Springer Tracts in Civil Engineering

Springer Tracts in Civil Engineering (STCE) publishes the latest developments in Civil Engineering - quickly, informally and in top quality. The series scope includes monographs, professional books, graduate textbooks and edited volumes, as well as outstanding PhD theses. Its goal is to cover all the main branches of civil engineering, both theoretical and applied, including:

- Construction and Structural Mechanics
- Building Materials
- Concrete, Steel and Timber Structures
- Geotechnical Engineering
- Earthquake Engineering
- Coastal Engineering; Ocean and Offshore Engineering
- Hydraulics, Hydrology and Water Resources Engineering
- Environmental Engineering and Sustainability
- Structural Health and Monitoring
- Surveying and Geographical Information Systems
- Heating, Ventilation and Air Conditioning (HVAC)
- Transportation and Traffic
- Risk Analysis
- Safety and Security

Indexed by Scopus

To submit a proposal or request further information, please contact:
Pierpaolo Riva at Pierpaolo.Riva@springer.com (Europe and Americas)
Wayne Hu at wayne.hu@springer.com (China)

Luís Bragança · Meri Cvetkovska ·
Viorel Ungureanu · Katerina Tsikaloudaki ·
Marianna Marchesi · Ana Andabaka ·
Ruben Paul Borg · Camila Cervantes
Editors

Shaping Circular Transitions in the Built Environment

From Barriers to Enablers

Springer

Editors
See next page

ISSN 2366-259X ISSN 2366-2603 (electronic)
Springer Tracts in Civil Engineering
ISBN 978-3-032-02833-4 ISBN 978-3-032-02834-1 (eBook)
https://doi.org/10.1007/978-3-032-02834-1

This work was supported by Associação iiSBE Portugal.

This Springer imprint is published by the registered company Springer Nature Switzerland AG
The registered company address is: Gewerbestrasse 11, 6330 Cham, Switzerland

Editors
Luís Bragança
Department of Civil Engineering
University of Minho
Guimarães, Portugal

Viorel Ungureanu
Department of Steel Structures
and Structural Mechanics
Polytechnic University of Timişoara
Timişoara, Romania

Marianna Marchesi
Welsh School of Architecture
Cardiff University
Cardiff, UK

Ruben Paul Borg
Faculty for the Built Environment
University of Malta
Msida, Malta

Meri Cvetkovska
Faculty of Civil Engineering
Ss. Cyril and Methodius University
in Skopje
Skopje, North Macedonia

Katerina Tsikaloudaki
Department of Civil Engineering
Aristotle University of Thessaloniki
Thessaloniki, Greece

Ana Andabaka
Faculty of Economics and Business
University of Zagreb
Zagreb, Croatia

Camila Cervantes
Department of Civil Engineering
University of Minho
Guimarães, Portugal

Preface

The transition towards a circular economy in the built environment is no longer a speculative ambition. It is a pressing necessity. Across the globe, the construction sector continues to exert tremendous pressure on natural resources, energy systems, and waste management infrastructures. At the same time, it holds untapped potential to become a transformative agent for sustainability, resilience, and socio-economic regeneration. This book, *Shaping Circular Transitions in the Built Environment—From Barriers to Enablers*, offers a timely and critical exploration of how that transformation can be effectively initiated and sustained.

The publication is structured around three core parts, each addressing essential dimensions of circularity in the built environment. Part I examines the current state of the art, identifying and mapping the technical barriers that hinder progress while highlighting key enablers, technological, economic, and behavioural that can catalyse change. These chapters reveal the complexity of circular implementation, where systemic innovation must overcome fragmented knowledge, outdated practices, and structural inertia.

Part II addresses the intricate and multi-layered policy landscape. Through a detailed analysis of EU-level directives and national policy frameworks, the book explores how circular economy goals can be aligned across governance levels. It offers practical insights to enhance coherence and integration, supporting the development of effective, context-sensitive instruments for accelerating the transition.

Final Part III of the book tackles a challenge often overlooked but fundamental to any successful transition: the harmonisation of language and concepts. Without a shared terminology and mutual understanding of key principles, stakeholders across academia, industry, and governance risk working in silos, duplicating efforts, or misaligning priorities. The comprehensive glossary presented here aims to unify discourse and enable effective communication and collaboration across disciplines, sectors, and national contexts. The glossary, which includes over 400 harmonised terms, is designed as a living document that can evolve with emerging regulatory frameworks and innovations in circular practice.

Additionally, the book places strong emphasis on the role of technical standards and implementation frameworks, recognising their central importance in operationalising circular practices. It not only analyses the state of standardisation in Europe and beyond, but also presents case studies and survey-based insights that reveal gaps, opportunities, and pathways for improvement. Several chapters also draw on practice insights from surveys, case studies, and cross-national comparisons, particularly in relation to the implementation of standards and frameworks. This evidence-based approach adds practical depth and reveals real-world dynamics in applying circular principles.

By combining critical analysis with actionable guidance, the chapters speak directly to professionals, policymakers, and researchers seeking to understand and influence circular transitions in practice. The book aims to support those designing and applying policies, tools, and business models grounded in circular principles, offering structured knowledge that is both reflective and ready to inform real-world implementation.

This collective effort, grounded in the outcomes of the COST Action CircularB, brings together a wealth of insights from researchers and practitioners across Europe and beyond. The diversity of perspectives reflected in the book enriches the analysis and ensures its relevance to different contexts, whether urban or rural, advanced or emerging economies.

We hope this publication will serve not only as a reference for current debates and future research, but also as a practical guide for those shaping the future of the built environment. In addition to academic audiences, this book will serve as a valuable resource for municipal planners, policymakers, consultants, and postgraduate students aiming to implement circular strategies in diverse built environments. As we move from barriers to enablers, from theory to practice, the vision of a regenerative, inclusive, and resilient built environment becomes more tangible, and more necessary than ever.

This book is the result of the collective endeavour of CircularB, which unites almost 500 experts from across 40 European countries. Their shared commitment to advancing the circular economy in the built environment was fundamental to the development of this volume. The insights presented across its chapters were shaped by interdisciplinary dialogue, rigorous exchange, and a collaborative spirit that exemplifies the COST Programme's mission of fostering excellence through networking. The Action's coordination and continuous knowledge-sharing enabled a coherent and structured reflection on the barriers, enablers, policies, and frameworks that define the circular transition in construction.

July 2025

Prof. Dr. Luís Bragança
Chair of COST Action
CA21103—Implementation of Circular
Economy in the Built Environment
(CircularB)
Department of Civil Engineering
University of Minho
Guimarães, Portugal

Acknowledgements

cost
EUROPEAN COOPERATION
IN SCIENCE & TECHNOLOGY

Funded by
the European Union

This publication is based upon work from COST Action CA21103 CircularB (Implementation of Circular Economy in the Built Environment, https://circularb.eu/), supported by COST (European Cooperation in Science and Technology).

COST (European Cooperation in Science and Technology) is a funding agency for research and innovation networks. Our Action helps connect research initiatives across Europe and enables scientists to grow their ideas by sharing them with their peers. This boosts their research, career and innovation.

www.cost.eu

circular B

Contents

Part I
Understanding the State of the Art and Key Drivers

Chapter 1
Technical Barriers and Challenges to Circular Strategies Implementation

Marianna Marchesi, Adriana Salles, Raluca Buzatu, Ferhat Karaca, Marilena De Simone, François Fohl, Elif Karacor, Stylianos Yiatros, Stella Tsoka, Genesis Camila Cervantes Puma, Christina Giarma, Katerina Tsikaloudaki, Jana Chovancová, Tatiana Ruchinskaya, Nadine Al-Bqour, Viorel Ungureanu, Alicia Zaragoza, Daniel Ferrandez, Ayfer Donmez Cavdar, and Emília Huttmanová

Abstract This chapter examines the primary technical barriers and challenges that impede the transition to a circular economy within the built environment. It highlights the complexities involved in integrating circular practices into the industry which traditionally relies on linear models, government and society. The analysis explores technical barriers at the urban, building and material levels of the built environment through a problem-based approach. Findings highlight a lack of standardized methodologies, inadequate design strategies for reuse and adaptability, insufficient data on material performance, and constraints in current construction practices and regulations. Furthermore, it underscores how technical fragmentation and limited interoperability among systems hinder the advancement of circularity. The chapter

M. Marchesi (✉) · N. Al-Bqour
Welsh School of Architecture, Cardiff University, Cardiff, UK
e-mail: marchesim@cardiff.ac.uk

A. Salles · G. C. C. Puma
ISISE, ARISE, Department of Civil Engineering, University of Minho, Guimarães, Portugal

R. Buzatu · V. Ungureanu
Department of Steel Structures and Structural Mechanics, Politehnica University Timisoara, Timisoara, Romania

F. Karaca
School of Engineering and Digital Sciences, Department of Civil and Environmental Engineering, Nazarbayev University, Astana, Kazakhstan

M. De Simone
Department of Environmental Engineering, University of Calabria, Rende, Italy

F. Fohl
ArcelorMittal Commercial RPS, Esch-Sur-Alzette, Luxembourg

E. Karacor
Landscape Architecture Department, Istanbul Technical University, Istanbul, Türkiye

L. Bragança et al. (eds.), *Shaping Circular Transitions in the Built Environment*, Springer Tracts in Civil Engineering, https://doi.org/10.1007/978-3-032-02834-1_1

also explores the potential benefits of digitalization, design for disassembly, and building information modelling (BIM) as a means to address these challenges. Ultimately, it emphasizes the need for coordinated efforts among stakeholders, supported by robust regulatory frameworks and innovation, to effectively implement circular economy principles in the construction industry.

Keywords Circular economy · Construction sector · Technical barriers · Challenges · Urban level · Building level · Material level

1.1 Introduction

The circular economy (CE) concept has been gaining momentum, as evidenced by the literature [1–3]. This is not just a passing trend but a crucial concept that academia, governments, policymakers, and institutions increasingly recognise and commit to. The urgency of adopting CE practices is underscored by global initiatives like the European Green Deal, which aims to achieve climate neutrality by 2050 through a "regenerative growth model" that emphasises sustainability and economic competitiveness [4]. Furthermore, the European Union has made significant strides in addressing the sustainability aspects of products through various initiatives and legislative frameworks, implemented on both a mandatory and voluntary basis [4]. This focus aligns with the principles of CE, and the transition from a linear "take-make-dispose" model to a circular one requires systemic change [5].

The shift towards a CE is a critical evolution in addressing the unsustainable practices of the traditional linear economy model. Unlike the traditional linear economy, the CE promotes a regenerative approach. Going beyond waste management and related practices, the concept of CE addresses challenges for the whole value chain,

S. Yiatros
Department of Civil Engineering and Geomatics, Cyprus University of Technology, Limassol, Cyprus

S. Tsoka · C. Giarma · K. Tsikaloudaki
Laboratory of Building Construction and Building Physics, School of Civil Engineering, Faculty of Engineering, Aristotle University of Thessaloniki, Thessaloniki, Greece

J. Chovancová · E. Huttmanová
Faculty of Management and Business, Department of Environmental Economy and Management, University of Prešov, Prešov, Slovakia

T. Ruchinskaya
TVR Design Consultancy, Willingham, Cambridge, UK

A. Zaragoza · D. Ferrandez
Escuela Técnica Superior de Edificación, Departamento de Tecnología de La Edificación, Universidad Politécnica de Madrid, Madrid, Spain

A. D. Cavdar
Department of Forest Industry Engineering, Karadeniz Technical University, Trabzon, Turkey

from design strategies to new business models, transitioning from ownership to service-based solutions, and keeping materials and products in the cycle, making them last longer [5].

The importance of adopting circular practices to tackle environmental impacts, resource shortages, and social and economic challenges cannot be overstated [3, 6, 7]. By minimising waste, optimising resource flows, and enabling transformative new patterns of production and consumption, CE can foster sustainability and well-being [8], being considered a powerful tool for achieving broader sustainability goals and a transformative model for sustainable development. Given its significant consumption of natural resources and energy, which accounts for approximately a third of the total energy consumed worldwide, the built environment plays a pivotal role in the transition to a CE and sustainable development [9]. This underscores the responsibility and potential impact of the built environment in this crucial transition.

Despite the clear benefits, the implementation of circular strategies in the built environment faces numerous challenges. The most prominent among these are technical barriers, which include difficulties in integrating circular principles into building design, construction, and material use. According to AlJaber et al. [10], technical barriers were the most frequently observed in the literature on the implementation of a CE in the building sector, followed by economic/market barriers, awareness-related barriers, implementation barriers, social barriers, and finally, support/promotion-related barriers. These technical barriers often arise from a lack of standardised methodologies, the limited availability of circular materials, and the complexity of retrofitting existing structures [11–13]. Addressing these challenges is critical not only for environmental sustainability but also for enhancing the economic viability and resilience of the construction sector.

1.1.1 Importance of Identifying Technical Barriers

Understanding and addressing technical barriers is crucial for advancing CE strategies within the built environment. These barriers often prevent the effective implementation of sustainable practices, such as the reuse and recycling of materials, energy efficiency, and waste reduction. By identifying these challenges, stakeholders, including policymakers, industry professionals, and researchers—can develop targeted solutions that facilitate the transition towards more sustainable and resilient building practices. This focus not only contributes to environmental goals, such as reducing carbon footprints and conserving resources, but also supports economic and social objectives by fostering innovation, creating jobs, and enhancing the overall quality of life.

Current research involves a comprehensive review of existing literature to pinpoint specific technical barriers. These barriers were categorised and quantified based on their frequency and relevance across different sources. The study also explored qualitative aspects to provide a deeper understanding of the contextual challenges associated with each barrier. The methodology was designed to ensure a thorough analysis, including quantitative data visualisation through diagrams and tables and qualitative insights supported by expert opinions and case studies. The insights from this report will help shape future policies and practices, ensuring a more sustainable and resource-efficient built environment.

The study employed a multi-step methodology to systematically analyse the technical barriers to circular strategies in the built environment. The process began with a quantitative data analysis, focusing on the prevalence and distribution of identified barriers across literature. Key steps include:

1. Identification and Categorisation: Barriers were identified from the literature and categorised based on their nature and impact on circular strategies. These categories included where and when the barriers occur, who is involved, and how they affect implementation.
2. Quantitative Analysis: A detailed quantitative analysis, including histograms and data tables, was conducted to quantify the frequency and relevance of each barrier category. This analysis provided a numerical basis for identifying the most critical barriers.
3. Qualitative Analysis: A qualitative review was undertaken alongside the quantitative data to explore the nuances and specific contexts in which these barriers arise. This involved a deeper examination of literature, supported by expert opinions and case studies.
4. Synthesis and Discussion: The findings from the quantitative and qualitative analyses were synthesised to form a comprehensive overview of the technical barriers. The discussion section of the report elaborates on these findings, offering recommendations and potential solutions to address the identified challenges.

This structured approach ensures a thorough understanding of the technical barriers to implementing a CE in the built environment. Based on the methodology proposed, this study intends to explore the following question: "What are the main technical barriers and challenges to implementing circular economy strategies in the built environment?" To this end, the questions proposed to be answered in this study are:

1. What are the technical barriers?
2. Where do the technical barriers occur?
3. When do the technical barriers occur?
4. Who is responsible for or involved in?
5. How does this impact the implementation of circular strategies in the built environment?

By answering these questions, this study could identify not only the main barriers but also the stakeholders, and implications for CE implementation. The insights gained from this study are essential for guiding future research, policy development, and practical applications in the sector.

1.2 Materials and Methods

The starting point of the literature review was the definition and specification of the context (built environment) and of the investigated subjects (technical barriers). The built environment was defined considering its levels, following the classification proposed by Pomponi and Moncaster [14] and Brincat [15]:

1. Urban level adapts the system boundary definition at a municipality/city/neighbourhood/urban agglomerate level.
2. Building level takes it to a building at the whole asset/system and components/products.
3. Material level focuses on the materials that are used within the construction.

The investigation of technical barriers was performed in parallel to all three levels of the built environment. A detailed description of technical barriers and their categorisation is provided in the next paragraph. Successively, the criteria for document collection and organisation are explained.

1.2.1 Technical Barriers—Definition and Classification

Technical barriers (or challenges) are described as challenges in science, engineering, industry, urban planning, and architecture that affect technicians, urbanists, architects, engineers, scientists, and entrepreneurs in the implementation of circular strategies in the built environment. They were analysed through the different levels of the built environment, considering this classification in categories [16]:

- information/data
- technical knowledge
- technical skills
- technical standards
- machines/equipment
- processes (such as assessment) and infrastructure
- methods/tools
- materials/products/building systems.

The technical barriers identified in the literature were framed according to the 5Ws approach to investigating the What, Who, Where, When, and hoW of a barrier since their answers are considered essential in information gathering and problem-solving. Specifically, this analysis can provide interesting findings for researchers, companies, and policymakers about where (layers based on the selected system boundaries), when (the life cycle stages) of the built environment, how barriers affect the implementation of a CE in the built environment (circular strategies), and who is responsible for or affected (stakeholders). Additional information addressing the questions of "why" was also included whenever available.

The 5Ws were detailed in Tables 1.1 and 1.2 based on the technical barrier classifications which were considered in this study. Within each barrier category, information related to identified barriers was collected in terms of building layers, stakeholders, life cycle phases, and circular strategies with the details of potential motivations when available. This information was elaborated for each reference using the templates of Tables 1.1 and 1.2.

The question "What" considered technical barriers identified in the literature review that fit the definition adopted in the study and then categorised according to the classification above and reported in the row "Technical barriers/challenges category and description" of Table 1.1.

The question "Who" focused on the stakeholders who are responsible for the barrier or mainly affected by the barrier. The analysis was based on the classification of stakeholders involved in a CE in the built environment defined by Munaro and Tavares [19] and reported below:

1. Clients: owners, users, consumers
2. Project professionals: project managers, designers, architects, engineers, facilities managers, investors, subcontractors, real estate agencies, builders, employees
3. Suppliers: manufacturers, process and service providers
4. Public: media, community representatives, neighbours, the press, the academy, pressure groups, civic institutions, visitors, the natural environment
5. Government: legal authorities, regional development agencies, civic institutions, government establishments
6. Others: non-governmental organisations (such as National or International Organisations for Standardisation).

Table 1.1 Template of "qualitative data collection"

	Data collected
Publication's reference	
Publication types (such as research paper, conference paper, book chapter, review, report, project deliverable, etc.)	
Brief abstract (max 100–150 words)	Aim Method Results
Technical barriers/challenges category and description	Lack of information/data
	Lack of technical knowledge
	Lack of technical skills
	Lack of technical standards
	Inadequate machines/equipment
	Inadequate processes and infrastructure
	Lack of methods/tools
	Lack of materials/products/building systems

Table 1.2 Template of "quantitative data collection"

What is the barrier category	Where it happens (when the barrier originates and when the barrier impacts)							When it happens						Who is responsible for/ involved in					
Technical barrier category	Building layer (a)							Building life cycle phase (b)						Stakeholders (c)					
	1	2	3	4	5	6	7	1	2	3	4	5	6	1	2	3	4	5	6
Lack of information/ data																			
Lack of technical knowledge																			
Lack of technical skills																			
Lack of technical standards																			
Inadequate machines/ equipment																			
Inadequate processes and infrastructure																			
Lack of methods/ tools																			
Lack of materials/ products/ building systems																			

What is the barrier category	How it affects the implementation of circular strategies in the built environment										Sources of data	Note
Technical barrier category	Circular strategies (d)										References include the barrier ref. n. reported in Table 1.1	
	1	2	3	4	5	6	7	8	9	10		
Lack of information/data												
Lack of technical knowledge												

(continued)

Table 1.2 (continued)

What is the barrier category	How it affects the implementation of circular strategies in the built environment										Sources of data	Note
Technical barrier category	Circular strategies (d)										References include the barrier ref. n. reported in Table 1.1	
	1	2	3	4	5	6	7	8	9	10		
Lack of technical skills												
Lack of technical standards												
Inadequate machines/ equipment												
Inadequate processes and infrastructure												
Lack of methods/ tools												
Lack of materials/ products/building systems												

(a) Building layers [17]: (1) site; (2) structure; (3) skin; (4) services; (5) space plan; (6) stuff; (7) system (district)
(b) Building life cycle phases (adapted from RIBA [18]; Brincat et al. [15]): (1) Strategic Definition and Briefing; (2) Design; (3) Manufacture; (4) Dismantling and Construction; (5) Use and Refurbishment; (6) End-of-life
(c) Stakeholders [19]: (1) Clients; (2) Project professionals; (3) Suppliers; (4) Public; (5) Government; (6) Others
(d) Circular Strategies [20]: (1) Refuse; (2) Rethink; (3) Reduce; (4) Reuse; (5) Repair; (6) Refurbish; (7) Remanufacture; (8) Repurpose; (9) Recycle; (10) Recover

This information was collected in the column "Stakeholders" of Table 1.2.

The question "Where" was related to building layers. According to Brand [21] and Zimmann et al. [17], the built environment can be framed in seven layers, and each of them holds specific functions with different expected service life as reported below:

1. Site is the location of the building (∞ yrs).
2. Structure is the building's load-bearing system, including the foundation and load-bearing elements (30–300 yrs).
3. Skin is the façade and building exterior (20–35 yrs).
4. Services are the pipes, wires, energy, heating/cooling and ventilation systems (15–30 yrs).
5. Space plan is the internal fit-out including walls and floors (10–30 yrs).

6. Stuff is the rest of the internal fit-out, including the furniture, lighting, and ICT (5–20 yrs).
7. System (district) is an additional layer that has been recently included with the intent to apply this approach beyond the scope of a building, in the context of a district and city [17].

Related information was collected in the column "Building layers" in Table 1.2.

With the question "When", we considered the building lifecycle model proposed by RIBA [18] and extended according to Brincat et al. [15] and reported below:

1. Strategic definition and Briefing
2. Design
3. Manufacture
4. Demolition and Construction
5. Use and Refurbishment
6. End-of-life.

Related information was collected in the column "Building life cycle stage" of Table 1.2.

Finally, regarding the question "How", identified technical barriers were related to the circular strategies affected by them, to identify evidence of their effect/impact on the CE implementation according to the framework proposed by Stijn and Gruis [20]. Stijn and Gruis [20] highlight that most of the design frameworks focused on the building level and based on Bocken et al. [22] and Potting et al. [23], they developed a circular design framework aiming to expand the approach to other levels and ensure that circularity is achieved within and beyond the designs' lifecycle. Their framework (reported in Table 1.3) intends to implement a systems approach in which components and systems are regarded from within their wider system environment. Since R0 and R1 strategies decrease the consumption of natural resources and materials applied in a product chain, by less product being needed for delivering the same function, they are generally also considered circularity strategies, even though they do not necessarily involve increasing the reuse of products and components, or reapplication of recycled materials.

1.2.2 Document Collection

The literature review search was performed using SCOPUS databases, looking at peer-reviewed journal articles, conference papers, book chapters and reports. Moreover, non-academic literature was collected, consisting of reports from European institutions, research projects, NGOs and industry by using their official websites. Only documents in English were searched (original and review articles published and in press) about the core themes, covering up to January 2024, when the searches were conducted. Excel spreadsheets were used to manage the references.

Table 1.3 Circular design framework applied to products [20]

Principle	Strategy	
Narrowing loops	RO refuse	Make the product redundant by abandoning its function and offering the same function with a radically different product
	R1 rethink	Make product use more intensive (e.g. through sharing products, or by putting multi-functional products on the market)
	R2 reduce	Increase efficiency in product manufacture or use by consuming fewer natural resources and materials
Slowing loops	R3 re-use	Re-use by another consumer of a discarded product that is still in good condition and fulfils its original function
	R4 repair	Repair and maintenance of defective products so they can be used with their original function
	R5 refurbish	Restore an old product and bring it up to date
	R6 remanufacture	Use parts of discarded products in a new product with the same function
	R7 repurpose	Use discarded products or their parts in a new product with a different function
Closing loops	R8 recycle	Process materials to obtain the same (high grade) or lower (low grade) quality
	R9 recover	Incineration of materials with energy recovery

The queries with the keywords for document searching were selected for each level as specified as follows:

Urban level queries: barrier* AND "circular cities" AND urban*; barrier* AND "urban circularity"; barrier* AND "circular cities" AND urban circularity*; "barriers" "challenges" AND "circular cities" AND "urban circularity"; barrier* AND "circular economy" AND "urban"). The results of the search showed 53 documents. After the first screening, 21 articles (scientific literature) were selected for the deep review.

Building level queries: barrier* AND "circular economy" AND building*; barrier* AND "circular economy" AND "built environment"; barrier* AND "circularity" AND "built environment"; barrier* AND "circularity" AND building*). A total of 119 documents were collected belonging to scientific literature, and a deeper screening validated 57 of them. The search of non-academic literature produced 32 documents. This number was reduced to 18 after the screening.

Material level queries: "technical barrier" AND "circularity" AND "built environment" AND "material" level; "technical barrier" AND "circular economy" AND "built environment" AND "material" level; "technical barrier*" AND "circular economy" AND material*; barrier* AND "circular economy" AND material*; "technical barrier*" AND circularity AND material*; "technical barrier*" AND circularity OR "circular economy" AND material*; "technical barrier*" AND "circular economy"

AND "built environment"). The results of the search showed 365 documents. To include only technical issues, the selection was limited to Engineering, Material Science, and Energy fields. After the first screening, 62 documents (scientific literature), research reports, book chapters, PhD thesis, and MSc thesis were selected for the deep review.

Following the initial search, our literature review proceeded through nine additional steps performed at each level of the built environment. They are listed and explained as follows:

1. *Excluding duplicates.* All documents were checked, and all duplicate documents were excluded.
2. *Title and keywords filter.* Each article's title and keywords were read, and documents not related to the search theme were excluded.
3. *Abstract filter.* The abstracts of the remaining articles were analysed, and documents not related to the theme were excluded.
4. *Retrieving full papers.* Full papers were downloaded. The ones not found available, using the access provided to the authors, were ruled out.
5. *Full Reading.* All remaining available documents were read in full to provide the theoretical basis for this review. After fully reading, papers not suitable for the purpose of this research, for reasons such as not having a strong relationship with the research theme, presenting blurred interpretations, presenting overly shallow approaches, or not significantly contributing to the topics/themes were excluded. The remaining articles compose the final portfolio. Moreover, the authors double-checked whether the downloaded documents were suitable for analysis and evaluation under the headings of material, building, and urban levels.
6. *Collecting data.* Qualitative Data were collected consistently using the template shown in Table 1.1.
7. *Comparing data.* Quantitative Data were summarised for each level using the template shown in Table 1.2.
8. *Analysis.* Data were analysed and discussed using a quantitative and qualitative approach for each level.
9. *Synthesis.* The final step was identifying patterns or faults in each barrier category at all built environment levels.

1.3 Results

1.3.1 Urban Level

1.3.1.1 Quantitative Analysis of the Literature

To understand, quantify, and identify the technical barriers, quantitative and qualitative approaches have been adopted. The first step is quantitative analysis, which identifies the number of barriers according to the analysed documents and how they are distributed among the eight categories.

Analysing the documents, of the 21 articles, 14 reported barriers under the category (6) *Inadequate processes and infrastructure*, which was the category most represented in the documents. In comparison, 12 articles reported barriers under the categories (1) *Lack of information/data*, (2) *Lack of technical knowledge*, and (4) *Lack of technical standards.* Only one article reported barriers under the category (5) *Inadequate machines/equipment*, as presented in Fig. 1.1.

In terms of the number of barriers presented in each category, the category (6) *Inadequate process and infrastructure* represents the category with the highest number of barriers reported, with 147 reported among the 21 articles. Although the categories (4) *Lack of technical standards*, (1) *Lack of information/data*, and (2) *Lack of technical knowledge* were reported in 12 articles, the number of barriers reported in each of these categories are, respectively, 98, 90, and 76, as shown in Fig. 1.2.

As detailed in the Materials and Methods session, the question "Where" is related to building layers (Site, Structure, Skin, Services, Space Plan, and Stuff). The System (District) was included as an additional layer, going beyond the building scope, to

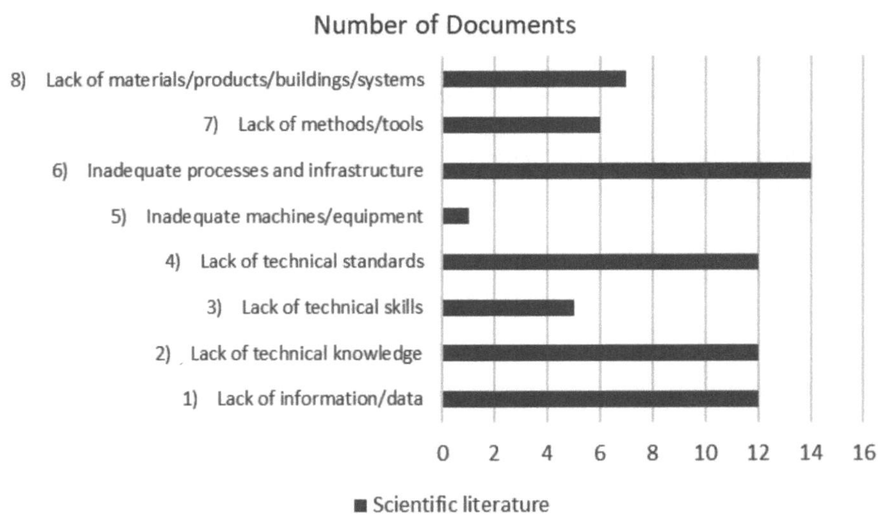

Fig. 1.1 Number of articles that reported the barrier category

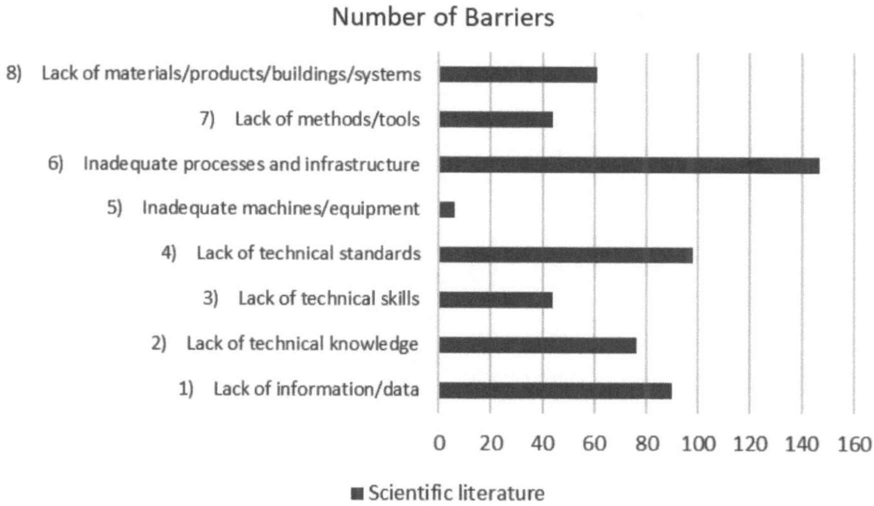

Fig. 1.2 Number of barriers reported per category

apply this approach to the urban-level context. In this regard, the barriers identified at the urban level were considered to occur only at the System (District) layer. However, not all barriers specifically addressed "Where", with only 62 barriers reported where they happen. For instance, category (6) Inadequate process and infrastructure is the most representative in the number of barriers reported, accounting for 147 barriers (Fig. 1.2). However, from these 147 barriers, only 15 specifically addressed where they happened. Figure 1.3 shows the number of barriers reported at the System (District) layer.

Analysing the life cycle phase of the examined barriers, most of them occur during the *Use and Refurbishment* phase, accounting for a total of 44 barriers reported during this phase. Followed by *Strategic Definition and Briefing*, accounting for 31 barriers, and *(Dismantling and) Construction* and *End-of-life* phases, with 28 barriers each. The life cycle phase with the lowest number of barriers is the *Manufacturing* phase, with 19 barriers reported.

Under the barriers categories, represented in Fig. 1.4, in category (6) *Inadequate process and infrastructure*, the *(Dismantling and) Construction* phase represents the highest number of barriers reported, with a total of 12 barriers reported, followed by *Use and Refurbishment*, with 11 barriers reported, and *End-of-life*, with 10 barriers reported.

The *Government* is the stakeholder reported as responsible for/involved in most of the barriers, being listed in 50 barriers within all categories. *Project professionals* and *Suppliers* are pointed out as the second and third stakeholders more responsible for or involved in the barriers. Looking at the categories presented in Fig. 1.5, (4) *Lack of standards* is the one with the most barriers reported, and *Government* and *Project professionals* are the group of stakeholders responsible for or involved in, accounting for 12 and 11 barriers, respectively.

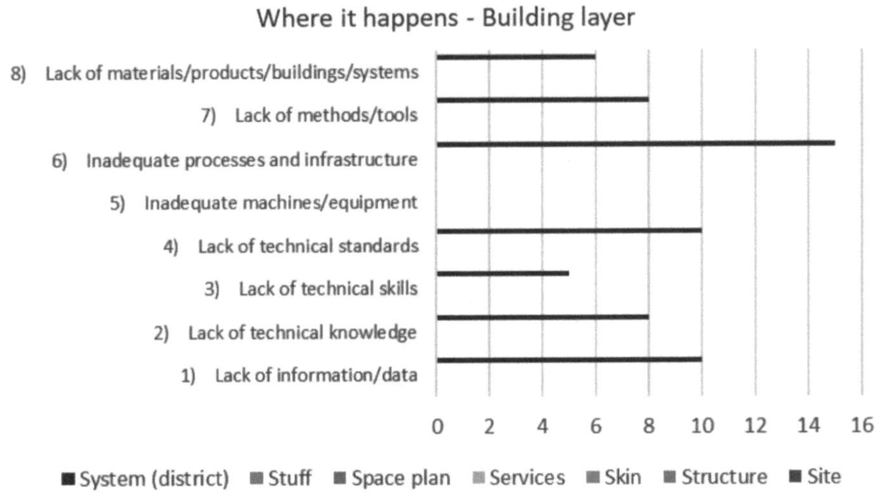

Fig. 1.3 Where the barriers happen

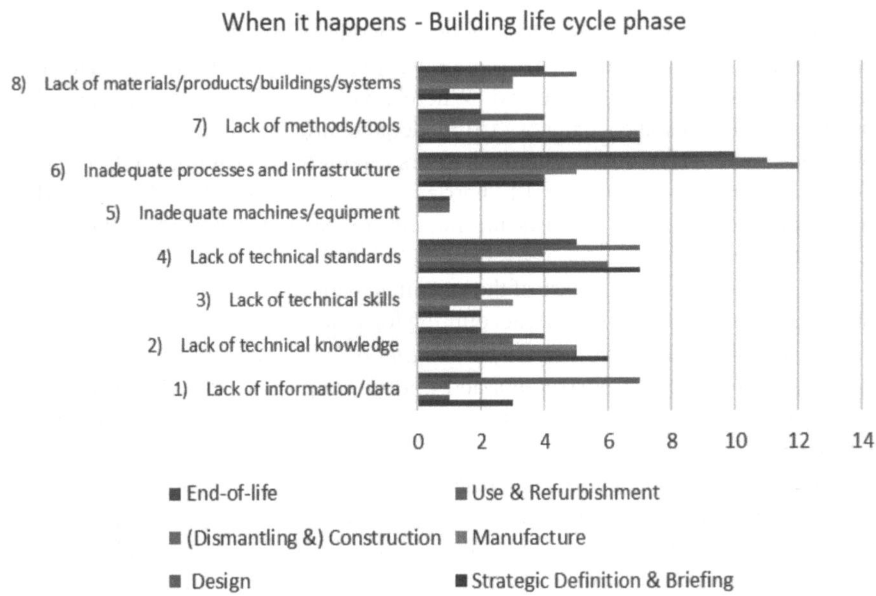

Fig. 1.4 When the barriers happen

Under category (6), *Inadequate process and infrastructure*, *Suppliers* are the group of stakeholders more related to the barriers, with 11 total. *Government* was also reported as responsible for/involved in 10 barriers of this category. In category (1) *Lack of information/data*, *Government* and *Project professionals* are the groups of

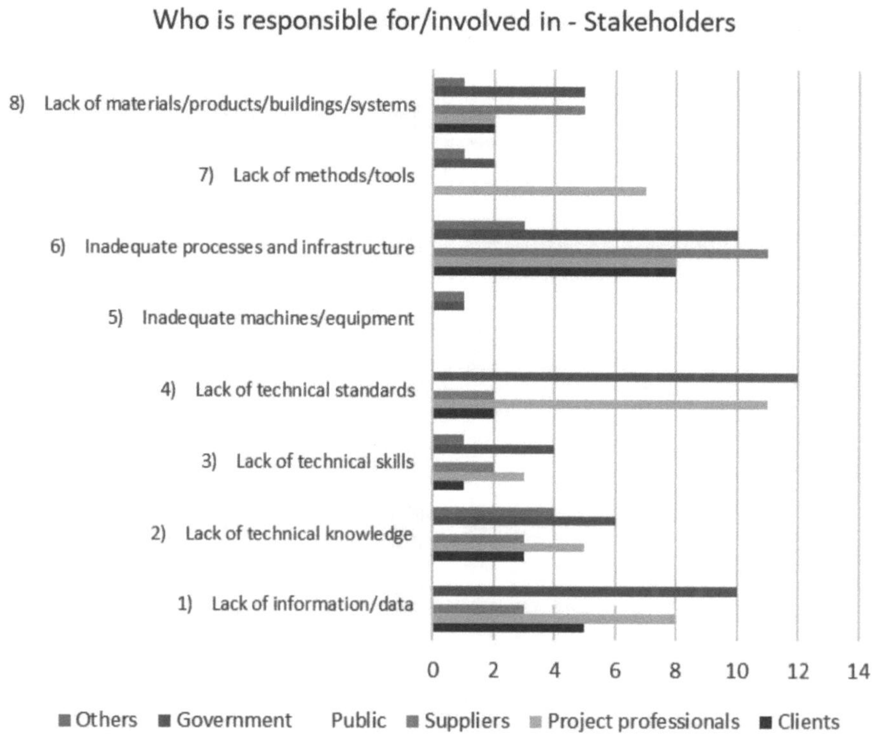

Fig. 1.5 Groups of stakeholders responsible for/involved in

stakeholders that are the most responsible for/involved in the barriers, accounting for 10 and 8, respectively.

None of the articles mentioned a relationship between the barriers and the circular strategy of R0 *Refuse.* At the urban level, most of the barriers affect R3 *Reuse,* followed by R8 *Recycle* and R7 *Repurposed* strategies. As illustrated in Fig. 1.6, (6) *Inadequate process and infrastructure* is the category where most of the barriers affect the circular strategies, with 13 barriers affecting R3 *Reuse* and 10 in R8 *Recycle.* (4) *Lack of standards* is the second category in terms of how the barriers affect the circular strategy implementation. Within this category, eight barriers affect both R3 *Reuse* and R8 *Recycle.*

It is important to notice that (1) *Lack of information/data* is the category of barrier that affects eight out of nine strategies in the circular strategy framework. It does not account for the most significant number of barriers, but it is the only category that affects almost all circular strategies.

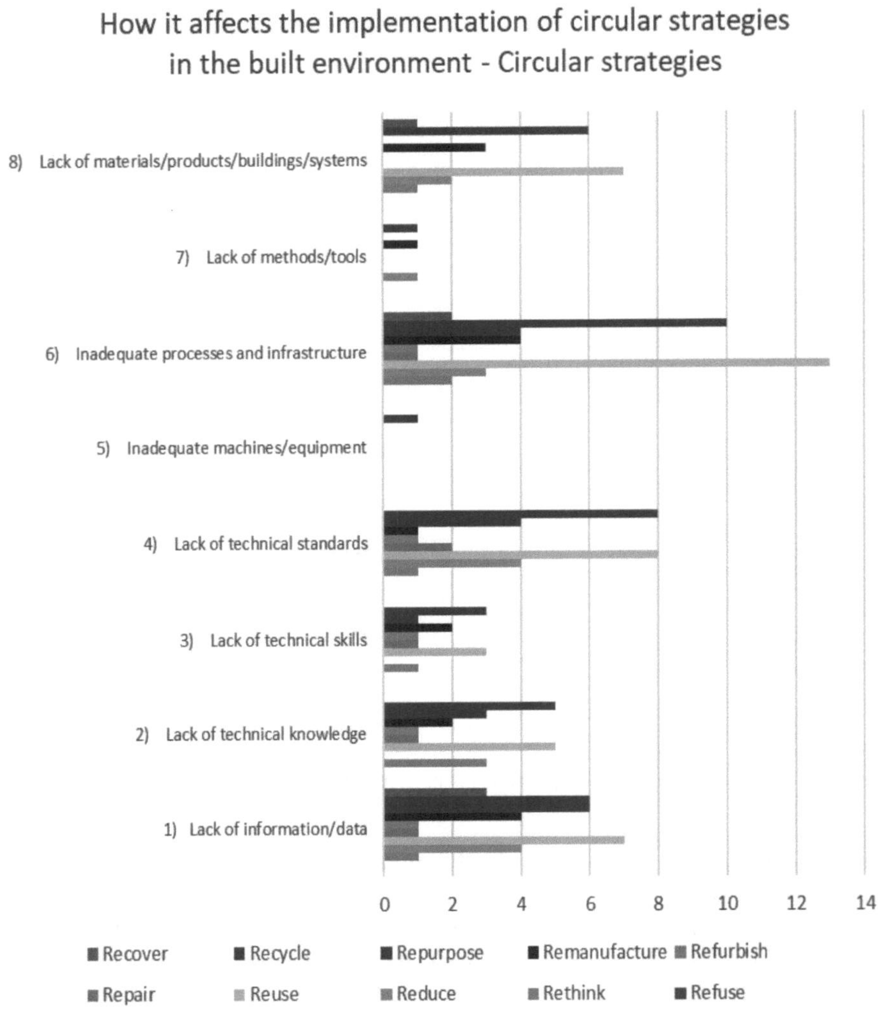

Fig. 1.6 How the barriers affect CE strategies

1.3.1.2 Collection of the List of Barriers in Each Category and Analysis of Relevance

This review synthesised findings from various studies and articles, highlighting the obstacles to adopting CE principles. The transition to a CE, particularly in urban environments and the construction sector, is fraught with numerous challenges.

After a comprehensive review of the articles, the identified barriers were systematically categorised into subcategories to provide a clearer understanding of the issues hindering the successful implementation of CE practices. These subcategories cover various challenges, from infrastructural inadequacies and technical limitations to

Table 1.4 Subcategories of barriers under Category 1 at the urban level

Subcategory	Count	Reference
Collection and sorting guidance	2	[24, 25]
Documentation and data quality	3	[19]
Data availability and accessibility	9	[10, 26–32]
Simulation and modeling	1	[19]
Knowledge transfer and integration	2	[33, 34]
Total	17	

production networks, logistics, and urban planning issues. This classification aims to facilitate targeted interventions and strategies to overcome these barriers and promote the adoption of CE practices.

Category 1: Lack of Information/Data

Implementing CE practices in urban environments and the construction sector hinges on the availability and quality of data and information. However, multiple studies have identified significant barriers in this regard. These barriers can be broadly categorized into five subcategories (Table 1.4). Each subcategory highlights specific issues that impede the efficient collection, management, and utilisation of data necessary for CE practices *Collection and Sorting Guidance.* This subcategory includes barriers related to the absence of clear guidelines for collecting and sorting construction and demolition waste (CDW). Proper guidance is crucial for efficient waste management and recycling processes.

Documentation and Data Quality. This subcategory addresses the lack of documentation for new and used building products, as well as issues with data quality and availability. Ensuring high-quality data is essential for the effective tracking and management of materials.

Data Availability and Accessibility. This subcategory includes barriers related to the availability and accessibility of data necessary for monitoring and planning CE activities. Limited data availability hinders effective decision-making and resource management.

Simulation and Modelling. This barrier pertains to the limitations of the simulation models used for planning and implementing CE practices. Incomplete models can lead to inadequate predictions and planning.

Knowledge Transfer and Integration. This subcategory includes barriers related to the transfer and integration of knowledge across different stakeholders and disciplines. Effective knowledge transfer is critical for scaling up CE practices and ensuring a comprehensive understanding among involved parties.

Table 1.5 Subcategories of barriers under Category 2 at the urban level

Subcategory	Count	Reference
Recycling and reuse technology	4	[24, 29, 35]
Technological expertise	3	[25, 30]
Design and disassembly	1	[27]
CE concept and principles	3	[27, 32, 36]
Production and manufacturing	1	[31]
Knowledge transfer and retention	2	[28, 31]
Total	14	

Category 2: Lack of Technical Knowledge

Successfully implementing CE practices in urban and construction settings requires robust technical knowledge across various domains. However, there are significant barriers related to technical knowledge that hinder these efforts. These barriers can be grouped into six subcategories (Table 1.5). Each subcategory captures specific challenges in acquiring, applying, and disseminating the technical expertise needed for CE practices.

Recycling and Reuse Technology. This subcategory focuses on the technological barriers related to recycling and reuse processes. Efficient recycling and reuse are crucial for the CE, but technological limitations can impede these processes.

Technological Expertise. This subcategory addresses the lack of specialized knowledge and skills required for implementing CE practices effectively. Technical expertise is essential for developing and applying innovative solutions in CE.

Design and Disassembly. This subcategory includes barriers related to product design and disassembly, which are critical for enabling recycling and reuse.

CE Concept and Principles. This subcategory focuses on the understanding and application of CE concepts and principles. Comprehensive knowledge of CE principles is vital for their effective implementation.

Production and Manufacturing. This subcategory addresses the technical knowledge barriers related to production and manufacturing processes. Efficient production methods are essential for minimising waste and promoting circularity.

Knowledge Transfer and Retention. This subcategory includes barriers related to the transfer and retention of technical knowledge among stakeholders. Effective knowledge transfer is critical for maintaining continuity and fostering innovation in CE practices.

Table 1.6 Subcategories of barriers under Category 3 at the urban level

Subcategory	Count	Reference
Human resources and knowledge of CE principles	2	[25, 28]
Access to business services and technical assistance	1	[37]
Public infrastructure and service providers' skills	2	[31, 34]
Total	5	

Category 3: Lack of Technical Skills

The successful implementation of CE principles and practices relies heavily on the availability of technical skills and expertise across various sectors. However, several studies have identified significant barriers related to the lack of technical skills, which can be categorised into three main subcategories (Table 1.6). Each subcategory highlights specific challenges that hinder the development and application of technical skills necessary for advancing CE initiatives.

Human Resources and Knowledge of CE Principles. This subcategory focuses on barriers to the availability of human resources with the necessary knowledge of CE principles. Understanding CE principles is essential for effectively integrating sustainable practices into urban and construction contexts.

Access to Business Services and Technical Assistance. This subcategory addresses barriers related to accessing business services and the technical assistance needed to implement CE practices. Adequate support services are crucial for overcoming technical challenges and fostering innovation in sustainable development.

Public Infrastructure and Service Providers' Skills. This subcategory includes barriers related to the technical skills and capabilities of public infrastructure and service providers. These entities play a vital role in implementing and maintaining CE initiatives on a broader scale.

Category 4: Lack of Technical Standards

The absence or inadequacy of technical standards across various domains hinders the implementation of CE practices in the built environment. The challenge presents itself in two primary subcategories (Table 1.7). These subcategories encompass specific barriers that impede the establishment of uniform guidelines and regulatory frameworks necessary for promoting sustainable practices and ensuring consistency in CE implementation.

Building Design and Construction Standards. This subcategory addresses barriers to the lack of technical standards in building design and construction practices conducive to CE principles. Practical standards are crucial for promoting eco-friendly building materials, efficient resource use, and sustainable construction techniques.

Table 1.7 Subcategories of barriers under Category 4 at the urban level

Subcategory	Count	Reference
Building design and construction standards	12	[19, 24, 26–28, 30–32, 34, 35, 38]
Regulatory framework and support	6	[31, 33, 36, 39]
Total	18	

Table 1.8 Subcategories of barriers under Category 5 at the urban level

Subcategory	Count	Reference
Inadequate recycling and waste processing technology	2	[24, 32]
Total	2	

Regulatory Framework Support. This subcategory focuses on barriers associated with the regulatory framework and government support required to establish and enforce technical standards for CE practices. Clear regulations and supportive policies are essential for incentivising compliance and fostering innovation in sustainable development.

Category 5: Inadequate Machines/Equipment

The practical implementation of CE principles in waste management and recycling heavily depends on the availability and adequacy of machines and equipment designed for these purposes. This challenge is categorised into one primary subcategory: Inadequate Recycling and Waste Processing Technology (Table 1.8). This subcategory highlights specific barriers that hinder the efficient processing and recycling of materials that are essential for CE practices.

Inadequate Recycling and Waste Processing Technology. This subcategory focuses on barriers related to the insufficiency and inadequacy of recycling and waste processing technology. Effective recycling and waste processing technologies are crucial for maximising material recovery and minimising waste in CE initiatives.

Category 6: Inadequate Processes and Infrastructure

The successful implementation of CE practices hinges on robust processes and infrastructure that support efficient resource management and sustainable development. This challenge is categorised into four main subcategories (Table 1.9). Each subcategory addresses specific barriers that impede the establishment and optimisation of processes and infrastructure necessary for advancing CE initiatives.

Infrastructure for Waste Management and Recycling. This subcategory focuses on barriers related to inadequate infrastructure for waste management and recycling.

Table 1.9 Subcategories of barriers under Category 6 at the urban level

Subcategory	Count	Reference
Infrastructure for waste management and recycling	8	[24, 25, 29, 31, 34, 35, 38]
Production networks and logistics	6	[26, 31, 35, 37, 38]
Urban planning	3	[27, 39, 40]
Challenges in implementing CE practices	4	[19, 26, 32, 33]
Total	21	

Adequate infrastructure is critical for facilitating the collection, sorting, and recycling of materials essential for CE practices.

Production Networks and Logistics. This subcategory addresses barriers associated with production networks and logistics that support CE practices. Efficient logistics and supply chain networks are essential for optimising resource flows and minimising waste in CE initiatives.

Urban Planning. This subcategory focuses on barriers related to urban planning practices that affect the implementation of CE principles. Integrated urban planning is essential for creating sustainable urban environments conducive to CE practices.

Challenges in Implementing CE Practices. This subcategory addresses broader challenges in implementing CE practices across different sectors. It encompasses barriers to policy, governance, and stakeholder engagement that are necessary to foster CE initiatives.

Category 7: Lack of Methods/Tools

The practical implementation of CE principles relies heavily on the availability and adequacy of methods and tools designed to support sustainable practices in urban planning, management systems, and design and construction. This challenge is categorised into three main subcategories (Table 1.10). Each subcategory addresses specific barriers that hinder the adoption and optimisation of methods and tools necessary for advancing CE initiatives.

Urban Planning. This subcategory focuses on barriers related to inadequate methods and tools in urban planning that hinder the integration of CE principles. Practical urban planning tools are essential for incorporating sustainable practices into urban development and promoting resource efficiency.

Management Systems. This subcategory addresses barriers associated with management systems lacking adequate methods and tools to support CE practices. Efficient management systems are crucial for optimising resource flows, tracking material use, and ensuring effective implementation of CE strategies. Design and Construction Tools.

Table 1.10 Subcategories of barriers under Category 7 at the urban level

Subcategory	Count	Reference
Urban planning	2	[19, 24]
Management systems	1	[24]
Design and construction tools	6	[10, 27, 34, 35]
Total	9	

Design and Construction Tools. This subcategory focuses on barriers related to inadequate methods and tools in design and construction that limit the implementation of CE principles. Effective design and construction tools are essential for promoting sustainable building practices, material reuse, and lifecycle assessments.

Category 8: Lack of Materials/Products/Buildings/Systems

The challenges related to the lack of materials, products, buildings, and systems for implementing CE principles are crucial barriers that impact the transition towards sustainable practices. These challenges are categorised into five subcategories (Table 1.11).

Material/Product/System Availability and Quality. This subcategory addresses barriers concerning the availability and quality of materials, products and systems essential for implementing CE practices. Ensuring a consistent supply of high-quality recycled or reclaimed materials, products, and systems is critical for sustainable construction and manufacturing processes.

Space and Storage Issues. This subcategory focuses on barriers related to spatial constraints and storage challenges that hinder the effective implementation of CE strategies. Adequate space for storing materials and products awaiting reuse or recycling is essential for optimising resource utilisation.

Cost Factors. This subcategory addresses barriers related to cost implications that deter the adoption of CE practices. High costs related to materials/products/systems procurement, processing, recovering and recycling can impede the economic viability of circular solutions.

Complexity and Composition. This subcategory focuses on barriers related to the complexity and composition of materials, products and systems used in circular practices. Understanding and managing the diverse compositions of materials and products are essential for effective recycling and reuse strategies.

Monitoring and Management. This subcategory addresses barriers to monitoring and managing material flows within the CE framework. Effective monitoring systems are crucial for tracking material usage, ensuring quality standards, and optimising resource efficiency.

Table 1.11 Subcategories of barriers under Category 8 at the urban level

Subcategory	Count	Reference
Material availability and quality	4	[31, 33]
Space and storage issues	4	[26, 37, 39]
Cost factors	2	[39]
Complexity and composition	1	[19]
Monitoring and management	1	[29]
Total	12	

Upon conducting a quantitative analysis of reported barriers from the literature, it is evident that *Inadequate processes and infrastructure* pose significant challenges to adopting CE practices in urban areas. These barriers highlight the urgent need for technological advancement, efficient resource management, and sustainable development. Bridging these skill gaps is crucial to fostering expertise within the industry and overcoming technical hurdles to CE adoption. Comprehensive infrastructure development, streamlined logistical systems, and integrated urban planning strategies are necessary to support sustainable material flows and enhance resource efficiency.

Furthermore, establishing robust *Technical Standards* is essential to ensuring consistency, reliability, and safety in materials reuse and recycling processes. This involves addressing challenges such as insufficient waste management and recycling infrastructure, fragmented production networks, logistical inefficiencies, and deficiencies in urban planning that integrate CE principles.

Moreover, the *lack of adequate urban planning tools*, management systems, and design and construction tools present additional barriers to effectively implementing CE practices. Enhancing the availability and functionality of these tools is critical for informed decision-making, optimising resource utilisation, and overcoming operational obstacles to sustainable development.

Lastly, challenges related to *limited availability and quality* of materials, spatial constraints, cost factors, material complexity, and inadequate monitoring and management practices underscore the need for strategic interventions. These include improving material availability, optimising spatial usage, mitigating cost implications, simplifying material compositions, and enhancing monitoring systems to manage sustainable resources effectively.

Furthermore, these multifaceted barriers require concerted efforts across technological innovation, policy development, and capacity-building initiatives. By overcoming these challenges, the transition towards a more sustainable urban environment can be accelerated, effectively promoting CE principles and practice.

1.3.1.3 Quantitative Data Analysis of Each Category

A summary of the collected quantitative data for each category is shown in Table 1.12.

Table 1.12 Quantitative data collection at the urban level

What is the barrier category	Where it happens							When it happens					
Technical barrier category	Building layer (a)							Building life cycle phase (b)					
	1	2	3	4	5	6	7	1	2	3	4	5	6
Lack of information/ data							10	3	1			7	2
Lack of technical knowledge							8	6	5	5	3	4	2
Lack of technical skills							5	2	1	3	2	5	2
Lack of technical standards							10	7	6	2	4	7	5
Inadequate machines/ equipment											1	1	1
Inadequate processes and infrastructure							15	4	4	5	13	11	10
Lack of methods/tools							8	7	7	1	2	4	2
Lack of materials/ products/ systems							6	2	1	3	3	5	4

What is the barrier category	Who is responsible for/ involved in						How it affects the implementation of circular strategies									
Technical barrier category	Stakeholders (c)						Circular strategies (e)									
	1	2	3	4	5	6	1	2	3	4	5	6	7	8	9	10
Lack of information/ data	5	8	3	7	10			1	4	7	1	1	4	6	6	3
Lack of technical knowledge	3	5	3	2	6	4		3		5	1	1	2	3	5	
Lack of technical skills	1	3	2	1	4	1		1		3	1	1	2	1	3	
Lack of technical standards	2	11	2	1	12			1	4	8	2	1	1	4	8	

(continued)

Table 1.12 (continued)

What is the barrier category	Who is responsible for/ involved in						How it affects the implementation of circular strategies									
Technical barrier category	Stakeholders (c)						Circular strategies (e)									
	1	2	3	4	5	6	1	2	3	4	5	6	7	8	9	10
Inadequate machines/ equipment					1	1									1	
Inadequate processes and infrastructure	8	8	11	6	10	3	0	2	3	13	1	1	4	4	10	2
Lack of methods/tools	0	7	0	0	2	1	0	0	1	0	0	0	1	0	1	0
Lack of materials/ products/ systems	2	2	5	2	5	1	0	1	2	6	0	0	3	0	5	1

(a) Building layers [17]: (1) site; (2) structure; (3) skin; (4) services; (5) space plan; (6) stuff; (7) system (district)
(b) Building life cycle phases (adapted from RIBA [18]; Brincat et al. [15]): (1) Strategic Definition and Briefing; (2) Design; (3) Manufacture; (4) (Dismantling and) Construction; (5) Use and Refurbishment; (6) End-of-life
(c) Stakeholders [19]: (1) Clients; (2) Project professionals; (3) Suppliers; (4) Public; (5) Government; (6) Others
(d) Circular Strategies [20]: (1) Refuse; (2) Rethink; (3) Reduce; (4) Reuse; (5) Repair; (6) Refurbish; (7) Remanufacture; (8) Repurpose; (9) Recycle; (10) Recover

Category 1: Lack of Information/Data

Where it happens. The studies addressing the lack of information/data barriers at the system (district) scale broadly describe these barriers under the headings "data capabilities and data availability" [26, 29], "limited knowledge" [10, 27, 31, 40], "the lack of integration among sources of information" [19, 24, 25, 28, 30, 33, 34], and "insufficient data" [32].

When it happens. Lack of information/data barrier at the system (district) scale has been discussed mostly in the "Use and Refurbishment" phase within the building life cycle phases [26, 27, 30, 31, 34]. However, dismantling and construction and end-of-life were evaluated in the scope of this barrier [27], as well.

Who is responsible for/involved in. The government, project professionals, and the public are primarily responsible for the lack of information/data barriers. However, clients and suppliers are also included in these responsible stakeholders [10, 24–31, 26, 27].

How it affects the implementation of circular strategies in the built environment. While it is emphasised in the literature that the lack of information/data barriers will mainly affect reuse, recycle, and repurpose [24–27, 26, 27], which are circular strategies in the built environment, it is stated that this barrier will also be effective on reduce, rethink, repair and refurbish [25, 26, 28, 30].

Category 2: Lack of Technical Knowledge

The identified technical barriers can be broadly categorised into several key knowledge gaps hindering the implementation of circular construction practices. The first one is "limited application knowledge". A lack of understanding of suitable circular construction technologies and how to effectively apply them to projects acts as a major barrier [30]. This encompasses understanding the design process for disassembly, as discussed in reference [27]. "Production and quality management challenges" is the second knowledge barrier identified. Insufficient knowledge in areas like production management and the high-quality manufacturing of recycled construction materials presents a significant hurdle [31]. "CE knowledge deficit" also provides a limited understanding of CE principles among stakeholders, including public infrastructure providers, construction professionals, and urban manufacturers, hindering widespread adoption [32]. "Scaling up challenges" are associated with the lack of knowledge required for scaling up circular construction businesses, and urban manufacturing operations pose a significant barrier [31]. The loss of traditional construction skills and the challenge of transmitting this knowledge to younger generations create a potential gap in the technical expertise needed for circular construction [28], and it can be listed as a "skills gap threatening knowledge transfer".

Immature recycling technologies [24] and a lack of technological expertise across various aspects of circular construction [25] pose significant technical challenges, and they can be listed as "technological immaturity and expertise shortages". The lack of this expertise further hinders the industry's ability to adapt to CE practices [25].

Where it happens. Analysing the identified skill gaps using the Zimmann et al. [17] levelling reveals their prevalence across various stages of urban circular construction. They are mostly linked to the district (system) level, but we can still link them to staff and service levels.

When it happens. The identified technical knowledge barriers can be mapped across the building life cycle phases, adapted from RIBA [18] and Brincat et al. [15]. In the initial stages (Strategic Definition and Briefing), understanding technical constraints [29], selecting suitable circular construction technologies and their applications [30], and integrating design for disassembly [27] are crucial. A commonly agreed-upon understanding of CE principles across the industry is necessary for effective project planning [27]. The design phase (Stage 2) requires further development of these elements, with a strong emphasis on technical expertise to ensure designs are compatible with circular construction practices [25]. Knowledge gaps in production management, high-quality manufacturing [31], and expertise in handling reused construction and demolition waste (CDW) [24] become critical during the manufacturing stage (Stage 3). Similarly, limitations in recycling technology [24] can impact the quality and viability of recovered materials. Some knowledge, like the principles of high-quality manufacturing [31], can be used in both the dismantling and construction stages (Stage 4). However, because of the limitations of current technologies for these steps in a circular framework, they need to be looked at separately [35]. The use and refurbishment stage (Stage 5) addresses the importance of stakeholder awareness of CE principles for building maintenance and potential reuse [32]. Preserving traditional construction skills alongside knowledge of reusing CDW [24] becomes vital during refurbishment projects [28]. Knowledge gaps in production management [31] and the limitations of recycling technology [24] can also impact decisions and options for material processing at the end-of-life stage (Stage 6). This mapping underscores the need to address these knowledge gaps throughout the entire building life cycle to successfully implement circular construction practices.

Who is responsible for/involved in. Our analysis of stakeholder responsibility for addressing technical knowledge gaps reveals a shared burden across various groups throughout the building life cycle [19]. Project professionals (2) play a crucial role in acquiring knowledge of suitable circular construction technologies and their application [30], while also collaborating with designers (2) to integrate design for disassembly principles and navigate the absence of a commonly agreed-upon CE concept [27]. Suppliers (3) need to address knowledge gaps in production management and high-quality manufacturing of recycled materials [31] to ensure a reliable supply chain. This knowledge gap is also relevant for government agencies (5) as they can play a role in setting standards and promoting the best practices. The loss of traditional construction skills and the need to equip younger generations [30] require collaboration between clients (1), the public (4), and the government (5) to ensure knowledge transfer and workforce development. Technical constraints [17] primarily impact the public (4) as they experience the limitations of existing infrastructure. Clients (1) and suppliers (3) share responsibility for addressing the lack of knowledge about reusing CDW [24]. Immature recycling technology [24] presents a challenge for clients (1), government (5), and other stakeholders (6) who may be involved in financing or promoting research and development efforts. Similarly, the lack of technological expertise [25] and the unavailability, which exacerbates CE adaptation [25], necessitate collaboration between suppliers (3) and other stakeholders (6) involved in technology development and implementation.

How it affects the implementation of circular strategies in the built environment. Our review reveals how technical knowledge barriers can affect the implementation of various circular construction strategies, as outlined by Stijn and Gruis [20]. Insufficient knowledge in production management and high-quality manufacturing [31] creates challenges for Reuse (R4), Remanufacture (R7), and Recycle (R9) strategies. These strategies rely on a robust system for processing and reintroducing materials back into the construction value chain, which requires skilled professionals and efficient production processes. The loss of traditional construction skills and the need to equip younger generations with the necessary knowledge [28] create hurdles for Rethink (R2), Repair (R5), Refurbish (R6), and Repurpose (R8) strategies. These approaches often involve extending the lifespan of existing buildings or components, requiring expertise in traditional construction methods alongside an understanding of how to adapt these skills for a circular approach [28]. The lack of knowledge about reusing construction and demolition waste (CDW) [41] affects not only Reuse (R4) but also Reduce (R3) strategies, as it limits the ability to divert waste from landfills. Immature recycling technology [24] poses challenges for Reuse (R4) and Repurpose (R8) strategies, as it can limit the quality and viability of recovered materials for these applications. Similarly, the lack of technological expertise and the unavailability of such expertise hinder CE adaptation [25] creating significant barriers to Reuse (R4) and Recycle (R9) strategies, which rely on efficient processing and reintegration of materials into the construction cycle. The lack of knowledge on design for disassembly [27] creates challenges for implementing Reuse (R4), Remanufacture (R7), Repurpose (R8), and Recycle (R9) strategies, as it hinders the ability to recover components and materials for future use.

Category 3: Lack of Technical Skills

Our literature review identified five key technical barriers under the "lack of technical skills" category for implementing circular construction principles in urban environments. These barriers can be broadly categorised as "skill gaps" and "knowledge deficits", impacting both the construction workforce and those involved in urban planning and infrastructure. One of the most critical challenges identified is "the lack of skills and expertise among public infrastructure and service providers" [34]. This includes knowledge of circular design principles, using recycled materials, and implementing construction methods that facilitate future disassembly and material recovery. Integrating circularity into urban construction projects becomes a major challenge without a skilled workforce equipped to implement these practices. A "lack of knowledge regarding scaling up production and understanding the contribution of urban manufacturing to the CE" presents another significant barrier [31]. Limited knowledge in this area hinders the development of robust and localised supply chains for recycled construction materials, a crucial element for achieving circularity in urban environments. "The potential loss of traditional construction skills and the need to equip young professionals with the necessary skills for circular construction" is a concern highlighted by another publication [28]. Ensuring a smooth transition and knowledge transfer between generations of construction professionals is vital to maintaining core construction skills while integrating them with circularity principles. It was also identified that "the lack of access to essential business-related services and technical assistance" as a barrier [37]. This includes access to financing, legal advice, and expertise in circular business models. We consider that having good access to or being able to access such services might be linked to other barrier categories (e.g. lack of methods/tools), however, we decided to relist it under the category of lack of technical skill, too, by considering its linkage with the professional skillsets among construction professionals needed to access those services. Without this support and skill set, construction companies and developers may struggle to navigate the complexities of implementing circular construction practices in their projects. "The shortage of human resources knowledgeable about CE principles across all stakeholders" involved in construction and urban planning hinders widespread adoption [25]. This technical barrier is also linked with other categories (e.g. it includes a lack of awareness and understanding among decision-makers, designers, and engineers), however, we decided to relist it here following the same logic given above.

Where it happens. Analysing the identified skill gaps using the Zimmann et al. [17] levelling reveals their prevalence across various stages of urban circular construction. They are all mostly linked to the district (system) level, but we can still link them to staff and service levels.

When it happens. Our review shows that skill gaps and knowledge deficits manifest across various stages of the urban circular construction lifecycle. During the strategic briefing and definition stage, two key challenges emerge a lack of expertise among public infrastructure providers [34] and limited knowledge regarding scaling up the production of recycled materials and understanding the role of urban manufacturing in the CE [31]. This knowledge gap concerning scaled-up production persists and impacts manufacturing processes as well [31]. Furthermore, the lack of access to essential business services and technical assistance likely hinders various aspects of manufacturing [37]. Interestingly, while the identified barriers don't explicitly point to specific knowledge deficits during dismantling, construction, and end-of-life stages, the previously mentioned challenges regarding production scaling and business services may also pose difficulties in these later phases [31, 37]. The literature review suggests that skill gaps and knowledge deficits are most frequently reported during the use and refurbishment stages, highlighting the need for further research to pinpoint the specific challenges faced during these crucial lifecycle phases.

Who is responsible for/involved in. While the lack of technical skills for circular construction cuts across nearly all stakeholders in the urban construction value chain, addressing these deficits requires a collaborative effort. The literature suggests a particular focus on construction professionals, who need training and knowledge to implement circular design and construction methods. Additionally, government intervention through policy changes, funding initiatives, and support for research and development can play a crucial role in creating a supportive environment for the widespread adoption of circular construction practices.

How it affects the implementation of circular strategies in the built environment. We mapped our analysis against the 10 Rs framework by Stijn and Gruis [20] to reveal how skill gaps and knowledge deficits can link to the implementation of various circular strategies in the built environment. The lack of knowledge regarding scaling up production and the limited understanding of urban manufacturing's role in the CE [31] pose challenges for strategies like Reuse (R4), Remanufacture (R7), and Recycle (R9). These strategies rely on a robust system for recovering, processing, and reintroducing materials back into the construction value chain, which requires a skilled workforce and a well-developed manufacturing sector. The loss of traditional skills and the need to equip young professionals with the necessary skills for circular construction [28] creates barriers to Rethink (R2), Repair (R5), Refurbish (R6), and Repurpose (R8) strategies. These strategies often involve extending the lifespan of existing buildings or components, requiring expertise in traditional construction methods alongside an understanding of how to adapt these skills for a circular approach. The lack of access to essential business services and technical assistance [37] and the shortage of human resources knowledgeable about CE principles [25] create significant barriers to Reuse (R4), Remanufacture (R7), and Recycle (R9) strategies. These strategies require not only technical expertise but also business acumen to navigate financing, logistics, and market complexities. A lack of access to support services and a shortage of personnel with the necessary knowledge can significantly hinder the economic viability and successful implementation of these circular construction strategies.

Category 4: Lack of Technical Standards

Where it happens. The studies addressing the lack of technical standards barrier at the system (district) scale indicate that these barriers are largely caused by the absence/inadequacy of standardised processes, certification, and guidelines [19, 24, 26, 27, 31, 32, 34, 36, 38, 39], not being up to date [31], the diversity of existing regulations and the difficulties they cause as a result of their development based on different spatial scales [28, 30, 33].

When it happens. The lack of information/data barriers at the system (district) scale has been discussed mostly under the headings of "Strategic Definition and Briefing" and "Use and Refurbishment" within the building life cycle phases [26, 28, 30, 31, 34–36, 36, 38]. However, design, manufacture, dismantling and construction and end-of-life phases are also evaluated within the scope of this barrier [19, 24].

Who is responsible for/involved in. Project professionals and the government are mostly held responsible for the Lack of technical standards barrier [19, 24, 26–28, 28, 30, 34–36, 35, 36]. However, clients [19, 28], suppliers [38, 39] and the public [28] are also included in these responsible stakeholders.

How it affects the implementation of circular strategies in the built environment. While it is emphasised in the literature that the lack of technical standards barrier will mostly affect reuse and recycling, which are circular strategies in the built environment [19, 24, 26, 27, 30, 38, 39], some studies state that this barrier also affects rethink, reduce, repair, refurbish, remanufacture, and repurpose [24, 26–28, 28, 38].

Category 5: Inadequate Machines/equipment

Researchers identified immature recycling technology as a major barrier [24].

When it happens. The main barrier was identified in the building life cycle phase: Dismantling and Construction, Use and Refurbishment, and End-of-life [24].

Who is responsible for/involved in. Government should put more emphasis on state-owned recycling centres because of their highest eco-efficiency. The government should encourage economic incentive measures such as shortening the loan application period and lowering the land rent for CDW management businesses. New market modes, such as the Public–Private-Partnership should also be encouraged to relieve the economic pressures on CDW treatment/recycling companies [24].

How it affects the implementation of circular strategies in the built environment. The main barrier affects the implementation of circular strategies in phase recycle, because the conditions for CDW reduction are not well developed (in China), mainly due to ineffective management systems, immature recycling technology, an underdeveloped market for recycled CDW products, and immature recycling market operation [24].

Category 6: Inadequate Processes and Infrastructure

The barriers in the category of Inadequate processes and infrastructure can be grouped as follows:

- **Waste collection and Classification Systems**: Low-quality collection and classification systems and Immature recycling technology [24].
- **Material Handling and Infrastructure**: Lack of pre- and post-use material handling systems [25], Resource-intensive infrastructure lock-in [25], Poorly developed infrastructure for waste management and recycling [10] and for support CE management [10], Insufficient or incompatible local processes and infrastructure [33], Non-integrated and refurbished infrastructure [29].
- **Integration and Networking**: Lack of integrated CDW processes, tools, and practices group [19], Lack of networking initiatives with legislative support [26], Limited integration of urban metabolism [27], Lack of reverse logistics infrastructure, Fragmentation of supply chains [35], Limited access to production networks [31, 37].
- **Spatial and Scaling Issues**: Lack of space for circular infrastructure [storage, reuse, recycling [31], Challenges in scaling up [37], Spatial precarity [39].
- **Planning and Urban Processes**: Neglecting people and ‘place-making’; poor connection and limited integration of place-making urban ‘processes’ and urban ‘metabolisms’ [27, 40].

Where it happens. Technical barriers related to Inadequate machines and equipment were mainly reported at the national (government) level [24, 25], local (municipal, urban) level, and organisational (company) level [19].

When it happens. The following phases of the building life cycle most frequently reveal inadequate machines and equipment:

- Dismantling and Construction [10, 19, 24–26, 31, 35, 37–39]
- Use and Refurbishment [25, 26, 29, 31, 34, 35, 37–39]
- End of life [19, 24, 26, 31, 35, 37–39].

Who is responsible for/involved in. Responsibility in this area was most frequently identified in the:

- Suppliers [25, 26, 29, 31, 35, 37–39], and Government group [10, 25, 26, 29, 31, 34, 35, 37, 39],
- Followed by clients [10, 19, 26, 29, 35, 38], and Project Professionals [10, 19, 26, 29, 35, 38].

How it affects the implementation of circular strategies in the built environment. The impact on circular strategies has most often been identified in the process of Reuse [19, 24, 26, 27, 29, 31, 34, 35, 37–39] and in the process of Recycle [19, 24, 26, 27, 29, 31, 37–39].

Category 7: Lack of Methods/tools

Literature identifies several key issues as the main barriers and challenges associated with the lack of methods and tools for implementing circular strategies. Firstly, urban systems fail to integrate circular design and thinking sufficiently, and professionals in the built environment lack the necessary design and logistics expertise [34]. There is also a notable lack of comprehensi CE metrics, tools, and design frameworks, which are essential for assessing and guiding the effectiveness of circular practices [10]. Similarly, according to [27], the lack of standardised methodologies for evaluating CE initiatives is a significant barrier, making it difficult to measure progress and ensure successful outcomes.

The integration of technology into design tools presents a particular challenge in the construction sector, as existing tools often fail to align with circular principles, underscoring the need for advanced design technologies [27]. Furthermore, the lack of enabling digital technologies, especially those that support real-time data tracking and resource management, hinders the implementation of circular strategies, especially in construction [35].

Finally, inappropriate urban planning and ineffective management systems exacerbate these challenges because they are not designed to support circular practices, resulting in resource inefficiencies and implementation issues [24].

Where it happens. Most reports at the system (district) level highlight the barriers and challenges associated with the lack of methods and tools for implementing circular strategies. This is evident in the literature, where numerous studies have highlighted these issues in the broader context of urban and district-level planning and management. For example, the lack of comprehensive CE metrics, tools, and design frameworks [10] and standardised methodologies [27] is a significant issue at the district level, as these are essential for assessing and guiding the effectiveness of circular practices across systems.

When it happens. The barriers and challenges related to the lack of methods and tools for implementing circular strategies occur most frequently in specific phases of the building life cycle. The literature review primarily identifies these phases as follows:

1. Strategic definition and briefing: This phase experiences a high number of barriers, with seven cases identified. During this phase, the lack of comprehensive CE metrics, tools, and design frameworks becomes apparent [10]. These tools are crucial for setting strategic goals and guiding the planning process towards circular practices. The absence of standardised methodologies [27] hinders the evaluation of CE initiatives, necessitating the early establishment of effective evaluation frameworks to track project progress and outcomes.
2. Design: Again, seven barriers were identified. As Hossain et al. [27] point out, the challenge of integrating technology into design tools is particularly relevant, as existing tools often lack alignment with circular principles, thereby hindering the integration of circular strategies. In addition, the lack of enabling digital technologies affects the ability to create designs that support real-time data tracking and resource management [35]. In addition, the lack of appropriate design and logistics expertise among built environment professionals becomes apparent at this stage [34].
3. Use and Refurbishment: Four barriers were identified at this stage. Specifically, the design of ineffective management systems and inappropriate urban planning presents significant challenges, as they do not facilitate the implementation of circular practices [24].

Who is responsible for/involved in. Project professionals and government agencies are primarily responsible for addressing the barriers and challenges associated with a lack of methods and tools for implementing circular strategies. Project professionals, including architects, engineers, designers, and construction managers, play a critical role in integrating CE principles into projects [27, 31, 34, 35]. Meanwhile, government agencies are essential in creating regulatory frameworks, providing incentives, and supporting the development of standardised methodologies for CE assessment [24].

How it affects the implementation of circular strategies in the built environment. The lack of methods and tools significantly impacts the implementation of circular strategies in the built environment, particularly strategies such as Reduce, Remanufacture and Recycle [24]. This lack of appropriate methods and tools undermines the potential to reduce resource consumption, remanufacture components, and recycle materials in the built environment.

Category 8: Lack of Materials/Products/Buildings/Systems

The main barriers and challenges in the category of lack of materials/products/buildings/systems for implementing circular strategies in the built environment are multifaceted and highlighted across the literature. Several studies have identified spatial challenges as significant obstacles to the implementation of circular strategies in the built environment. One of the main issues is the scarcity of affordable sites and spatial precarity, which limit the availability of suitable locations for circular practices. This includes a shortage of affordable industrial land within urban areas and reduced space for urban manufacturing, restricting the potential for CE practices in cities [37, 39]. Additionally, the limited space to stock disassembled materials before reuse impacts the efficiency of material recovery and reuse processes [26].

Moreover, material-related issues present substantial barriers. The low cost of new materials discourages the use of recycled materials, compounded by the high cost of labor required for deconstruction and material recovery [39]. There is also limited work and interest in high-value or niche products, and a limited availability of quality recycled material from local sources [31]. Furthermore, the complexity of materials and building compositions, involving several layers and modifications over their lifespan, makes disassembly and recycling challenging [19].

Monitoring and managing urban resource flows is also a complex task that poses significant challenges for implementing circular strategies [29].

Where it happens. The barriers related to the lack of materials, products, buildings, and systems for implementing circular strategies are mainly reported at the system (district) level. This is evident in the literature, which highlights challenges such as the scarcity of affordable sites, limited industrial land within urban areas, and insufficient space for urban manufacturing and storing disassembled materials [26, 37, 39]. These barriers are particularly pronounced in urban districts, where spatial and resource constraints are more acute.

When it happens. The barriers related to the lack of materials, products, buildings, and systems for implementing circular strategies occur at various phases of the building life cycle. They can be prioritised based on their occurrence as follows:

1. Use and Refurbishment: This phase encounters the highest number of barriers (5), including limited availability of quality recycled materials, high labor costs, scarcity of affordable sites, spatial precarity, and limited space for storing disassembled materials [4, 26, 37, 39].
2. End-of-life: The end-of-life phase faces four barriers, such as high labor costs for deconstruction, scarcity of space to store disassembled materials, and difficulties in managing complex material compositions [4, 26, 37, 39].
3. Manufacture: The manufacturing phase is affected by three barriers, including limited availability of quality recycled materials, high labor costs, and lack of affordable industrial land within urban areas [31, 37, 39].

4. (Dismantling and) Construction: This phase also encounters three barriers, including the scarcity of affordable sites, limited space for urban manufacturing, and high labor costs [31, 37, 39].
5. Strategic Definition and Briefing: This initial phase has two barriers, including the scarcity of affordable sites and challenges in monitoring and managing urban resource flows [29, 39].
6. Design: The design phase encounters a challenge due to the restricted availability of high-quality recycled materials [39].

Who is responsible for/involved in. The responsibility for addressing barriers to implementing circular strategies involves multiple stakeholders, prioritised by the number of barriers they are involved in. Project professionals and the public are each involved in five barriers, such as managing design challenges and influencing market demand for circular practices [4, 26, 37, 39]. Clients and suppliers handle two barriers each, including the scarcity of affordable sites and providing quality recycled materials [26, 39]. The government is responsible for one barrier, primarily related to creating supportive policies [39].

How it affects the implementation of circular strategies in the built environment. The barriers to implementing circular strategies in the built environment affect various strategies to differing extents. The most impacted strategy is Reuse, with seven barriers, such as limited space to store disassembled materials and the scarcity of quality recycled materials [19, 26, 29, 31, 37, 39]. Recycling follows six barriers, including the complexity of materials [19, 26, 29, 31, 37, 39]. Remanufacturing faces three barriers, such as the limited availability of quality recycled materials and high labour costs [31, 37, 39]. Reduce encounters two barriers [39]. Rethink is affected by one barrier, with spatial precarity limiting suitable locations for circular practices [39], and Recover also has one barrier [29].

1.3.2 Building Level

1.3.2.1 Quantitative Analysis of Literature

The first step in the elaboration of technical barriers provided by the literature at the building level focused on a quantitative approach to understanding the number and typology of consulted documents. Moreover, a quantification of the identified barriers in total and each category was extracted. Figure 1.7 shows the number of documents providing the different barriers distinguished between scientific and non-academic literature including reports (research, technical, project deliverable) provided by European Institutions website (such as CORDIS, JRC, EPC, OPE) and global and national Associations ad Cooperatives of designers, engineers, consultants, companies (such as ARUP, Circle, UKGBC).

Lack of information/data (48 documents) and *lack of technical standards* (45) were the most consistent barrier categories, followed by *lack of technical knowledge*

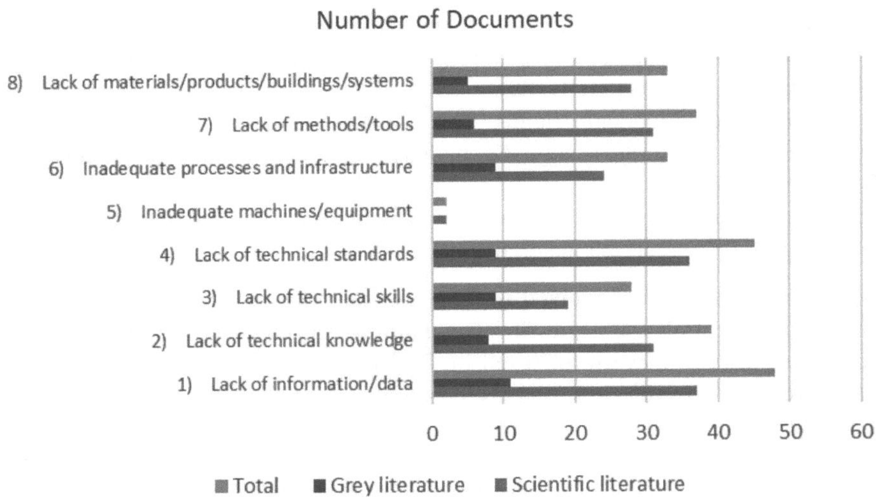

Fig. 1.7 Total consulted documents, in academic (called scientific literature) and non-academic literature (called grey literature), providing information about the different categories of technical barriers

(39) and *lack of methods/tools* (37). The remaining categories present a comparable amount of related documents, except for *inadequate machines/equipment* which were recorded in only two documents. The scientific literature prevails in all the categories.

The number of barriers discovered in terms of multiple and variation in each category is presented in Fig. 1.8.

The highest number of barriers is recorded in *methods/tools* (89) and *information/ data* (73), and they are mainly found in scientific literature. A significant number of identified barriers are also related to *materials/products/buildings/systems* (68), *technical standards* (65), *and technical knowledge* (60). Figure 1.9 shows the building layers in which the barriers are found.

The lack of methods and tools is primarily recorded at the structure, skin and services layers. A similar pattern is observed in the lack of information and data. This pattern is a result of the most frequent barriers. In other cases, such as lack of materials/products/buildings/systems and lack of technical standards, obstacles to the implementation of the CE were also consistently recorded in stuff. Figure 1.10 shows the building life cycle phases affected by each barrier.

It is clear how the design phase is influenced by the lack of methods, standards, skills knowledge, and data. Also, dismantling and construction and end-of-life detection have a significant influence on technical barriers. The stakeholders responsible for the technical barriers are presented in Fig. 1.11.

Project professionals and suppliers have a crucial role. In addition, the importance of governments is evident in the development and implementation of technical standards. The circular strategies affected by technical barriers are shown in Fig. 1.12.

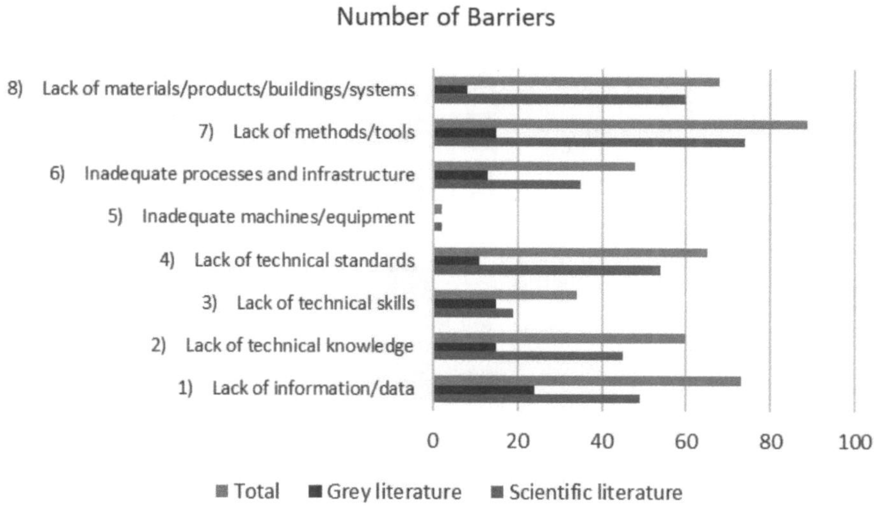

Fig. 1.8 Total barriers, and barriers provided by scientific and non-academic literature in the different categories

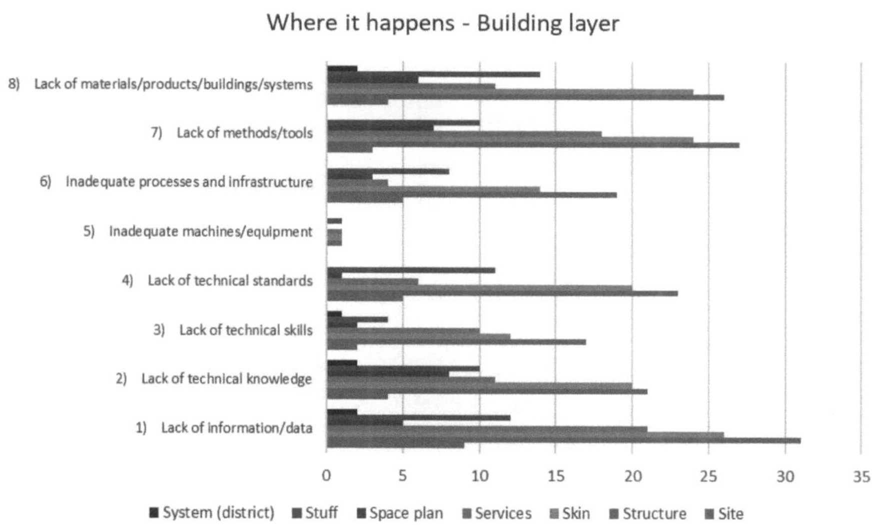

Fig. 1.9 Building layers in which technical barriers were recorded

Reuse and recycling are the circular strategies more affected by technical barriers, as already noticed at the urban level. Also, recovery shows a significant impact from the technical barriers.

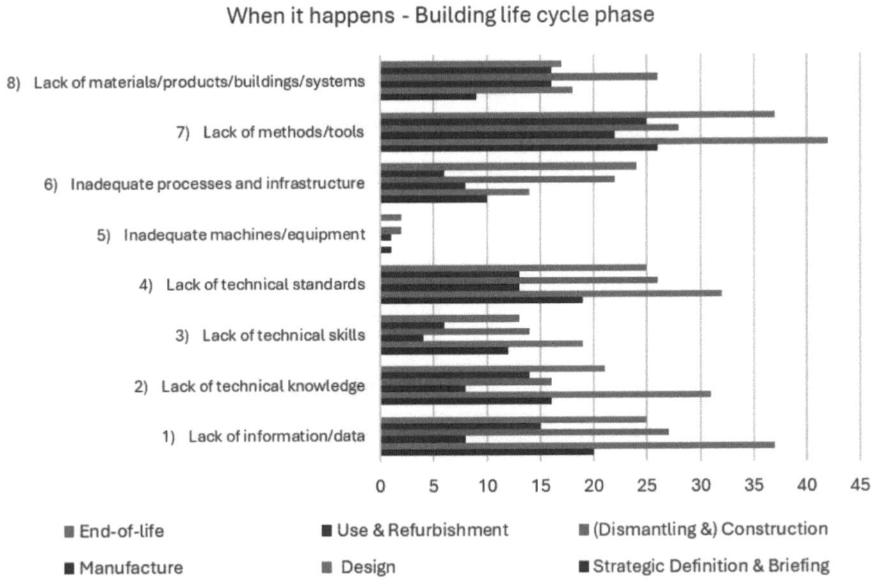

Fig. 1.10 Building life cycle stages in the different categories of technical barriers

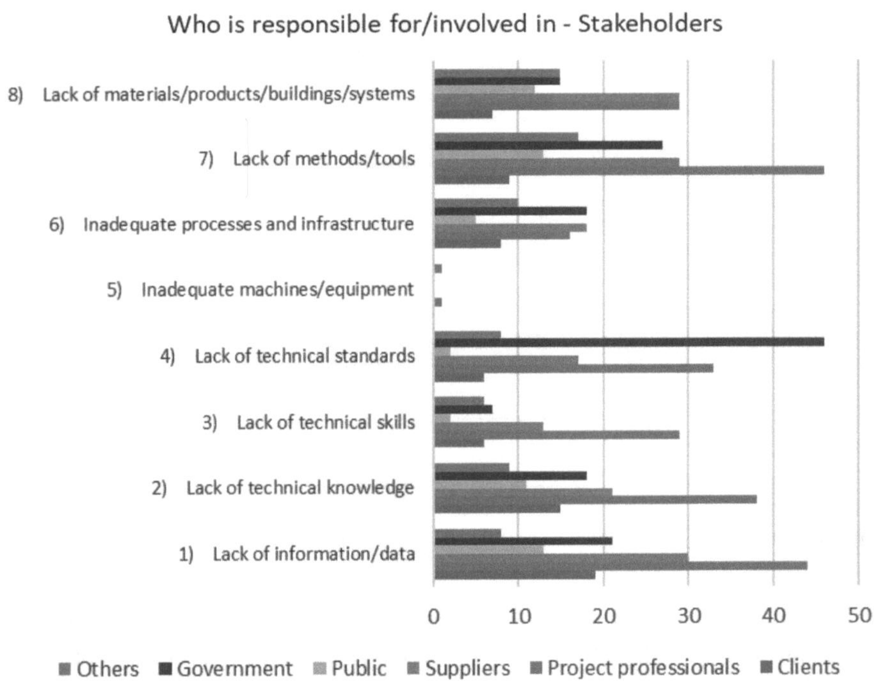

Fig. 1.11 Stakeholders involved in the different categories of technical barriers

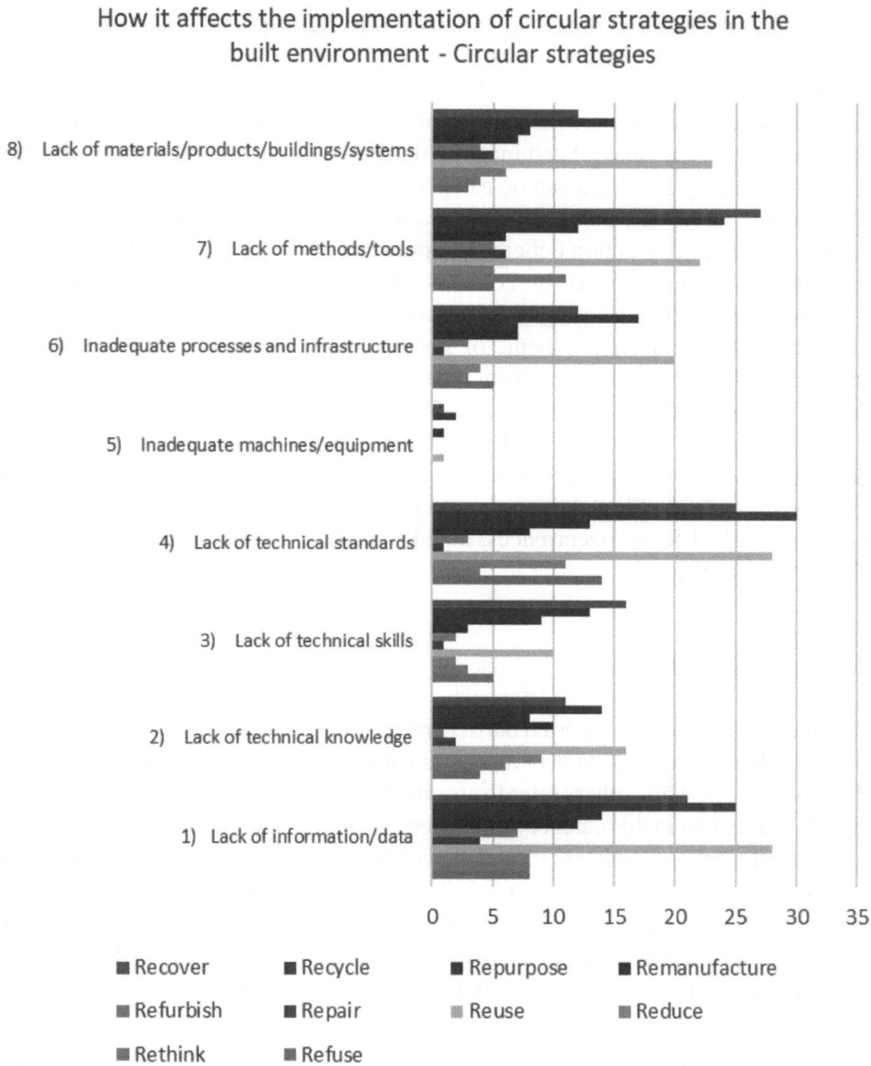

Fig. 1.12 Technical barriers categories affecting circular strategies

1.3.2.2 Collection of the List of Barriers in Each Category, and Analysis of Relevance

Category 1: Lack of Information / Data

The list of barriers in the category was produced by looking at the collected data and annotating them in Table 1.13. Repetitions or similarities were removed. In the case of a large number of barriers, subcategories were identified. Then, the more relevant barriers and subcategories are listed. Finally, a brief description of the barriers in the category and their relevance is provided.

Through the quantitative analysis of reported barriers from the collected literature, it was revealed that issues such as the lack of information on indexing material stock, tracking reused materials, related material databases, inaccuracies, and the absence of data on existing buildings can be categorised under a single theme: the lack and inaccuracy of information about the material stock. This theme emerged as the most prevalent type of information deficiency, appearing 38 times at the building level in the identified literature [6, 10, 19, 26, 34, 41–62].

Furthermore, within the category of lack of data/information, 15 technical barriers can be categorised under the theme of sustainability assessment. This theme was identified as the second most prevalent type of information deficiency at the building scale in the analysed literature [6, 53, 57–59, 62–70]. The theme of disassembly and technical information was noted 12 times [43, 53, 55, 57–59, 61, 71–73]. The lack of stakeholder awareness was reported five times among the extracted barriers within the first category [68, 74–77]. Additional themes, including digital, data interoperability, and property, had fewer occurrences, noted four, four, and two times respectively [19, 55, 65, 78–80].

Category 2: Lack of Technical Knowledge

A detailed analysis of the reported barriers revealed that they are of different levels: a considerable portion of them is of a general nature (e.g. "Lack of effective technical CE knowledge"), while others refer to specific fields of technical knowledge. Also, it is worth noting that in this category of barriers, issues related to business approaches/ market conditions and societal/cultural views have been included. In the context of the present analysis, none of the reported barriers in the category of "Lack of technical knowledge" has been excluded from the considered list, although some of them could have been classified in other major categories of barriers as well. The main subcategories into which the barriers belonging to the category of "Lack of technical knowledge" are grouped as follows (Table 1.14):

(i) General: adopted also in cases where more than one axes of design and construction are highlighted to the same degree (e.g., "Technical difficulties during the design and construction process").
(ii) Design-related issues (DfD, circular design, etc.).
(iii) Materials and resources (incl. also waste, reuse of materials and components, etc.).
(iv) Sustainability (also when issues such as carbon-intensive processes and or LCA are concerned).
(v) BIM and digital background.
(vi) Construction-related issues.
(vii) Business/market/economy-associated parameters (mostly economic/financial concerns).

Table 1.13 Collection of the list of barriers in the lack of information/data category and analysis of their relevance at the building level

Lack of information/data	Quantitative analysis
Stakeholders' awareness	*5*
Lack of awareness among stakeholders	
Lack of awareness and familiarity among professionals and the public (e.g. clients)	
Lack of awareness across the board (public/industry/government)	
Lack of awareness and culture in the industry	
Lack of data, reusing, mining, stock	*38*
Lack of information on the quality of reused materials	
Lack of information on the mechanical properties of the recovered materials	
Scarcity of data	
Disassembly, technical info	*12*
Technical perceptions	
Unavailability of information on technical features in the deconstruction process	
Unavailability of disassembly information	
Sustainability assessment	*15*
Lack of broadly recognized methodologies for calculating the embodied carbon of circular technical solutions	
Assessment of environmental benefits	
Digital	*4*
Lack or overload of digital information related to products and materials	
Lack of digital data sources to feed repositories such as DBL that can promote circularity	
In the design phase, maintenance providers use more and more IoT/Condition-Based maintenance approaches, therefore requesting more data from systems on actual performance to plan maintenance. Important data/information points are missing, most of the missing are at the start of maintenance cycles and are added later	
Data Interoperability	*4*
Data exchange and interoperability issues	
Lack of interoperability between systems and data	
Property	*2*
Property owners do not have (direct) access to supply information channels of reused products, components and materials	
Limited access to future property users in the design phase	

(viii) Perceptual/cultural/societal issues (when views, readiness for change, and opinions are, among others, addressed).

It is worth noting that the aforementioned subcategories have been arbitrarily determined based on the described content of the barriers. The axis of this determination is the field of technical knowledge ("what") and not the stakeholder involved in the lack ("who").

It should be pointed out that some of the barriers could have been classified into more than one subcategory. In each one of those cases, the axis that appeared to be the determining one regarding the lack of knowledge for a specific field prevailed, and the respective subcategory encompassed the barrier. For example, although the barrier "Necessity of a key-expert-operator (one person or one company) that can manage and follow the entire circular building process. LCA sustainability experts" does have a component of a general nature, it also contains a specific reference to LCA; this the reason why it is classified under the "Sustainability" subcategory.

In any case, the "General" subcategory seems to be the prevalent one, with relevant barriers being cited 21 times [19, 34, 49, 50, 52, 56–58, 61, 66–68, 72, 81–85]. It is interesting to note that a major group in this subcategory consists of very general statements revolving around the lack of knowledge/experience/awareness/training for building circularity, CE, etc.

Category 3: Lack of Technical Skills

A list of identified barriers in the lack of technical category, and their subcategory relevance, is shown in Table 1.15.

Through the quantitative analysis of reported barriers from the collected literature, it was revealed that issues such as lack of people with the right skills for the CE [74, 86], lack of skills [34, 45, 51, 61, 77, 80, 81, 83, 85, 87, 88], lack of systemic CE education and training for supply chain members [51, 62, 72, 73], lack of training and knowledge for professionals [48, 65, 80], lack of skills in design and implementation [19, 44, 55–57], lack of awareness and communication skills [68], educational barriers [58, 59, 68] can be categorised under a single theme: lack of technical skills.

This theme emerged as the most prevalent type of information deficiency, appearing 30 times at the building level in the identified literature [19, 34, 45, 48, 51, 55–59, 61, 62, 65, 68, 72–74, 77, 80, 81, 83, 85–89].

Category 4: Lack of Technical Standards

The lack of technical standards as a barrier is mentioned 64 times in the reference documents. These mentions can be broadly grouped into four themes (Table 1.16): (i) lack of standardisation/guidelines, (ii) lack of legal policies, (iii) sustainability quality assurance and (iv) lack of flexibility.

Table 1.14 Collection of the list of barriers in the lack of technical knowledge category and analysis of their relevance at the building level

Lack of technical knowledge	Quantitative analysis
General	*21*
Complexity/confused incentives	
Sectoral barriers: 1-lack of bandwidth compounded by no coherent vision	
Lack of effective technical CE knowledge	
Lack of employee knowledge and awareness about how to integrate practices in the industry. Lack of organisational capabilities and in-house knowledge about CE	
Lack of knowledge of the appropriate technologies and how to apply them, especially in the integration of receiving systems and reverse logistics	
Lack of knowledge retention and learning networks	
Lack of expertise and knowledge in circularity in construction	
Lack of knowledge, training, and experts in CE	
Lack of awareness and knowledge of building circularity	
Poor stakeholder experience (to the extent where project professionals are concerned)	
Lack of understanding/awareness/information/technology	
Practitioners and supply chain with inadequate technical knowledge	
Political/citizen ignorance of technical aspects when making decisions	
Technical difficulties during the design and construction process	
Lack of awareness, knowledge and experience with the CE	
Lack of architects and developers with the required knowledge and willingness to adapt	
Stakeholders throughout the value chain remain insufficiently familiar with how CE principles do or could operate in the built environment	
Knowledge gap between stakeholders: designers and engineers sometimes lack the practical expertise of contractors, demolition contractors, and deconstruction engineers, and vice versa	
Matching standardisation with a variety of end uses	
Lack of knowledge-sharing of best practices in the building environment and translation to different contexts, lack of large-scale demonstration projects and experience with new technologies	
Different levels of knowledge and awareness in the value chain	
Design-related issues (design for disassembly, etc.)	*11*
Lack of practical application (actual practice) of DfMA (Design for Manufacture and Assembly) approach in construction (Studies of DfMA in construction heavily relied on simulations rather than the actual construction)	
Lack in comparison for structures constructed using DfMA and DfD (Design for Disassembly) with structures built with conventional approaches	

(continued)

Table 1.14 (continued)

Lack of technical knowledge	Quantitative analysis
Lack of execution of sustainability assessment studies to evaluate the impact of DfMA and DfD	
Most of DfMA in OSC (off-site construction) studies focuses on the structural components only	
The DfD guidelines are intended only for cold-formed steel structures	
A small body of knowledge addresses the actual usage of DfD in construction	
Lack of leadership, cost, and schedule considerations in the university curriculum for building design	
Limited circular design (incl. lack of know-how people)	
Lack of knowledge on circular design principles, and on the carbon and resource intensity of the linear economy	
Lack of expanded knowledge on circular design practices	
Limited experience with circular design within a built environment	
Materials and resources	*8*
Architects have misconceptions about waste management hierarchy, rarely find objective reasons to reuse building components and have conflicting views about DfD and resiliency	
Not enough people with specific knowledge about reused materials	
Lack of a waste management professional	
Unknown causes of design waste, perception of waste	
Lack of knowledge of reclaimed materials	
Difficulties in understanding and developing EPDs (Environmental Product Declarations)	
Lack of knowledge on reuse and recycling techniques and the capacity to implement	
Design teams are not aware or open to exploring the reuse of existing building products and installations (no quick visibility and transparency on secondary products and materials that are available and meet the required quality criteria)	
Sustainability	*6*
Lack of technical knowledge of sustainable practices	
Few building professionals have the skills to plan and deliver a carbon–neutral building	
Real estate valuation processes do not—at present—fully or reliably account for the value of sustainability, neither in terms of energy consumption nor material circularity	
Insufficient carbon literacy	
Necessity of a key expert operator (one person or one company) that can manage and follow the entire circular building process. LCA sustainability experts	

(continued)

Table 1.14 (continued)

Lack of technical knowledge	Quantitative analysis
Knowledge gaps: There are knowledge gaps regarding the long-term effectiveness of nature-based and regenerative solutions for climate change mitigation and adaptation. Current research is still inconclusive and incomplete in a series of aspects, including the effectiveness of nature-based solutions (for instance, as regards positive effects on human health and wellbeing or the comparative merits of various approaches in the long term)	
BIM and digital background	*3*
Lack of BIM specialists	
Insufficient knowledge of 3D modelling, building information modelling (BIM) and digital twins (DT)	
Lack of technical knowledge in use and capability of BIM throughout the construction cycle	
Construction-related issues	*2*
Lack the appropriate knowledge of construction workers (for example dry construction methods)	
Lack of knowledge regarding traditional joinery disassembly and reassembly	
Business/market/economy issues	*3*
Lack of business active models	
Lack of knowledge on how a CE can be beneficial financially	
Small projects include a lack of budgets to afford some of the specialised consulting	
Perceptual/cultural/societal issues	*6*
Short term perspective	
Conservative mindset	
Lack of acceptance and demand	
Three main uncertainties regarding the applicability and feasibility of PSS have been identified to date: the readiness of companies to adopt them, the readiness of consumers to accept them, and their environmental implications	
The user chooses what he is willing to exchange	
In the operation phase, some architect bureaus operate in a niche market where they first check what is already available before starting the actual design	

The theme of the lack of standardisation and guidelines in CE, DfD, DfMA and BIM was the most prevalent with the large number of appearances (38) in the literature, followed closely by the lack of legal policies (32), whether to consistently regulate or incentivise CE or DfD. The theme of sustainability and quality assurance, which refers to the lack of relevant indicators and other priorities, is found in 21 mentions, while the theme of the lack of flexibility in existing codes and regulations appeared only on 6 occasions.

Table 1.15 Collection of the list of barriers in the lack of technical skills category and analysis of their relevance at the building level

Lack of technical skills	Quantitative analysis
Lack of people with the right skills for CE	*2*
Lack of technical skills in the private sector on CE innovative building projects	
Lack of the necessary capabilities and skills for EOL value extraction	
Lack of skills	*12*
Lack of understanding of LCC, and on top of that, factoring in design for disassembly	
Not enough people with the skills for harvesting reused materials	
Lack of skilled tradesmen	
Lack of educational/training programs for industry employees	
Lack of skilled tradesmen	
Lack of standardised practices	
Lack of leadership for a CE: leadership has played a critical part in the delivery of existing circular built environment projects	
Lack of systemic CE education and training for supply chain members	*4*
Lack of knowledge and training on the environmental benefits and how to design a building for deconstruction of the design team	
Lack of expertise, knowledge flow, or in-house skill sets to repair and remanufacture	
Lack of training, skills and education support. Lack of documentation to support competence improvement	
Lack of skills and expertise	
Lack of basic digital skills and e-literacy	
The workforce is largely trained to work in a linear manner	
Lack of training and knowledge for professionals	*3*
Lack of instruction of decision-makers in an explicit and robust manner	
Lack of training of professionals	
Lack of understanding/skills in applying the principles of the CE	
Lack of information technology (IT) and AI professionals	
Lack of skills in design and implementation	*5*
Lack of adequate skills and training, including the limited use of Information and communication technologies (ICTs)	
Barriers for BIM application, such as the upfront cost of implementation, risks of adoption linked to IP and copyright laws, licensing issues between collaborating parties, and lack of technical skills and experience to drive implementation within organisations	
In the design phase, a barrier is still the classical view of the supplier/customer relationship from a property owner's point of view	

(continued)

Table 1.15 (continued)

Lack of technical skills	Quantitative analysis
Design as a barrier to future refurbishment: design that does not focus on the ability to be adapted, or partially/fully disassembled	
Design as a barrier to future refurbishment: design that does not focus on the ability to be adapted, or partially/fully disassembled	
Change of the traditional skill set of waste management on construction sites	
Lack of awareness and communication skills	*1*
Lack of awareness and skills across the value chain	
Lack of detail within requirements and uncommon language (not specific enough, and terms unfamiliar to the local sector)	
Lack of communication and limited awareness on the financial benefits of CE	
Lack of educational programs	*3*
Limited attention to CE in architecture studies and design	
Lack of education programs in circular thinking for primary and secondary school, vocational and higher technical education	

Category 5: Lack of Machines and Equipment

A list of identified barriers in the lack of machines and equipment category and their sub-category relevance are shown in Table 1.17.

Through the quantitative analysis of reported barriers from the collected literature, it was revealed that issues such as the need for infrastructure and equipment to perform deconstruction for a CE [47] and lack of suitable equipment for the process of deconstruction and material reclamation [50], can be categorised under a single theme: lack of infrastructure and equipment.

This theme emerged as the most prevalent type of information deficiency, appearing 2 times at the building level in the identified literature [47, 50].

Category 6: Inadequate Processes and Infrastructure

A list of identified barriers in the inadequate processes and infrastructure category, and their sub-category relevance, is shown in Table 1.18.

Following the analysis of the literature in the category "Inadequate processes and infrastructure", a total of six barriers have been identified. The lack of adequate logistic systems and spaces stood out as the most important barrier, appearing 22 times in the literature consulted. Within this barrier, the lack of space for the storage of materials was identified as the most relevant, in addition to other barriers such as the lack of space for the processing of recovered materials, a deficient logistic system for their supply, as well as an underdeveloped bank system of recovered materials [10, 26, 38, 47–49, 51, 61, 62, 73, 83, 94, 95].

Table 1.16 Collection of the list of barriers in the lack of technical standards category and analysis of their relevance at the building level

Lack of technical standards	Quantitative analysis
Lack of standardisation/guidelines	*38*
Lack of standardisation and/or guidelines in CE	[6, 10, 19, 26, 44–47, 47, 49–52, 55, 58, 59, 67, 68, 72–75, 78, 80–83, 86, 89, 90]
Lack of standardisation and/or guidelines in DfD	
Lack of standardisation and/or guidelines in DfMA	
Lack of BIM standards	
Lack of legal policies	*32*
Lack of legal policies to standardize DfD/CE	[6, 10, 26, 41, 45, 46, 50, 51, 53, 55, 58, 59, 65, 68, 69, 71, 73, 77, 81–83, 83, 90–93]
Lack of legal policies to incentivize DfD/CE	
Lack of a consistent regulatory framework for CE	
Sustainability and quality assurance	*21*
Lack of indicators (traceability, certification, modularity components)	[6, 10, 19, 26, 45, 47, 51, 55, 58, 59, 67–69, 71, 72, 80, 89, 90]
High priority in other issues	
Lack of flexibility	*6*
Lack of flexibility in (existing) codes and standards	[10, 45, 52, 66, 87]
Lack of flexibility in regulations for reused/salvaged materials	

Table 1.17 Collection of the list of barriers in the lack of machines and equipment category and analysis of their relevance at the building level

Lack of machines and equipment	Quantitative analysis
Lack of machines and equipment	2
Need for infrastructure and equipment to perform deconstruction	
Lack of suitable equipment for the process of deconstruction and material reclamation	

Category 7: Lack of Methods/Tools

A list of identified barriers in the lack of methods/tools category, and their sub-category relevance, is shown in Table 1.19.

In the category "lack of methods/tools", up to eight different barriers have been identified, the most significant barrier being the lack of CE approaches during design. This barrier was mentioned up to 28 times in the literature. Within this barrier, the lack of integration of holistic design processes and the integration of new technologies to assist design were considered the main problems. Other issues identified were incompatibility with future deconstruction of buildings, lack of adaptability design, as well as the lack of optimised repurposing plans for deconstruct and demolition, the complexity and constraints of circular design, and lack of continuous project implementation, monitoring and maintenance [10, 19, 26, 34, 43, 46–49, 55, 62, 65, 72, 79, 80, 90, 93, 94].

Category 8: Lack of Materials/Products/Building Systems

A list of identified barriers in the lack of materials/products/building systems, and their sub-category relevance, is shown in Table 1.20.

Within the category of lack of materials/products/building systems, the most relevant barrier is related to the availability of materials and products, mentioning a total of 28 times in the literature. Within this barrier, the quality and quantity of recycled materials and components are presented as the main sub-barrier. Other identified shortcomings were the availability of materials and alternatives, the lack of bio-based materials and eco-innovation, the lack of demountable elements on the market, the need for critical raw materials and the industrialisation of natural materials [6, 10, 26, 34, 38, 45, 47, 50, 56, 57, 61–63, 73, 86, 89, 93, 94].

Table 1.18 Collection of the list of barriers in the inadequate processes and infrastructure category and analysis of their relevance at the building level

Inadequate processes and infrastructure	Quantitative analysis
Lack of infrastructure for CE (general)	*6*
Lack of infrastructure for recycling	
Lack of infrastructure for CE management	
Lack of institutional infrastructure for CE	
Lack of supporting infrastructure	
Lack of adequate logistic system and spaces	*22*
Lack of or undeveloped logistic system	
Lack of storage facilities	
Lack of material bank/market	
Supply chain integration	
Lack of space for process-recovered components	
Material recovery	*12*
Lack of recycling plants	
Lack of adequate processes and technology to disassemble	
Lack of advanced recycling technology	
Presence of materials unsuitable for reuse	
Incomplete of performance optimization in waste generation and extending asset life	
Processing waste productions	
Connectivity and collaboration	*8*
Lack of digital infrastructure for connectivity	
Lack of information exchange system	
Lack of collaboration between stakeholders	
Professionals involved in the initial phases are not present at the end-of-life stage	
Linear construction	*4*
Fragmented and linear construction industry models	
Misalignment between business planning cycles and built environment asset life cycles	
Other	*7*
Outdated technology	
Lack of support between private and public sectors	
Increased construction process frames	
Jobsite physical constraints	
Performance assessment issues	

Table 1.19 Collection of the list of barriers in the lack of methods/tools category and analysis of their relevance at the building level

Lack of methods/tools	Quantitative analysis
Lack of adequate information tools	*15*
Lack of an adequate database	
Lack of adequate information management	
Disregard of circular economic implications	
Lack of information tools	
Lack of data-driven and empirically validated assessment models	
Difficulty in building data management	
Lack of methods to obtain material information	
Lack of CE design approaches	*28*
Lack of integration of holistic design approaches	
Restriction in design with reuse components	
Lack of CE strategies integration and other emerging digital technologies	
Complex design requirements	
Lack of ongoing enforcement, monitoring and maintenance of projects	
Design for deconstruction incompatibility	
Lack of adaptability design	
Lack of optimised repurposing plans for deconstructing and demolition	
Lack of coordination, management and traceability	*9*
Lack of coordination and management throughout the project life cycle	
Lack of collaboration tools use and development	
Need additional transport	
Lack of traceability systems	
Partial communication between certifications' metrics and goals	
Change in work procedures. Interoperability and compatibility issues	
Lack of adequate logistics	
Lack of guidelines, indicators and metrics	*22*
Lack of complete CE indicators and metrics	
Lack of CE guidelines	
Lack of reliable assurance for reuse products	
Lack of a tool for CE principles	
Unification of the CE concept	
Subjective assessment	
Ambiguity in weighting and scoring	
Aligning policies and technical themes in certification credits	
Lack of implementation of material passports in terms of legislation	
Lack of decision-making protocols	

(continued)

Table 1.19 (continued)

Lack of methods/tools	Quantitative analysis
Inadequate or complex digital tools	*4*
Complexity of software	
Incorrect use of BIM (Building Information Modeling)	
Lack of large-scale digital models in renovation	
Use of property software and protocols	
Lack of tools and methods	*4*
Lack of CE tools	
Lack of relevant tools for material reuse	
Lack of holistic and adequate methods	
Complexity in projects	*2*
Complex buildings	
Complexity of disassembly	
Other	*8*
Uncertainty about the future uses and the end-of-life phase	
Lack of innovation in processes during the building life cycle	
Walmart effect	
Material recovery	
The necessity of refurbishing	
Individual buildings	
Focus on recycling and waste management	

1.3.2.3 Quantitative Data Analysis of Each Category

Category 1: Lack of Information/Data

Results in this category were analyzed by looking at general quantitative data for this barrier category, and summarized in Table 1.21.

Where it happens. Within the most relevant types of barriers, (lack of data on reusing and material stock) the structure and skin are the building layers mostly affected by these barriers [10, 19, 41, 42, 44, 47, 49, 51–55, 53, 54]. The barriers related to sustainability assessment were reported on the layers of structure, skin, and service [53, 67, 69, 70]. In terms of lack of data related to stakeholder awareness, these barriers were mainly reported at the layers of site and structure [68, 74–77]. Regarding the lack of data on disassembly and technical info, the layers of structure and skin were mainly affected by the type of lack of Data/Information barrier [45, 55, 71, 72]. In terms of data deficiency in digital applications, the layer of services was addressed as the most affected layer by this type of barrier [55, 79, 80]. The structure and skin layers were also influenced by the lack of information [10, 19, 26, 41, 42, 44, 45, 47, 50–52, 52, 54, 55, 60, 62, 67, 70–72, 71, 72, 74, 79–81, 79]

Table 1.20 Collection of the list of barriers in the lack of materials/products/building systems category and analysis of their relevance at the building level

Lack of materials/products/buildings/systems	Quantitative analysis
Design/material/building complexity	*19*
Building and design complexity	
Material complexity	
Project uniqueness	
Building rigidity	
Additional procedures for material recovery	
Availability of materials and products	*27*
Quality and quantity of recycled materials and components	
Lack of availability of materials	
Lack of bio-based materials	
Lack of eco-innovation products	
Lack of material alternatives	
Industrialization of natural materials	
Lack of removable elements on the market	
Need for critical raw materials	
Lack of CE design and metrics	*5*
Lack of CE design and metrics	
Lack of design for disassembly	
Market	*3*
Mismatch of demand and supply of reused materials	
Lack of CE business models	
Undeveloped secondary material market	
Structural issues	*3*
Poor structural integrity	
Structural over-dimensioning when using salvaged materials	
Lack of technology and tools	*6*
Lack of collaboration tools	
Technology readiness	
Lack of solutions to satisfy each performance requirement	
Lack of evaluation of the reuse potential of secondary materials	
Other	*7*
Long product life cycles (building and materials)	
Lack of appropriate construction and demolition waste management	
Adequate incentives	
Lack of building maintenance	

Table 1.21 Quantitative data analysis—lack of information/data at the building level

"Lack of technical knowledge"	Where it happens							When it happens					
	Building layer (a)							Building life cycle phase (b)					
	1	2	3	4	5	6	7	1	2	3	4	5	6
Total nr of occurrences in all the reported cases	*7*	*30*	*29*	*20*	*3*	*9*	*1*	*23*	*40*	*7*	*26*	*14*	*27*
Lack of information/data on reusing and material stock	1	16	17	11	0	4	0	6	19	4	14	6	12
Lack of information on sustainability	0	2	3	2	1	1	0	5	7	3	3	2	4
Lack of information on disassembly and technical info	2	4	3	1	0	1	0	2	5	0	4	1	3
Lack of information on stakeholders' awareness	4	4	2	2	2	2	1	4	3	0	3	4	4
Lack of information on digital applications	0	2	2	3	0	0	0	3	2	0	1	1	2
Lack of information on the interoperability of data	0	1	1	1	0	1	0	3	4	0	1	0	2
Lack of information on the property	0	1	0	0	0	0	0	0	0	0	0	0	0

"Lack of technical knowledge"	Who is responsible for/involved in						How it affects the implementation of circular strategies in the built environment									
	Stakeholders (c)						Circular strategies (d)									
	1	2	3	4	5	6	1	2	3	4	5	6	7	8	9	10
Total nr of occurrences in all the reported cases	*20*	*52*	*30*	*14*	*20*	*7*	*7*	*4*	*5*	*42*	*6*	*8*	*13*	*14*	*23*	*20*
Lack of information/data on reusing and material stock	9	24	16	6	10	3	2	0	0	19	2	3	8	7	12	11
Lack of information on sustainability	2	9	5	2	3	1	0	2	2	4	1	2	2	1	1	2
Lack of information on disassembly and technical info	3	7	3	1	0	0	1	0	1	7	1	1	1	2	3	1
Lack of information on stakeholders' awareness	4	3	3	3	4	1	3	1	1	3	1	1	1	3	3	3
Lack of information on digital applications	1	2	1	1	1	1	1	1	1	2	1	1	1	8	2	2

(continued)

Table 1.21 (continued)

"Lack of technical knowledge"	Who is responsible for/ involved in						How it affects the implementation of circular strategies in the built environment									
	Stakeholders (c)						Circular strategies (d)									
	1	2	3	4	5	6	1	2	3	4	5	6	7	8	9	10
Lack of information on the interoperability of data	1	3	1	0	2	1	0	0	0	0	0	0	0	0	3	1
Lack of information on the property	0	0	1	1	0	0	0	0	0	1	0	0	0	0	0	0

(a) Building layers [17]: (1) site; (2) structure; (3) skin; (4) services; (5) space plan; (6) stuff; (7) system (district)
(b) Building life cycle phases (adapted from RIBA [18]; Brincat et al. [15]): (1) Strategic Definition and Briefing; (2) Design; (3) Manufacture; (4) (Dismantling and) Construction; (5) Use and Refurbishment; (6) End-of-life
(c) Stakeholders [19]: (1) Clients; (2) Project professionals; (3) Suppliers; (4) Public; (5) Government; (6) Others
(d) Circular Strategies [20]: (1) Refuse; (2) Rethink; (3) Reduce; (4) Reuse; (5) Repair; (6) Refurbish; (7) Remanufacture; (8) Repurpose; (9) Recycle; (10) Recover

followed by the service layer [10, 26, 44, 45, 49, 51, 60, 62, 72, 74, 75, 79, 81], and the space plan and system layers with (3 and 1) occurrences respectively [6, 45, 74].

When it happens. Within the theme of lack of information related to reusing and material stock, the design, construction, and end-of-life phases at the phases mostly affected [6, 26, 34, 41, 44, 45, 48, 49, 62]. The same pattern appeared in the barriers listed under the theme of sustainability assessment [62–64]. The barriers listed under the theme of lack of data related to disassembly and technical information mainly affect the design and construction [45, 71–73]. Furthermore, it should be noted that the barriers related to the lack of stakeholders' awareness mainly affect the earlier and later stages of the building lifecycle: the strategic definition and briefing, use and refurbishment, and end-of-life [68, 74–77]. Also, the barriers related to the lack of data in digital applications showed influencing the stage of strategic definition and briefing [55, 79, 80]. In general, the technical barriers in this category mostly affect the design stage followed by construction stage [6, 10, 19, 26, 41, 44–46, 46, 48, 62–64, 63, 64, 74–77, 74].

Who is responsible for/involved in. Within the most prevalent type of barriers in this category, the lack of data related to reusing and material stock, project professionals and the public are mostly affected by the lack of information/data. Clients and suppliers are also influenced [26, 42, 45, 46, 75, 81, 97, 98]. This pattern was also noticed in the technical barriers listed under the themes of lack of data related to sustainability and disassemble and technical information [45, 57, 59, 61, 62, 64, 65, 68–73, 70]. Project professionals resulted in being affected by barriers listed under the themes of lack of digital information and lack of data interoperability [78–80].

The barriers listed under the theme of lack of data related to stakeholder awareness impacted clients, project professionals, and government [68, 74–77]. In general, at the most, the project professionals were the stakeholder that is affected most from the barriers in the category of lack of information and data [10, 19, 26, 34, 41, 42, 44–46, 48–50, 55–57, 59–62, 57, 59–61, 71–81].

How it affects the implementation of circular strategies in the built environment. Within the most prevalent type of barriers presented by the theme of lack of data related to reusing and material stock, it was observed that this type of barriers affects the implementation of circular strategies of reuse, recycling and repurposing, respectively [42, 75, 81, 97, 98]. Also, it was noted that the barriers listed under the second and third themes, the lack of data related to sustainability and disassembly and technical info, reusing is the most affected circular strategy [45, 53, 55, 57, 59, 62, 67, 72]. In general, among all representative themes, the reusing was the most affected circular strategy by the barriers extracted from the literature related to this category [10, 19, 26, 34, 45, 51–55, 53–55, 57, 59, 60, 62, 75–77, 72, 75, 76].

Category 2: Lack of Technical Knowledge

The barriers reported to belong to the "Lack of technical knowledge" category are 60 in total (incl. potential double references of the same barrier in different sources). However, only 53 of them have specific information reported with regard to the various fields examined in the context of the analysis (1. Where it happens; 2. When it happens; 3. Who is responsible; 4. How do they affect the implementation of circular strategies in the built environment). Furthermore, it is interesting to note that the information detected for each barrier does not necessarily cover the whole spectrum of the four aforementioned fields. In fact, information about the first section exists for 25 cases, while the respective numbers for the second, third and fourth sections are 41, 44 and 27, respectively. This information, as well as the number of identified instances in the whole population of the barriers listed in this category, are presented in Table 1.22.

The quantitative information regarding the barriers included in each one of them is shown in Table 1.23.

Where it happens. A first indication derived from the above Table is that the building layers most affected are the "structure" and "skin", with 21 [19, 26, 46, 49, 50, 52–54, 54, 68, 72, 73, 75, 78, 81, 82, 87, 89] and 20 occurrences [19, 26, 46, 49, 50, 52, 53, 68, 72, 73, 75, 78, 81, 82, 100], respectively, out of the 25 cases with existing data for this section. The less intensely appearing are the "site" [68, 75, 78, 81] and "district" [75, 78].

When it happens. Regarding the building life cycle stage, it seems that most of the stages appear with similar frequencies, apart from the "Design" stage, which is the most represented one with 31 occurrences out of the 41 cases with existing data for this section [19, 26, 34, 44, 46, 49, 50, 53, 54, 56–58, 58, 61, 72, 75, 78, 81–83, 82,

Table 1.22 Quantitative data analysis summary—lack of technical knowledge at the building level

What is the barrier category	Where it happens							When it happens					
Technical barrier category	Building layer (a)							Building life cycle phase (b)					
	1	2	3	4	5	6	7	1	2	3	4	5	6
Total nr. of occurrences in all the reported cases	4	21	20	11	8	10	2	16	31	8	16	14	21
Nr. of cases with information in each specific section {(a), (b), (c), (d)}	25							41					
Number of cases with information in ANY one of the specific sections	53												
Total nr. of cases reported	60												

What is the barrier category	Who is responsible for/ involved in						How it affects the implementation of circular strategies in the built environment									
Technical barrier category	Stakeholders (c)						Circular strategies (d)									
	1	2	3	4	5	6	1	2	3	4	5	6	7	8	9	10
Total nr. of occurrences in all the reported cases	15	38	21	11	18	9	4	6	9	16	2	1	10	8	14	11
Nr. of cases with information in each specific section {(a), (b), (c), (d)}	44						27									

(continued)

Table 1.22 (continued)

What is the barrier category	Who is responsible for/ involved in						How it affects the implementation of circular strategies in the built environment									
Technical barrier category	Stakeholders (c)						Circular strategies (d)									
	1	2	3	4	5	6	1	2	3	4	5	6	7	8	9	10
Number of cases with information in ANY one of the specific sections																
Total nr. of cases reported																

(a) Building layers [17]: (1) site; (2) structure; (3) skin; (4) services; (5) space plan; (6) stuff; (7) system (district)
(b) Building life cycle phases (adapted from RIBA [18]; Brincat et al. [15]): (1) Strategic Definition and Briefing; (2) Design; (3) Manufacture; (4) (Dismantling and) Construction; (5) Use and Refurbishment; (6) End-of-life
(c) Stakeholders [19]: (1) Clients; (2) Project professionals; (3) Suppliers; (4) Public; (5) Government; (6) Others
(d) Circular Strategies [20]: (1) Refuse; (2) Rethink; (3) Reduce; (4) Reuse; (5) Repair; (6) Refurbish; (7) Remanufacture; (8) Repurpose; (9) Recycle; (10) Recover

83, 87] and the "Manufacture" stage, which is the most underrepresented one [26, 44, 49, 73, 75, 78, 83].

Who is responsible for/involved in. The stakeholders who seem to be more involved in the identified barriers are the project professionals [19, 26, 34, 44, 46, 49, 50, 53–57, 55–57, 61, 64, 68, 72, 73, 81–83, 78, 81–83, 85]. References are made not only to designers but also to those involved in the construction and in other types of activities.

How it affects the implementation of circular strategies in the built environment. The circular strategies mostly affected are "Reuse" [19, 34, 49, 50, 53–56, 55, 56, 72, 75, 78, 81, 82, 87] and "Recycle" [19, 26, 34, 46, 48, 49, 53, 54, 61, 72, 75, 78, 81, 100], while only one occurrence has been reported for "Refurbish" [78].

Category 3: Lack of Technical Skills

Results in this category were analysed by looking at general quantitative data for this category, and summarized in Table 1.24.

Where it happens. The studies addressing the lack of technical skills barriers largely described these barriers at structure [34, 44, 45, 61, 72–74, 74, 77, 80, 81, 88] and skin building layers [44, 45, 61, 72, 73, 77] under subheadings lack of skills, lack of

Table 1.23 Detailed quantitative data analysis—lack of technical knowledge at the building level

"Lack of technical knowledge"	Where it happens							When it happens					
	Building layer (a)							Building life cycle phase (b)					
	1	2	3	4	5	6	7	1	2	3	4	5	6
Issues of a general nature	2	8	8	4	1	2	0	3	11	2	7	1	4
Design-related issues	0	5	7	4	5	4	0	8	10	0	0	6	6
Materials and resources	0	4	1	0	0	1	0	0	4	1	4	3	5

"Lack of technical knowledge"	Who is responsible for/ involved in						How it affects the implementation of circular strategies in the built environment									
	Stakeholders (c)						Circular strategies (d)									
	1	2	3	4	5	6	1	2	3	4	5	6	7	8	9	10
Issues of a general nature	4	13	6	0	4	1	2	4	1	7	1	0	4	4	6	6
Design-related issues	0	2	0	0	1	1	0	0	1	1	0	0	1	0	1	0
Materials and resources	2	7	3	2	3	0	0	1	1	4	0	0	2	0	4	3

(a) Building layers [17]: (1) site; (2) structure; (3) skin; (4) services; (5) space plan; (6) stuff; (7) system (district).
(b) Building life cycle phases (adapted from RIBA [18]; Brincat et al. [15]): (1) Strategic Definition and Briefing; (2) Design; (3) Manufacture; (4) (Dismantling and) Construction; (5) Use and Refurbishment; (6) End-of-life
(c) Stakeholders [19]: (1) Clients; (2) Project professionals; (3) Suppliers; (4) Public; (5) Government; (6) Others
(d) Circular Strategies [20]: (1) Refuse; (2) Rethink; (3) Reduce; (4) Reuse; (5) Repair; (6) Refurbish; (7) Remanufacture; (8) Repurpose; (9) Recycle; (10) Recover

systemic CE education and training for supply chain members and lack of training and knowledge for professionals.

When it happens. The lack of technical skills barriers at the structure and skin layers has been discussed mostly in "Strategic definition and briefing" [34, 44, 62, 65, 72, 74, 80, 81, 86, 88] "Design" [34, 44, 45, 55–58, 57, 72–74, 72–74, 77, 80, 81], "Dismantling and construction" phases.

Who is responsible for/involved in. The government [62, 65, 72, 77, 80, 83, 85], project professionals [34, 44, 45, 45, 51, 55–57, 57, 59, 61, 62, 65, 72–74, 73, 74, 77, 80, 81, 87–89] and suppliers [34, 55, 56, 59, 68, 72, 85, 86] are mostly responsible or affected by the lack of technical skills.

How it affects the implementation of circular strategies in the built environment. The lack of technical skills barrier mostly affects Refuse [72, 74, 77, 81, 88], Reuse [62, 72, 74, 77, 80, 81, 87–89], Repurpose [45, 56, 57, 72, 77, 81, 87–89], Recycle [34, 45, 62, 72, 73, 77, 80, 81, 86–89] and Recover [34, 45, 62, 72–74, 74, 77, 80, 86–89].

Table 1.24 Quantitative data analysis—lack of technical skills at the building level

What is the barrier category	Where it happens							When it happens					
Technical barrier category	Building layer (a)							Building life cycle phase (b)					
	1	2	3	4	5	6	7	1	2	3	4	5	6
Lack of technical skills		13	7	3	2	2		11	15	4	13	10	11

What is the barrier category	Who is responsible for/ involved in						How it affects the implementation of circular strategies in the built environment									
Technical barrier category	Stakeholders (c)						Circular strategies (d)									
	1	2	3	4	5	6	1	2	3	4	5	6	7	8	9	10
Lack of technical skills	6	24	7	2	6	6	5	3	2	8	1	2	3	10	12	14

(a) Building layers [17]: (1) site; (2) structure; (3) skin; (4) services; (5) space plan; (6) stuff; (7) system (district)
(b) Building life cycle phases (adapted from RIBA [18]; Brincat et al. [15]): (1) Strategic Definition and Briefing; (2) Design; (3) Manufacture; (4) (Dismantling and) Construction; (5) Use and Refurbishment; (6) End-of-life
(c) Stakeholders [19]: (1) Clients; (2) Project professionals; (3) Suppliers; (4) Public; (5) Government; (6) Others
(d) Circular Strategies [20]: (1) Refuse; (2) Rethink; (3) Reduce; (4) Reuse; (5) Repair; (6) Refurbish; (7) Remanufacture; (8) Repurpose; (9) Recycle; (10) Recover

Category 4: Lack of Technical Standards

Results in this category were analysed by looking at general quantitative data for this category, and summarized in Table 1.25.

Where it happens. The dominant discussion on the need for technical standards lies mostly in "Structure" and "Skin", with "Stuff" to a lesser extent. In terms of where these barriers take place, the most occurrences of the lack of standardisation and guidelines are observed in "Structure" with 13 occurrences [6, 10, 19, 45, 52, 72, 74, 75, 80–82, 82, 86] and "Skin" with 12 occurrences [6, 10, 19, 51, 52, 72, 74, 75, 80–82, 82] and further down the list is "stuff" with 6 occurrences [6, 19, 26, 47, 50, 73]. A similar pattern is observed for the lack of policies and consistent regulations with 8 occurrences for "Structure" [10, 41, 45, 71, 77, 82] and nine for "Skin" [10, 41, 45, 51, 71, 77, 82] with five occurrences for "Stuff" [10, 26, 50, 65, 73]. The pattern also continues with Sustainability and Quality assurance with 7 occurrences for "Structure" [6, 10, 19, 71, 72, 80], eight occurrences for "Skin" [6, 10, 19, 51, 71, 72, 80] and five occurrences for "Stuff" [6, 19, 26, 47]. The fourth and smaller theme on the lack of flexibility seems to be taking place in four occurrences in the "Structure" [10, 45, 52, 87] and in three occurrences respectively in "Skin" and "Services" [10, 45, 52].

Table 1.25 Quantitative data analysis—lack of technical standards at the building level

What is the barrier category	Where it happens							When it happens					
Technical barrier category	Building layer (a)							Building life cycle phase (b)					
	1	2	3	4	5	6	7	1	2	3	4	5	6
Lack of technical standards	5	22	20	6	1	11		19	31	13	26	13	25

What is the barrier category	Who is responsible for/ involved in						How it affects the implementation of circular strategies in the built environment									
Technical barrier category	Stakeholders (c)						Circular strategies (e)									
	1	2	3	4	5	6	1	2	3	4	5	6	7	8	9	10
Lack of technical standards	6	32	17	2	46	8	14	4	11	27	1	3	8	13	30	25

(a) Building layers [17]: (1) site; (2) structure; (3) skin; (4) services; (5) space plan; (6) stuff; (7) system (district)
(b) Building life cycle phases (adapted from RIBA [18]; Brincat et al. [15]): (1) Strategic Definition and Briefing; (2) Design; (3) Manufacture; (4) (Dismantling and) Construction; (5) Use and Refurbishment; (6) End-of-life
(c) Stakeholders [19]: (1) Clients; (2) Project professionals; (3) Suppliers; (4) Public; (5) Government; (6) Others
(d) Circular Strategies [20]: (1) Refuse; (2) Rethink; (3) Reduce; (4) Reuse; (5) Repair; (6) Refurbish; (7) Remanufacture; (8) Repurpose; (9) Recycle; (10) Recover

When it happens. The lack of technical standards seems to be affecting all life cycle phases considerably in almost all themes examined here. The two more pronounced phases with the largest number of occurences was Design with 31 occurences [10, 26, 41, 44–46, 46, 49, 50, 65, 71–75, 72–75, 77, 78, 80], while Dismantling and Construction [6, 10, 19, 26, 41, 44–47, 46, 47, 50, 71–75, 71–74] and End-of-Life [6, 10, 26, 41, 45, 46, 50, 52, 67, 69, 71–75, 73–75, 77, 81] followed with 26 occurrences each. This is expected as DfD as well as the actual construction/deconstruction phases require standards and guidelines. A sizable number of occurrences (19) was observed for the strategic definition and Briefing phases [6, 10, 19, 41, 44–46, 46, 49, 50, 65, 67, 71, 73, 74, 80, 92], which can be attributed to the novelty of CE and the absence of key legal policies and technical standards which hinder CE flagship projects.

Who is responsible for/involved in. On the issue of stakeholders involved, the grand majority of the mentions consider the government as an important stakeholder with 46 occurences [6, 6, 10, 19, 26, 41, 44–47, 49–53, 49, 71–75, 52, 80–83, 71–75]. This is expected as when we refer to government, this would include regulatory organisations at different levels of government as well as standardisation entities and agencies. Another key stakeholder is the project professional (32 occurrences) [6, 6, 10, 19, 26, 41, 45–47, 47, 49, 71–75, 72–75, 77, 78, 91–93] who influences standards but is also the main stakeholder that has to comply with them. Third in importance are the manufacturers (17 occurrences) [6, 26, 41, 45–47, 47, 71, 72, 74, 75, 78,

Table 1.26 Quantitative data analysis—Lack of machines and equipment at the building level

What is the barrier category	Where it happens							When it happens					
Technical barrier category	Building layer (a)							Building life cycle phase (b)					
	1	2	3	4	5	6	7	1	2	3	4	5	6
Lack of infrastructure and equipment		1	1	1		1		1		1	2		2

What is the barrier category	Who is responsible for/ involved in						How it affects the implementation of circular strategies in the built environment									
Technical barrier category	Stakeholders (c)						Circular strategies (e)									
	1	2	3	4	5	6	1	2	3	4	5	6	7	8	9	10
Lack of infrastructure and equipment		1								1			1		2	1

(a) Building layers [17]: (1) site; (2) structure; (3) skin; (4) services; (5) space plan; (6) stuff; (7) system (district)
(b) Building life cycle phases (adapted from RIBA [18]; Brincat et al. [15]): (1) Strategic Definition and Briefing; (2) Design; (3) Manufacture; (4) (Dismantling and) Construction; (5) Use and Refurbishment; (6) End-of-life
(c) Stakeholders [19]: (1) Clients; (2) Project professionals; (3) Suppliers; (4) Public; (5) Government; (6) Others
(d) Circular Strategies [20]: (1) Refuse; (2) Rethink; (3) Reduce; (4) Reuse; (5) Repair; (6) Refurbish; (7) Remanufacture; (8) Repurpose; (9) Recycle; (10) Recover

81] along similar lines as the project professionals. This pattern was evident in all 4 themes.

How it affects the implementation of circular strategies in the built environment. In terms of circular strategies the lack of technical standards seems to mostly affect recycling (30 occurrences) followed by reuse (27 occurrences) and recovery (25 occurrences), these three being ranked in the same way across all 4 themes [6, 10, 19, 26, 41, 45, 46, 50–52, 52, 59, 67–69, 71–75, 71–75, 77, 80].

Category 5: Lack of Machines and Equipment

Results in each category were analysed by looking at the general quantitative data for this category, and summarized inTable 1.26.

Where it happens. The studies addressing the lack of infrastructure and equipment described these barriers at structure [50], skin [50] and services [50] layers under the sub-heading "need for infrastructure and equipment to perform deconstruction".

When it happens. The lack of infrastructure and equipment barriers at structure, skin, services, and building layers has been discussed mostly in "Strategic Definition and Briefing" [50] "Design" [50], "Dismantling and construction" phases [47, 50] and "End of Life" [47, 50].

Table 1.27 Quantitative data analysis—inadequate processes and infrastructure at the building level

What is the barrier category	Where it happens							When it happens					
Technical barrier category	Building layer (a)							Building life cycle phase (b)					
	1	2	3	4	5	6	7	1	2	3	4	5	6
Inadequate processes and infrastructure	5	19	14	4	3	8	0	10	14	8	22	6	24

What is the barrier category	Who is responsible for/ involved in						How it affects the implementation of circular strategies in the built environment									
Technical barrier category	Stakeholders (c)						Circular strategies (e)									
	1	2	3	4	5	6	1	2	3	4	5	6	7	8	9	10
Inadequate processes and infrastructure	8	16	18	5	18	10	5	3	4	20	1	3	7	7	17	12

(a) Building layers [17]: (1) site; (2) structure; (3) skin; (4) services; (5) space plan; (6) stuff; (7) system (district)
(b) Building life cycle phases (adapted from RIBA [18]; Brincat et al. [15]): (1) Strategic Definition and Briefing; (2) Design; (3) Manufacture; (4) (Dismantling and) Construction; (5) Use and Refurbishment; (6) End-of-life
(c) Stakeholders [19]: (1) Clients; (2) Project professionals; (3) Suppliers; (4) Public; (5) Government; (6) Others
(d) Circular Strategies [20]: (1) Refuse; (2) Rethink; (3) Reduce; (4) Reuse; (5) Repair; (6) Refurbish; (7) Remanufacture; (8) Repurpose; (9) Recycle; (10) Recover

Who is responsible for/involved in. The project professionals [50] are mostly affected by the lack of infrastructure and equipment.

How it affects the implementation of circular strategies in the built environment. The lack of infrastructure and equipment mostly affects Reuse [50], Remanufacture [50], Recycling [47, 50] and Recovery strategies [47].

Category 6: Inadequate Processes and Infrastructure

Results in each category were analysed by looking at the general quantitative data for this category, and summarized in Table 1.27.

Where it happens. The barriers in the inadequate processes and infrastructure category refer mainly to the structure of buildings [10, 19, 41, 42, 47, 50, 51, 53, 61, 62, 71, 80, 81, 88, 93]. In addition, a significant number of barriers have been identified in building services [10, 19, 41, 47, 50, 51, 61, 62, 71, 80, 81].

When it happens. Most barriers related to inadequate processes and infrastructure appear at the end-of-life of buildings [19, 41, 46–50, 48–50, 53, 57, 61, 73, 93–95]. On the other hand, the dismantling and construction phase was the second stage where the highest number of barriers was detected [10, 19, 26, 41, 47–49, 49, 56, 59, 61, 62, 71, 73, 81, 85, 88, 88, 94].

Who is responsible for/involved in. Suppliers [26, 41, 47–49, 49, 56, 71–73, 72, 73, 81, 83] and government [10, 19, 26, 46–50, 48–50, 53, 61, 62, 72, 73, 80] were identified as the main actors responsible for inadequate construction processes and infrastructure. Project professionals [10, 41, 42, 46, 46, 56, 57, 71–73, 80–82, 81] were also heavily implicated in this barrier, although to a lesser extent than the previous ones.

How it affects the implementation of circular strategies in the built environment. According to the literature, inadequate processes and infrastructure mainly affect the reuse strategy [10, 26, 38, 42, 49, 51, 53, 57, 59, 71, 72, 83, 88, 95, 99]. With a high but lower impact than the previous one is the recycling strategy [10, 19, 26, 38, 41, 48–50, 50, 53, 59, 61, 62, 72, 73, 80, 81].

Category 7: Lack of Methods/Tools

Results in each category were analysed by looking at the general quantitative data for this category and summarized in Table 1.28.

Where it happens. The study showed that barriers related to the lack of methods/tools occur mostly at the building structure level [10, 34, 46–49, 48, 49, 51, 62, 65, 70, 72, 79–81, 79]. It is also prominent at skin level, although to a much lesser extent [10, 19, 34, 46–49, 48, 49, 51, 62, 65, 72, 79–81].

When it happens. The lack of methods/tools mainly happens at the building design stage [10, 19, 34, 43, 45–49, 47, 60–62, 55, 72–75, 78–81, 72–75]. The end-of-life stage was also significantly identified for this barrier [6, 10, 26, 43, 45–47, 47, 49, 55, 60, 72–74, 73, 74, 79].

Who is responsible for/involved in. The project professionals were identified as the key actors involved in or responsible for the barriers related to the lack of methods/tools, well above the other stakeholders [19, 26, 34, 43, 45–49, 47, 48, 60–62, 56, 60–62, 62, 65, 72, 73, 75, 78].

How it affects the implementation of circular strategies in the built environment. The literature review highlights recovery as the main strategy affected by barriers related to lack of methods/tools [10, 26, 34, 46, 47, 49, 60–62, 62, 72–75, 79–81, 75, 79], closely followed by recycling [10, 19, 26, 34, 46, 49, 55, 60–62, 62, 65, 72, 74, 75, 79, 81, 94] and reuse [10, 19, 26, 34, 47, 49, 55, 61, 62, 72, 74, 75, 79, 93, 94, 102].

Table 1.28 Quantitative data analysis—lack of methods/tools at the building level

What is the barrier category	Where it happens							When it happens					
Technical barrier category	Building layer (a)							Building life cycle phase (b)					
	1	2	3	4	5	6	7	1	2	3	4	5	6
Lack of methods/tools	3	27	24	18	7	10	0	26	42	22	28	25	37

What is the barrier category	Who is responsible for/ involved in						How it affects the implementation of circular strategies in the built environment									
Technical barrier category	Stakeholders (c)						Circular strategies (e)									
	1	2	3	4	5	6	1	2	3	4	5	6	7	8	9	10
Lack of methods/ tools	9	46	29	13	27	17	5	11	5	22	6	5	6	12	24	27

(a) Building layers [17]: (1) site; (2) structure; (3) skin; (4) services; (5) space plan; (6) stuff; (7) system (district)
(b) Building life cycle phases (adapted from RIBA [18]; Brincat et al. [15]): (1) Strategic Definition and Briefing; (2) Design; (3) Manufacture; (4) (Dismantling and) Construction; (5) Use and Refurbishment; (6) End-of-life
(c) Stakeholders [19]: (1) Clients; (2) Project professionals; (3) Suppliers; (4) Public; (5) Government; (6) Others
(d) Circular Strategies [20]: (1) Refuse; (2) Rethink; (3) Reduce; (4) Reuse; (5) Repair; (6) Refurbish; (7) Remanufacture; (8) Repurpose; (9) Recycle; (10) Recover

Category 8: Lack of Materials/Products/Buildings/Systems

Results in each category were analysed by looking at general quantitative data for this category, and summarized in Table 1.29.

Where it happens. The studies addressing the lack of materials/products/buildings/systems are mostly at the structure level [6, 10, 19, 34, 38, 44, 45, 47, 51, 55, 62, 73, 82, 87, 89, 93, 100], and to a slightly lower extent at building skin level [6, 10, 19, 26, 34, 44, 45, 47, 51, 62, 63, 73, 82, 89, 93, 96, 100].

When it happens. This barrier category occurs mainly during the dismantling and construction phase [6, 10, 19, 26, 34, 44, 45, 47, 50, 61, 62, 73, 81, 87]. Other phases mentioned, but less important, were design and the end of life.

Who is responsible for/involved in. Both project professionals and suppliers are equally responsible for the lack of materials/products/constructions/systems. Others such as the public and government are much less involved [6, 10, 19, 26, 34, 38, 44–47, 46, 47, 61–63, 61–63, 73, 81, 86, 87, 89].

How it affects the implementation of circular strategies in the built environment. The lack of materials/products/buildings/systems affects mostly the reuse strategy, far above the other approaches [6, 10, 26, 34, 45, 47, 56, 57, 62, 73, 82, 87, 89, 93,

Table 1.29 Quantitative data analysis—lack of materials/products/building systems at the building level

What is the barrier category	Where it happens							When it happens					
Technical barrier category	Building layer (a)							Building life cycle phase (b)					
	1	2	3	4	5	6	7	1	2	3	4	5	6
Lack of materials/ products/ buildings/ systems	4	26	24	11	6	14	2	9	18	16	26	16	17

What is the barrier category	Who is responsible for/ involved in						How it affects the implementation of circular strategies in the built environment									
Technical barrier category	Stakeholders (c)						Circular strategies (e)									
	1	2	3	4	5	6	1	2	3	4	5	6	7	8	9	10
Lack of materials/ products/ buildings/ systems	7	29	29	12	15	15	3	4	6	23	5	4	7	8	15	12

(a) Building layers [17]: (1) site; (2) structure; (3) skin; (4) services; (5) space plan; (6) stuff; (7) system (district)
(b) Building life cycle phases (adapted from RIBA [18]; Brincat et al. [15]): (1) Strategic Definition and Briefing; (2) Design; (3) Manufacture; (4) (Dismantling and) Construction; (5) Use and Refurbishment; (6) End-of-life
(c) Stakeholders [19]: (1) Clients; (2) Project professionals; (3) Suppliers; (4) Public; (5) Government; (6) Others
(d) Circular Strategies [20]: (1) Refuse; (2) Rethink; (3)Reduce; (4) Reuse; (5) Repair; (6) Refurbish; (7) Remanufacture; (8) Repurpose; (9) Recycle; (10) Recover

100]. Other strategies, such as recycling and recovery, are affected to a much lesser extent.

1.3.3 Material Level

1.3.3.1 Quantitative Analysis of Literature

The prevalence of identified barriers was quantified in total and for each category. Figure 1.13 presents the document frequency of the different barriers, as identified in the scientific and non-academic literature at the material level.

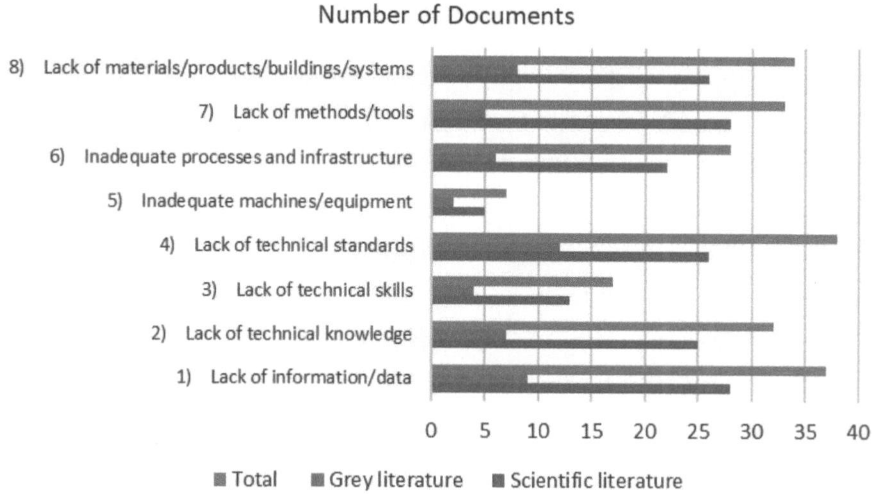

Fig. 1.13 Total consulted documents, in academic (called scientific literature) and non-academic literature (called grey literature), providing information about the different categories of technical barriers at the material level

Among the analysed 62 articles, 38 reported barriers under the category (4) *Lack of technical standards,* which formed the category most represented in the documents. *Lack of information/data* was the next most consistent barrier category (reported in 37 documents), followed by *Lack of materials/products/buildings/systems* (reported in 34 documents) and *Lack of methods/tools* (reported in 33 documents). Scientific literature was the dominant source for all reported barriers.

The main categories of technical challenges and barriers were established by quantifying the barriers that appeared across the reviewed literature, at the material level. The highest number of barriers was reported for the *Lack of materials/products/buildings/systems* category: 92 barriers across both scientific and non-academic literature, as illustrated in Fig. 1.14.

The second prominent category among the number of barriers reported was *Lack of information/data,* with 63 barriers reported in total, followed by *Lack of methods/tools*, with 59 challenges and barriers reported across the reviewed literature. *Inadequate machines/equipment* was the category found to have the least technical barriers encountered in the construction sector, at the material level.

Challenges and barriers were reported in six categories (according to Ref. [17]): site, structure, skin, services, space plan and stuff. For the material level, technical barriers and challenges were mainly reported in terms of building layers in the Structure layer for all barrier categories, followed by the Skin layer. Notably, challenges related to the space plan layer were mentioned the least frequently. Document frequency of these challenges is illustrated in Fig. 1.15, categorised by the building layers' level.

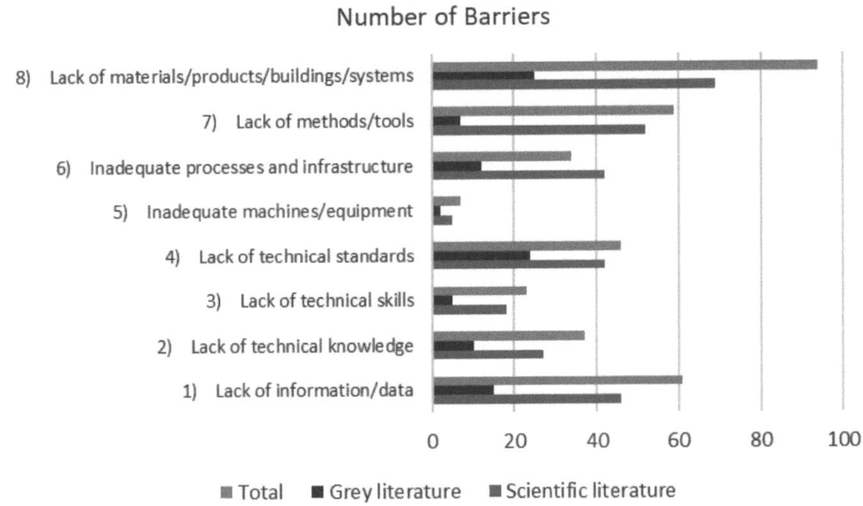

Fig. 1.14 Main categories of technical barriers, for the material level, by occurrence

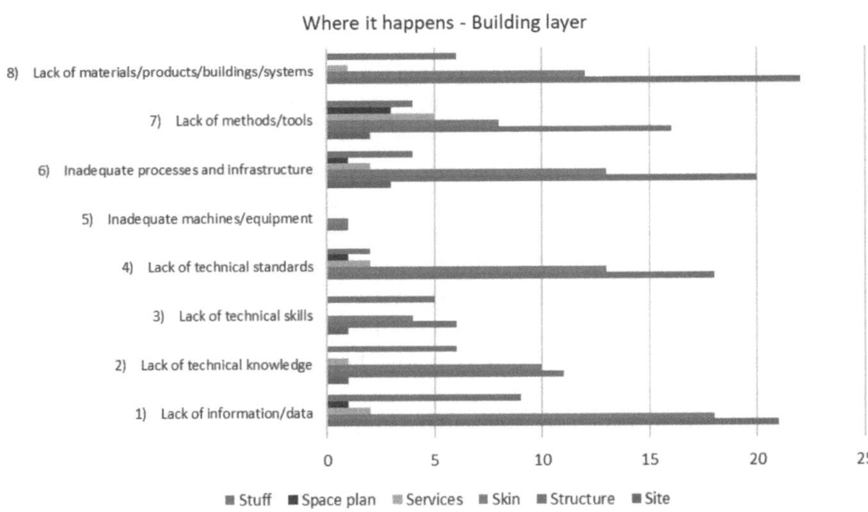

Fig. 1.15 Challenges and technical barriers at the material level regarding the layers of the building where they are reported

Regarding the building life cycle phase, the challenges and technical barriers were quantified for their occurrence in six categories (adapted from Royal Institute of British Architects [18]; Council et al. [15]): *Strategic Definition and Briefing, Design, Manufacture, (Dismantling and) Construction, Use and Refurbishment* and *End-of-life*, as shown in Fig. 1.16.

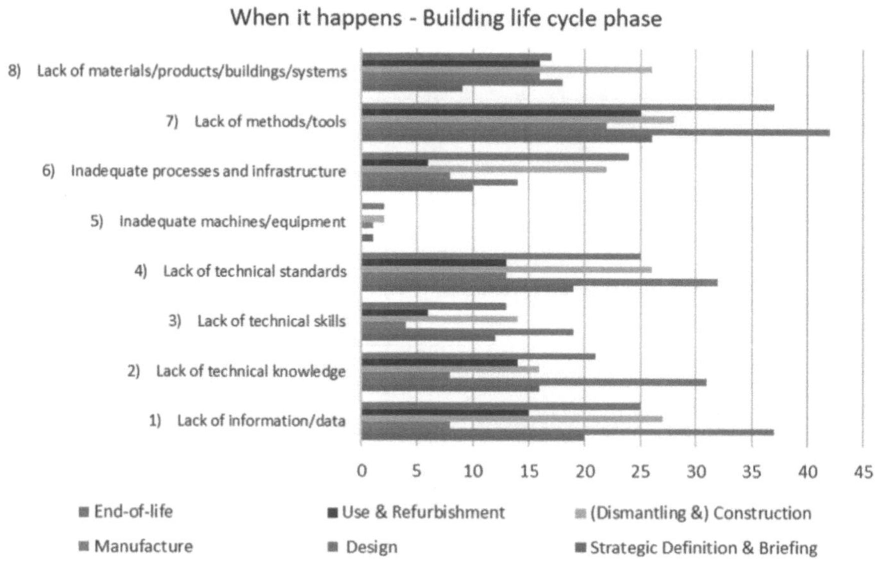

Fig. 1.16 Challenges and technical barriers for the material level, regarding their occurrence in the building life cycle phase

Design and *End-of-life* were the most prevalent building life cycle phases when barriers occurred, affected by a *Lack of methods/tools, a Lack of information/data. Lack of technical standards* and *Lack of technical knowledge.* The second most prevalent building life cycle phase when barriers occurred was *(Dismantling and) Construction*, affected mostly by the *Lack of methods/tools,* the *Lack of information/ data,* the *Lack of technical standards* and by the *Lack of materials/products/buildings/ systems.* The remaining categories present a comparable occurrence for the barriers and challenges reported.

Following [19], those responsible for or involved in addressing technical challenges at the material level were classified into six stakeholder groups (illustrated in Fig. 1.17). These groups include clients (those commissioning the project), project professionals (such as architects and engineers), suppliers (providing materials), the public (potentially impacted by the project), the government (with relevant regulations), and others not included in these categories.

By far the most frequently mentioned stakeholder in the studies examined was the *Government*, which was mostly responsible for *the lack of technical standards.* The following stakeholders involved in/responsible for technical barriers were found to be the *Project professionals*, who registered the highest occurrence in all technical barrier categories, except for the *lack of technical standards.* The *Public and Other* (those not included in the first five categories of stakeholders) were found to be the least responsible for/involved in the technical barriers encountered in the construction sector, at the material level.

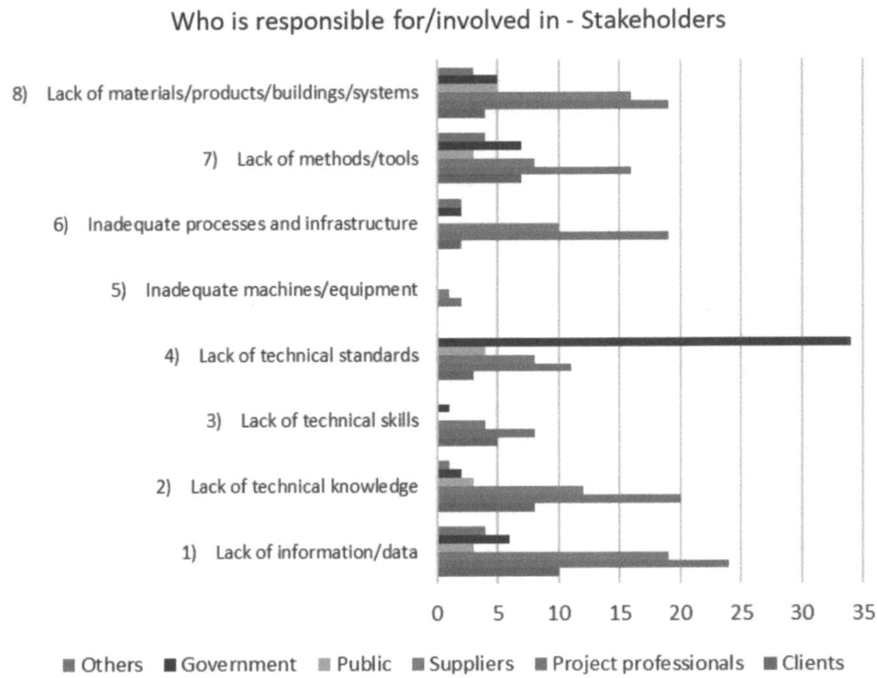

Fig. 1.17 Stakeholders responsible for/involved in technical barriers, for the material level

The research classified CE strategies in the built environment into 10 categories (according to van Stijn and Gruis [20]): *Refuse, Rethink, Reduce, Reuse, Repair, Refurbish, Remanufacture, Repurpose, Recycle, Recover*. Figure 1.18 explores how technical barriers affect the implementation of each strategy.

Reuse and *Recycle* were the most affected circular strategies by the technical barriers, followed by *Repurpose* and *Repair*. These strategies were affected mostly by barriers like *Lack of materials/products/buildings/systems, Inadequate processes and infrastructure, Lack of information/data* or *Lack of methods/tools.*

1.3.3.2 Collection of the List of Barriers in Each Category and Analysis of Relevance

A comprehensive analysis of existing research was conducted to identify obstacles hindering the implementation of CE principles. The findings underscore the significant challenges associated with transitioning to a CE for the material level in the built environment.

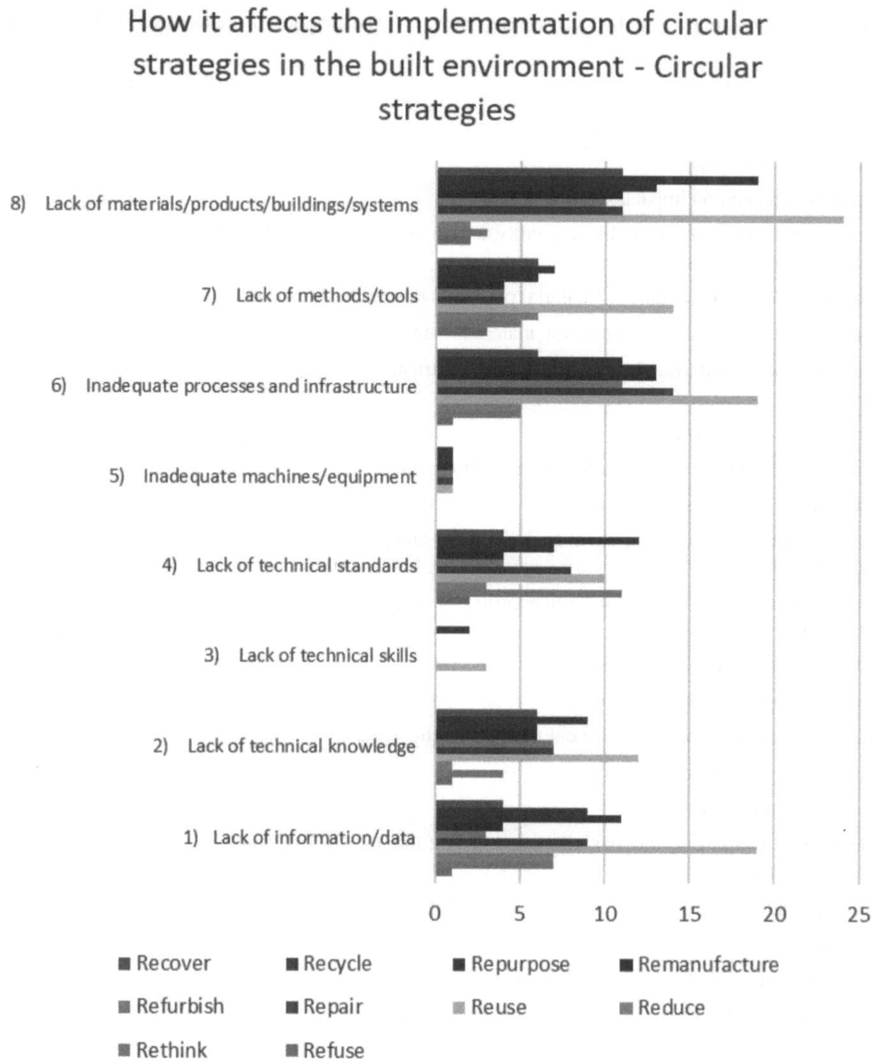

Fig. 1.18 Circular strategies at the material level affected by the technical barriers

Category 1: Lack of Information/Data

A list of identified barriers in the lack of information/data category, and their subcategory relevance, is shown in Table 1.30.

A quantitative analysis of reported barriers on Lack of information/data revealed a critical subcategory: Data Availability, Accessibility, and Accuracy. This subcategory, encompassing issues like unreliable data, lack of Material Passport updates, and

Table 1.30 Quantitative analysis of reported barriers to the lack of information/data at the material level

Lack of information/data	Quantitative analysis
Data availability, accessibility and accuracy	*33*
Lack of reliable data	
Lack of data, e.g. on impacts	
Lack of data about the availability, quality, and quantity of salvaged building components	
Lack of information update on material passports during the lifespan	
Lack of detailed data on construction materials—information gap	
Lack of quality and availability of data (privacy, trust, ownership, access)	
Traceability and certification	*14*
Lack of information on the quality of reused materials	
Certifications and responsibility for reused products	
Lack of traceability of component characteristics and performance in the value chain	
Lack of information on the mechanical properties of the recovered materials	
Lacking certification/traceability	
Lack of knowledge and awareness	*11*
Lack of knowledge and engagement throughout the value chain	
Awareness of recycled materials	
Lack of interest and awareness	
Inadequate awareness, understanding, and insight into the CE in C&D waste management	
Lack of understanding of the circular loops of MCPs	
Sustainability	*3*
Lack of monitoring the benefits of circular development	
Information on the durability issues of reused products	

limited information on salvaged components, emerged as the most frequent information deficiency at the material level, appearing 33 times across the reviewed literature [6, 7, 19, 34, 49, 90, 97, 104–123].

The next mostly mentioned technical barrier subcategory in *Lack of information/data* category was *Traceability and Certification*, which occurred 14 times across the reviewed literature [10, 49, 104, 117, 118, 122, 124–128]. This subcategory involves challenges and technical barriers at the material level, such as lacking certification/traceability, certifications and responsibility of reused products, lack of traceability of components' characteristics and performance in the value chain or, lack of information on the quality of reused materials.

A similar occurrence to *Traceability and Certification* was reported for *Lack of Knowledge and Awareness* for the material level. This subcategory appeared 11 times across the reviewed literature [10, 48, 62, 81, 115, 117, 120, 129–131] and was related to challenges and barriers such as lack of interest and awareness, inadequate awareness, understanding and insight into CE in C&D waste management, awareness of recycled materials or Low environmental awareness.

The *Sustainability* subcategory, encompassing issues like a lack of monitoring benefits of circular development or information regarding durability issues of reused products, occurred three times across the reviewed literature [34, 132].

It is worth mentioning that, sometimes, different challenges and issues belonging to the same subcategory were reported within the same literature reference.

Category 2: Lack of Technical Knowledge

A list of identified barriers in the lack of technical knowledge category, and their sub-category relevance, is shown in Table 1.31.

Through a deep bibliographical review, a total of 36 barriers were reported to be directly associated with the lack of technical knowledge. Some of them (six) expressed a general lack of knowledge on circularity, sustainability and green building issues with regard to the material level [6, 48, 50, 97, 107, 108, 110, 133–137]. Within the same framework, some indicated a lack on knowledge throughout the supply and value chain [34, 81, 113, 120] and others indicated in more detail the stage suffering more from the lack of technical knowledge. More specifically, many denoted the design stage [10, 19, 97, 120, 123, 124], but most of them concerned the end of life processes, i.e. demolition, disassembly, reuse and recycle [6, 48, 50, 97, 107, 108, 110, 133–137]. The lack of technical knowledge on the materials properties, tracking and reuse potential was also ranked among the most important barriers [19, 104, 106, 114, 116, 121, 129, 138].

Category 3: Lack of Technical Skills

A list of identified barriers in the lack of technical skills category, and their sub-category relevance, is shown in Table 1.32.

Through a deep bibliographical review, a total of 19 barriers were identified as a lack of technical skills. Most of them (nine) were not thoroughly explained and expressed a general lack of skills or competence, associated with CE, green building risk factors, or decision making [49, 50, 81, 104, 106, 108, 109, 114, 129]. In more detailed indications, some specified lack technical skills throughout the supply chain [106, 113, 114, 121], at the design [34, 120, 121] and at the end-of-life processes, i.e. demolition, deconstruction and reuse [97, 137].

Table 1.31 Quantitative analysis of reported barriers on lack of technical knowledge at the material level

Lack of technical knowledge	Quantitative analysis
Lack of awareness—general knowledge—familiarity with CE, green building or sustainability concepts	*6*
Lack of investigation of the relationship between GB risks and project outcomes	
Lack of exploration in cross-country or developing countries	
Lack of expertise and knowledge in circularity in construction	
Lack of knowledge: circular management and building, environmental impact and certification	
Unaware of the opportunities available to save carbon through the reuse approach	
Lack of awareness, interest, knowledge and engagement. Difficulties in delivering CE projects without support or a business case	
Lack of knowledge of the value and supply chain	*4*
Lack of knowledge and engagement throughout the value chain	
Lack of proper training and development programs for the members of supply chain	
Lack of skills and expertise (amongst public infrastructure and service providers)	
Lack of knowledge and engagement throughout the value chain	
Lack of knowledge on the design stage	*6*
Design team buy-in	
Lack of adequate information in building design	
Lack of effective green building design development	
Design-related issues: matching the design of the new building with the strength of the recovered elements	
Insufficient use and development of CE-focused design and collaboration tools, information, and metrics	
Limited experience with circular design within the built environment	
Lack of knowledge on the materials' properties	*8*
Materials knowledge	
Knowledge on adopting material tracking systems	
Accurate information regarding materials/tracking in SC towards recycling is not available	
Materials knowledge	
Difficulties in understanding and developing EPDs	
The perception and reaction to the geopolymer cement representing a green solution to the construction industry remains uncertain	
Lack of monitoring quality conditions	

(continued)

Table 1.31 (continued)

Lack of technical knowledge	Quantitative analysis
Limited or unknown information on the technical performance of the current asset, e.g. loading capacity of the structural frame, energy performance of the façade	
Lack of knowledge on the end-of-life procedures	*12*
Knowledge on reuse of construction materials	
Knowledge on reuse of construction materials	
Needing policymakers to enforce the use of recycled/salvaged materials to a certain extent	
Lack of knowledge and an underdeveloped market reduces the capability of CDW reduction in a CE system	
Knowledge of reclaimed materials: quantities of material recovered, identifying reclaimed components, material data availability; material data update	
Negative experiences of designers with the usage of most of the remanufactured products	
Lack of disassembly guides	
Confusion between reuse and recycling	
Impact of upfront demolition must not be ignored (impact on LCA of new project to be discussed)	
Lack of technical solutions for effective disassembly and reconnecting methods of the reused wood components	
Unknown of the most influential criteria for the reuse strategy	
Lack of knowledge on how to reuse/recycle treated wood	

Category 4: Lack of Technical Standards

The quantitative analysis of reported barriers on *Lack of Technical Standards* revealed that the most critical subcategory is the *Lack of Consistent Regulatory Framework*, appearing 25 times across the literature [10, 49, 50, 62, 81, 104, 105, 108, 109, 115, 119, 121, 122, 124, 129, 134, 135, 137, 139–141]. This subcategory involves a lack of new regulations on circularity but also contradictions between existing regulations, a lack of flexibility and a lack of obligatory use of recycled materials on building projects.

The next mostly mentioned technical barrier subcategory was Lack of *Standardisation*, which occurred 13 times across the reviewed literature [10, 49, 104, 117, 118, 122, 124–128]. This subcategory mainly involves challenges of lacking standards on the quality of the reused materials (Table 1.33).

The subcategory of *Complexity in Certification Processes* appeared six times across the reviewed literature [90, 107, 110, 114, 133, 142] and was related to challenges and barriers, such as complex certification and testing processes for recovered materials and guidelines on the number of reuse cycles.

Table 1.32 Quantitative analysis of reported barriers on lack of technical skills at the material level

Lack of technical skills	Quantitative analysis
On a general level	*9*
Lack of skills	
Inconsistency in identifying GB risk factors	
Technical competence of the users	
Not ready for radical changes to adopt CE principles	
To make an accurate decision	
Lack of skills by employees in CE	
On the supply chain	*4*
Adopting material tracking systems	
Lack of CE skills by employees in the supply chain	
Make the right decision in SC (supply chain) to implement CE in the most efficient way	
Difficulties finding someone to take the products and materials on	
At the design process	*3*
Lack of circular thinking in the design process	
Planning constraints	
Difficulty of measuring circularity (for the Netherlands)	
At the end of life processes	*3*
Complexities associated with reusing large-sized, complex-shaped, and compositionally diverse composite products	
Lack of awareness of the demolition crews (for steel reuse in Canada)	
Damage during deconstruction (for steel reuse in Canada)	

Finally, the *End-of-Life Policies* subcategory, encompassing issues like a consistent framework on the processes at the end-of-life of a material so as to be reused, occurred two times across the reviewed literature [6, 128].

Category 5: Inadequate Machines/Equipment

A list of identified barriers in the inadequate machines/equipment category, and their sub-category relevance, is shown in Table 1.34.

Compared to other barriers on the material level, inadequate machines and equipment are rarely mentioned as barriers in the analysed literature. In total, this barrier is cited 10 times (in [50, 104, 113, 120, 129, 130, 137, 143]). Nevertheless, the technological level has its importance: the lack of technologies, technological platforms, recycling processes as well as the slow technological evolution are relevant

Table 1.33 Quantitative analysis of reported barriers on the *lack of technical standards* at the material level

Lack of technical standards	Quantitative analysis
Lack of a consistent regulatory framework	*25*
Lack of regulations/legislations	
Lack of flexibility in building codes and regulations	
Contradictions between regulations	
Lack of mandatory reuse regulations for recycled products	
Inconsistent application of existing regulations	
Lack of legislation regarding the properties of the reclaimed materials	
Lack of standardisation	*13*
Material quality specifications for reuse	
Lack of standards for reused and recycled building products	
Complexity of certification processes	*6*
Complex certification and testing processes for recovered materials	
Lack of easy-to-adapt certification schemes for various materials	
Guidelines for testing secondary materials and the number of reuse cycles	
Lack of end-of-life policies	*2*
Ambiguous or inadequate end-of-life policies	
End-of-life policies only for materials considered waste	

Table 1.34 Quantitative analysis of reported barriers on inadequate machines/equipment at the material level

Inadequate machines/equipment	Quantitative analysis
Technological level	*10*
Lack of technologies/inadequate technologies	
Lack of technological platform	
Lack of technological evolution	
Lack of recycling process	

barriers for the implementation of CE in the built environment according to the identified literature. Especially Morel et al. [62] and Mahpour [129] identified inadequate machines and equipment as a barrier.

Category 6: Inadequate Processes and Infrastructure

A list of identified barriers in the lack of the inadequate processes and infrastructures category, and their sub-category relevance, is shown in Table 1.35.

Table 1.35 Quantitative analysis of reported barriers on inadequate processes and infrastructures at the material level

Inadequate processes and infrastructure	Quantitative analysis
Inadequate processes and infrastructure of material recovery	*10*
Material recoverability (selection, harmfulness, lifespan)	
Treatment technologies for secondary materials (concrete, steel, wood, glass, e.g.)	
Uncertainties/Unmet quality in recycling processes	
Inadequate infrastructure of the secondary raw material industry	
The lack of processing and retail hubs for salvaged lumbers	
Needed a fragile system and specific skill sets to deconstruct of the system	
Lack of sorting/insufficient storage/site constraints	*11*
Lack of sorting and storage spaces	
Storage of recovered materials	
Location of material recovery facilities	
Inappropriate waste collection services and recycling infrastructure	
On-site sorting to ensure that the product stream is relatively pure for reuse of wood	
CE-focused design	*3*
Insufficient use or development of designs based on CE principles	
Inefficient methods reduce the profitability of circular business cases	
Complexity of the design process	
Lack of facilities/additional complexity	*4*
Lack of facilities for reclaimed materials	
Inadequate processes of composite construction/joining technique	
Barriers for reuse/recycling of drywall are the lack of gypsum recycling facilities	
Market	*8*
Lack of structured market (bank materials, lack of platform, storage)	
Lack of reuse markets/supply chains	
Supply and demand matching	
Challenges of take-back from other companies	
Underdeveloped market for salvaged components/market for prefabrication is heavily dependent on imports	
Having no clear entry point into the market for fine recycled aggregates and recycled fines	
Lack of info/technology	*6*
Lack of information about the existing structure and materials	
Material tracking system or passport	
Inadequate CE infrastructure to support CE management	

(continued)

Table 1.35 (continued)

Inadequate processes and infrastructure	Quantitative analysis
Lack of infrastructure for implementing circularity in construction	
Lack of institutional infrastructure based on the linear economy model	
Sustainability	*3*
Lack of demonstrating a real impact on reducing carbon emissions	
Lack of principles towards construction design sustainability	
Content of hazardous substances in existing building products	
Disassembly	*2*
Existing built environment are not built for disassembly	
Difficulty in dismantling wood	
Cost	*4*
High cost needed for the collection and storage of reused elements	
Increased cost of recycling and recovering materials	
The expense of hiring a structural engineer to certify the products	

On the material level, the barrier of inadequate processes and infrastructure was identified 51 times through the bibliographical review [6, 10, 49, 50, 81, 97, 104, 106–108, 113, 114, 117, 121, 124, 125, 129, 131, 133–135, 137–139, 143–147]. Main categories within this barrier are inadequate processes and infrastructure of material recovery (10 mentions) [50, 104, 114, 121, 131, 133, 134, 143], the lack of sorting and insufficient storage (site constraints) with 11 citations [10, 97, 104, 124, 129, 135, 137, 139, 144], and market related issues with eight mentions [6, 49, 50, 104, 113, 124, 145, 146]. Market-related issues consist mainly of a lack of reuse markets, supply and demand matching and in general, a lack of developed markets for salvaged building materials. The identified references also reveal a lack of facilities and technologies [10, 49, 50, 97, 106, 117, 124, 137], CE focused design issues [49, 81, 107] as well as roadblocks related to sustainability [108, 125, 147], disassembly [49, 137] and cost [97, 131, 137, 138].

Category 7: Lack of Methods/Tools

A list of identified barriers in the lack of methods/tools category, and their sub-category relevance, is shown in Table 1.36.

Barriers due to the lack of methods and tools are reported 56 times in the identified literature; hence, it is a relevant barrier to the implementation of CE. The quantitative analysis has shown that it is most often about the lack of tools, design tools and metrics [6, 10, 19, 34, 49, 90, 97, 104, 105, 109, 114, 115, 120, 121, 125, 129, 132, 140, 148, 149]. This goes hand in hand with difficulties in adapting CE principles to already existing tools and methods. Other relevant identified categories are the lack

Table 1.36 Quantitative analysis of reported barriers to lack of methods and tools at the material level

Lack of method/tool	Quantitative analysis
Awareness of material recovery by stakeholders/producers/company holders	*6*
Lack of bandwidth compounded by no coherent vision	
Low quantity of secondary construction materials	
Applications of reuse in new construction applications are yet to be explored	
No coordination between all stakeholders in the value chain	
Lack of tool/design or adopting existing method/tool	*23*
Difficulties of the programming and design phases of adaptation for construction material	
Difficulties of changing the culture away from traditional design towards adopting flexibility	
Lack of procedures to monitor the progress toward CE transition	
Lack design tools for reusing timber or for green design	
Limited studies adopted the latest methods to assess risks	
Lack of technical assessment methods of recycled building materials	
Lack of detailed methodology for the selection of materials according to the CE principles	
Lack of digital/sensor methodologies	
Limited or lack of CE metrics/tools/design	
Lack of performance analysis and design tools for evaluating end-of-life scenarios of buildings	
Certification/Standardization/Classification	*8*
Lack of Certification	
Lack of standardisation/classification	
Lack of tools for identifying salvaged materials	
Complexity of the recycling processes	*4*
Complexity of materials and building composition for recycling processes	
Lack of project specification	
Time constraints	
Longer required time for the deconstruction of the building materials	
Testing methods	1
Required destructive testing	
Building information modelling	2
Insufficient data management (data unavailability in accessibility; data compliance as—as-built-accuracy; data requirements; data exchange and interoperability issues)	
Difficulty in modelling of continuous loops of materials	

(continued)

Table 1.36 (continued)

Lack of method/tool	Quantitative analysis
Disassembly	*5*
Complex disassembly shorts service life	
Technical challenges regarding material recovery	
Inaccessible joints	
Lack of standards for Design for Disassembly (DfD) tools	
The high operational costs	
Environmental impact	*2*
Overemphasizing recycling and non-environment friendly methods	
Governance, regulatory frameworks and political landscape	*1*
Inadequate current strategies and frameworks (cradle to cradle, industrial ecology) overlooking the underlying social, political, economic and technical aspects	

of awareness for material recovery by all the stakeholders [81, 97, 110, 120, 133, 149] and a lack of certification/standardisation and classification [6, 7, 10, 19, 104]. Furthermore, time constraints [124, 131, 136, 145] and the complex implementation of CE, sometimes linked to high cost and hurdles to recycling [7, 19, 124, 139] are cited as barriers in the identified literature. Cited rather rarely are the lack of technical knowledge and skills [81], lack of testing methods [124], insufficient implementation of BIM [7, 48], high cost [50], issues related to the disassembly [120, 134], environmental impacts [129] and regulatory frameworks [117].

Category 8: Lack of Materials/Products/Building Systems

A list of identified barriers in the lack of materials/products/building systems category, and their sub-category relevance, is shown in Table 1.37.

The quantitative analysis of reported barriers on *Lack of materials/products/buildings/systems* revealed that the subcategory *Lack of performance and quality assurance for reused/recovered materials* is the most frequent barrier reported across literature, for the material level. This subcategory, involves issues like low volume and inconsistent quality of the recovered materials, durability issues of reused products, use of polluted or low recoverability materials, low performance and/or the absence of documentation on performance of the recovered materials, hazardous or contaminated materials etc., appearing 34 times across the reviewed literature [6, 7, 10, 49, 50, 104, 105, 108, 111, 114, 124, 126, 127, 133, 134, 136, 137, 141, 148, 150].

The next three technical barrier subcategories in the Lack *of materials/products/buildings/systems* category recorded a similar number of mentions across the reviewed literature. *Lack of systems*, which occurred 14 times across the reviewed

Table 1.37 Subcategories of barriers under Category 8 at the material level

Lack of materials/products/buildings/systems	Quantitative analysis
Lack of performance and quality assurance for reused/recovered materials	*34*
Quality of materials at end-of-life (EOL)	
Use of polluted or low recoverability materials	
Ability to deliver high-quality remanufactured products	
Contamination—coatings with (now) banned chemicals	
Durability issues of reused products	
Low performance and/or the absence of documentation on performance	
Reusable goods with a good lifetime: at the time of the re-use, they are out of date and standard	
Maintaining the quality of products made from recovered materials	
Low volume and inconsistent quality	
Lacking quality assurance	
Lack of systems	*14*
Most data storage and transfer happen through external documentation	
QR codes are not a sufficient system for embedded or encapsulated materials, such as cast-in-place concrete or insulation behind an exterior wall	
Taking back products for reuse becomes a tough task for service providers due to the legal problems of retaining the sold product	
No standardization of diversity	
No space for storage	
Lack of a producer-based responsibility system in the production of construction materials	
Complexity of material recovery	*13*
Technical challenges related to material recovery	
Complexity of materials and building composition (several layers and modifications during its lifespan)	
Difficulty in recovering materials, hazardous components	
Complexity and the irreversible nature of structural connections	
Difficult to manage the dismantling and reusing, or recycling	
Lack of available materials/products for reuse	*12*
Supply chain dynamics and availability of reused materials	
Limited availability of reuse products	
Limited stock availability	
Other limitations of materials/products/buildings/systems	*16*
Low level of application of recycled products	
Using finite recyclable construction materials	
Increasing material diversity and the number of composite materials and structures	

(continued)

Table 1.37 (continued)

Lack of materials/products/buildings/systems	Quantitative analysis
Demolition projects range depending on the age of structures	
Recycled product may have unknown strengths and quality properties	
Long product life cycle	
Building component lifespan	
Design constraint for reclaimed material use	
Determination of future use of secondary materials	
The trend of using inexpensive and non-durable materials/components in construction	

literature [81, 105, 106, 114, 115, 120, 121, 129, 134, 139, 150], involved technical barriers and challenges, such as lack of producer-based responsibility system in production of construction materials, no space for storage, lack of clearly defined national goals, targets, and visions to move toward CE, no standardisation of diversity or issues like the fact that most data storage and transfer happen through external documentation.

Similar occurrence to the *Lack of systems* was reported for the *Complexity of material recovery* and for the *Lack of available materials/products for reuse,* at the material level. *Complexity of material recovery* appeared 13 times across the reviewed literature [10, 19, 48, 62, 110, 114, 120, 125, 131, 136, 137, 148] and was related to challenges such as difficulty to recover materials, building complexity/building composition—several layers, delay in deconstruction due to the complexity and irreversible nature of structural connections or difficulty to manage the dismantling and reusing or recycling. *Lack of available materials/products for reuse* appeared 12 times across the reviewed literature [6, 97, 104, 114, 120, 124, 131, 145, 150] and was related to challenges, such as limited availability of reuse products, supply chain dynamics and availability of reused materials, certain materials not widely available in every region of the world, inadequate flexibility on section sizes of reuse steel, or the location/availability/quantity of the recovered materials.

Barriers that could not be grouped in the above subcategories of the *Lack of materials/products/ buildings/systems* category were assigned to the *Other limitations of materials/products/buildings/systems* subcategory. This subcategory encloses barriers like low level of application of recycled products, usage of finitely recyclable construction materials, increasing material diversity and number of composite materials and structures, demolition projects ranges depending on the age of structures, long product life cycle, building component lifespan, design constraints for reclaimed material use and the trend of using inexpensive and non-durable materials/ components in construction. *Other limitations of materials/products/buildings/systems* occurred 16 times across the reviewed literature [6, 48, 50, 97, 105, 111, 114, 115, 126, 127, 129, 133, 137, 141, 149].

As recorded in all previous technical barrier categories, in some cases, different challenges and issues belonging to the same subcategory were reported within the same literature reference.

1.3.3.3 Quantitative Data Analysis of Each Category

A summary of the collected quantitative data for each category is shown in Table 1.38.

Category 1: Lack of Information/Data

Where it happens. Regarding the building layers, the studies addressing the lack of information/data barriers at the material level reported that these barriers affect mostly the structure [6, 7, 19, 34, 90, 97, 109, 111, 112, 115–119, 122, 123, 125, 128, 132, 141] and the skin [6, 7, 19, 34, 90, 97, 106, 111, 112, 116, 118, 119, 122, 128, 132, 141] of a building, while at the other end of the spectrum is the space plan of a project mentioned once [112] and the site (not mentioned at all), which are the least affected by barriers regarding lack of information/data.

When it happens. Lack of information/data barriers at the material level has been evaluated as affecting mostly the Design [6, 7, 19, 34, 49, 90, 97, 112, 116, 122, 123, 125, 132] and End-of-life phases [6, 90, 97, 109, 111, 112, 114, 116–118, 118, 122, 125, 126, 128] within the building life cycle phases, followed by (Dismantling and) Construction [6, 97, 109, 112, 114, 115, 117, 119, 125, 128, 132], Strategic Definition and Briefing of the project [49, 90, 97, 108, 112, 118, 122, 123, 129] and Use and Refurbishment building phase [97, 112, 114, 117, 118, 120, 129, 132].

Who is responsible for/involved in. Considering the stakeholders involved in/responsible for the technical barriers, project professionals [6, 7, 19, 34, 49, 50, 90, 109, 111–113, 118–123, 119–123, 125] and suppliers [6, 7, 34, 49, 50, 90, 111–113, 113, 116, 120, 122, 123, 126, 128, 129, 132] are considered mostly responsible for the lack of information/data barrier at the material level. The following stakeholders are Clients [50, 112, 113, 120, 121, 125, 129, 131, 136, 141], who influence the lack of information/data barrier at the material level. The rest of the stakeholders seem to represent minor interest in the literature regarding the influence on lack of information/data barrier at the material level: Government [19, 50, 118, 120, 132, 141], Others—not included in other categories [110, 126, 128, 141] and Public [50, 120, 141].

How it affects the implementation of circular strategies in the built environment. The reviewed literature emphasizes that the lack of information/data barrier significantly impacts circular strategies in the built environment at the material level. This includes reuse [6, 10, 49, 90, 97, 108, 109, 111, 112, 117–119, 122, 123, 125, 132, 141], repurpose [49, 90, 97, 111, 112, 117–119, 122, 123, 132] and repair [6, 97, 111, 117–119, 121, 122, 131]. However, the impact extends, affecting also strategies

Table 1.38 Quantitative data collection for the material level involving technical barrier categories

What is the barrier category	Where it happens						When it happens					
Technical barrier category	Building layer (a)						Building life cycle phase (b)					
	1	2	3	4	5	6	1	2	3	4	5	6
Lack of information/data	–	21	18	2	1	9	9	15	5	11	8	15
Lack of technical knowledge	1	11	10	1	–	6	14	19	10	13	11	9
Lack of technical skills	1	6	4	–	–	5	3	5	3	5	3	7
Lack of technical standards	–	18	13	2	1	2	9	14	8	5	6	11
Inadequate machine/equipment		1	1	–	–	–	–	–	1	3	2	2
Inadequate processes and infrastructure	3	20	13	2	1	4	5	12	9	11	7	16
Lack of methods/tools	2	16	8	5	3	4	11	20	7	16	8	12
Lack of materials/products/systems	–	22	12	1	–	6	5	14	9	16	11	21

What is the barrier category	Who is responsible for/involved in						How it affects the implementation of circular strategies in the built environment									
Technical barrier category	Stakeholders (c)						Circular strategies (e)									
	1	2	3	4	5	6	1	2	3	4	5	6	7	8	9	10
Lack of information/data	10	24	19	3	6	4	1	7	7	19	9	3	4	11	9	4
Lack of technical knowledge	8	20	12	3	2	1	1	4	1	12	7	7	6	6	9	6
Lack of technical skills	5	8	4	–	1	–	–	–	–	3	–	–	–	–	2	–
Lack of technical standards	3	11	8	4	34	–	2	11	3	10	8	4	4	7	12	4
Inadequate machine/equipment	–	2	1	–	–	–	–	–	–	–	1	1	1	1	1	–
Inadequate processes and infrastructure	2	19	10	–	2	2	1	5	5	19	14	12	13	13	11	6

(continued)

Table 1.38 (continued)

What is the barrier category	Who is responsible for/ involved in						How it affects the implementation of circular strategies in the built environment									
Technical barrier category	Stakeholders (c)						Circular strategies (e)									
	1	2	3	4	5	6	1	2	3	4	5	6	7	8	9	10
Lack of methods/tools	7	16	8	3	7	4	3	5	6	14	4	4	4	6	7	6
Lack of materials/ products/ systems	4	19	16	5	5	3	2	3	2	24	11	10	11	13	19	11

(a) Building layers [17]: (1) site; (2) structure; (3) skin; (4) services; (5) space plan; (6) stuff; (7) system (district)
(b) Building life cycle phases (adapted from Council et al. [15]; Royal Institute of British Architects [18]): (1) Strategic Definition and Briefing; (2) Design; (3) Manufacture; (4) (Dismantling and) Construction; (5) Use and Refurbishment; (6) End-of-life
(c) Stakeholders [19]: (1) Clients; (2) Project professionals; (3) Suppliers; (4) Public; (5) Government; (6) Others
(d) Circular Strategies [20]: (1) Refuse; (2) Rethink; (3) Reduce; (4) Reuse; (5) Repair; (6) Refurbish; (7) Remanufacture; (8) Repurpose; (9) Recycle; (10) Recover

like reducing [49, 90, 109, 112, 116, 126, 132], rethinking [7, 19, 34, 112, 116, 118, 123] recover [49, 112, 117, 132] and remanufacture [49, 97, 111, 117].

Category 2: Lack of Technical Knowledge

Where it happens. Concerning the building layers, the studies highlighting the lack of technical knowledge at the material level have found that these barriers primarily impact the structure [6, 10, 19, 97, 116, 123, 133, 135, 137, 138, 141] and the skin [6, 19, 97, 106, 116, 123, 135, 137, 138, 141], as well as the "stuff" of a building [6, 106, 114, 116, 129, 135]. The layers that were barely reported in the studied bibliography referred to the site [135], the services [19] and the space plan.

When it happens. Lack of technical knowledge barriers at the material level have been evaluated as affecting every phase of the building life cycle. As the most prevalent phase the Design was emerged [6, 10, 19, 34, 48–50, 50, 81, 90, 97, 114, 116, 120, 133–135, 134, 135], followed by the Strategic Definition and Briefing [10, 19, 48, 49, 81, 97, 107, 108, 114, 116, 133–135, 135], (Dismantling and) Construction [6, 34, 48–50, 50, 81, 105, 114, 133–135, 134] and Use and Refurbishment [34, 48, 49, 81, 109, 110, 114, 133–135, 135]. The phases of Manufacture [48–50, 50, 81, 114, 133–135, 134] and End of life were also widely addressed [6, 48, 49, 81, 109, 110, 114, 134, 137].

Who is responsible for/involved in. Considering the stakeholders, project professionals [6, 10, 19, 34, 48, 49, 81, 97, 109, 110, 113, 114, 116, 120, 123, 133, 135, 137, 138, 141] and suppliers [6, 19, 48, 81, 108–110, 110, 113, 114, 120, 123] are primarily held responsible for the lack of technical knowledge barrier at the material level, followed by clients [6, 48, 81, 113, 114, 133, 135, 141]. Other stakeholders appear to be of minor interest in literature concerning their influence on this issue, including the Public [109, 120, 141], government [19, 141] and others [141].

How it affects the implementation of circular strategies in the built environment. The reviewed literature emphasises that the lack of technical knowledge barrier significantly impacts circular strategies in the built environment at the material level. This includes reuse [10, 48, 97, 107, 121, 123, 124, 133, 136–138, 138], recycle [6, 10, 48, 108, 121, 123, 135–137], repair [6, 10, 48, 121, 123, 136, 138] and Refurbish [10, 48, 121, 123, 135, 136, 138]. The lack of technical knowledge also affects Remanufacture [10, 48, 121, 123, 135, 136], Repurpose [10, 48, 121, 123, 135, 136] and Recover [10, 48, 121, 123, 135, 136]. The strategies Refuse and Reduce were the least indicated, in Refs. [120] and [10], respectively.

Category 3: Lack of Technical Skills

Where it happens. Concerning the building layers, the studies highlight the lack of technical skills at the structure [49, 97, 108, 109, 137] the skin [97, 106, 137], as well as the "stuff" of a building [50, 106, 120, 121, 129]. The layers that were barely reported in the studied bibliography referred to the site, the services and the space plan.

When it happens. At the material level, the Lack of technical skills has been identified as a barrier affecting each phase of the building life cycle. End of life has been referenced the most [49, 81, 97, 108, 114, 137], followed by Design [34, 49, 81, 108, 114] and Dismantling and Construction [81, 97, 108, 114]. The phases of Strategic definition [81, 108, 114], Manufacture [81, 108, 114], and Use-Refurbishment were also addressed [81, 108, 114].

Who is responsible for/involved in. Considering the stakeholders, project professionals [34, 81, 108, 113, 114, 137, 140], clients [50, 81, 104, 108, 114] and suppliers [81, 108, 114] are identified as mostly associated with the lack of technical skills barrier at the material level.

How it affects the implementation of circular strategies in the built environment. The reviewed literature does not emphasise on the impact that the lack of technical skills barrier has on the implementation of circular strategies in the built environment at the material level. The only reported strategies are reuse [49, 137] and recycle [137].

Category 4: Lack of Technical Standards

Where it happens. Regarding the building layers, the studies addressing the lack of technical standards barrier at the material level reported that it mostly affects the structure (18 refs) [6, 7, 97, 116, 118, 119, 122, 128, 133, 135, 137, 138, 140–142, 146, 147] and the skin (13 refs) of a building [6, 97, 116, 118, 119, 122, 128, 135, 137, 138, 140, 141]. On the other hand, very limited studies mentioned that the barrier affects the services (two refs) ([6, 140]), space plan (one ref) [140] and staff (2 refs) ([6, 34]). The site layer is not related to this barrier.

When it happens. The lack of technical standards barrier at the material level has been found to be critical at the Design (14 refs) [81, 97, 107, 116, 118, 119, 124, 128, 135, 138, 140, 141] and End-of-life phases within the building life cycle (11 refs) [6, 81, 97, 107, 118, 128, 133, 135, 137, 139, 140]. Nine studies have also evaluated the lack of technical standards as an important barrier at the Strategic Definition and Briefing Phase [81, 97, 118, 119, 128, 135, 138, 140, 141], followed by Manufacture (eight refs) [81, 107, 135, 137, 140, 142, 146, 147] Use and Refurbishment (six refs) [81, 107, 108, 135, 139, 140] and Dismantling and Construction (five refs) [81, 107, 135, 137, 140].

Who is responsible for/involved in. In terms of the stakeholders involved, the Government is considered the most responsible one for the lack of technical standards barrier at the material level (34 refs) [6, 7, 50, 81, 97, 104, 105, 107–110, 114–116, 118–121, 116, 118, 119, 134–139, 128, 129, 131, 134]. The Project Professionals (11 refs) [6, 7, 10, 81, 97, 116, 118, 133, 135, 140, 146] are also considered responsible or highly involved in the barrier. The rest of the stakeholders seem to have lower influence on the lack of technical data barrier at the material level: Suppliers (eight refs), [10, 81, 90, 107, 140, 142, 146, 147], Public (four refs), [97, 107, 128, 141], Clients (three refs), [10, 107, 140], Others—not included in other categories (0 refs).

How it affects the implementation of circular strategies in the built environment. The reviewed literature emphasises that the lack of technical standards barrier has a considerable impact on implementing circular strategies in the built environment at the material level. 12 refs mentioned that it affects the Recycle strategy [6, 10, 90, 118, 119, 122, 128, 133, 135, 138, 142, 146], followed by Rethink (11 refs), [7, 10, 116, 118, 119, 135, 138, 140, 142, 146, 147], Reuse (10 refs), [6, 10, 90, 97, 118, 122, 124, 128, 137, 140] and Repair (eight refs), [6, 10, 49, 90, 122, 128, 133, 140]. The circular strategies Refurbish [6, 10, 128, 140], Remanufacture [6, 10, 128, 140], and Recover [10, 90, 122, 128] were mentioned in four refs each, as influenced by the lack of technical standards barrier.

A large number of studies examining technical barriers at the material level, suggested that the lack of a consistent regulatory framework [49, 50, 81, 104, 105, 109, 115, 116, 134] along with inadequate standardisation procedures for recovered materials [6, 97, 114, 118, 120, 131, 136] play a key role in the application of CE in construction.

Category 5: Inadequate Machines/Equipment

Where it happens. The examined literature that deals with the barrier of inadequate machines and equipment rarely identifies relevant building layers. Structure and skin are identified once each [137]. The site, services, space plan and stuff are not related to this barrier.

When it happens. No specific life cycle phase stands out. The barrier of Inadequate machines and equipment on the material level is discussed in the following phases: Manufacture, Dismantling and Construction, Use and Refurbishment, End-of-life [113, 130, 137, 143]. This barrier does not occur in the life cycle stages of strategic definition and design.

Who is responsible for/involved in. Project professionals and suppliers are identified as responsible for the lack of adequate machines and equipment [120, 129, 137]. Clients, the public, governments and others are not met in the bibliographical review.

How it affects the implementation of circular strategies in the built environment. No clear conclusion can be drawn by analysing the identified literature. The barrier, Inadequate machines and equipment, doesn't seem to affect one CE strategy, particularly at a wide extent. The lack of adequate machines and equipment affects the following CE strategies, while each of them is addressed only once: Reuse, Repair, Refurbish, Remanufacture, Repurpose, Recycle [137]. The barrier is not influencing the refuse, rethinking, reducing and recover strategies.

Category 6: Inadequate Processes and Infrastructure

Where it happens. On a material level, the barrier of inadequate processes and infrastructure mainly takes place in the structure of a building (20 refs) followed closely by the skin (13 refs) [6, 10, 49, 97, 106, 117, 121, 124, 125, 131, 133–138, 143, 145–147]. The other building layers, namely the site, services, space plan and stuff, are rarely addressed.

When it happens. The bibliographical review shows that the lack of processes and infrastructure has an impact mainly on the end-of-life of a building [6, 10, 49, 50, 108, 113, 124, 125, 135, 137, 138, 143, 144, 146, 147]. Besides, the design as well as the construction and dismantling stage, are life cycle phases that often deal with the lack of adequate processes and infrastructure. Furthermore, the strategic definition, manufacture and the use and refurbishment phase are mentioned as relevant life cycle phases in the identified literature (with 5, 9 and 7 citations respectively).

Who is responsible for/involved in. Mostly responsible for/involved in the lack of processes and infrastructure are project professionals and suppliers, with 19 and 10 citations respectively [6, 10, 49, 50, 97, 106, 108, 113, 124, 125, 135, 137, 138, 143, 144, 146, 147]. However, the clients, the government and other stakeholders are also cited as responsible. The quantitative analysis did not reveal that the public is involved or responsible for this particular barrier.

How it affects the implementation of circular strategies in the built environment. Reuse is the most affected CE strategy when analysing the lack of adequate processes and infrastructures. It is cited 19 times in total in the following references [10, 49, 97, 117, 121, 124, 125, 131, 133, 135, 137, 138, 143, 145, 147]. Besides the reuse strategy, it could be identified that the Repair, Refurbish, Remanufacture, Repurpose and Recycle strategies are severely impacted.

Category 7: Lack of Methods/Tools

Where it happens. In line with barrier six, the structure is the building layer that is the most impacted (cited 16 times) by the lack of methods and tools, followed by the skin layer (cited eight times). The following references discuss structure and skin as relevant building layers: [6, 10, 19, 90, 109, 110, 112, 124, 125, 132–134, 139, 140, 148, 149]. Some references cite the site, services, space plan and stuff as impacted by the lack of methods and tools, but the occurrence per building is quite low, which shows a limited impact from the lack of methods and tools.

When it happens. According to the identified literature, the lack of methods and tools mainly occurs in the design (20 citations) and dismantling and construction (16 citations) life cycle stages. Those are referenced 36 times in: [7, 34, 48, 49, 81, 90, 97, 109, 112, 114, 115, 117, 124, 125, 129, 131–133, 140, 148]. Not to be neglected is the impact on the strategic definition, manufacture, use and refurbishment and end-of-life stages, as those are cited several times with 11, seven, eight and 12 citations respectively.

Who is responsible for/involved in. Project professionals are mostly responsible for the lack of methods and tools (16 mentions), which is discussed in the following literature: [6, 7, 19, 34, 48, 49, 81, 90, 97, 109, 112, 124, 125, 132, 140, 148]. However, clients, suppliers and governments are also involved. Considering the lack of methods and tools, the Public and others are not often met as involved stakeholders in the identified literature.

How it affects the implementation of circular strategies in the built environment. The lack of methods and tools mainly affects the reuse strategy, with 14 mentions in the identified literature. This is addressed in the following references: [6, 7, 19, 97, 110, 112, 117, 124, 125, 131, 132, 140, 145]. All of the other circular strategies are also impacted by the lack of methods and tools, but rather on a moderate level.

Category 8: Lack of Materials/Products/Buildings/Systems

Where it happens. Studies addressing the Lack of materials/products/buildings/systems barriers at the material level reported that, regarding the building layers, these barriers affect mainly the structure [6, 7, 10, 48–50, 97, 105, 111, 119, 121, 124, 125, 131, 133, 136, 137, 141, 148, 150] and the skin [6, 7, 10, 106, 111, 119, 137, 141, 149, 150] of a building, while the other layers of the building (space plan, site, services) are rarely addressed and therefore, appear not to be affected by this barrier.

When it happens. Lack of materials/products/buildings/systems barriers at the material level has been evaluated as affecting mostly the End-of-life [6, 7, 10, 48, 50, 81, 111, 114, 115, 120, 125, 126, 133, 137, 139, 148, 149] and (Dismantling and) Construction [10, 50, 97, 104, 114, 115, 120, 124, 125, 127, 129, 131, 136, 145, 148, 149] within the building life cycle phases, followed by Design [6, 10, 48–50, 50, 97, 106, 114, 115, 119, 120, 141, 148] and Use and Refurbishment of the project [48–50, 50, 114, 115, 119, 120, 131, 136, 139].

Who is responsible for/involved in. The project professionals [6, 10, 48, 50, 97, 111, 114, 119, 125, 129, 137, 139, 141, 148, 150] and the suppliers [6, 7, 48, 50, 97, 106, 111, 114, 115, 119, 120, 124, 126, 141, 148, 150] are mostly responsible for the Lack of materials/products/buildings/systems barrier. Public [97, 115, 120, 126, 139], clients [115, 136, 139, 141] and government [81, 97, 114, 115, 126] are also included in the responsible stakeholders, but they seem to represent minor interest in the literature regarding the influence on Lack of materials/products/buildings/systems barrier, at the material level, than the suppliers and project professionals.

How it affects the implementation of circular strategies in the built environment. The reviewed literature emphasises that the Lack of materials/products/buildings/systems barrier significantly impacts circular strategies in the built environment at the material level. The most affected circular strategies are: reuse [6, 7, 10, 49, 97, 110, 111, 114, 119, 121, 124, 125, 127, 129, 131, 133, 134, 136, 137, 145] and recycle [6, 7, 10, 48, 50, 110, 111, 114, 115, 126, 129, 134, 137, 148, 149], followed by repurpose [7, 10, 48, 50, 114, 125, 129, 137, 148], repair [7, 10, 114, 119, 129, 134, 137], remanufacture [7, 10, 114, 119, 125, 129, 137], recover [7, 10, 48, 50, 97, 121, 125, 126, 136, 148, 149] and refurbish [7, 10, 110, 114, 129, 137], with the last four circular strategies being affected similarly. As recorded in all previous technical barrier categories, in some cases, different challenges and issues belonging to the same subcategory were reported within the same literature reference.

1.4 Discussion

Looking at general patterns, we observed that the lack of standards and the lack of data are both the most relevant barrier categories at all levels. They both appear as the most recurring barrier categories among the first three at all levels. The urban and building levels also report the lack of knowledge as a recurrent type of barrier, while at the building and material levels, it is acknowledged that the lack of approaches, methods and metrics is a common challenge. Additionally, we observed specific challenges: the lack of adequate infrastructure is recognized as the main barrier category at the urban level; the lack of materials, products and systems is reported as a relevant challenge at the material level.

Also, we noticed concentration on the skin and the structure layers at the building and material level, while building layers which have a shorter life in a building—the services, space plan and stuff—are less discussed. Considering, for example, that an interior door is highly re-useable, it is surprising that the space plan lacks research on applications of the 10R approach for a CE. Also, it is worth noticing that, in these layers, the implementation of material/product/systems passports is most beneficial. So, the lack of research or literature on applications in these building layers is unexpected despite evidence of their relevance.

The life cycle stages that result in being mostly affected by barriers at all levels are the construction and the end-of-life stages. At the urban level, refurbishment is also reported as a recurring stage in the literature on barriers and challenges, while the design phase is frequently acknowledged at the building and material levels. In terms of strategies for a CE, we noted a concentration on the Reduce, Reuse, and Recycling strategies across the 3 levels of the built environment. This trend may be because the 3R approach has been widely discussed and adopted for many years, but it may have limited people's thinking. The 10R approach widens the opportunities for sustainable practice while removing the false idea that Recycling is the ultimate aim or the justification to provide for sustainability. This pattern is consistent with results from another study, which reported that the construction industry is focused on recycling and reuse as the main "R" strategies, while the manufacturing industry tends to achieve higher levels of circularity due to broader implementation of remanufacturing and industrial symbiosis [151]. Finally, we observed that professionals, suppliers and governments are consistently documented as stakeholders mostly affected by or responsible for technical barriers at all levels.

Each level has discussed below trends, patterns and gaps observed in the study implemented within their level.

1.4.1 Urban Level

At the urban level, the barriers are reported mainly under the category of *Inadequate processes and infrastructure,* followed by categories of *Lack of technical standards*, *Lack of information/data*, and *Lack of technical knowledge*. This can be related to the characteristics of the urban environment, where the *Government* is mainly responsible for the processes of CE implementation. Moreover, *Project Professionals* and *Suppliers* were identified as the other stakeholders involved in the process, being directly affected by the barriers encountered in *Inadequate processes and infrastructure, Lack of technical standards*, *Lack of information/data*, and *Lack of technical knowledge*. Consequently, the barriers are reported mostly during the *Use and Refurbishment, Strategic Definition and Briefing, (Dismantling and) Construction,* and *End-of-life* phases. Another characteristic of the urban level is how it affects the

CE strategies, with most barriers affecting *Reuse* and *Recycle* strategies. In terms of "Where", as explained in the previous section, the barriers are primarily reported at the District (System) layer.

Analysing category by category, the barriers under the *Lack of information/data* category have been identified as "data capabilities and data availability", "limited knowledge", "the lack of integration among sources of information", and "insufficient data". Within the life cycle, they are primarily reported in the *Use and Refurbishment* phase. The *Government* is the main stakeholder responsible for this, followed by Project *Professionals.* The literature has shown that the barriers related to the *Lack of information/data* mainly affect *Reuse, Recycl*e and *Repurpose* CE strategies.

The second category, *Lack of technical knowledge*, presents barriers related to the knowledge gap in understanding the CE concept, such as "limited application knowledge", "production and quality management challenges", "CE knowledge deficit", "scaling up challenges", "skills gap threatening knowledge transfer", and "technological immaturity and expertise shortage". They are mostly linked to the district (system) level; however, some can be connected to staff and service levels. Manly reported in the *Strategic Definition and Briefing, Design,* and *Manufacturing* phases, they can also be mapped across the different life cycle phases, underscoring the need to address these knowledge gaps throughout the entire life cycle to implement circular construction practices successfully. The *Government* plays an essential role in setting standards and promoting best practices, ensuring knowledge transfer and workforce development, and may be involved in financing or promoting research and development efforts. Thus, this is the main stakeholder involved in this category. These barriers can affect the implementation of many CE strategies, especially *Reuse, Recycle,* and Repurpose. For instance, insufficient knowledge in production management and high-quality manufacturing creates challenges for *Reuse* and *Recycle* strategies. The loss of traditional construction skills directly affects the Repurpose strategy. The lack of knowledge about reusing construction and demolition waste (CDW) and immature recycling technology pose challenges to *Reuse* and *Repurpose* strategies. Similarly, the lack of technological expertise and the unavailability of such expertise hinder CE adaptation and create significant barriers to *reuse* and *recycle* strategies.

Five key technical barriers were identified under the *Lack of technical skills* category, namely "the lack of skills and expertise among public infrastructure and service providers", "the lack of knowledge regarding scaling up production and the contribution of urban manufacturing to the CE", "the potential loss of traditional construction skills", "the lack of access to essential business-related services and technical assistance", and "the shortage of human resources knowledgeable about CE principles across all stakeholders". They are all mainly related to the district (system) level but can also be related to staff and services levels. Even though the skill gaps and knowledge deficits manifest across various stages of the urban circular construction lifecycle, they are predominantly during the *Strategic Definition and Briefing* and *Manufacture* stages. The literature suggests a particular focus on *Construction Professionals* as the leading group of stakeholders responsible for or affected by the lack of technical skills. Additionally, *Government* intervention can play a crucial role in creating a supportive environment for the adoption of circular construction practices. The barriers identified in this category pose challenges for CE strategies like *Reuse, Remanufacture*, and *Recycle.*

Under the category of *Lack of technical standards,* barriers are primarily caused by the "absence/inadequacy of standardised processes", "certification and guidelines", which frequently are not updated, "the diversity of existing regulations", and consequently, the different levels of development and spatial scales causing difficulties in their understanding and correct implementation. These barriers were mainly discussed during the life cycle phases of *Strategic Definition and Briefing* and *Use and Refurbishment.* Due to their intrinsic characteristics, the *Government* and *Project Professionals* are the groups responsible for or are affected by these barriers. In terms of CE strategies, the barriers under the *Lack of technical standards* category affect the majority of *Reuse* and *Recycle* strategies.

Inadequate machines/equipment is the category with the lowest number of articles reporting it, appearing in only one article. The primary barrier identified in this category is the "immature recycling technology". The life cycle phases related to it are *Dismantling and Construction, Use and Refurbishment,* and *End-of-life*; consequently, the CE strategy most affected by this category is *Recycle.* The *Government* is mainly responsible for overcoming this barrier by encouraging economic incentive measures and recycling centres.

In terms of *Inadequate processes and infrastructure,* the barriers identified can be classified into five main groups: (1) *Waste collection and Classification Systems,* in which "Low-quality collection and classification systems", "Immature recycling technology" are the main barriers identified; (2) *Material Handling and Infrastructure,* with "Lack of pre and post-use material handling systems", "Resource-intensive infrastructure lock-in", "Poorly developed infrastructure for waste management and recycling and for support CE management", "Insufficient or incompatible local processes and infrastructure", and "Non-integrated and refurbished infrastructure" as the main barriers; (3) *Integration and Networking,* grouping "Lack of integrated CDW processes, tools, and practices group", "Lack of networking initiatives with legislative support", "Limited integration of urban metabolism", "Lack of reverse logistics infrastructure", "Fragmentation of supply chains", and "Limited access to production networks"; (4) *Spatial and Scaling Issues,* with "Lack of space for circular infrastructure", "Challenges in scaling up", "Spatial precarity"; and finally, (5) *Planning and Urban Processes,* where the barriers identified are "Neglecting people and 'place-making'" and "Poor connection and limited integration of place-making urban 'processes' and urban 'metabolisms'". These barriers can be classified into three different levels: the national (Government) level, local (municipal, urban) level, and organizational (company) level. Hence, they occur mainly during *Dismantling and Construction, Use and Refurbishment,* and *End-of-life* phases. *Suppliers* and the *Government* are the main ones responsible for these barriers, followed by *Clients,* and *Project Professionals.* The impact on CE strategies has most often been identified in *Reuse* and the process of *recycling* strategies.

In the *Lack of methods/tools,* the main barriers and challenges identified in the literature include the "lack of appropriate design" and "logistics expertise", this is because circular design and thinking are not sufficiently embedded in urban systems. Furthermore, there is a lack of comprehensive CE metrics, tools and design frameworks, as well as a lack of standardised methodologies for evaluating CE initiatives, making it challenging to implement and measure progress and ensure successful outcomes. The literature also indicates a lack of enabling digital technologies, especially those that support real-time data tracking and resource management. These barriers and challenges are primarily reported in the broader context of urban and district-level planning and management. They occur most frequently in specific life cycle phases, such as *Strategic definition and briefing*, *Design*, and *Use and Refurbishment*. *Project professionals*, including architects, engineers, designers, and construction managers, are the group most affected by these barriers. Meanwhile, *Government* agencies are responsible for overcoming the barriers once they are responsible for creating regulatory frameworks, providing incentives, and supporting the development of standardised methodologies for CE assessment. The lack of methods and tools significantly impacts the implementation of circular strategies, particularly strategies such as *Reduce*, *Remanufacture* and *Recycle*.

Finally, under the last category of barriers, *Lack of materials/products/building systems*, spatial challenges have been reported as significant barriers to implementing circular strategies in the built environment. The scarcity of affordable sites and spatial precarity can limit the availability of suitable locations for circular practices. Moreover, material-related issues present substantial barriers associated with the low cost of new materials versus the use of recycled materials. Also, monitoring and managing urban resource flows is a complex task posing significant challenges for circular strategy implementation. Those barriers occurred during the *Use and Refurbishment* phase, highlighting in this phase the limited availability of quality recycled materials, high labour costs, scarcity of affordable sites, spatial precarity, and limited space for storing disassembled materials and during the *End-of-life* phase, in which high labour costs for deconstruction, shortage of space to store disassembled materials, and difficulties in managing complex material compositions are reported. *Project professionals* and the *public* are the stakeholders most involved in or affected by these barriers. They are involved in barriers such as managing design challenges and influencing market demand for circular practices. Ultimately, the most impacted CE strategy is *Reuse*, with barriers such as limited space to store disassembled materials and the scarcity of quality recycled materials.

1.4.2 Building Level

The implementation of CE strategies in buildings is mainly affected by a lack of information/data and a lack of technical standards followed by a lack of technical knowledge and a lack of methods/tools. These barriers limit the implementation of CE strategies mainly in the structure, skin and service layers of buildings. Regarding the CE implementation phases, the design, dismantling and construction and end-of-life are mainly affected involving project professionals and suppliers and limiting them in the application of the reuse, recovery and recycling strategies. Governments can have a crucial role in the implementation of technical standards to remove these barriers and support the reuse, recovery and recycling of materials, components and systems in buildings.

The *lack of information and data* appears frequently in literature as the first common barrier category. The barriers reported in this category can be listed in several themes. The most prevalent barrier theme is the lack of data related to reusing and material stock, followed by sustainability and disassembly and lack of technical info. In general, throughout this category, the barriers affect project professionals in the implementation of reusing as a circular strategy at the design stage in the layers of structure and skin. From the quantitative analysis, we can deduce that this pattern is shown in the lack of data related to reusing and material stock, the lack of data related to sustainability, and the lack of data related to sustainability. Lack of data related to stakeholder awareness shows a completely different pattern. These types of barriers affect clients and the government in implementing the circular strategies of reusing, repurposing, recycling, and recovering on the site and structure layers at the strategic definition and briefing, the use and refurbishment, and the end-of-life stages. The barriers listed under the theme of lack of data related to digital applications are less explored in the literature. Data showed these barriers mainly affect project professionals in implementing circular strategies of reusing, recycling, and recovering of the service layer in the stages of strategic definition and briefing, and design.

In the *technical knowledge-related barrier* section, the prevalent barriers that were determined in the previous steps are further discussed, so that the main trends in their temporal and spatial dimensions, as well as in their impacts and involved stakeholders, are detected.

The main subcategories of technical knowledge-related barriers are:

- issues of general nature,
- design-related issues and
- materials and resources.

The barriers of general nature, including among others knowledge and awareness gaps among the involved parties, lack of knowledge on circularity principles, concepts and relevant technologies, difficulties and complexities of various levels, happen at almost all building layers except for district; the layers that seem to be more frequently reported are "Skin" and "Structure". As expected, the barriers belonging to this subcategory concern all the examined building life-cycle stages; the stage of "Design" is the one that is by far the most involved one, followed by "Services". With the "Project Professionals" being the leading party among the stakeholders, these barriers have been reported to affect the implementation of the vast majority of the examined circularity strategies; "Reuse" is the one that seems to be mostly affected, with "Recycle" and "Recover" being rather close to it.

The design-related issues (e.g., lack of knowledge on circular design principles, including Design for Disassembly and Design for Manufacture and Assembly) are mainly detected, concerning building layers, at the "skin", with the "structure" and "space plan" following closely. "Services" and "stuff" are also clearly represented. The main life cycle stages involved are primarily "Design" and secondarily "Strategic Definition and Briefing" (the other stages are present to varying degrees), while out of the stakeholders participating (reported for a few cases), the project professionals prevail. The circularity principles affected are "Reduce", "Reuse", "Remanufacture" and "Recycle".

Finally, the barriers included in the third subcategory "Materials and Resources" (among which misconceptions about waste management hierarchy, lack of knowledge about reused and reclaimed materials, as well as about reuse and recycling techniques, are listed) seem to mainly concern the layer of "Structure". The most intensely reported life-cycle stage is the "End of life", closely followed by the "Dismantling and Construction" and the "Design". The most frequently mentioned stakeholder group are by far the "Project professionals". The reported circular principles are "Rethink", "Reduce", "Reuse", "Remanufacture", "Recycle" and "Recover", with "Reuse" and "Recycle" being the ones that are more strongly represented.

In the *Lack of Technical Skills*, through the quantitative analysis of reported barriers from the collected literature, it was revealed that issues such as lack of people with the right skills for a CE, lack of skills, lack of systemic CE education and training for supply chain members, lack of training and knowledge for professionals, lack of skills in design and implementation, lack of awareness and communication skills, educational barriers were categorised under one theme: Lack of technical skills.

This theme emerged as the most prevalent type of information deficiency, appearing 30 times at the building level in the identified literature.

The theme lack of technical skills emerges as the most significant barrier to the implementation of CE practices. This theme encompasses a variety of specific issues that can be categorized as reported in the Table 1.39.

Quantitative analysis underscores a widespread recognition of the lack of technical skills as a barrier in various stages and layers of the built environment. The studies collectively indicate that addressing these barriers requires coordinated efforts from government bodies, project professionals, and suppliers. The effective implementation of circular strategies such as refuse, reuse, repurpose, recycle, and recovery is significantly hindered by these technical skill deficiencies, highlighting an urgent need for enhanced education and training programs tailored to the principles of CE. Findings are reported on the Table 1.40:

The integration of quantitative and qualitative data offers a comprehensive understanding of the lack of technical skills as a relevant barrier category to CE implementation. The quantitative analysis provides a clear picture of the prevalence and distribution of these barriers, while the qualitative insights explain the specific issues and their broader implications. This combined approach enables a more effective identification of intervention points and strategies to address these barriers and promote the successful adoption of CE practices in the built environment.

Lack of technical skills often arises due to several factors, including lack of education and training, complexity of technology, innovation and adoption lag, cost and resource constraints and regulatory and policy environment, which may not

Table 1.39 Lack of technical skills categorization

Lack of people with the right skills for CE
Lack of people with the right skills for CE: This barrier highlights a general shortage of individuals who possess the necessary competencies for CE initiatives, indicating a fundamental gap in the workforce's capabilities *Lack of the necessary capabilities and skills to EOL value extraction*: This barrier points to the specific skills needed to manage the end-of-life phase of products, a critical aspect of CE
Lack of skills
Lack of skills: A broad and pervasive issue, this barrier indicates a general deficiency in the required skills across the board *Lack of technical skills in the private sector on CE innovative building projects*: This specific barrier underlines the gap in technical expertise necessary for innovative CE projects in the private sector *Not enough people with skills for harvesting reused materials*: This highlights a niche but essential skill set required for effective material reuse *Lack of skilled tradesmen*: A recurring barrier that shows a shortage of skilled labor necessary for CE projects *Lack of educational/training programs for industry employees*: Points to an insufficiency in structured training programs that could help bridge the skill gap *Lack of expertise, knowledge flow, or in-house skill sets to repair and remanufacture*: This indicates a specific shortage in repair and remanufacturing skills within organisations *Lack of skills and expertise*: A general barrier highlighting a broad lack of necessary expertise *Lack of standardized practices*: Without standardized practices, it's challenging to develop and implement consistent skill sets *Lack of understanding/skills in applying the principles of the CE*: Highlights a fundamental gap in the comprehension and application of CE principles *Lack of basic digital skills and e-literacy*: This barrier reflects the need for foundational digital competencies, which are crucial in modern CE practices *Lack of information technology (IT) and AI professionals*: Indicates a gap in advanced technical skills, particularly in IT and AI, which are increasingly important in CE *Lack of leadership for a CE*: Leadership skills are essential for driving CE initiatives, and this barrier points to a gap at the management level
Lack of systemic CE education and training for supply chain members
Lack of systemic CE education and training for supply chain members: Reflects a gap in comprehensive educational programs tailored for the supply chain *Lack of knowledge and training on the environmental benefits and how to design a building for deconstruction of the design team*: Specific training gaps in designing for deconstruction and environmental benefits *Lack of training, skills, and education support. Lack of documentation to support competence improvement*: Points to a lack of support and resources necessary for skill enhancement *Lack of understanding/skills in applying the principles of the CE:* Re-emphasises the need for better education on CE principles
Lack of training and knowledge for professionals
Lack of instruction of decision-makers in an explicit and robust manner: Decision-makers require clear and robust training to implement CE effectively *Lack of training of professionals:* A general barrier pointing to insufficient professional development programs *Lack of information technology (IT) and AI professionals:* Advanced technical skills are crucial, and this gap affects the ability to leverage IT and AI in CE

(continued)

Table 1.39 (continued)

Lack of design and implementation skills
Professionals are commonly not trained to design NZCB (net zero carbon buildings): This points to a specific training gap in designing sustainable buildings Lack of adequate skills and training, including the limited use of Information and communication technologies (ICTs): Reflects a broader issue in technical and ICT skills Barriers for BIM application such as upfront cost of implementation, risks of adoption linked to IP and copyright laws, licensing issues between collaborating parties and lack of technical skills and experience to drive implementation within organisations: Highlights specific technical barriers in implementing Building Information Modeling (BIM) In the design phase, a barrier is still the classical view of supplier/customer relationship from a property owner point-of-view: Points to a need for rethinking traditional relationships and incorporating CE principles *Design as barrier to future refurbishment: design that does not focus on the ability to be adapted, or partially/fully disassembled*: Reflects a gap in design thinking and skills *Change of the traditional skill set of waste management on construction sites*: Indicates a need for updating waste management skills to align with CE
Lack of awareness and communication skills
Lack of awareness and skills across the value chain: Highlights the need for broader awareness and skills development throughout the entire value chain *Lack of Communication and limited awareness on the financial benefits of CE*: Reflects a need for better communication skills and understanding of CE's financial benefits
Lack of educational programs
Limited attention to CE in architecture studies and design: Indicates a gap in architectural education focused on CE *Lack of education programs in circular thinking for primary and secondary school, vocational and higher technical education*: Reflects a need for integrating CE education at all levels *The workforce is largely trained to work linearly*: highlighting the need to shift from linear to circular training programs *Lack of detail within requirements and uncommon language (not specific enough and terms unfamiliar to the local sector)*: Indicates a need for clear and standardised educational materials

support or incentivize the adoption of CE technologies, creating barriers to their implementation.

Lack of technical skills can slow adoption rates due to a lack of technical know-how, leading to missed economic opportunities. They can also have an environmental impact, as the benefits of reduced resource consumption and waste generation may not be realised. Additionally, organisations that cannot innovate and adapt to CE practices may lose competitiveness in their markets over time.

To advance the CE, it is essential to invest in education and training programmes that focus on CE principles and the necessary technical skills. Collaboration between academia, industry, and government should be fostered to share knowledge, best practices, and innovations in circular technologies. Supporting research and development is crucial to innovating new technologies and methodologies that simplify and enhance CE practices. Additionally, establishing supportive regulatory frameworks and policies will incentivise the adoption of these practices. Providing financial

Table 1.40 Finding summary about lack of technical skills

Prevalence at structure and skin layers
The barriers are more frequently reported at the structure layer (13 instances) than at the skin layer (7 instances)
Key project phases affected
Design phase is the most affected with 15 instances Strategic definition and briefing, dismantling and construction, and end-of-life phases also show significant impacts with 11, 13, and 11 instances, respectively
Stakeholders Involved
Project professionals are the most frequently cited group (24 instances), indicating their critical role in addressing these barriers Governments and suppliers are also key stakeholders, with 6 and 7 instances, respectively
Impact on Circular Strategies
The barrier affects all circular strategies, with the highest impact on recovery (14 instances) and recycling (12 instances) Refuse (5 instances), reuse (8 instances), and repurpose (10 instances) are also significantly impacted

support through incentives, grants, or subsidies can help organizations overcome the initial costs associated with implementing CE technologies.

Non-technical barriers that should be acknowledged include: (1) Lack of Awareness and Understanding; (2) Policy and Regulatory Hurdles; (3) Market Barriers; (4) Financial Constraints; (5) Cultural and Behavioural Factors.

The lack of technical skills is a significant barrier to the implementation of CE practices in the built environment. Addressing this barrier requires a multifaceted approach, including enhanced educational programs, targeted professional training, supportive policies, and improved communication and collaboration across the value chain. By focusing on these areas, we can develop the necessary skills to drive the transition to a more sustainable and CE.

The *lack of international/regional/national standards and guidelines* appears frequently in literature as a common barrier category. It seems to affect both design and (de)construction waste management, whether it is about rules for design for disassembly and adaptability (ISO 20887 not ready at the time), circular design or circular construction waste management. There is also mention of the lack of flexibility in the existing codes and regulations across the spectrum of circular design and waste management. Although the European Waste management protocol offers incentives for reuse and circularity, it seems that this has not percolated (or communicated) to national member states' legislation in an effective way. This is coupled with the absence of reliable procedures for evaluating reuse materials in terms of quality and durability, as well as logistics/accounting pertaining to their stock and subsequent use. Numerous of these barrier mentions are currently considered to be addressed by the aforementioned ISO, as well as a new EN standard currently under work by CEN TC350 Subcommittee 1, titled "CE in the construction sector" with delegates from different European member states from industry, public sector and academia.

In the *lack of machines and equipment*, through the quantitative analysis of reported barriers from the collected literature, it was revealed that the need for infrastructure and equipment to perform deconstruction for CE and the lack of suitable equipment for the process of deconstruction and material reclamation can be categorised under the lack of machines and equipment category (Table 1.41).

This theme emerged as the most prevalent type of information deficiency, appearing 2 times at the building level in the identified literature.

Barriers in this category are uniformly present across structure, skin, and services layers, indicating a widespread issue affecting the entire building lifecycle. This barrier category is particularly prominent during the dismantling and construction and end-of-life phases. This underscores the importance of having the right infrastructure and equipment at crucial stages of the building lifecycle to facilitate CE practices. Reuse, remanufacture, recycle, and recover strategies are all significantly impacted by the lack of infrastructure and equipment. This indicates that the barrier affects multiple aspects of CE, hindering the overall effectiveness of circular strategies.

Reuse, remanufacture, recycle, and recover strategies are all significantly impacted by the lack of infrastructure and equipment. Project professionals play a crucial role in mitigating this barrier by advocating for and investing in the necessary infrastructure and equipment. Addressing this barrier will enhance the ability to reuse,

Table 1.41 Lack of machines and equipment categorization

Need for infrastructure and equipment to perform deconstruction for CE
Core Requirement for CE Practices: Infrastructure and equipment are foundational to executing deconstruction in a manner that maximizes material recovery and reuse. Without proper facilities, even the most skilled workforce cannot efficiently deconstruct and repurposed materials *Impact on Material Flow*: The absence of dedicated infrastructure can disrupt the flow of materials through the supply chain, leading to inefficiencies and increased waste. Effective CE implementation requires a seamless process where materials can be easily processed and redistributed *Investment Needs*: Establishing the necessary infrastructure requires significant investment. This barrier highlights the financial and logistical challenges that organisations and governments face in creating a supportive environment for CE practices
Lack of suitable equipment for the process of deconstruction and material reclamation
Technical Capability: Suitable equipment is essential for efficient and effective deconstruction. Advanced tools and machinery can enhance the precision and speed of deconstruction activities, facilitating higher recovery rates of reusable materials *Safety and Efficiency*: The right equipment ensures that deconstruction processes are not only efficient but also safe. Improper or inadequate equipment can pose risks to workers and may result in lower-quality reclaimed materials *Innovation and Adaptation*: This barrier highlights the need for continuous innovation in deconstruction technology. As CE practices evolve, so must the equipment used, adapting to new materials, methods, and sustainability standards

remanufacture, recycle, and recover materials, thereby promoting sustainability and efficiency in the built environment.

The main technological barrier encountered in the category of *inadequate processes and infrastructure* at the building level was the lack of adequate logistic systems and spaces. This barrier emerged predominantly over the others, with specific reference to the lack of space for the storage of recovered materials in other buildings. Therefore, the main challenges occur when the end-of-life phase of the building begins, mainly in elements related to the structure of the building. In this case, suppliers are the main agents involved, affecting the development of reuse actions.

In the *lack of methods and tools*, the lack of CE approaches in the design of buildings stands out as the main barrier related to the lack of methods/tools. Mostly, the lack of design approaches that consider the whole life cycle of buildings, and the integration of new technologies during this process are especially relevant during building design. Therefore, project professionals are the main actors involved in this barrier, affecting considerable recovery strategies.

Relating to the *lack of materials/products/buildings/systems*, the quality and quantity of recycled materials and components mostly appear as the main technical barrier in terms of lack of materials/products/constructions/systems. This barrier, which considerably affects the building structure, occurs during the dismantling and construction phase. Project professionals and suppliers are principally responsible for this obstacle, affecting the implementation of the reuse strategy over other approaches.

1.4.3 Material Level

At the material level, technical barriers to CE implementation predominantly pertain to a *Lack of technical standards.* Subsequently, challenges related to *Lack of information/data, Lack of materials/products/buildings/systems*, and *Lack of methods/tools* are encountered. This can be attributed to the characteristics of the material level within the built environment, where the *Government* is considered to be the main responsible for the implementation of CE processes, followed by *Project Professionals* and *Suppliers* who were identified as the other stakeholders involved in the process/responsible for the technical barriers that impede the implementation of CE in the built environment at the material level. As a result, the barriers are reported mainly during the *Design, End-of-life,* and *(Dismantling and) Construction phases,* while *Reuse* and *Recycle* were the most affected circular strategies by the technical barriers, followed by *Repurpose* and *Repair*. These strategies were primarily hindered by barriers like a *Lack of materials/products/buildings/systems, Inadequate processes and infrastructure, Lack of information/data* or *Lack of methods/tools.*

The barriers under the *Lack of information/data* category have been classified as “Data Availability, Accessibility and Accuracy”, “Traceability and Certification”, “the Lack of Knowledge and Awareness”, and “Sustainability related issues”. Regarding the building layers, the studies addressing the *lack of information/data* barriers at the material level reported that these barriers affect mostly the *Structure* and the *Skin* of a building, and within the life cycle, they are mostly reported in the *Design* and *End-of-life* phases. *Project Professionals,* followed by *suppliers,* are the main stakeholders involved in/responsible for these technical barriers, at the material level. The literature has shown that the barriers related to the *Lack of information/data* mostly affect *Reuse, Repurpose, Recycle*, and *Repair* CE strategies. Addressing the data gap is crucial to overcoming these challenges and unlocking the full potential of circularity in the construction sector. Some solutions to overcome the *lack of information/data* barrier in circular construction at the material level could be "Transparency through Technology and Collaboration" solutions, "Education and Regulation" and "Data-driven Decision Making and Innovation".

Five technical barriers subcategories were identified in the category of *Lack of technical knowledge*, specifically "Lack of knowledge on the end-of-life procedures", "Lack of knowledge on the materials properties", "Lack of knowledge on the design stage", "Lack of awareness—general knowledge—familiarity with CE, green building or sustainability concepts", and "Lack of knowledge through the value and supply chain". These barriers primarily impact the *Structure* and the *Skin* of a building, and they are the most prevalent in the *Design, Strategic Definition and Briefing,* and *(Dismantling and) Construction* phases. Taking into account the stakeholders, *project professionals* and *suppliers* are primarily held responsible for the lack of technical knowledge barriers at the material level. According to the reviewed literature, the *lack of technical knowledge* barrier significantly impacts circular strategies in the built environment at the material level. This includes, especially *Reuse, Recycle, Repair,* and *Refurbish. Lack of technical knowledge* hinders the adoption of CE practices, leading to a poor and inefficient supply chain, reducing the penetration of reused materials on the market, and triggering the risk of material availability. The consequences influence the whole life cycle of construction with wider and significant impacts. To align with circular economy strategies, all stakeholders must adapt their practices and cultivate new knowledge and skills. Concurrently, governments and policy makers should prioritise research and knowledge development on material flows to enhance their control and inform strategic decision-making. Integrating CE principles into university curricula for architects, engineers, and construction professionals can foster a new generation of practitioners equipped with the knowledge and skills to navigate a circular built environment.

The third barrier category, *Lack of technical skills,* refers to the lack of the necessary expertise of individuals or organizations to effectively design, implement, or manage circular systems. The barriers reported in the scientific literature related to the *Lack of technical skills* could be divided as "the lack of technical skills on a general level", "the lack of technical skills in the supply chain", "the lack of technical skills in the design process", and "the lack of technical skills in the end-of-life processes". They were mainly related to *Structure* and *Skin,* concerning the building levels, but the skill gaps and knowledge deficits were reported to affect each phase of the building life cycle at the material level, predominantly during the *End-of-life* and closely followed by *Design, Dismantling and Construction, Manufacture* and

Use-Refurbishment stages. The literature primarily identifies *Construction Professionals* as the key stakeholder group responsible for addressing the *lack of technical skills.* Additionally, *Suppliers* and *Clients* also play a role in being responsible for the *Lack of technical skills* in the implementation of the CE in the built environment at the material level. The barriers identified in this category pose challenges to CE at the material level for strategies such as *Reuse* and *Recycle.* Following the *lack of technical knowledge,* this barrier can be overcome through education, research, and innovation.

The barriers reported in the reviewed literature related to the category *Lack of technical standards,* could be divided as "Lack of consistent regulatory framework", "Lack of standardisation", "Complexity of certification processes", and "Lack of end-of-life policies". Regarding the building layers, the studies addressing the *lack of technical standards* barrier at the material level reported that it mainly affects the *structure* and the *skin.* These barriers have been found to be critical at the *Design* and *End-of-life* phases within the building life cycle, closely followed by the *Strategic Definition and Briefing* phase and *Manufacture.* The *Government* is held responsible for the vast majority of the *Lack of technical standards* at the material level, followed by *Project Professionals* who are directly affected by these barriers. The reviewed literature emphasises that the lack of technical standards barrier has a considerable impact on circular strategies in the built environment at the material level, affecting more or less in the same manner the *Recycle, Rethink, Reuse,* and *Repair* strategies. The vast majority of the analysed studies highlight the importance of Governmental actions toward the establishment of specific regulatory frameworks and technical standards on the reuse and recovery of materials in the construction industry. Incentivising measures are also expected to encourage local governments to develop their territory according to the principles of the CE and switch to the use of materials in line with CE principles.

Inadequate machines/equipment is the category with the lowest number of studies reporting it, appearing in seven studies. Barriers identified in this category, such as "Lack of technologies/inadequate technologies", "Lack of technological platform", "Lack of recycling process" or "Lack of technology evolution" affect the *skin* and the *structure* within the building layers. No specific life cycle phase stands out as being affected by this barrier: *Inadequate machines and equipment* at the material level are discussed in the following phases: *Manufacture, Dismantling and Construction, Use and Refurbishment, End-of-Life. Project professionals* and *suppliers* are identified as responsible for the lack of adequate machines and equipment in the built environment. However, no clear conclusion can be drawn by analysing the identified literature in terms of how this barrier affects the implementation of circular strategies in the built environment: each of the solutions, *Reuse, Repair, Refurbish, Remanufacture, Repurpose,* and *Recycle,* is addressed only once. To overcome the barrier of *inadequate machines/equipment, there* should be increased research, development, and technological innovation, while advanced technologies should be applied to process C&D waste management.

In terms of *Inadequate processes and infrastructure,* the barriers identified can be classified into several main groups: "Lack of sorting/insufficient storage/site constraints", "Inadequate processes and infrastructure of Material Recovery", "Market", "Lack of info/technology", "Lack of facilities/additional complexity", "CE-focused design", "Sustainability" and "Disassembly". At the material level, the barrier of *inadequate processes and infrastructure* mainly affects the *structure* of a building, followed closely by the *skin* of the building. *Project professionals* and *suppliers* are the main ones responsible for these barriers, which have impacts mainly at the *end-of-life* of a building. Besides, the *design* and the *construction and dismantling* stages are life cycle phases that often deal with the lack of adequate processes and infrastructure. The impact on CE strategies driven by *Inadequate processes and infrastructure* barriers has most often been identified in *Reuse* strategies. To effectively address the deficiencies in material recovery processes and infrastructure, a holistic framework for embedding CE principles within the built environment is essential: the company culture should be ready to apply innovative methods effectively, the governments should support the construction companies economically, interaction between stakeholders should be more effective to create links between demolition contractors and stockists, ineffective C&D waste processes should be addressed by advanced technologies and by promoting upcycling and reusing strategies. Incentives, such as tax breaks, subsidies, and financing options specifically designed for circular construction projects, can help offset initial costs and create a more level playing field. Furthermore, promoting successful case studies and highlighting the economic benefits of circularity can inspire greater adoption of CE-oriented processes and infrastructure.

In the *Lack of methods/tools*, the main barriers and challenges identified in the literature include the subcategories "Lack of tool/design or Adopting to existing method/tool", "Certification/ Standardisation/Classification", "Awareness of material recovery by stakeholders/producers/company owners", "Complexity of the recycling processes", "Disassembly". The literature also indicates a lack of enabling digital technologies, such as BIM and interoperability issues, a lack of non-destructive testing methods, and issues related to environmental impact. In line with all the previous barriers, the *structure* is the building layer that is the most affected by the *lack of methods and tools*, followed by the *skin* layer. Barriers occur more frequently in the *Design* and *Dismantling and Construction* stages. *Project professionals* are primarily responsible for the *lack of methods and tools* in the environment built at the material level. However, *clients, suppliers,* and *governments* are also involved. The *lack of methods and tools* impacts the implementation of all circular strategies, with the greatest impact on the Reduce strategy. The lack of tools and hurdles to adapt to existing methods and tools are discussed in the identified literature, and ways to overcome these barriers are addressed. There is a need for widespread BIM use, design guidelines for circular buildings and related technologies for tools, material passports and regulatory frameworks. With the support of researchers, all other stakeholders in the value chain have the power to overcome technological barriers, such as the lack of digital technologies that help to make deconstruction more efficient and cost-effective, while governments should support the construction industry financially.

In terms of *Lack of materials/products/buildings/systems,* the barriers identified can be classified into five main groups: “Lack of performance and quality assurance for reused/recovered materials”, “Lack of systems”, “Complexity of material recovery”, “Lack of available materials/products for reuse” and “Other limitations of materials/products/buildings/systems” that encompasses limitations such as “Low level of application of recycled products”, “Long product life cycle”, “Design constraint for reclaimed material use”, “The trend of using inexpensive and non-durable materials/components in construction” or “Increasing material diversity and number of composite materials and structures”. Studies addressing the *Lack of materials/products/buildings/systems* barriers at the material level reported that, regarding the building layers, these barriers affect mainly the *structure* and the *skin* of a building, while the other layers of the building are rarely addressed and therefore, appear not to be affected by this barrier. These barriers have been evaluated as affecting in a high manner the *End-of-life, (Dismantling and) Construction, Design* and *Use and Refurbishment* of the project, showing that the scarcity of materials and products suitable for circular construction stems from a complex web of design limitations, knowledge gaps, economic factors, and practical challenges in deconstruction and material recovery. *Project professionals* and *suppliers* are mostly responsible stakeholders for the *Lack of materials/products/buildings/systems* barrier category. The CE strategies most affected by the *Lack of materials/products/buildings/systems* are the *Reuse* and *Recycle* strategies, with barriers such as “low volume and inconsistent quality of the recovered materials”, “no space for the storage of disassembled materials”, “difficulty to recover materials”, “hazardous components” or “design constraint for reclaimed material use”. Solutions to overcome the *Lack of materials/products/buildings/systems* mentioned in the literature were related to the need for research, development and innovation, education and awareness-raising efforts, financial incentives, regulatory frameworks and collaboration throughout the entire supply chain.

1.5 Conclusions

This study investigated technical challenges that affect the implementation of circular strategies in the built environment through a literature review. The investigation was conducted by examining barriers identified in the academic and non-academic literature at various levels (urban, building, and material) of the built environment.

We observed general patterns in (1) the types of barrier categories that most affected the implementation of circular strategies in the built environment, (2) the life cycle phases and (3) building layers in which technical challenges are observed, (4) the circular strategies mainly implemented and (5) the stakeholders involved or responsible for them. The lack of standards and the lack of data are both the most recurrent barrier categories at all levels. The life cycle stages that result in being mostly affected by barriers at all levels are the construction and the end-of-life stages. In the adoption of circular strategies in the built environment, the focus is placed on the 3R approach (Reduce, Reuse, Recycle). It highlights that while the 3R approach has been widely adopted, it may have limited broader thinking on sustainability. The 10R approach, which includes additional strategies like remanufacturing, offers more opportunities for sustainability and shifts focus away from recycling as the ultimate goal. Additionally, we noted a lack of focus on building layers with shorter lifespans (e.g., services, space plans), where strategies like material passports could be highly beneficial. Finally, we observed that professionals, suppliers and governments are consistently documented as stakeholders mostly affected by or responsible for technical barriers at all levels.

In the literature review focused on **urban-related technical barriers** to implementing circular strategies, the analysed studies and papers identified a variety of barriers. To enhance our understanding, these barriers were categorized into eight distinct groups.

- The most prominent category was *Inadequate processes and infrastructure.* Literature provides numerous evidence on challenges associated with the infrastructure for waste management and recycling. Additionally, the absence of efficient production and supply-chain networks and logistics that support CE practices were frequently noted as a significant technical barrier within this category.
- *Lack of technical standards* is another significant barrier. Technical standards are particularly missing in building design and construction, to promote eco-friendly building materials, efficient resource use, and sustainable construction techniques. There were also identified the barriers associated with the regulatory framework and government support.
- A substantial number of studies identified barriers related to the *Lack of information/data.* From the urban perspective, availability and quality of data and information are prerequisites for efficiently implementing CE strategies and practices in the construction sector. Barriers such as absence of clear guidelines for collecting and sorting construction and demolition waste (CDW), lack of documentation for new and used building products and issues with the quality and availability of data, as well as transfer and integration of knowledge across different stakeholders and disciplines, were the most widely reported barriers in this category.
- Several barriers were identified in the subcategory *Lack of technical knowledge.* Technical knowledge is especially vital in areas such as recycling and reuse processes, design and disassembly of products and production and manufacturing processes. The technological expertise and skills and knowledge transfer in this area are crucial for innovation, development and upscaling CE practices across the sector.
- Among all the barriers identified, the primary stakeholder highlighted was the government, followed by project professionals. However, suppliers and clients also play an important role as stakeholders.
- The most affected CE strategies at the urban level are Reuse, followed by Recycle and Repurpose. Relevant stakeholders should concentrate their efforts to mitigate these barriers and facilitate wider and more efficient implementation of these strategies in the building sector.

At the building level, we have identified the following range of CE barriers:

- The implementation of CE strategies in buildings is mainly affected by the *lack of information/data, technical standards, technical knowledge and methods/tools*. Because of these barriers, project professionals and suppliers are mainly affected in the application of the reuse, recovery and recycling strategies during the design, dismantling and construction and end-of-life phases.
- Regarding *lack of data*, barriers relate to reusing and material stock followed by lack of data for sustainability assessment and lack of technical info for disassembly. These barriers mainly affect the implementation of reusing on the building structure and skin.
- In the *technical standard barrier* category, the main barriers concern the lack of standardisation/guidelines, lack of legal policies, and sustainability quality assurance affecting professionals in recycling, reuse and recovery during the design and (de)construction waste management phases. Also, the absence of reliable procedures for evaluating reused materials limits the reuse.
- Regarding the *technical knowledge* category, the main barriers reported are of general nature, followed by the design-related- and the materials and resources-related ones. The barriers occur mainly at the skin and structure layers and most predominantly at the design stage; all circular strategies are affected, with reuse and recycling holding the lead.
- The main barrier in the *lack of methods/tools* category is the lack of CE approaches in buildings' design. The design stage of building structure and skin is affected by the lack of design approaches considering buildings' whole life cycle, and new technologies' integration in it. Professionals are mainly involved; recovery, recycling and reuse strategies are considerably affected.
- Insufficient investment in *infrastructure and equipment*, coupled with a deficit in training and expertise among professionals, significantly contributes to these challenges. Existing infrastructure often supports linear economy practices, which do not align with the requirements related to the implementation circular strategies.
- Technical barriers have multifaceted negative impacts, including slow adoption rates, missed economic opportunities, environmental impact, and loss of competitiveness. Governments and organisations need to prioritise investments in specialised infrastructure and tools tailored to support CE practices. This includes developing facilities and acquiring equipment capable of handling the complexities involved in reclaiming and repurposing materials. There is also a critical need to focus on education and training, along with substantial investment in research and development. Furthermore, collaboration across sectors is paramount. By fostering partnerships between academia, industry, and government entities, stakeholders can share valuable insights, best practices, and innovations in CE technologies.
- Non-technical barriers also need to be considered. Regulatory barriers, such as the lack of supportive policies and regulations, can impede the adoption of circular practices. Financial barriers, including high initial costs and a lack of financial incentives, can deter organisations from investing in CE practices. Cultural and

behavioural barriers, such as resistance to change and limited awareness of the benefits associated with CE practices, can prevent their widespread adoption. Market barriers, including limited demand for reclaimed materials and products, can undermine the economic viability of circular strategies.

At the material level, the extensive literature review and the deep analysis of the bibliographical sources from the scientific and non-academic literature revealed the following:

- The *Lack of technical standards* has been identified as the most significant barrier for implementing the CE principles in the built environment at the material level, followed by the *Lack of information/data*, *Lack of materials/products/building systems* and *Lack of methods/tools*. The technical barriers at the material level were mainly associated with the building structure and skin, and they affect mostly the Design and End-of-life building life cycle phases, with significant impacts on the implementation of the circular strategies, Reuse, Recycle, Repurpose and Repair. Among the stakeholders that are involved in the implementation of circular strategies at the material level, the Government was considered as the primary responsible for the technical barriers, especially with regard to the lack of technical standards, followed by the project professionals and the suppliers.
- The *lack of technical standards* embraced barriers that are related to the lack of consistent regulatory framework, i.e. absence of regulations on circularity, but also contradictions between existing regulations, limited flexibility and enforcement of using recycled materials on building projects and standardisation, as well as with the lack of standardisation, mainly as regards the quality and the properties of the reused materials. The lack of information/data was associated with the data availability, accessibility and accuracy, as well as the materials' traceability and certification. Both of these two major barrier categories at the material level affect primarily the Design and End of Life phases within the building life cycle.
- In the general barrier represented by the *Lack of materials/products/building systems* it was indicated that issues like the low availability and inconsistent quality of the recovered materials, the durability of reused products, the use of polluted materials or materials of low recoverability, the low performance and/or the absence of documentation on the performance of the recovered materials, etc., hinder the wide implementation of CE strategies on the built environment. Moreover, the *Lack of systems,* involving merely storage space and logistics, was discerned among the more distinguished categories of this specific technical barrier. The building life cycle phases End of life and Dismantling and Construction are primarily affected by the Lack of materials/products/building systems, but also by the Lack of methods and metrics.

Recommendations for stakeholders:

- Effective implementation of CE practices is often hindered by inadequate infrastructure. Therefore, investments in development, modernisation and extending supply-chain networks tailored to CE practices in the built environment are highly recommended.

- Technical standards and supportive policies should be developed and enforced to promote eco-friendly building materials, efficient resource use and sustainable construction techniques.
- The relevant stakeholders should address the lack of quality and accessibility of data on CDW management. Establishing clear guidelines for collecting, sorting and documenting CDW for example, via Data sharing platforms, can facilitate more circular material use.
- Support collaborative initiatives between stakeholders, encourage partnership and knowledge sharing. Also, partnership among academia, policy-makers and building professionals should be encouraged to drive innovations and overcome barriers in CE implementation.
- Governments can have a crucial role in removing technical barriers at the building level and supporting the reuse, recovery and recycling of materials, components and systems in buildings. Government should introduce incentives, subsidies and initiate development of policies that support economic viability of CE practices.
- While the ISO, as well as a new EN standard currently under work, may help to overcome technical standard barriers, governments can also play a decisive role in promoting technical standards and the European Waste Management protocol in national legislations.
- Technical knowledge gaps intersect the whole chain of stakeholders involved in CE implementation in the construction sector. Training activities organised by various actors in the field tailored to the specific needs of stakeholders, knowledge sharing, inclusion of relevant courses in university curricula/development of educational programs and emphasis on the development of good examples/best practice case-studies on various aspects of the required knowledge (e.g., product design for disassembly) can provide the basis for actual advancements.
- The lack of methods and tools dictates the need for the development of practical guidelines, standards and policies for various processes in the building sector (e.g. reclaimed materials, DfD), as well as of design-support tools f. Also, the creation of tools and the development of indicators/models for the assessment and monitoring of circularity in buildings will play an important role, along with the development of accessible databases and the use of new (digital) technologies.

The systematic bibliographic research not only provided the technical barriers that hinder the implementation of CE in the built environment at the material level, but also indicated recommendations and solutions to overcome them and promote the transition from the linear construction scheme towards a more circular approach. The recommendations are applicable to material, urban and building levels, with key insights focusing on:

- Education and awareness raising among all stakeholders to drive change across all levels of industry and society, e.g. through the integration of CE principles into educational programmes and lifelong courses, the training and upgrading of the workers' skills, the initiation of public campaigns, etc.

- Support of research and innovation in relevant fields, in order to further develop and exploit technological advancements, such as Industry 4, radio frequency identification (RFID) material tagging and other digital tools, e.g. BIM, AI, digital twins, neural networks, etc.
- The establishment of relevant regulations, standards and codes to promote circular construction, along with the integration of contractual requirements for the design, construction and procurement, aligned with CE principles,
- The development of cost-effective and trusted certification protocols and testing processes for assessing the materials' properties, thus fostering their recovery and reuse potential, in consort with the development of design guidelines, multi-criteria decision procedures, material passports and well-designed indicators,
- The provision of financial incentives, such as increasing demolition and landfill fees, combined with subsidies, tax breaks, and low-interest loans for companies implementing circular design and business models, as well as increasing the financial value of reused materials, e.g. through advanced sorting or higher inclusion rates in new products,
- Strengthening collaboration throughout the entire supply chain in order to bring together clients, suppliers, contractors, and designers to ensure that circular principles are embedded throughout the project lifecycle.

The limitations of the study are related to the lack of a robust scientific approach when research investigates barriers. This is evident in the varying methods of investigation and interpretation adopted in the bibliographic references examined. Findings of different levels of generality, as well as of different focusing and target audiences, were dealt with, leading together with the involvement of several researchers of different backgrounds in this study, to the integration of a degree of subjectivity into the analysis. This issue runs through several steps of the process and intersects all the levels examined. Beginning from the initial phase of the barrier classification into the identified categories (lack of data, lack of technical standards, etc.), through the reporting of their dimensions and relevant aspects (where/when/who is involved/ how is circularity implementation affected), and reaching the final stage of identifying subcategories of barriers within each main category, fragments of subjectivity in the adopted approaches and assumptions can be detected, which are due also to the different depth of interpretation in the reviewed sources. However, despite the inaccuracies that may have affected the analysis in its details, general trends have indeed been identified, and the central conclusions are based on a solid basis. Moreover, the non-academic literature provided less detailed information; therefore it limitedly complemented the scientific literature on this topic.

References

1. Çimen Ö (2021) Construction and built environment in circular economy: a comprehensive literature review. J Clean Prod 305:127180. https://doi.org/10.1016/j.jclepro.2021.127180

2. Murray A, Skene K, Haynes K (2017) The circular economy: an interdisciplinary exploration of the concept and application in a global context. J Bus Ethics 140(3):369–380. https://doi.org/10.1007/s10551-015-2693-2
3. Munaro MR, Tavares SF, Bragança L (2020) Towards circular and more sustainable buildings: a systematic literature review on the circular economy in the built environment. J Clean Prod 260:121134. https://doi.org/10.1016/j.jclepro.2020.121134
4. European Commission. Directorate General for Communication (2020) Circular economy action plan: for a cleaner and more competitive Europe. Publications Office of the European Union, Luxembourg. https://doi.org/10.2779/717149
5. European Environment Agency (2024) Accelerating the circular economy in Europe: state and outlook 2024. Publications Office of the European Union, Luxembourg. https://doi.org/10.2800/055236
6. Cruz Rios F, Grau D, Bilec M (2021) Barriers and enablers to circular building design in the US: an empirical study. J Constr Eng Manag 147(10):04021117. https://doi.org/10.1061/(ASCE)CO.1943-7862.0002109
7. Selman A, Nørkjær Gade A (2020) Barriers of incorporating circular economy in building design—in a Danish context. Unpublished report
8. Geissdoerfer M, Savaget P, Bocken NMP, Hultink EJ (2017) The circular economy—a new sustainability paradigm? J Clean Prod 143:757–768. https://doi.org/10.1016/j.jclepro.2016.12.048
9. International Energy Agency (2013) Transition to sustainable buildings: strategies and opportunities to 2050. OECD Publishing, Paris. https://www.oecd-ilibrary.org/energy/transition-to-sustainable-buildings_9789264202955-en
10. AlJaber A, Martinez-Vazquez P, Baniotopoulos C (2023) Barriers and enablers to the adoption of circular economy concept in the building sector: a systematic literature review. Buildings 13(11):2778. https://doi.org/10.3390/buildings13112778
11. Ellen MacArthur Foundation (2013) Towards the circular economy: economic and business rationale for an accelerated transition. [Online]. Available: https://www.ellenmacarthurfoundation.org/towards-the-circular-economy-vol-1-an-economic-and-business-rationale-for-an
12. European Environment Agency (2016) Circular economy in Europe: developing the knowledge base. Publications Office of the European Union, Luxembourg. [Online]. Available: https://www.eea.europa.eu/publications/circular-economy-in-europe
13. World Economic Forum (2014) Towards the circular economy: accelerating the scale-up across global supply chains. [Online]. Available: https://www3.weforum.org/docs/WEF_ENV_TowardsCircularEconomy_Report_2014.pdf
14. Pomponi F, Moncaster A (2017) Circular economy for the built environment: a research framework. J Clean Prod 143:710–718. https://doi.org/10.1016/j.jclepro.2016.12.055
15. European Innovation Council and SMEs Executive Agency et al. (2023) Study on measuring the application of circular approaches in the construction industry ecosystem: final study. Publications Office of the European Union.https://doi.org/10.2826/488711
16. Cambridge University Press. Cambridge Dictionary|English Dictionary, Translations & Thesaurus. [Online]. Available: https://dictionary.cambridge.org/
17. Zimmann R, O'Brien H, Hargrave J, Morrell M (2016) Circular economy in the built environment. Arup. [Online]. Available: https://www.arup.com/insights/circular-economy-in-the-built-environment/
18. Royal Institute of British Architects (2020) RIBA plan of work. [Online]. Available: https://www.architecture.com/knowledge-and-resources/resources-landing-page/riba-plan-of-work#available-resources
19. Munaro MR, Tavares SF (2023) A review on barriers, drivers, and stakeholders towards the circular economy: the construction sector perspective. Cleaner Responsib Consum 8:100107. https://doi.org/10.1016/j.clrc.2023.100107
20. van Stijn A, Gruis V (2019) Towards a circular built environment: an integral design tool for circular building components. Smart Sustain Built Environ 9(4):635–653. https://doi.org/10.1108/SASBE-05-2019-0063

21. Brand S (1995) How buildings learn: what happens after they're built. Penguin Books, New York. [Online]. Available: http://archive.org/details/howbuildingslear00bran
22. Bocken NMP, de Pauw I, Bakker C, van der Grinten B (2016) Product design and business model strategies for a circular economy. J Ind Prod Eng 33(5):308–320. https://doi.org/10.1080/21681015.2016.1172124
23. Potting J, Hekkert MP, Worrell E, Hanemaaijer A (2017) Circular economy: measuring innovation in the product chain. PBL Netherlands Environmental Assessment Agency. [Online]. Available: http://www.pbl.nl/sites/default/files/cms/publicaties/pbl-2016-circular-economy-measuring-innovation-in-product-chains-2544.pdf
24. Huang B, Wang X, Kua H, Geng Y, Bleischwitz R, Ren J (2018) Construction and demolition waste management in China through the 3R principle. Resour Conserv Recycl 129:36–44. https://doi.org/10.1016/j.resconrec.2017.09.029
25. Abdulai SF, Nani G, Taiwo R, Antwi-Afari P, Zayed T, Sojobi AO (2024) Modelling the relationship between circular economy barriers and drivers for sustainable construction industry. Build Environ 254:111388. https://doi.org/10.1016/j.buildenv.2024.111388
26. Giorgi S, Lavagna M, Wang K, Osmani M, Liu G, Campioli A (2022) Drivers and barriers towards circular economy in the building sector: stakeholder interviews and analysis of five European countries' policies and practices. J Clean Prod 336:130395. https://doi.org/10.1016/j.jclepro.2022.130395
27. Hossain MdU, Ng ST, Antwi-Afari P, Amor B (2020) Circular economy and the construction industry: existing trends, challenges and prospective framework for sustainable construction. Renew Sustain Energy Rev 130:109948. https://doi.org/10.1016/j.rser.2020.109948
28. Pintossi N, Ikiz Kaya D, Pereira Roders A (2021) Assessing cultural heritage adaptive reuse practices: multi-scale challenges and solutions in Rijeka. Sustainability 13(7):3603. https://doi.org/10.3390/su13073603
29. Williams J (2019) Circular cities: challenges to implementing looping actions. Sustainability 11(2):423. https://doi.org/10.3390/su11020423
30. Campbell-Johnston K, Cate JT, Elfering-Petrovic M, Gupta J (2019) City level circular transitions: barriers and limits in Amsterdam, Utrecht and The Hague. J Clean Prod 235:1232–1239. https://doi.org/10.1016/j.jclepro.2019.06.106
31. Tsui T, Peck D, Geldermans B, Van Timmeren A (2020) The role of urban manufacturing for a circular economy in cities. Sustainability 13(1):23. https://doi.org/10.3390/su13010023
32. Obersteg A et al (2019) Urban regions shifting to circular economy: understanding challenges for new ways of governance. Urban Planning 4(3):19–31. https://doi.org/10.17645/up.v4i3.2158
33. Dąbrowski M, Varjú V, Amenta L (2019) Transferring circular economy solutions across differentiated territories: understanding and overcoming the barriers for knowledge transfer. Urban Planning 4(3):52–62. https://doi.org/10.17645/up.v4i3.2162
34. Williams J (2022) Challenges to implementing circular development—lessons from London. Int J Urban Sustain Dev 14(1):287–303. https://doi.org/10.1080/19463138.2022.2103822
35. Wuni IY (2022) Mapping the barriers to circular economy adoption in the construction industry: a systematic review, Pareto analysis, and mitigation strategy map. Build Environ 223:109453. https://doi.org/10.1016/j.buildenv.2022.109453
36. Dorst H, Van Der Jagt A, Runhaar H, Raven R (2021) Structural conditions for the wider uptake of urban nature-based solutions—a conceptual framework. Cities 116:103283. https://doi.org/10.1016/j.cities.2021.103283
37. A+BE (2023) Spatial approaches to a circular economy. A+BE|Archit Built Environ 2023(20). https://doi.org/10.7480/ABE.2023.20
38. Mhatre P, Gedam VV, Unnikrishnan S, Raut RD (2023) Circular economy adoption barriers in built environment—a case of emerging economy. J Clean Prod 392:136201. https://doi.org/10.1016/j.jclepro.2023.136201
39. Verga GC, Khan AZ (2022) Space matters: barriers and enablers for embedding urban circularity practices in the Brussels Capital Region. Front Built Environ 8:810049. https://doi.org/10.3389/fbuil.2022.810049

40. Webb R et al (2018) Sustainable urban systems: co-design and framing for transformation. Ambio 47(1):57–77. https://doi.org/10.1007/s13280-017-0934-6
41. Tleuken A, Torgautov B, Zhanabayev A, Turkyilmaz A, Mustafa M, Karaca F (2022) Design for deconstruction and disassembly: barriers, opportunities, and practices in developing economies of Central Asia. Procedia CIRP 106:15–20. https://doi.org/10.1016/j.procir.2022.02.148
42. Rakhshan K, Morel J-C, Daneshkhah A (2021) Predicting the technical reusability of load-bearing building components: a probabilistic approach towards developing a circular economy framework. J Build Eng 42:102791. https://doi.org/10.1016/j.jobe.2021.102791
43. Rahla KM, Bragança L, Mateus R (2019) Obstacles and barriers for measuring building's circularity. IOP Conf Ser Earth Environ Sci 225:012058. https://doi.org/10.1088/1755-1315/225/1/012058
44. Tirelli D, Besana D (2023) Moving toward net zero carbon buildings to face global warming: a narrative review. Buildings 13(3):684. https://doi.org/10.3390/buildings13030684
45. Eray E, Sanchez B, Haas C (2019) Usage of interface management system in adaptive reuse of buildings. Buildings 9(5):105. https://doi.org/10.3390/buildings9050105
46. Roxas CLC et al (2023) Design for manufacturing and assembly (DfMA) and design for deconstruction (DfD) in the construction industry: challenges, trends and developments. Buildings 13(5):1164. https://doi.org/10.3390/buildings13051164
47. Rakhshan K, Morel J-C, Alaka H, Charef R (2020) Components reuse in the building sector—a systematic review. Waste Manag Res 38(4):347–370. https://doi.org/10.1177/0734242X20910463
48. Charef R, Emmitt S (2021) Uses of building information modelling for overcoming barriers to a circular economy. J Clean Prod 285:124854. https://doi.org/10.1016/j.jclepro.2020.124854
49. Allam A, Nik-Bakht M (2023) Barriers to circularity in construction: an analysis of experts' perspectives. In: Proceedings 2023 European conference on computing in construction and the 40th international CIB W78 conference, pp 1–10. https://doi.org/10.35490/EC3.2023.234
50. Durdyev S, Koc K, Tleuken A, Budayan C, Ekmekcioğlu Ö, Karaca F (2023) Barriers to circular economy implementation in the construction industry: causal assessment model. Environ Dev Sustain. https://doi.org/10.1007/s10668-023-04061-8
51. AlJaber A, Alasmari E, Martinez-Vazquez P, Baniotopoulos C (2023) Life cycle cost in circular economy of buildings by applying building information modeling (BIM): a state of the art. Buildings 13(7):1858. https://doi.org/10.3390/buildings13071858
52. Guerra BC, Leite F (2021) Circular economy in the construction industry: an overview of United States stakeholders' awareness, major challenges, and enablers. Resour Conserv Recycl 170:105617. https://doi.org/10.1016/j.resconrec.2021.105617
53. Hossain MS, Khatun M (2021) A qualitative-based study on barriers to change from linear business model to circular economy model in built environment—evidence from Bangladesh. Circ Econ Sust 1(3):799–813. https://doi.org/10.1007/s43615-021-00050-z
54. European Commission. Joint Research Centre (2023) Techno-economic and environmental assessment of construction and demolition waste management in the European Union: status quo and prospective potential. Publications Office of the European Union, Luxembourg. https://doi.org/10.2760/721895
55. BIMcert (2019) BIMcert—1. Construction skills, 2. Energy efficiency, 3. Regulating supply chains, 4. Tackling climate change. EU. https://doi.org/10.3030/785155
56. UK Green Building Council (2022) System enablers for a circular economy. UK GBC. [Online]. Available: https://ukgbc.org/resources/system-enablers-for-a-circular-economy/
57. UK Green Building Council (2019) Circular economy guidance for construction clients: how to practically apply circular economy principles at the project brief stage. [Online]. Available: https://ukgbc.org/resources/circular-economy-guidance-for-construction-clients-how-to-practically-apply-circular-economy-principles-at-the-project-brief-stage/
58. Circle Economy, WBCSD (2018) Barriers and recommendations to scale the circular built environment

59. Circle Economy, Metabolic (2022) The circularity gap report: built environment—The Netherlands. [Online]. Available: https://www.circle-economy.com/resources/circularity-gap-report-built-environment-the-netherlands
60. Cervo H, Ferrasse J-H, Descales B, Van Eetvelde G (2020) Blueprint: a methodology facilitating data exchanges to enhance the detection of industrial symbiosis opportunities—application to a refinery. Chem Eng Sci 211:115254. https://doi.org/10.1016/j.ces.2019.115254
61. Assmann IR, Rosati F, Morioka SN (2023) Determinants of circular business model adoption—a systematic literature review. Bus Strat Env 32(8):6008–6028. https://doi.org/10.1002/bse.3470
62. Charef R, Morel J-C, Rakhshan K (2021) Barriers to implementing the circular economy in the construction industry: a critical review. Sustainability 13(23):12989. https://doi.org/10.3390/su132312989
63. Azcárate-Aguerre JF, Den Heijer AC, Arkesteijn MH, Vergara d'Alençon LM, Klein T (2023) Facades-as-a-service: systemic managerial, financial, and governance innovation to enable a circular economy for buildings. Lessons learnt from a full-scale pilot project in the Netherlands. Front Built Environ 9:1084078. https://doi.org/10.3389/fbuil.2023.1084078
64. Azcárate-Aguerre JF, Conci M, Zils M, Hopkinson P, Klein T (2022) Building energy retrofit-as-a-service: a total value of ownership assessment methodology to support whole life-cycle building circularity and decarbonisation. Constr Manag Econ 40(9):676–689. https://doi.org/10.1080/01446193.2022.2094434
65. Han D, Kalantari M, Rajabifard A (2024) The development of an integrated BIM-based visual demolition waste management planning system for sustainability-oriented decision-making. J Environ Manag 351:119856. https://doi.org/10.1016/j.jenvman.2023.119856
66. Sala Benites H, Osmond P, Prasad D (2022) Inquiry on perceptions and practices of built environment professionals regarding regenerative and circular approaches. Buildings 13(1):63. https://doi.org/10.3390/buildings13010063
67. Çetin S, Gruis V, Straub A (2021) Towards circular social housing: an exploration of practices, barriers, and enablers. Sustainability 13(4):2100. https://doi.org/10.3390/su13042100
68. Executive Agency for Small and Medium Sized Enterprises, Deloitte, Building Research Establishment, In Extenso Innovation Croissance (2021) Study on circular economy principles for buildings' design: final report. Publications Office of the European Union, Luxembourg. https://doi.org/10.2826/3602
69. Carra G, Magdani N (2017) Circular business models for the built environment. Arup. [Online]. Available: https://www.arup.com/insights/circular-economy-in-the-built-environment/
70. Agirre I (2017) REnnovates: flexibility activated zero energy districts. Mondragon/LKS
71. Condotta M, Zatta E (2021) Reuse of building elements in the architectural practice and the European regulatory context: inconsistencies and possible improvements. J Clean Prod 318:128413. https://doi.org/10.1016/j.jclepro.2021.128413
72. Oluleye BI, Chan DWM, Olawumi TO, Saka AB (2023) Assessment of symmetries and asymmetries on barriers to circular economy adoption in the construction industry towards zero waste: a survey of international experts. Build Environ 228:109885. https://doi.org/10.1016/j.buildenv.2022.109885
73. Munaro MR, Tavares SF (2023) Design for adaptability and disassembly: guidelines for building deconstruction. Constr Innov. https://doi.org/10.1108/CI-10-2022-0266
74. Torgautov B, Zhanabayev A, Tleuken A, Turkyilmaz A, Mustafa M, Karaca F (2021) Circular economy: challenges and opportunities in the construction sector of Kazakhstan. Buildings 11(11):501. https://doi.org/10.3390/buildings11110501
75. Talpur BD, Liuzzi S, Rubino C, Cannavale A, Martellotta F (2023) Life cycle assessment and circular building design in South Asian countries: a review of the current state of the art and research potentials. Buildings 13(12):3045. https://doi.org/10.3390/buildings13123045
76. Zuofa T, Ochieng EG, Ode-Ichakpa I (2023) An evaluation of determinants influencing the adoption of circular economy principles in Nigerian construction SMEs. Build Res Inf 51(1):69–84. https://doi.org/10.1080/09613218.2022.2142496

77. De Bruijn TJNM, Entrop AG (2019) Developing a transition agenda towards a circular economy: the Dutch case of the province of Overijssel. IOP Conf Ser Earth Environ Sci 297(1):012012. https://doi.org/10.1088/1755-1315/297/1/012012
78. Takyi-Annan GE, Zhang H (2023) Assessing the impact of overcoming BIM implementation barriers on BIM usage frequency and circular economy in the project lifecycle using Partial Least Squares Structural Equation Modelling (PLS-SEM) analysis. Energy Build 295:113329. https://doi.org/10.1016/j.enbuild.2023.113329
79. Gómez-Gil M, Espinosa-Fernández A, López-Mesa B (2022) Contribution of new digital technologies to the digital building logbook. Buildings 12(12):2129. https://doi.org/10.3390/buildings12122129
80. Hedberg A, Šipka S, Bjerkem J (n.d.) Creating a digital roadmap for a circular economy. [No further publication information provided.]
81. Hart J, Adams K, Giesekam J, Tingley DD, Pomponi F (2019) Barriers and drivers in a circular economy: the case of the built environment. Procedia CIRP 80:619–624. https://doi.org/10.1016/j.procir.2018.12.015
82. Izquierdo RS, Soliu I, Migliaccio GC (2024) Enablers and barriers to implementation of circular economy practices in the built environment: an exploratory study. J Leg Aff Dispute Resolut Eng Constr 16(2):05023009. https://doi.org/10.1061/JLADAH.LADR-1094
83. Weerakoon P, Thayaparan M (2023) Analysis of barriers towards implementing circular economy in Sri Lankan built environment. In: 11th world construction symposium—2023. Ceylon Institute of Builders, Sri Lanka, pp 162–173. https://doi.org/10.31705/WCS.2023.14
84. Calheiros CSC, Stefanakis AI (2021) Green roofs towards circular and resilient cities. Circ Econ Sust 1(1):395–411. https://doi.org/10.1007/s43615-021-00033-0
85. Acharya D, Boyd R, Finch O (2018) From principles to practices: first steps towards a circular built environment. Arup, Ellen MacArthur Foundation
86. SYSTEMIQ, Ellen MacArthur Foundation (2017) Achieving "Growth Within". [No publisher specified]
87. Kanters J (2020) Circular building design: an analysis of barriers and drivers for a circular building sector. Buildings 10(4):77. https://doi.org/10.3390/buildings10040077
88. Owojori OM, Okoro C (2022) The private sector role as a key supporting stakeholder towards circular economy in the built environment: a scientometric and content analysis. Buildings 12(5):695. https://doi.org/10.3390/buildings12050695
89. Cruz-Rios F, Grau D (2020) Design for disassembly: an analysis of the practice (or lack thereof) in the United States. In: Construction research congress 2020, Tempe, Arizona, pp. 992–1000. American Society of Civil Engineers. https://doi.org/10.1061/9780784482889.105
90. World Green Building Council (2023) The circular built environment playbook. [Online]. Available: https://worldgbc.org/article/circular-built-environment-play-book
91. Mackenbach S, Zeller JC, Osebold R (2020) A roadmap towards circularity—modular construction as a tool for circular economy in the built environment. IOP Conf Ser Earth Environ Sci 588(5):052027. https://doi.org/10.1088/1755-1315/588/5/052027
92. Bertino G, Rose G, Kisser J (2021) Drivers and barriers for implementation and international transferability of sustainable pop-up living systems. Circ Econ Sust 1(3):935–965. https://doi.org/10.1007/s43615-021-00063-8
93. European Commission. Directorate General for Research and Innovation (2023) ERA industrial technology roadmap for circular technologies and business models in the textile, construction and energy-intensive industries. Publications Office of the European Union, Luxembourg. https://doi.org/10.2777/188014
94. Ababio BK, Lu W (2023) Barriers and enablers of circular economy in construction: a multi-system perspective towards the development of a practical framework. Constr Manag Econ 41(1):3–21. https://doi.org/10.1080/01446193.2022.2135750
95. UK Green Building Council (2022) Insights on how circular economy principles can impact carbon and value. [Online]. Available: https://ukgbc.org/resources/how-circular-economy-principles-can-impact-carbon-and-value/

96. Azcarate-Aguerre JF, Klein T, Konstantinou T, Veerman M (2022) Facades-as-a-service: the role of technology in the circular servitisation of the building envelope. Appl Sci 12(3):1267. https://doi.org/10.3390/app12031267
97. Kanyilmaz A, Birhane M, Fishwick R, Del Castillo C (2023) Reuse of steel in the construction industry: challenges and opportunities. Int J Steel Struct 23(5):1399–1416. https://doi.org/10.1007/s13296-023-00778-4
98. Bank LC, Arias FR, Gentry TR, Al-Haddad T, Tasistro-Hart B, Chen J-F (2019) Structural analysis of FRP parts from waste wind turbine blades for building reuse applications. In: Zingoni A (ed) Advances in engineering materials, structures and systems: innovations, mechanics and applications, 1st edn. CRC Press, pp 1520–1524. https://doi.org/10.1201/9780429426506-262
99. UK Green Building Council (2020) Circular economy innovation insights: reuse and products as a service. [Online]. Available: https://ukgbc.org/resources/circular-economy-implementation-packs-for-products-as-a-service-and-reuse/
100. Fang D, Berglund-Brown J, Iwakuni D, Mueller C (2023) Carbon and craft: lessons from the deconstruction, relocation, and reuse of a traditional Japanese house's timber structure. J Phys Conf Ser 2600(19):192002. https://doi.org/10.1088/1742-6596/2600/19/192002
101. Circle Economy (2017) A future-proof built environment: putting circular business models into practice
102. Atapattu AMDS, Hadiwattage C, Perera BAKS, Rajaratnam D (2024) Employing criteria scoring matrix in appraising the economic return of transcending to a circular built environment. Smart Sustain Built Environ 13(2):267–290. https://doi.org/10.1108/SASBE-04-2023-0077
103. Scialpi G, Declercq J (2023) Adaptability in healthcare buildings: a perspective through Joseph Bracops Hospital. Front Med Technol 5:1199581. https://doi.org/10.3389/fmedt.2023.1199581
104. Morel J-C, Charef R, Hamard E, Fabbri A, Beckett C, Bui Q-B (2021) Earth as construction material in the circular economy context: practitioner perspectives on barriers to overcome. Philos Trans R Soc B 376(1834):20200182. https://doi.org/10.1098/rstb.2020.0182
105. Alagusic M, Milovanovic B, Pecur IB (n.d.) Recycled aggregate concrete—sustainable use of construction and demolition waste and reduction of energy consumption. [No publication details available]
106. Byers BS, Cheriyamulla S, Ewason J, Hall DM, De Wolf C (2022) Using engraved QR codes to connect building components to materials passports for circular construction. In: 2022 European conference on computing in construction. https://doi.org/10.35490/EC3.2022.226
107. Christensen TB, Johansen MR, Buchard MV, Glarborg CN (2022) Closing the material loops for construction and demolition waste: the circular economy on the island Bornholm, Denmark. Resour Conserv Recycl Adv 15:200104. https://doi.org/10.1016/j.rcradv.2022.200104
108. Dsilva J, Zarmukhambetova S, Locke J (2023) Assessment of building materials in the construction sector: a case study using life cycle assessment approach to achieve the circular economy. Heliyon 9(10):e20404. https://doi.org/10.1016/j.heliyon.2023.e20404
109. Nguyen HD, Macchion L (2023) Risk management in green building: a review of the current state of research and future directions. Environ Dev Sustain 25(3):2136–2172. https://doi.org/10.1007/s10668-022-02168-y
110. Ginga CP, Ongpeng JMC, Daly MKM (2020) Circular economy on construction and demolition waste: a literature review on material recovery and production. Materials 13(13):2970. https://doi.org/10.3390/ma13132970
111. Lederer J, Gassner A, Kleemann F, Fellner J (2020) Potentials for a circular economy of mineral construction materials and demolition waste in urban areas: a case study from Vienna. Resour Conserv Recycl 161:104942. https://doi.org/10.1016/j.resconrec.2020.104942
112. Yi T (2023) Research on building material design in circular economy based on improved neural network. Soft Comput 27(14):10369–10378. https://doi.org/10.1007/s00500-023-08302-w

113. Bilal M, Khan KIA, Thaheem MJ, Nasir AR (2020) Current state and barriers to the circular economy in the building sector: towards a mitigation framework. J Clean Prod 276:123250. https://doi.org/10.1016/j.jclepro.2020.123250
114. Govindan K, Hasanagic M (2018) A systematic review on drivers, barriers, and practices towards circular economy: a supply chain perspective. Int J Prod Res 56(1–2):278–311. https://doi.org/10.1080/00207543.2017.1402141
115. Liu J, Wu P, Jiang Y, Wang X (2021) Explore potential barriers of applying circular economy in construction and demolition waste recycling. J Clean Prod 326:129400. https://doi.org/10.1016/j.jclepro.2021.129400
116. Andersen BH, Salvetti G, Komkova A (2021) Circular economic modelling—barriers and challenges throughout the value circle. In: Cunha VMCF, Rezazadeh M, Gowda C (eds) Proceedings of the 3rd RILEM spring convention and conference (RSCC 2020), RILEM Bookseries, vol 35. Springer, Cham, pp 121–130. https://doi.org/10.1007/978-3-030-76543-9_12
117. Iacovidou E, Hahladakis JN, Purnell P (2021) A systems thinking approach to understanding the challenges of achieving the circular economy. Environ Sci Pollut Res 28(19):24785–24806. https://doi.org/10.1007/s11356-020-11725-9
118. Rootzén J, Karlsson I, Johnsson F, Kadefors A, Uppenberg S (2020) Supply-chain collective action towards zero CO_2 emissions in infrastructure construction: mapping barriers and opportunities. IOP Conf Ser Earth Environ Sci 588(4):042064. https://doi.org/10.1088/1755-1315/588/4/042064
119. Nian S, Pham T, Haas C, Ibrahim N, Yoon D, Bregman H (2022) A functional demonstration of adaptive reuse of waste into modular assemblies for structural applications: the case of bicycle frames. J Clean Prod 348:131162. https://doi.org/10.1016/j.jclepro.2022.131162
120. Springvloed P (2021) The drivers and barriers for a circular economy in the built environment of the Netherlands. Master thesis. https://studenttheses.uu.nl/handle/20.500.12932/39916
121. Surgenor A, Winch R, Wakiwaka J, Finch O (2019) Circular economy guidance for construction clients: how to practically apply circular economy principles at the project brief stage. UK Green Building Council. https://ukgbc.org/resources/circular-economy-guidance-for-construction-clients-how-to-practically-apply-circular-economy-principles-at-the-project-brief-stage/
122. Larsson K, Gammelsæter E (n.d.) Circular materials in infrastructure. Sweco Group. https://www.swecogroup.com/urban-insight/circularity/report-circular-materials-in-infrastructure/
123. Thelen D et al (2018) Scaling the circular built environment: pathways for business and government. WBCSD. https://www.wbcsd.org/resources/scaling-the-circular-built-environment-pathways-for-business-and-government/
124. Densley Tingley D, Cooper S, Cullen J (2017) Understanding and overcoming the barriers to structural steel reuse, a UK perspective. J Clean Prod 148:642–652. https://doi.org/10.1016/j.jclepro.2017.02.006
125. Selvaraj S, Chan T-M (2024) Recommendations for implementing circular economy in construction: direct reuse of steel structures. J Constr Steel Res 214:108439. https://doi.org/10.1016/j.jcsr.2023.108439
126. Klotz M, Haupt M, Hellweg S (2022) Limited utilization options for secondary plastics may restrict their circularity. Waste Manag 141:251–270. https://doi.org/10.1016/j.wasman.2022.01.002
127. Environmental Protection Agency (2009) Recover your resources—reduce, reuse, and recycle construction and demolition materials at land revitalization projects. https://www.epa.gov/sites/default/files/2015-09/documents/cdbrochure.pdf
128. Wahlström M, Hradil P, Sansom M (2019) D1.3 Report on the legal barriers and opportunities—PROGRESS
129. Mahpour A (2018) Prioritizing barriers to adopt circular economy in construction and demolition waste management. Resour Conserv Recycl 134:216–227. https://doi.org/10.1016/j.resconrec.2018.01.026

130. Khadim N, Agliata R, Marino A, Thaheem MJ, Mollo L (2022) Critical review of nano and micro-level building circularity indicators and frameworks. J Clean Prod 357:131859. https://doi.org/10.1016/j.jclepro.2022.131859
131. Forrest J (2021) The feasibility of recycling and reusing building materials found in single-family homes built after 1970 in Metro Vancouver. Metro Vancouver & University of British Columbia. https://sustain.ubc.ca/sites/default/files/2021-010_Feasibility%20of%20recycling%20and%20reusing_Forrest.pdf
132. Vefago LHM, Avellaneda J (2012) The environmental impact of recycled concrete. https://www.semanticscholar.org/paper/The-Environmental-Impact-of-Recycled-Concrete-Vefago/f362116a22ff212c0c4a5b03087ebebdafbc8580
133. Mhatre P, Gedam VV, Unnikrishnan S (2024) Management insights for reuse of materials in a circular built environment. Waste Manag Res 42(5):396–405. https://doi.org/10.1177/0734242X231187570
134. Geboes E, Galle W, De Temmerman N (2023) Make or break the loop: a cross-practitioners review of glass circularity. Glass Struct Eng 8(2):193–210. https://doi.org/10.1007/s40940-022-00211-y
135. Schuster S, Geier S (2022) CircularWOOD—towards circularity in timber construction in the German context. IOP Conf Ser Earth Environ Sci 1078(1):012030. https://doi.org/10.1088/1755-1315/1078/1/012030
136. Deweerdt M, Mertens M (2020) A guide for identifying the reuse potential of construction products. European Regional Development Fund—Interreg North-West Europe, Deliverable 1.2 of Activity 1 of the Work Package 2 (WP T2)
137. Nakajima S, Russell M (eds) (2014) Barriers for deconstruction and reuse/recycling of construction materials. CIB Publication 397. Delft (Netherlands). http://www.irbnet.de/daten/iconda/CIB_DC29788.pdf
138. Papamichael I, Voukkali I, Loizia P, Zorpas AA (2023) Construction and demolition waste framework of circular economy: a mini review. Waste Manag Res 41(12):1728–1740. https://doi.org/10.1177/0734242X231190804
139. Yu ATW, Wong I, Mok KSH (2021) Effectiveness and barriers of pre-refurbishment auditing for refurbishment and renovation waste management. Environ Challenges 5:100231. https://doi.org/10.1016/j.envc.2021.100231
140. Wanaguru K, Mallawaarachchi H, Vijerathne D (2022) Circular Economy (CE) based material selection: development of a CE-based '10R' evaluation framework for building construction projects in Sri Lanka. In: Proceedings of 10th world construction symposium 2022, BEMRU, University of Moratuwa, pp 208–219. https://doi.org/10.31705/WCS.2022.18
141. Galkina D (2022) Circular economy in the construction industry: barriers and drivers. MSc thesis, Politecnico di Milano. https://www.politesi.polimi.it/retrieve/c4476e67-b329-490a-9914-9c8b856e4fec/10759691_Galkina%20Darya_Thesis.pdf
142. Low Carbon Concrete Group (2022) Low carbon concrete Routemap. Institution of Civil Engineers. https://www.ice.org.uk/media/200i0yqd/2022-04-26-low-carbon-concrete-routemap-final_rev.pdf
143. Zhang C et al (2023) Life cycle assessment of material footprint in recycling: a case of concrete recycling. Waste Manag 155:311–319. https://doi.org/10.1016/j.wasman.2022.10.035
144. Gooroochurn M (2022) Circular homes—an energy-water-materials nexus for community climate engagement and action at grassroots level in the built environment. Results Eng 15:100548. https://doi.org/10.1016/j.rineng.2022.100548
145. Leversha D (2022) Delivering steel's circular economy potential. The Institution of Structural Engineers. https://www.istructe.org/journal/volumes/volume-100-(2022)/issue-9/delivering-steel-circular-economy-potential/
146. Pacheco J, De BJ, Lamperti TM (2023) Use of recycled aggregates in concrete: opportunities for upscaling in Europe. Publications Office of the European Union, Luxembourg. JRC131294. https://doi.org/10.2760/144802
147. Høibye L, Sand H (2018) Circular economy in the Nordic construction sector: identification and assessment of potential policy instruments that can accelerate a transition toward a

circular economy. Nordisk Ministerråd. https://urn.kb.se/resolve?urn=urn:nbn:se:norden:org:diva-5166
148. Stanton A (2021) End-of-life and reusing mass timber. Timber Tectonics in the Digital Age. https://timbertectonics.com/2021/12/14/end-of-life-and-reusing-mass-timber/
149. Klewe T, Völker T, Landmann M, Wilsch G, Kruschwitz S (n.d.) Laser-based sorting of construction and demolition waste for the circular economy
150. Keena N et al (2022) A life-cycle approach to investigate the potential of novel biobased construction materials toward a circular built environment. Energies 15(19):7239. https://doi.org/10.3390/en15197239
151. Yang Y, Guan J, Nwaogu JM, Chan APC, Chi H, Luk CWH (2022) Attaining higher levels of circularity in construction: scientometric review and cross-industry exploration. J Clean Prod 375:133934. https://doi.org/10.1016/j.jclepro.2022.133934

Chapter 2
Drivers for Circular Economy in the Built Environment

Kenza Belkhiri, Alina Mitache, Raluca Buzatu, Viorel Ungureanu, and Katerina Tsikaloudaki

Abstract This work explores the drivers for circular economy principles integration in the built environment. It highlights the reuse of materials through urban mining and the role of material passports in supporting efficient resource management. It addresses challenges in tracking and valuing materials throughout the building lifecycle and highlights digital tools such as Building Information Modelling (BIM) or sensor technologies that support material identification, reuse, and recycling. The research additionally explores the interaction between smart city technologies and circular economy approaches, demonstrating how data-driven systems can optimize material flows and promote sustainable urban development. Furthermore, it considers the potential and barriers to adopting circular construction practices in emerging markets, underscoring the need for tailored policies, education, and innovative business models. By integrating technological innovation, policy frameworks, and collaborative approaches, this work outlines pathways to enhance resource efficiency, reduce waste, and foster resilient and sustainable built environments globally.

Keywords Circular economy · Built environment · Policy drivers · Sustainability · Digital tools

---

K. Belkhiri · A. Mitache · R. Buzatu · V. Ungureanu (✉)
Department of Steel Structures and Structural Mechanics, Politehnica University Timisoara, Timisoara, Romania
e-mail: viorel.ungureanu@upt.ro

V. Ungureanu
Laboratory of Steel Structures, CCTFA, Romanian Academy, Timisoara Branch, Timisoara, Romania

K. Tsikaloudaki
School of Civil Engineering, Aristotle University of Thessaloniki, Thessaloniki, Greece

L. Bragança et al. (eds.), *Shaping Circular Transitions in the Built Environment*, Springer Tracts in Civil Engineering, https://doi.org/10.1007/978-3-032-02834-1_2

2.1 Introduction

The built environment significantly contributes to the depletion of global resources and the creation of waste, being responsible for about 40% of worldwide energy consumption and 30% of greenhouse gas emissions [1]. The concept of the circular economy (CE) aims to disconnect economic growth from resource consumption by promoting reuse, recycling, and regeneration of materials and systems [2]. The implementation of CE principles in the built environment acts as a strategic response to sustainability challenges. This chapter explores the key drivers that accelerate the transition to a circular economy in the built environment and synthesizes insights from academic research, policy frameworks, and industry practices.

Conventional linear models are unviable, especially given the considerable environmental impact of the construction industry, as it accounts for a large portion of the world's energy consumption and greenhouse gas production [3]. The CE framework seeks to close material loops and to promote strategies such as reuse, recycling, and extension of product lifespans, reducing waste and conserving resources [4].

A core motivation for adopting circular principles in the built environment is the urgent need to address climate change and resource scarcity. The sector's contribution to global emissions and resource depletion has led to industry-wide calls for more sustainable practices, the Paris Agreement and related international policies serving as catalysts for change [3]. Regulatory frameworks and policy instruments play a key role in navigating the industry toward circularity, as well as facilitating the adoption of circular business models and practices [5].

However, the effectiveness of these policies often depends on their ability to move beyond end-of-pipe solutions and foster systemic change throughout the entire building lifecycle [4, 6].

Economic considerations are equally significant in shaping the transition to a circular built environment. The high costs associated with sustainable construction often deter clients and developers from prioritizing circular solutions, especially when these involve larger upfront investments. Funding mechanisms and economic incentives are therefore essential to bridge the affordability gap and make circular construction more accessible [7]. The market acceptance of circular business models is also influenced by political and economic conditions, highlighting the need for coordinated action across multiple stakeholders [5]. Environmental imperatives further reinforce the case for circularity. The use of hazardous substances in construction materials exposes both human health and ecosystems to risk. The circular economy offers pathways to reduce these risks by encouraging the selection of safer materials and the recovery of valuable resources at the end of a building's life [8].

However, the focus of current research and policy often remains limited to recycling and material savings, with less attention given to innovative design concepts such as adaptability, flexibility, and dismantlability, which are necessary to realizing the full potential of CE in the built environment [6]. Social and behavioural factors

Fig. 2.1 Semantic map of circular built environment drivers

also influence the adoption of circular practices. Environmental, economic, and regulatory challenges often cover barriers, such as lack of awareness, resistance to change, and insufficient technical knowledge [8]. Overcoming these barriers requires more technological innovation and cultural and organizational changes in the construction industry [7, 9]. The integration of digital tools, such as Building Information Modelling (BIM), can facilitate the implementation of CE principles by enabling better resource management, waste reduction, and lifecycle optimization, tailored to specific project and client needs, as highlighted in Fig. 2.1 [10].

Transparency and documentation are increasingly recognized as vital components of sustainable construction. The growing need for detailed information about material flows, environmental effects, and lifecycle performance has driven the creation and adoption of environmental product declarations (EPDs) and similar certification systems, which serve to measure and clearly convey the sustainability attributes of buildings and construction materials. However, the dominant emphasis on energy and carbon metrics may overlook other important environmental indicators, potentially shifting burdens from one impact category to another. A more holistic approach is needed to ensure that circular strategies deliver comprehensive environmental benefits across multiple dimensions [11]. Although notable progress has been achieved in enhancing awareness and establishing policy directions, the broad application of circular economy principles remains incomplete, necessitating continuous efforts from policymakers, industry stakeholders, and researchers [5, 6].

2.2 Key Drivers for Circular Economy in the Built Environment

2.2.1 Regulatory and Policy Drivers

Governments and international organizations are introducing legislation and policies to insert circularity into construction practices. Governments are also adopting enabling environments for circular practices through targeted funding programs, public–private partnerships, regulatory mandates, and financial incentives.

2.2.2 Government Legislation and Incentives

The Circular Economy Action Plan of the European Union [12] offers a framework that prioritizes sustainable product design, waste prevention, and reuse of construction materials (see Fig. 2.2). Additional frameworks—such as the Waste Framework Directive, and national strategies like the Netherlands' "Circular Economy by 2050"—serve as foundational instruments for advancing circularity in the construction sector [13, 14].

These policies mandate actions including design for deconstruction, material reuse, and pre-demolition audits, thereby reducing landfill disposal and enabling material recovery. Building codes and certifications such as BREEAM and LEED are increasingly incorporating CE metrics, such as material efficiency and adaptability. The EU Taxonomy Regulation further defines criteria for economic activities that contribute to sustainability, including those within circular construction [15].

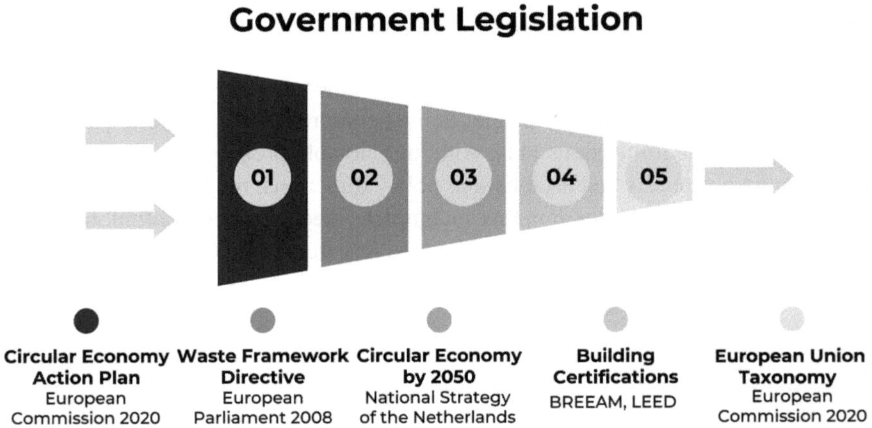

Fig. 2.2 Key policy and regulatory milestones supporting the transition to a circular economy in the built environment

Similarly, national-level building regulations and green public procurement (GPP) policies are evolving to mandate circular design principles [15]. Notable examples include national roadmaps like Finland's strategy to become a circular leader by 2035 and the broader EU Circular Economy Action Plan, both of which emphasize resource efficiency and sustainable construction practices [16, 17].

Government legislation and incentives are fundamental drivers for advancing the circular economy within the built environment. Regulatory frameworks, including updated standards and certifications, play a central role in shaping industry practices by mandating or encouraging the adoption of circular principles.

Without clear legislative direction of government policies and regulatory instruments, stakeholders can lack the motivation or clarity needed to integrate circular practices into their operations [18]. Regulatory pressure, when present, forces industry actors to adapt their production and business models, ensuring compliance with evolving sustainability requirements [7]. This pressure can be a catalyst for innovation, as companies are encouraged to develop new solutions that align with legislative expectations.

The economic incentives provided by governments further reinforce the adoption of circular economy strategies. These incentives can take various forms, such as funding schemes, tax benefits, or assurance programs for reused and recycled products.

The availability of government funding is often a decisive factor in enabling companies to invest in circular solutions, particularly when the immediate financial returns are uncertain or when the business case for renewable or recycled materials is less favourable compared to conventional alternatives based on fossils [7]. According to Munaro and Tavares [18], the establishment of marketplaces for material circularity and the implementation of incentive schemes for circular products are effective mechanisms for encouraging both producers and consumers to participate in circular value chains.

Sector-specific policies are also necessary, the shift to a circular economy demands tailored strategies to tackle the specific challenges and seize the opportunities present in each sector or value chain. A generic one-size-fits-all approach is insufficient given the diversity of materials, processes, and stakeholders involved in the built environment. By developing targeted policies, governments can more effectively support the scaling up of circular initiatives and encourage collaboration across municipal, provincial, and national levels.

In summary, government legislation and incentives serve as both direct and indirect devices for the adoption of the circular economy in the built environment. They provide the necessary regulatory pressure, financial support, and sector-specific guidance to drive systemic change, while also promoting stakeholder engagement, knowledge sharing, and innovation [7, 14, 18].

2.2.2.1 International Agreements and Standards

International agreements and standards significantly influence the implementation of circular economy principles in policy. At the European level, the ambition to implement the circular economy package of the European Commission, along with relevant Council conclusions, demonstrates a coordinated effort to set clear expectations for member states. The alignment of national strategies with international directives ensures that the transition toward circularity is not isolated but surrounded by a broader context of global sustainability objectives. The Netherlands, for example, has articulated its commitment to a circular economy by 2050, explicitly referring to the preservation of natural capital and the prioritization of renewable and widely available raw materials. This national ambition is closely linked to international agreements, as it seeks to optimize resource deployment and minimize environmental and health risks.

The integration of these standards into national policy frameworks is essential for maintaining consistency and comparability across borders, facilitating cross-national collaboration and innovation, thus the lack of clear standards and legislation, both at the national and international levels, it is still identified as a barrier to the implementation of circular practices in the construction sector [19]. The establishment and adoption of robust standards can address ambiguities, particularly in areas such as end-of-waste regulations, which have been highlighted as significant challenges by industry stakeholders [4]. By providing clarity and predictability, international standards reduce regulatory uncertainty and support circular solutions across the built environment.

Furthermore, the influence of international agreements extends to the encouragement of technological, social, and systemic innovations. These agreements often set ambitious targets for resource conservation and environmental protection, prompting countries to develop and implement innovative solutions that align with global best practices. The resulting policy frameworks create economic opportunities, reduce dependency on scarce raw materials, and contribute to a cleaner environment, as observed in the Dutch context.

The interaction between international agreements and national policy actions accelerates the adoption of circular economy principles and ensures that the built environment evolves in a manner that is both sustainable and resilient. The literature also categorizes regulatory drivers as both soft (institutional and social) and hard (technical and economic), with international agreements and standards often bridging these categories by providing both normative guidance and technical specifications [20]. This dual function increases the effectiveness of policy interventions, as it addresses not only the structural aspects of circularity but also the cultural and institutional dimensions that influence stakeholder behaviour. In summary, international agreements and standards are important in guiding the regulatory and policy drivers for circular economy implementation in the built environment. They establish a foundation for harmonized action, reduce barriers related to regulatory ambiguity, and provide an environment for innovation and collaboration across national boundaries [4, 19].

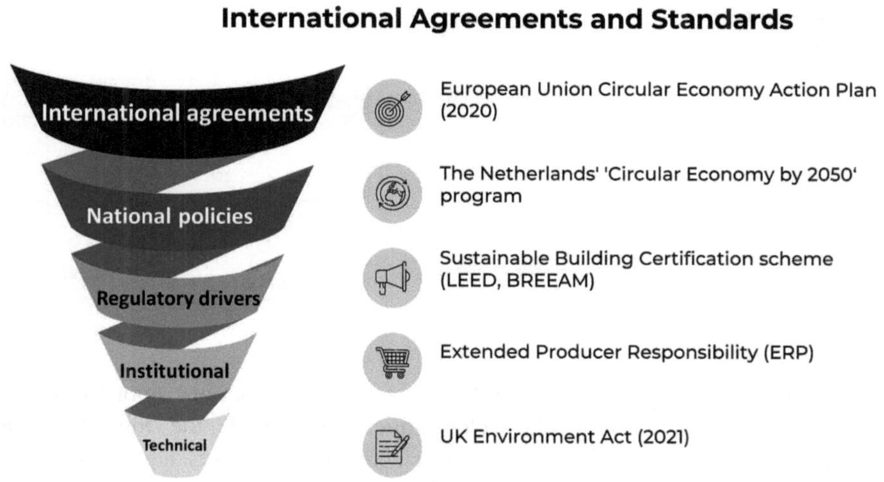

Fig. 2.3 International agreements and standards

Comparative International and Sectoral Drivers

Policies and the multilevel governance in the Nordic countries (Denmark, Finland, Sweden, Norway) are the drivers of CE implementation in the construction sector which coordinates national waste management strategies, local CE roadmaps, and aligns public procurement guidelines to promote innovation and circular practices (see Fig. 2.3) [21].

Illustrative examples include the UK Environment Act [22], which sets targets for improving resource efficiency and mandates the reuse of construction materials. The Netherlands' 'Circular Economy by 2050' includes Amsterdam's Circular Strategy, which sets the ambitious target to reduce the use of virgin raw materials by 50% by 2030 and may also be considered as an illustrative example. The city requires circular procurement for public projects and facilitates collaboration through platforms dedicated to material reuse. A notable case is the Buiksloterham neighborhood, where deconstructed building materials were reused in new developments [23].

As an impact, these initiatives have adopted systemic collaboration between municipalities, architects, engineers, and contractors, encouraging innovation within supportive regulatory frameworks.

2.2.2.2 Local Policy Innovations

Local policy innovations are the fundamental drivers for promoting the circular economy in the built environment. Municipal and regional authorities hold a distinctive capacity to customize regulatory frameworks and incentive schemes to align with local conditions. A key strategy involves introducing circular economy goals

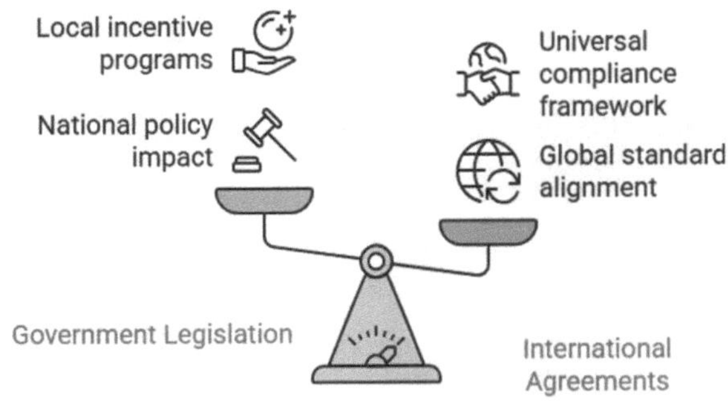

Fig. 2.4 Regulatory and policy drivers

into local innovation policies, to coordinate the efforts and actions among research institutions, businesses, and public sector entities. For example, the Dutch National Research Agenda has incorporated circular economy objectives into its strategic pathways, helping local governments and businesses to recognize opportunities for circular solutions. Additionally, the development of resources such as the Natural Capital Atlas illustrates how local policies can support the discovery of novel business models and revenue opportunities based on natural circularity. Local policy initiatives aim to establish cross-sectoral collaborations between industries, academia, and government organizations. For example, sector-specific efforts in agri-food, logistics, horticulture, and water management receive funding through innovation grants and collaborative platforms. Such measures not only stimulate research and development activities but also promote new partnerships and facilitate knowledge sharing, both essential for moving beyond linear practices (see Fig. 2.4).

Another important aspect of local policy innovation involves establishing certification frameworks and sustainability standards linked to municipal or regional contexts. While global certifications often serve as market motivators for innovative buildings, locally developed schemes can offer more context-sensitive incentives for circular activities. These regional certification systems may be especially valuable in areas where international standards are less common or where environmental and social circumstances demand personalised solutions. By making these local certifications visible, market interest in circular construction and refurbishment can increase, forcing local stakeholders to follow sustainable innovations [19].

Local authorities play a key role in promoting digital tools and decision support frameworks, essential for circular design and construction. The integration of technologies such as BIM, material passports, and sustainability certification platforms at the municipal level has the potential to modify conventional construction practices by implementing lifecycle considerations into both project planning and implementation. Even if national policies are slow to encourage reversible design approaches, local governments can serve as pioneers and advocates for these digital solutions [14].

The implementation of new technologies is often facilitated by local incentives, training programs, and demonstration initiatives. By defining resource use and acquisition criteria, local governments can influence building practices, promoting the reuse of excess materials and minimizing waste generation. Such policy measures rely on lifecycle assessments and other analytical instruments that provide evidence for informed decision-making [7].

Local governments often lead initiatives such as campaigns, workshops, and knowledge sharing events that help build community engagement and strengthen the capabilities of industry professionals. These educational activities are integrated in regulatory strategies that usually require or encourage circular practices. In summary, innovative local policies advance the circular economy in the built environment by developing regulatory frameworks, encouraging cross-sector partnerships, establishing regionally appropriate certification schemes, facilitating digital advances, and enhancing resource efficiency. These locally measures refer to specific regional needs and support the transition toward sustainability in construction and urban development [7, 14, 19, 20].

2.2.3 *Economic Drivers*

Economic drivers have the role in advancing the implementation of the circular economy in the construction industry. They not only improve cost savings and resource efficiency but also offer financial incentives for companies adopting circular practices. Moreover, sustainability-linked finance mechanisms are attracting investment into circular projects, supporting the business case for CE implementation [3, 18, 24–26]. These economic factors also contribute to increasing the market demand for sustainable buildings [13].

2.2.3.1 Cost Savings and Resource Efficiency

CE practices improve resource efficiency through material recycling strategies and reduce costs associated with waste management [24]. Practices such as design for disassembly, material reuse, and whole-life costing contribute to long-term economic efficiency, especially in construction projects with extended lifecycles [7, 16]. Buildings designed for disassembly or adaptability can retain value through material recovery and reuse of components [11].

New circular business models—such as product-as-a-service, leasing programs, and take-back systems—are attracting new customer segments, generating revenue from waste streams, and contributing to global resource efficiency [4, 13, 24, 27]. In addition, technological innovations and digitalization can also influence costs reduction, which can improve profit margins [5, 25].

Green finance mechanisms and Environmental, Social, and Governance frameworks (ESG) are incentivizing investments in sustainable and circular real estate projects [4]. These approaches not only sustain brand reputation and stakeholder trust but also unlock new revenue streams and align with ESG-aligned value creation strategies [14, 27]. CE enables differentiation in the marketplace and aligns with long-term value creation models, especially for family-owned construction firms seeking legacy continuity [7, 17].

Implementing Green Methods and Technologies (GMTs) improves resource efficiency and reduces operational costs related to water and energy consumption, which are key enablers of net zero carbon goals [28]. Furthermore, Whole Life Costing (WLC) and environmentally informed valuation models highlight the long-term financial advantages of resource-efficient design and operation [13].

Fiscal instruments including Eco modulated Extended Producer Responsibility (EPR) programs, subsidies, and tax incentives for recycled content-reduce financial risk and improve the economic feasibility of CE practices [20, 29]. For instance, eliminating double VAT on recycled materials can make circular models more competitive with conventional linear systems [20, 29].

Additional incentives, such as reduced VAT on refurbishment projects, pricing mechanisms to discourage non-traceable materials, and increasing material scarcity, can further stimulate circular practices [5, 7, 14]. The development of a well-constructed business case demonstrating these economic advantages is a fundamental driver for adopting circular economy principles [13].

Cost savings and resource efficiency stand out as economic drivers in promoting the adoption of circular economy principles in the built environment. The 'reduce costs' pressure, both short-term and long-term, requires organizations to adopt strategies that minimizes resource consumption and waste generation.

Organizations allocate financial resources towards sustainability and circular initiatives to reveal opportunities for cost reduction. Justified by their potential for long-term economic benefits, despite their significant initial costs, investment projects exploring new sustainable business models are becoming dominant. Consequently, the availability of funding is vital, particularly during early implementation stages, to ensure circular initiatives become financially feasible [7]. Such financial support not only facilitates economic feasibility but also aligns with institutional frameworks and broader policy objectives. Moreover, market incentives play a central role in promoting resource efficiency. By internalizing social and environmental costs into product pricing, companies are encouraged to optimize material flows, reduce waste, and increase the use of recycled and secondary materials. This alignment of economic and sustainability objectives creates favourable conditions for circular practices.

The literature underlines the value of resource efficiency in reducing dependency on primary resources, thereby mitigating risks associated with supply chain disruptions. Improving resource efficiency significantly lowers costs and increases economic resilience, particularly in import-dependent economies. Additionally, energy efficiency is recognized as an important factor, with potential cost savings derived from energy recovery processes within waste streams. Achieving such resource efficiency demands shifts in business models and operational strategies, stimulating organizations to innovate, diversify existing business practices, or adopt external circular business models to realize these economic benefits [16]. However, several barriers hinder this transition, including high initial investment costs, uncertain short-term economic returns, and the necessity for collaborative innovation across supply chains [8]. Addressing these obstacles typically requires multi-level interventions, such as infrastructural support and resource-sharing platforms, further aligning business strategies with global sustainability goals [26, 30].

Ultimately, achievements such as of cost savings and resource efficiency through circular economy practices depends on a combination of supportive policy frameworks, robust market mechanisms, and committed organizational leadership. When these elements converge, circular initiatives can deliver economic and environmental outcomes, transforming conventional construction practices into sustainable and resilient systems [26].

Strengths, Weaknesses, and Mechanisms for Support of Savings and Resource Efficiency

Cost savings and resource efficiency represent a fundamental economic driver by directly reducing operational expenses and alleviating supply chain vulnerabilities, particularly in regions dependent on imported raw materials. Investments in circular business models, such as material banks and energy recovery systems, yield substantial lifecycle savings and enhance resilience against resource price volatility. In addition, policy frameworks that integrate environmental costs, such as carbon pricing, effectively align economic incentives with overarching sustainability objectives, fostering greater adoption of circular practices.

Nevertheless, several weaknesses hinder the full realization of these economic advantages. A notable challenge is the absence of comprehensive quantitative evidence, including clear return-on-investment analyses and detailed case studies, which limits the practical demonstration of potential savings. Moreover, the substantial initial expenses and the ambiguity surrounding short-term financial gains are frequently viewed as major obstacles, particularly for small and medium-sized businesses. The potential for scalable pilot projects remains largely untapped, which presents a missed opportunity for validating circular economy approaches.

Supporting mechanisms are essential to address these challenges. Reuse strategies, involving reclaimed steel or recycled concrete, effectively lower material procurement and disposal costs, with industrial symbiosis networks reportedly achieving cost reductions between 20 and 30% through cross-sectoral material exchanges. Circular design methods, such as passive heating and modular construction, contribute significantly to energy efficiency, further boosting the overall return on investment over the lifecycle of buildings [30]. Additionally, resource efficiency acts as a safeguard against volatility in virgin material prices, strengthening supply chain stability and resilience [26].

To further support cost savings and resource efficiency, integrating lifecycle costing into procurement practices can ensure a focus on long-term economic gains rather than immediate cost considerations. Providing tax incentives to stimulate cross-sector material exchanges and industrial symbiosis can also encourage broader collaboration and the adoption of resource-efficient practices across the industry.

2.2.3.2 Market Demand for Sustainable Buildings

The market demand for sustainable buildings has emerged as a significant economic driver in the transition toward a circular economy within the built environment. Increased environmental awareness and shifting societal values have led to a transformation in consumer preferences, and stakeholders across the construction value chain are increasingly recognizing the long-term value of sustainable and circular practices. This growing demand is fuelled not only by end users seeking healthier, more resource-efficient living and working environments, but also by investors and developers who view sustainable buildings as assets that offer enhanced market valuation and reduced operational risk [18, 19].

Economic considerations play a central role in shaping this demand. The increasing interest of clients in environmentally responsible construction serves as a powerful market pull for circular solutions [31]. Furthermore, the availability of financial incentives, ranging from tax incentives and subsidies to favourable lending conditions, has further strengthened the business case for sustainable construction [5, 13]. Market pressures are further intensified by increasing raw material costs and volatility, investor expectations of ESG-aligned assets, and the emergence of financial instruments tied to environmental performance.

A critical component enabling this market shift is the development of robust markets for secondary materials. As emphasized by Munaro and Tavares [18], the creation of physical and digital marketplaces for reclaimed and recycled construction materials—supported by inventory systems and transparent tracking—plays a central role in operationalizing buildings as material banks. These infrastructures not only facilitate the flow of materials but also reduce the dependence on virgin resources and enable the construction sector to respond effectively to market demands for sustainability.

The dissemination of information about the availability and quality of secondary materials improves decision-making throughout the value chain and contributes to the normalization of circular solutions. In addition, financial motivations such as cost savings from resource efficiency and waste minimization further bolster this market demand. The traditional linear construction model based on extraction, use, and disposal is increasingly perceived as both environmentally and economically untenable [8]. In response, clients and developers are actively pursuing alternatives that offer reduced lifecycle costs, lower carbon footprints, and alignment with evolving regulatory frameworks.

Sustainable buildings additionally offer companies a strategic edge in competition, allowing them to attract environmentally conscious customers, strengthen brand reputation, and improve market positioning [19, 26]. Circular business models such as product-as-a-service, leasing frameworks, and materials-sharing platforms have gained traction by enabling resource recovery and reuse, thus supporting both economic and environmental objectives [5, 25].

Research has demonstrated that integrating circular principles into building design and construction significantly reduces material usage and waste production, thereby enhancing the economic feasibility of sustainable buildings [14, 32]. Furthermore, sustainable buildings are increasingly perceived as instruments for achieving broader social objectives, including corporate social responsibility and compliance with the United Nations Sustainable Development Goals [33]. Stakeholders are thus motivated not solely by direct financial returns, but also by reputational gains and access to new funding opportunities favouring circular practices.

Education and awareness efforts are also integral to market development. By improving the understanding of the benefits associated with circular and regenerative construction practices, such initiatives help build a culture of innovation and sustainability within the sector [19]. However, several challenges remain that may dampen demand. Misconceptions about the performance and quality of recycled materials persist, and there is a notable absence of universally accepted protocols for their certification and integration into new projects [32].

Overcoming these barriers requires a coordinated approach that involves policymakers, industry practitioners, and academic institutions. Establishing trust, developing standardized regulatory frameworks, and investing in professional training are essential to support the broader adoption of circular economy principles [11, 19]. According to Pomponi and Moncaster [11], interdisciplinary collaboration is particularly important in addressing the systemic complexity of market transformation and in ensuring that the built environment can fully harness the potential of circular economy strategies. In conclusion, the market demand for sustainable buildings is shaped by a dynamic interplay of economic, social, and regulatory factors. The expansion of secondary material markets, the pursuit of cost efficiency, the alignment with sustainability frameworks, and the increasing awareness of stakeholders collectively are fostering a shift towards circular building practices. This trend reflects a wider transition within the construction sector toward resource-conscious, resilient, and forward-looking development models.

Strengths and Weaknesses, and Mechanisms for Support of the Market Demand for Sustainable Buildings

Among the primary strengths driving the market demand for sustainable buildings is the dual pull of consumers and investors. Health-conscious end-users increasingly seek environmentally responsible spaces, while ESG-oriented investors prioritize assets aligned with sustainable development goals. This convergence of consumer and institutional demand creates significant momentum for circular adoption. Another strength lies in the emergence of transparent and functional markets for secondary materials. Platforms dedicated to reclaimed construction components reduce the reliance on virgin resources and promote market transparency, facilitating informed decision-making in the sector.

However, key weaknesses persist. Negative perceptions surrounding the reliability and performance of recycled materials continue to inhibit widespread acceptance of the market. Furthermore, the lack of standardized certification systems, such as environmental product declarations (EPDs), hinders the validation and scalability of circular solutions. The demand for sustainable construction also varies between regions due to cultural and economic disparities, resulting in fragmented adoption patterns and inconsistent implementation.

To address these limitations, several supportive mechanisms are increasingly employed. Premium pricing associated with certified green buildings, such as those with LEED or BREEAM labels, demonstrates the economic value of circular performance, encouraging developer participation. Corporate sustainability commitments, particularly among ESG-driven investors and tenants, create a strong 'pull' effect for circular supply chains. Additionally, public procurement policies that mandate recycled content, such as the EU Level(s) framework, serve to institutionalize demand and establish guaranteed markets for secondary materials.

Strategic interventions to further strengthen market demand include the expansion of green certification schemes that validate circular performance and the launch of public awareness campaigns to educate consumers on the environmental, economic, and health benefits of circular building solutions. These combined efforts, which span policy, finance, and communication, create an enabling environment for sustainable construction, positioning the built environment as a cornerstone in the broader transition toward a circular economy.

Public procurement policies that prioritize circular economy principles can successfully drive consistent demand for sustainable and innovative building solutions [5, 18]. Additionally, long-term asset value increases—driven by operational cost savings-further incentivize sustainable design choices [3]. By emphasizing these benefits, supply chain stakeholders can influence client awareness and increase demand for sustainable construction [18]. In some jurisdictions, sustainability certifications for buildings are mandatory, further accelerating market alignment with circular construction practices [14].

Case Study: Madaster Platform

Madaster functions as a digital material registry, allowing buildings to be documented as 'material banks', tracking the origin, quality, and value of construction components. Using the platform, empowers the asset owners to plan for disassembly and monetize recovered materials. As an impact, the approach facilitates the development of circular asset portfolios and supports end-of-life material valorisation, while it also helps developers comply with regulations related to circular construction and material reuse. Madaster has been widely adopted in the Netherlands and is expanding to other countries such as Germany, Switzerland, and Norway. The platform proves to be a powerful enabler of circular construction, offering a structured, data-driven approach to material management that aligns environmental goals with economic and regulatory frameworks.

2.2.3.3 Financial Incentives and Green Financing. Exploring the Role of Economic Policies in Promoting Circular Construction Practices

Access to green financing has emerged as a key economic enabler in the transition toward circular economy (CE) practices within the built environment (see Fig. 2.5). Dedicated financial instruments, such as green loans, grants, and sustainability-linked investment schemes, play a critical role in enhancing the feasibility, scalability, and long-term viability of circular construction initiatives. In a sector where decisions are often constrained by short-term financial considerations and profitability metrics, access to capital is a determining factor in whether circular strategies can go beyond conceptual design into implementation.

As argued by Ahrens [7], financial support functions not only as a strategic economic lever but also as a foundational prerequisite for companies considering engaging in circular construction. In the absence of well-structured financial incentives or institutional support, circular initiatives can be deemed unviable, despite their evident long-term environmental and cost-saving benefits. The economic rationale for green financing is closely intertwined with broader imperatives such as resource efficiency, reducing lifecycle costs, and mitigation of environmental externalities.

This underscores the dual role of green financing: as both a catalyst for innovation and a mechanism to lower the perceived financial risk of adopting circular models.

The importance of financial access affects both large firms and small to medium-sized enterprises (SMEs). Circular construction projects demand longer payback periods, involve higher upfront capital expenditures, and carry operational uncertainties distinct from those associated with conventional linear projects. These conditions necessitate the development of bespoke financial products that align with the specific economic logic and lifecycle characteristics of circular models.

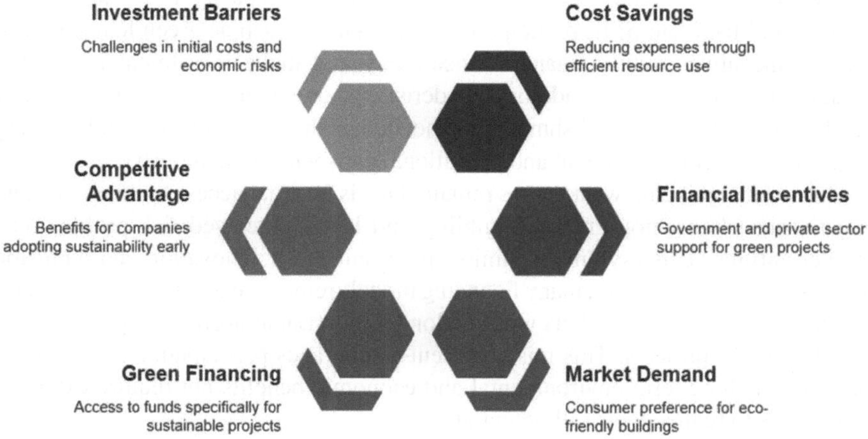

Fig. 2.5 Economic drivers of sustainable construction

Beyond financial incentives, green financing contributes to sectoral transformation through support for capacity building and knowledge dissemination.

Munaro and Tavares [18] observe that technological limitations and knowledge gaps continue to impede effective deployment of green funds, particularly among organizations lacking prior experience with circular approaches. This highlights the need for tailored financial tools that address the heterogeneous needs of industry actors, alongside targeted investment in managerial and technical upskilling. Adabre et al. [33] further emphasize that alignment between financial incentives and stakeholder priorities is critical to securing sustained engagement and impact.

When properly designed, funding mechanisms can simultaneously promote economic efficiency and environmental integrity. Instruments that reward resource-efficient design, encourage recycled content, and penalize wasteful practices reinforce the broader goals of circularity and sustainable development. This integration not only addresses immediate financial barriers but also drives systemic transformation by fostering collaboration across sectors and enabling new business models to emerge and scale.

In summary, green financing operates as a multidimensional driver that underpins the economic feasibility of circular economy strategies in the built environment. It facilitates the mitigation of upfront costs, enables strategic investment in skills and infrastructure, and aligns capital flows with long-term sustainability goals [7, 18, 20, 33]. However, its effectiveness depends on its integration with supportive policy frameworks, robust institutional structures, and broad-based stakeholder participation.

Strengths, Weaknesses, and Mechanisms for Support of Financial Incentives and Green Financing

A key strength of green financing lies in its catalytic capacity to derisk circular projects and facilitate SME participation. Mechanisms such as green loans, grants, and tax incentives reduce financial uncertainty and make investment in circular models more attractive. In addition to derisking, green finance enables capacity building through the establishment of knowledge-sharing platforms and training programs that bridge technical and operational gaps within the industry.

However, significant weaknesses remain. One is unequal access to financing, and larger firms often enjoy greater eligibility and lower perceived risk profiles than smaller entities. This asymmetry limits opportunities for innovation and inclusion across the sector. Moreover, many financing models remain geared toward short-term returns, a mindset that conflicts with the longer investment horizons typically associated with CE projects. This misalignment discourages participation in initiatives that promise long-term environmental and economic benefits but require extended timeframes to realize their full potential.

To support the scaling of green financing, several mechanisms have proven effective. Grant schemes, green bonds, and low-interest loans help offset the initial capital required for circular economy pilots, such as modular construction or the implementation of material passports [7]. Fiscal instruments, including tax deductions for the use of recycled materials or levies on waste disposal, further guide investment toward sustainable practices. Blended finance models, particularly those involving public–private partnerships (PPPs), pool resources and distribute risk, making large-scale infrastructure such as material recovery hubs more feasible.

Strategic recommendations for enhancing the impact of green financing include the establishment of venture capital funds explicitly targeting circular start-ups and the implementation of extended producer responsibility (EPR) schemes, which allocate funding for recycling infrastructure through producer levies. Together, these interventions not only address funding gaps but also embed circularity into the economic fabric of the built environment.

2.2.3.4 Competitive Advantage for Early Adopters

Adopting circular economy (CE) principles early in the construction sector offers a significant competitive edge, especially as markets become increasingly eco-conscious and diverse. Companies that implement circular business models (CBMs) ahead of others can present unique value propositions—such as utilizing recycled materials, employing low-impact building methods, and showcasing measurable environmental performance—that set them apart in a competitive market. These firms not only attract sustainability-oriented clients but also cultivate long-term customer loyalty by aligning their services with evolving societal and regulatory expectations [5, 7].

Economic incentives play a significant role in motivating early adoption. Companies that integrate CE strategies often generate additional revenue streams through activities such as material recovery, product take-back schemes, and refurbishment services. These initiatives reduce waste disposal costs, improve resource efficiency, and create new opportunities for profitability, particularly in the context of increasing material scarcity and regulatory stringency [5]. Thus, early movers are strategically positioned to respond to the growing demand for sustainable construction, leveraging their operational readiness to capture market share and influence client expectations [20].

Beyond operational efficiency, early adopters frequently aim to be recognized as industry pioneers, setting standards for innovation and sustainability. This pursuit of leadership improves organizational visibility and brand reputation, enabling greater access to capital, skilled talent, and strategic partnerships. In a setting where environmental, social, and governance (ESG) criteria are becoming more significant, the capacity to demonstrate environmental responsibility and progressive business practices holds particular value [7]. Although market entry by additional actors may eventually dilute the initial advantage held by early adopters, those who embed circularity at the core of their operations are more likely to sustain leadership. By institutionalizing continuous improvement, optimizing processes, and cultivating stakeholder networks, early adopters are well prepared to adapt to policy changes, client demands, and technological developments [5, 7].

Collaboration with clients and project partners further enhances competitive positioning. Engaging stakeholders from the earliest stages—through co-design processes, industry-client workshops, or integrated delivery models—ensures that circular principles are reflected in project outcomes. This partnership also enables the joint development of innovative solutions and the exchange of knowledge, improving the company's standing as a reliable ally in sustainable development [14, 20].

The regulatory context also plays a vital role. Governments and policy frameworks are increasingly rewarding circular practices through incentives, standards, and public procurement requirements. Early adopters, having already integrated these principles, are better positioned to comply with emerging mandates and can use their compliance as a strategic asset. This alignment opens pathways to preferential contracts, pilot project funding, and enhanced market access [33].

In essence, the competitive advantage derived from the early adoption of CE practices is multifaceted. It includes enhanced brand differentiation, access to niche and emerging markets, operational cost savings, the creation of new revenue streams, and elevated stakeholder trust. These benefits are reinforced by a supportive policy landscape and a growing imperative for the construction industry to meet sustainability commitments [5, 7, 20, 33].

Strengths, Weaknesses, and Mechanisms for Support of Competitive Advantage for Early Adopters

The competitive edge gained by early adopters of circular economy strategies is underpinned by several strengths. Foremost among these advantages is the capacity to differentiate products and services through the implementation of circular business models, such as modular construction and material passports, which appeal to investors driven by ESG and environmentally conscious clients. Furthermore, these firms diversify their income by transforming waste into valuable inputs and offering circular services, thus improving their resilience in volatile markets [33].

As CE practices become mainstream, the uniqueness of early adoption can be eroded unless reinforced by sustained innovation and continuous differentiation. Furthermore, the collaborative ethos that underpins circular transitions can paradoxically challenge competitive positioning. Sharing knowledge and best practices, while essential for scaling circularity, can blur proprietary advantages and lead to competitive parity.

Several mechanisms can help maintain and amplify competitive advantage. First-mover benefits allow early adopters to claim leadership in niche markets such as net zero retrofits and to form partnerships with clients prioritizing ESG compliance. Strategic branding of circular innovations—such as leasing models for fixtures or biodegradable building materials—enhances visibility and customer loyalty. Additionally, monetizing waste streams or offering services like selective deconstruction contributes to long-term value creation [16].

To support these dynamics, strategic initiatives are needed. The establishment of incubators for circular business models can accelerate the piloting and scaling of innovative practices. Collaborative research and development (R&D) initiatives across sectors further enable companies to share risks and rewards while driving the evolution of cutting-edge solutions.

Ultimately, while economic drivers alone are insufficient for a full transition to CE, they serve as indispensable enablers when aligned with regulatory foresight and active stakeholder collaboration. Redirecting financial flows through mechanisms such as cost transparency, market creation, and risk-sharing initiatives can unlock significant economic value. As the Ellen MacArthur Foundation has estimated, the construction sector could harness up to $1.8 trillion in global circular opportunities by 2030 if CE is embraced not as a compliance burden but as a strategic engine for resilience, innovation and value creation.

2.2.4 Environmental Drivers

The global construction industry is highly dependent on finite resources, contributing substantially to greenhouse gas emissions, environmental pollution, landfill overflow, and the depletion of natural ecosystems [34]. In this context, environmental drivers (see Fig. 2.6) play an important role in the advancement of sustainability within the built environment. They support the transition toward circular practices and directly contribute to broader goals such as climate change mitigation and the pursuit of net-zero carbon targets [10, 24].

2.2.4.1 Climate Change Mitigation

Mitigating climate change and preserving biodiversity are increasingly recognized as central motivations to advance circular construction practices. These practices,

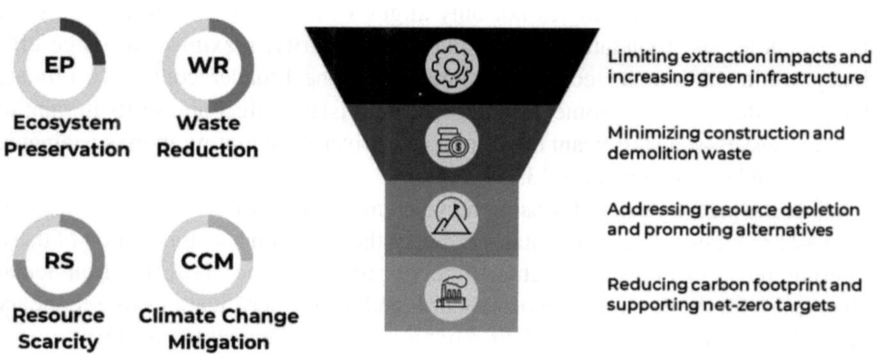

Fig. 2.6 Environmental drivers in construction

particularly through the reduction of embodied carbon through material reuse and lifecycle-oriented design, are essential in achieving sustainability goals in the built environment [34]. Escalating concerns about environmental degradation, resource depletion, and climatic instability continue to catalyse the integration of the principles of circular economy (CE) in the construction sector, which aims to minimize environmental externalities while fostering long-term sustainability [24].

The construction industry, responsible for a substantial share of global carbon emissions and a significant proportion of greenhouse gas (GHG) emissions, plays a critical role in this transition [3, 19, 33, 35]. Increased environmental awareness has intensified efforts to reduce emissions, aligning industry practices with net zero commitments [7, 36]. CE strategies, such as material reuse, modular construction, and passive design, substantially reduce embodied carbon and support emissions reduction targets [14]. These approaches not only limit resource consumption and waste generation, but also directly mitigate environmental burdens [3, 24, 27].

The regulatory landscape, including environmental legislation and climate policies, exerts considerable influence by mandating reduced environmental impacts across construction activities [36]. In particular, policy instruments such as Nearly Zero-Energy Building (NZEB) standards challenge designers and developers to achieve exceptional energy performance in alignment with sustainability objectives [32]. Circular building typologies, utilizing on-site renewable energy systems, are capable of reaching energy neutrality or even generating surplus energy, thus reducing the reliance on fossil fuels.

Material selection is equally critical to achieving these goals. The adoption of low-embodied-energy materials contributes significantly to reducing lifecycle energy use, while improvements in thermal insulation further optimize energy efficiency during the operational phase of buildings [32].

The European Green Deal reinforces these objectives by designating the construction sector as a key focus area for economic and ecological transition, emphasizing the adoption of cleaner and more circular building processes [21, 37]. In this context, the integration of recycled content and enhanced recycling mechanisms within the building sector are essential strategies to achieve substantial emission reductions.

The pursuit of energy-neutral and energy-positive buildings through the deployment of renewable technologies not only aligns with environmental mandates but also embodies the foundational principles of circularity: maximize resource efficiency while minimizing ecological harm [32]. The broader context of climate change, evidenced in phenomena such as the heat island effect and shifts in rainfall patterns, serves as a significant environmental driver for adopting more sustainable and resilient building practices [33].

While the primary driver for using recycled materials is often resource efficiency and waste reduction (environmental concerns), the selection and application of these materials must also ensure structural integrity and resilience [10]. Environmental drivers push for the development of standards and quality certifications for secondary materials, which is essential for their reliable use in resilient structures [10].

The negative environmental effects of buildings and infrastructure on ecosystems drive the need for a more holistic and sustainable approach to the built environment

[33]. This inherently links to sustainable urban planning, with the aim of minimize environmental impact on a larger scale and thus reduce vulnerability [19].

Concepts such as "regenerative sustainability" consider the built environment as a whole and aim for self-regenerating social and ecological systems, addressing broader environmental vulnerabilities [19].

Case Study: ReCreate Project (EU)

The ReCreate Project, funded by the European Union's Horizon 2020 program, exemplifies a pioneering approach to circular construction through direct reuse of structural concrete elements from existing buildings. Instead of employing energy-intensive recycling techniques, the project focuses on the selective deconstruction, transportation, and reintegration of precast concrete components, such as beams, slabs, and columns, into new constructions.

By preserving the structural integrity of these elements, ReCreate significantly reduces the environmental impact typically associated with conventional demolition and material reprocessing. The project has demonstrated the potential to achieve up to 80% reduction in carbon emissions compared to the use of virgin concrete. The initiative operates in multiple European countries, including Finland, Sweden, Germany, and the Netherlands, serving as a cross-national model of best practice in circular urban development. Integrates digital innovations such as material passports and Building Information Modelling (BIM) to support the traceability, assessment, and planning required for the reuse of structural components.

ReCreate contributes to the operationalization of the urban mining concept by treating existing buildings as material banks. In doing so, it underscores the viability of large-scale structural reuse as a means of reducing embodied carbon, enhancing material efficiency, and aligning construction practices with climate and circular economy objectives.

2.2.4.2 Resource Scarcity and Waste Reduction

Environmental drivers play a significant role in addressing resource scarcity and waste reduction in the construction industry by highlighting the negative consequences of the traditional linear "take, make, and dispose" model [13, 33, 38]. Recognition of these environmental impacts acts as a primary motivator for change toward a circular economy (CE) [24].

Material reuse is a primary step from a linear to a circular economy, reducing the consumption of raw materials [6]. The depletion of virgin materials, especially in the construction sector, combined with the escalating waste generation, has caused a shift towards material recovery, recycling centres and waste diversion policies [39, 40]. Construction and demolition activities generate over 800 million tonnes of waste annually in the EU alone. Circular strategies address waste overload and resource depletion through material efficiency, traceability, and reuse infrastructures [13, 20].

The circular economy emphasizes the reuse, refurbishment, and recycling of building materials and components. This reduces the need for new raw material extraction and processing, which are significant sources of embodied carbon [10]. Circular building design considers materials and components capable of sustaining several lifecycles, accentuating durability and resilience to maximize their potential reuse [32].

Addressing Depletion of Natural Resources

A central objective of the circular economy (CE) is to maintain materials in continuous circulation, thus minimizing dependence on virgin resource extraction. The imperative to reduce the environmental impacts associated with raw material procurement serves as a critical motivator for the adoption of CE strategies [38]. Concerns about resource scarcity and the ecological burden of conventional construction materials have accelerated interest in more sustainable alternatives, such as biobased and recycled inputs [24, 26]. The CE model promotes the substitution of non-renewable material and energy sources with regenerative, circular options that can be reintegrated into production cycles [5]. Steel recycling, for example, offers a compelling opportunity to reduce energy use and bypass the environmental costs of extraction and long-distance transport [37].

Furthermore, the pursuit of circular practices is often driven by the potential for economic gains and improved environmental performance [13]. Increasing resource scarcity further incentivizes the development of innovative technologies capable of recovering waste materials for reintegration into construction value chains [35].

Minimising Construction and Demolition Waste

Environmental considerations, such as reducing pollutant generation and conserving landfill capacity, have increasingly placed deconstruction as a more sustainable alternative to conventional demolition methods [41]. Although deconstruction often involves higher upfront costs and greater labour input, its potential to significantly minimize waste and facilitate the reuse of materials constitutes a compelling environmental incentive [21, 41]. A fundamental strategy within the circular economy is to embed design-for-disassembly principles at the early stages of building conception, thereby enhancing the ability to recover components efficiently at the end of their lifecycle [8, 32]. Moreover, prolonging building lifespans through renovation and repair, as opposed to premature demolition, aligns closely with CE objectives focused on resource conservation and environmental performance [21].

The global proliferation of construction and demolition waste (CDW) and its associated ecological repercussions underscore the urgency of advancing recycling and upcycling initiatives [10, 33]. These environmental pressures serve as key drivers for scaling circular practices, including material recovery, processing, and reuse [24, 35, 38]. Redirecting CDW away from landfills through targeted recycling strategies delivers both economic and environmental benefits [42]. In particular, upcycling is emerging as an innovative approach that enables high-value reuse, contributing to improved material quality in new constructions [6]. Furthermore, the environmental imperative to reduce waste generation has spurred technological advancements in recycling processes, with increasing emphasis placed on the ecological compatibility and performance of recycled construction products [18].

Tackling Landfill Overload

Environmental concerns regarding overburdened landfills, the escalation of illegal waste dumping, and growing public opposition to ecological degradation have become key drivers in promoting the diversion of construction waste from disposal sites [24]. In this context, the circular economy (CE) framework advocates for a fundamental shift towards minimizing waste generation at its source [5]. Regulatory instruments, such as landfill bans and the imposition of higher landfill taxation, serve to incentivize stakeholders to explore more sustainable waste management alternatives, notably material recycling and recovery practices [14].

These environmental imperatives further reinforce the development of closed-loop systems for building materials, where waste is reconceptualized as a valuable input for new products or processes, thus minimizing the demand for landfill space [5]. The transition away from an 'end-of-life' paradigm towards one centred on continual material recovery, both within individual systems and across sectors, supports the goal of CE of maintaining resources in circulation at their highest possible value [35]. This includes the strategic reuse of end-of-life building components and the systematic deconstruction of materials for reintegration into new construction cycles [13].

Preserving Ecosystems and Biodiversity

Environmental awareness of the harmful effects of resource extraction on ecosystems and biodiversity drives the need to limit the environmental impacts [24, 26, 38]. By reducing the reliance on virgin materials through CE strategies, the pressure on natural habitats due to mining and deforestation is reduced [37, 38]. The concept of regenerative sustainability aims at an overall systemic consideration of the ecological system, moving away from degenerative linear input–output models [19].

Although not explicitly detailed as a direct outcome of environmental drivers in the context of CE in construction, the broader environmental movement and understanding of ecological degradation in urban areas support the improvement of green infrastructure [43]. The absence of vegetation, for instance, diminishes ecosystems' ability to sequester carbon dioxide and moderate temperature fluctuations [43]. Regenerative neighbourhoods also consider the ecology of place and habitat exchange [8, 19]. The move towards sustainable urban planning, often driven by environmental concerns, inherently aims to minimize the overall environmental impact of the built environment [19].

In summary, economic drivers, such as the potential for cost reduction, resource efficiency, access to financial incentives, competitive advantages, and growing market demand for sustainable options, are key in promoting the adoption of circular economy principles and driving the market toward more sustainable building practices [3, 18, 24–26].

2.2.5 Technological Drivers

2.2.5.1 Innovations in Material Reuse and Recycling

Technological advancements support circularity through improved design, planning, and tracking of materials. Technology enables better predictive maintenance and efficient management of the building lifecycle. Innovations in digital technologies are facilitating circular economy practices in the built environment. Building Information Modelling (BIM), GIS, IoT sensors, LCA, digital twins, and material passports enhance transparency in material flows and enable better planning for reuse and recycling [17, 39, 44]. These technologies improve traceability, inform design decisions, and help manage resource flows throughout the building lifecycle [14].

With the growing pressure within the construction sector regarding the necessity to reduce its environmental footprint, innovations in material reuse and recycling have been some of the best enablers of the circular economy. Such new developments aim to retain material value through minimal extraction of virgin resources and minimize the generation of construction and demolition waste. Inside the sphere of technological drivers, allocating resources towards science and technology is regarded as crucial for the implementation of the CE, along with the adoption of cutting-edge technologies [45].

Mechanical recycling is vital for recovering construction materials, significantly reducing waste, and minimizing environmental impacts associated with resource extraction by enabling closed-loop processes, playing an important role in the recycling of heavy construction materials such as concrete and steel. Modern crushing, magnetic separation, and robotic sorting offer efficient recovery of steel bars and aggregates from demolished buildings. Recycled concrete aggregates (RCAs) can be used in road base and non-structural concrete or new concrete mixes. Moreover, steel is 100% recyclable and can be recycled as many times as required without loss of quality [46].

Chemical recycling represents a more advanced and transformative alternative to traditional mechanical recycling, particularly in the processing of complex plastic and composite materials. This method breaks down polymers at the molecular level, enabling the recovery of high-quality feedstocks suitable for the production of new materials. Applications in the construction sector include the recycling of intricate plastic components such as polystyrene insulation and PVC pipe. One notable technique is pyrolysis, through which synthetic crude oil is extracted from plastic waste and subsequently refined into new high-value products. This approach effectively closes the material loop by avoiding downcycling and preserving the quality and utility of recycled materials [47].

Chemical recycling thus offers a promising solution to address the challenges associated with complex material streams, ensuring high-grade reuse, and reducing waste. The broader adoption of advanced recycling and upcycling technologies, including chemical recycling and robotic systems for selective demolition, further supports the circular treatment of composite materials and reinforces the transition toward a more sustainable construction industry [26].

Upcycling refers to the process by which waste materials or low-value end-of-life products are transformed into items of significantly higher value and utility. Within the construction sector, upcycling can involve repurposing old concrete blocks into terrazzo-like surfaces or the use of reclaimed wood to introduce distinctive architectural or design elements. Such practices not only preserve the embodied energy of materials but also retain and enhance their cultural and aesthetic significance [48]. An illustrative example is the reuse of discarded glass bottles that, rather than being sent to landfill, can be converted into decorative building components such as tiles or façade panels. Emerging technologies have enabled the processing of coloured glass waste into sophisticated architectural finishes, exemplifying the intersection of sustainability and design innovation [49]. Therefore, upcycling encourages both material circularity and creative reinvention, reducing the volume of construction waste while simultaneously adding cultural, aesthetic, and economic value to the built environment.

Biobased materials are increasingly at the forefront of circular construction because of their renewable nature, their capacity to sequester carbon, and their biodegradability. Notable innovations in this domain include hempcrete, a composite derived from hemp fibres and lime, and mycelium-based materials, which are grown from fungal root systems. These materials exemplify the environmental benefits of circularity: mycelium has been developed into insulation panels and bricks with excellent thermal performance, while hempcrete offers essential breathability for low carbon wall assemblies [50]. Their integration into building systems reflects a broader shift toward low-impact, regenerative construction practices.

The adoption of CE principles also varies by material type. In the case of wood-based construction, the sector benefits from a cultural and operational alignment with circularity, including a strong emphasis on bio-based design and cascading use. This alignment has supported active engagement with circular strategies such as modular building approaches and remanufacturing, refurbishment, and construction [36]. As a result, wood-based construction has emerged as a leader in circular innovation, bolstered by favourable regulatory and market environments. In contrast, the plastic-based construction sector faces more complex challenges. It remains predominantly dependent on downstream circular strategies, particularly mechanical recycling and energy recovery. Progress in this area is often driven by external pressures, including regulatory mandates, improved systems for waste stream traceability, and industrial symbiosis initiatives aimed at interindustry collaboration [36]. However, the sector's advancement is limited by fragmented governance structures and higher technical barriers, which limit the scalability and integration of CE solutions.

Disassembling and reusing building parts is another pillar of material circularity. Modular construction techniques allow prefabricated components to be assembled and taken apart without substantial degradation and reused later. This results in less construction time, reduced waste, and, subsequently, improved adaptability of the building to possible future change or relocation [51].

Material passports, powered by blockchains, also play a critical role in safeguarding material transparency for reuse. These digital assets keep records of the provenance, quality, and composition of materials, ensuring that those extracted from disassembled buildings can be reused with sustainability credentials backed by assessment. Through the application of these digital solutions, a more systematic and scalable approach to design for disassembly will find increasing support [52]. Material passports initiate a paradigm shift from waste management to resource management in the built environment. Material passports enable better planning for deconstruction and reuse by documenting material specifications and histories, enhancing transparency, and facilitating material recovery at the end of a building's lifecycle. Material banks and secondary markets for reclaimed building components would further economically justify these short-lived investment buildings [53].

Technological advancements are critical in promoting circular economy practices, particularly in the construction sector. These innovations improve material tracking, improve recycling processes, and increase the reusability of materials, ultimately reducing environmental footprints and waste generation.

One pressing issue is the scalability of these innovative technologies, and their implementation across various scales of projects can be challenging due to costs, complexity, or lack of expertise. There may be a gap in the regulatory frameworks that support these innovations; without appropriate policies that encourage the adoption of such technologies, progress may stagnate.

Alongside the material advances, emerging technologies in modular construction and robotic deconstruction are accelerating the construction industry's capacity to adopt circular economy (CE) principles on a larger scale [7, 36]. These innovations improve both the efficiency and precision of material recovery and reuse, making circular strategies increasingly viable in practice. While the discussion primarily focuses on the technological aspects, the social dimension deserves more emphasis. Enhancing public understanding and acceptance can greatly influence the success of such initiatives.

2.2.5.2 Digital Tools: BIM, IoT, and Lifecycle Tracking

Building Information Modelling (BIM) and digital twin technologies are defining the future-the influence of real-time material tracking on monitoring the lifecycle of building components and the optimization of disassembly processes [54]. BIM models carry detailed metadata on materials' origin, recyclability, and assembly techniques, so that at the end of a building's life, components can be reliably recovered and reused. BIM is a baseline for digital collection to embed circular principles in sign-to-build work. Through BIM architecture, engineering professionals can model buildings in 3D while integrating material data, performance metrics, and lifecycle analyses into the design process. Circular design principles-such as modularity, disassembly, and reuse-become computable within the BIM environment and can provide simulation and optimization opportunities to minimize waste and improve resource use [55].

BIM is also instrumental in supporting Life Cycle Assessment (LCA) and it allows users to model the environmental impact due to the time-compatible use of building materials. This includes emissions from production, maintenance, and demolition, allowing an informed selection of relevant materials and construction methods based on sustainability criteria [56]. BIM improves design efficiency, minimizes waste and optimizes the use of materials by enabling collaborative planning and visualization, leading to more sustainable building practices.

Life Cycle Assessment (LCA) represents a thorough method for assessing the environmental effects of buildings across their full lifespan, incorporating phases from construction and operation to demolition and disposal. One of its key indicators is the quantification of global warming potential (GWP), expressed in metric tons of CO_2-equivalent (tCO_2eq), which enables the assessment of greenhouse gas emissions associated with each phase of the building's lifecycle. By offering a detailed environmental profile, LCA supports informed decision-making regarding material selection and design strategies, thereby promoting sustainability throughout the building process. Integration of BIM significantly enhances the effectiveness of LCA by enabling greater transparency and data reporting across all stages of the lifecycle. This capability gives BIM a distinct advantage over conventional approaches. When coupled with Building Energy Modelling (BEM), BIM facilitates advanced energy performance simulations that are not only more time efficient, but also capable of producing highly detailed and complex system models.

Blockchain technology is emerging as an efficient means for traceability and transparency in material supply chains through multi-enterprise material track and trace-involving end-to-end records of sourcing, processing, and installation of a material or component enabling tamperproof and secure recording of where, when and how a material is or was sourced, processed, and installed [47].

Through the application of these digital solutions, a more systematic and scalable approach to design for disassembly will find increasing support. Third-party verification systems, blockchain-based material tracking, and enhanced protocols for data-sharing will help foster some transparency and credibility in reporting sustainability metrics [57]. The major obstacle is the capital investment upfront in designing the buildings to be disassembled and the requirement of specialized knowledge and training within the construction industry. With advances in technologies, buildable information modelling or model-based management systems, and digital twins, as well as systems for tracking materials based on blockchain, will be important in recovering and reusing building components for the benefits of Design for Disassembly.

A digital twin refers to a dynamic virtual model of a physical building that continuously receives real-time data streams. This technology enables the simulation of the behaviour of a building based on its material composition, environmental conditions captured through Internet of Things (IoT) devices, and other operational variables over time. As such, digital twins provide a solid foundation for intelligent asset management and long-term strategic planning, particularly in the context of retrofitting and maintenance. They facilitate predictive interventions aimed at minimizing waste and avoiding premature system failures.

To accelerate the adoption of this technology within the construction industry, it is essential to standardize modular building components and integrate digital twin systems into lifecycle analysis processes. Digital twins not only mirror the physical structure in real time, but also align closely with its performance metrics, offering a synchronized perspective on building functionality. From a circular economy point of view, digital twin technology is particularly compelling, as it consolidates critical data for resource efficiency and connects circularity goals with broader objectives in architecture, engineering, construction and operations (AECO) [58].

The implementation of digital twins enhances operational performance and resource management by allowing for simulation and analysis of building's behaviour in real-time, supporting continuous improvement. In lifecycle management, digital twins enable the implementation of circular economy concepts by optimizing the redesign, recycling, and end-of-life processes of products and buildings in these phases. This lifecycle-wide approach also promotes cost efficiency due to the fact that digital twins minimize waste and optimize resource use, resulting in tremendous savings during both construction and operational periods [59]. Moreover, the advanced monitoring and forecasting capabilities of electronic twins further allow better control over assets and operational processes and help predict failure and ensure timely intervention to avoid risk [60]. Digital twins also enable organizations to better understand and assess customers to improve their products through a better understanding of consumer behaviour and preferences [61]. Such innovations may also open new avenues where organizations today can innovate and explore new products or services [62]. With digital twins in asset management, there is a way to see the real-time condition of assets so that changes can be easily captured or adapted to ensure the overall efficiencies of operations.

Digital twinning is, in fact, much more about the Internet of Things (IoT): the networks connecting all sensors, actuators, and displays in a building to become 'smart', that is to automate certain functions such as opening doors and regulating its ventilation systems. Beyond this local automation, IoT also records data that establish the history and overall performance of the building, thus ensuring global goals and constraints affect local operations. The IoT is not just an enabler, it is the necessity because the digital twin assumes an IoT-laden building with bidirectional communication and synchronization, including feedback to its users and operators [62].

Integrating digital tools along the lifecycle flow could embed CE principles into the built environment (see Fig. 2.7). It may not be sufficient for an isolated tool adopting the principles of digital. A lifecycle perspective immediately reveals that digital technologies are never merely facilitators but transform the scene toward a very different paradigm of circular and sustainable construction [55].

Furthermore, the IoT could bring tariff regulation to a new level by providing real-time information on resource usage and waste management to regulators, allowing them to establish more accurate and fair pricing reflecting actual costs and environmental impacts [63]. The IoT also monitors the quality of manufactured goods and the shipping of recycling materials. In the regulatory context, IoT interventions in recycling will also be necessary, including IoT standards for devices, securing data privacy and security, and creating protocols for data sharing between stakeholders [55].

Meanwhile, it would be advisable to invest in infrastructure to boost IoT in recycling. In addition, stimulating collaborations between academia and industry, possibly with the aid of innovation competitions, could lead to new IoT developments in recycling. Strengthening public–private partnerships would synergize in resource mobilization and skill sharing to support innovations aimed at IoT adoption in the recycling area. The overarching implementation of these IoT-related policies would not only directly lead to jobs being created in the recycling industry, but also indirectly rejuvenate the market for recycled materials across European countries, thus showcasing how this technology would transform the recycling industry [63]. Integration of Building Information Modelling (BIM), the Internet of Things (IoT) and effective lifecycle tracking is essential to improve efficiency and sustainability in construction. These digital tools facilitate better planning, monitoring and management of building projects, enabling the collection and analysis of real-time data.

Consequently, they support informed decision making, optimize resource management, and promote the principles of circular economy by ensuring that materials are used effectively throughout their lifecycle. For example, the integration of these digital tools can be complex and costly, presenting barriers for smaller firms or projects with limited budgets. There is a risk of overreliance on technology, which could overshadow traditional practices that contribute to sustainable outcomes. Data security and privacy concerns are paramount, particularly with respect to the extensive data collection involved with IoT devices. Ensuring robust cybersecurity measures and user consent is essential to maintain trust among stakeholders.

Robotics and prefabrication methods support CE through improved quality control, disassembly readiness, and modularity, essential for future reuse and adaptability [13]. Automated manufacturing methods and robotics improve the production of modular components, contributing to quicker construction times and reduced material waste. Specifically, robotics can aid in the fabrication of modules, assist in transporting them to the job site, and streamline the assembly process on-site.

Robotics has also actively grown in the construction industry for waste sorting, such as construction and demolition waste (C&D waste) [64]. Automated deconstruction is rapidly emerging as a viable alternative to conventional demolition. Robotic

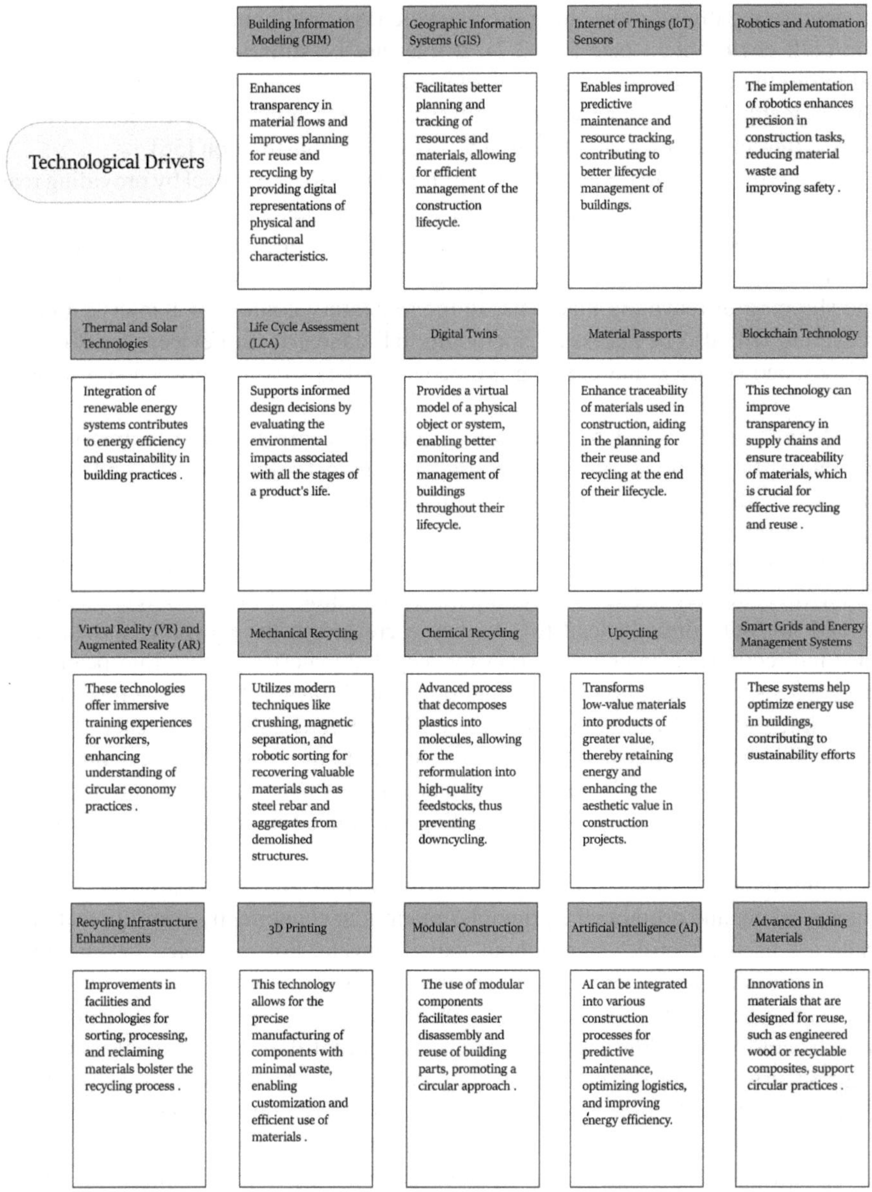

Fig. 2.7 Technological drivers

systems, such as drones and mechanical arms, can be programmed to perform precise, targeted operations on structures, allowing efficient recovery of valuable materials such as steel beams, timber panels, and façade components.

Despite the presence of automated waste sorting systems that segregate through some physical properties, manual sorting is inevitable at the end of the line [65]. However, the high labour cost associated with manual sorting, as well as its low efficiency, becomes deterrent to recycling and reuse. The promises that artificial intelligence (AI) holds as a possible solution in waste identification arise from its successes in other application scenarios [66]. AI can discover patterns and predict events, while deep learning (DL) techniques can use visual and chemical properties of waste objects to identify them [67].

2.2.5.3 Automation and Robotics

Therefore, robotic systems could be deployed to sort materials according to their appropriate categories after identifying waste through robots. C&D wastes are usually heavier than other wastes, therefore, they require robotic systems that have higher operational payloads and are more capable of performing the separation task efficiently. This not only maintains material flows within the deconstruction stage but also assists toward reducing landfilling and enabling secondary raw material markets in the construction sector when compared to traditional demolition techniques that generate a mixed waste stream [56].

While implementation of robotics and automation in the recycling and construction sectors enhance efficiency by improving material sorting and recovery, reducing labour costs, and minimizing environmental impact, there are pertinent challenges that need to be addressed: one major concern is the high initial investment required for deploying robotic systems, which may be prohibitive for smaller companies or those with limited financial resources. Additionally, there remains a reliance on manual sorting at the end of the recycling process, indicating that existing robotic solutions may not yet be fully capable or integrated into the overall workflow. Another concern related to the implementation of robotics and automation is the potential job displacement caused by it; while efficiency gains are desirable, they may result in reduced employment opportunities in manual labour roles, raising ethical concerns about the impact on the workforce.

Additive Manufacturing, namely 3D Printing, is revolutionizing the construction by enabling the production of custom-made building components on-demand in an environmentally friendly manner. For example, walls, furniture, or insulation panels can be printed with recycled plastics or construction waste powders. Manufacturing technology has advanced to the point where digital 3D printing models can produce complex building components directly on-site, eliminating the need for additional labour costs. Additive manufacturing is increasingly seen as a key enabler of the circular economy, not only because it allows for customized production, but also because it supports effective lifecycle management of the components it creates [55]. 3D printing helps in minimizing material wastage by ideally putting only what is necessary for making the particular desired component and leaving off-cuts and surplus with traditional subtractive methods. Furthermore, this increases the complexity of the resulting geometries for energy performance and material efficiency, thus minimizing the environmental impact during the entire lifecycle of a building [68].

Additive manufacturing is the futuristic technology that makes automation of manufacturing the most complex architectural components possible without any increased labour costs [69]. The actual mechanical process of 3D printing buildings or components is carried out based on a digital model whereby layers of material are laid down, one on top of the other, until the entire object comes into being [70].

Prefabrication refers to the off-site production of building components or building units that are later assembled on-site. This approach becomes particularly efficient when paired with modular design, enabling rapid construction and straightforward disassembly. In modular construction, components or entire building modules are manufactured in controlled factory environments and then transported to the construction site for assembly. This method not only improves efficiency and reduces material waste but also enhances the potential for disassembly and reuse. All these together, Design for Disassembly, Modularity, Material Selection, and Standardization, set the stage for circular economy strategies in modular construction.

Today's modular architecture is the product of prefabricated components and modules that can be combined and built on site, thereby accelerating the pace of building and improving flexibility and resource efficiency [71]. A major principle of modular design is that standardized parts can be used in different configurations to achieve various design requirements. This flexibility is fundamental to modular success, meaning that modular structures may be appropriate for any type of building, from homes to large commercial and industrial developments.

Modular parts disassembled and reassembled allow extensions or alterations in the future to maximize the life of buildings [72]. Artificial intelligence integrates well with modular construction operations. AI programs can filter through all data transmitted by IoT devices and even manage pattern matching and decision-making using predictive models [73].

Additionally, modular construction is highly efficient not only for above-mentioned initial assembling but also for further alteration, expansion, and relocation of spaces. All of this ultimately expands their life span and helps in retaining materials within the built environment. Because modular architecture uses prefabricated, interchangeable parts, homes can adapt to the changing needs of their occupants. Realignable layouts, mobile walls, and plug-and-play utility systems allow homeowners to modify their areas without having to move. Furthermore, modularity aligns housing design with long-term environmental goals by facilitating component reuse and reducing material waste [74].

To support different strategy implementations for Design for Disassembly, evaluation frameworks for modular construction development are increasingly being populated by researchers and industry practitioners [54]. Therefore, the development of holistic evaluation instruments should drive improvement, which ultimately ensures greater trust in the market. In addition to that, regulatory obstacles and acceptance in the marketplace are other important hindrances, as persistent building codes and regulations may not fully support the consequential requirements of circular modular construction. Therefore, it is of paramount importance to foster the necessary regulatory reforms, policy support, and collaborative action from the industry to ease widespread adoption [75]. Although the ecological and long-term economic benefits of modular construction through material reuse and cost optimization are evident, high initial costs and lack of standardization still pose a major challenge.

The future of circular modular construction strongly depends on the advancement of emerging digital tools, innovative materials, and collaborative frameworks that facilitate Design for Disassembly and reuse strategies. Furthermore, with the seamless integration of these in-built environment-inducing changes towards sustainable and resource-efficient living, these aspects would ultimately help buildings shift from linear to circular waste minimization systems optimized materials and a fully considered lifecycle for every component [76].

However, the arrival of smart materials-shaped memory alloys and polymers-introduced self-disassembling components into the realm of modular construction. Specifically, these materials can change their properties under predetermined conditions (such as temperature changes) to help automatically detach a building component without much manual effort [22]. Reversible BIM modelling embeds information on the recycling, reusability, and structural dependencies of modular components to accommodate disassembly. While the core focus of BIM is the construction and maintenance of buildings, Reversible BIM gives paramount importance to deconstruction, interchangeable components, and circularity of materials. With this integration, buildings respond positively to change, spaces become adapted for alternative use, and selected materials can be recovered, strengthening the argument for adaptive modular architecture in a circular urban context. Smart materials are critical in modern construction, as they offer self-healing properties, adaptive response to environmental changes, and improved performance with superior strength-weight ratios.

2.2.5.4 Emerging Technologies for Sustainability

Artificial intelligence (AI) facilitates the combination of data-driven design and material selection by weighing cost, environmental impact, structural performance, and availability. Machine-learning algorithms allow the construction team to test design scenarios and discover combinations of materials and construction methods that minimize waste and emissions and maximize reuse. AI can also be used to predict when components may fail based on their performance and reduce failures through early intervention or through retrofits. This leaves structures functionally sound for longer and prevents premature demolition or material wastage [77]. Advances in artificial intelligence and automation are pivotal for the future of intelligent modular systems. Advanced AI design tools enable architects and engineers to design optimized modular configurations that maximize space use. When it comes to construction, automated manufacturing methods and robotics will be useful to improve accuracy, quantify material waste, and speed up modular piece construction [78].

The lifecycle performance of modular buildings includes four main phases: the production phase, the construction phase, the use phase (also known as occupancy or operation) and the end-of-life phase. In the construction phase of modular buildings, specific tasks involve module fabrication, transportation of modules to the job site, and assembly on site. These activities differ from those of traditional buildings. Throughout all phases of a building's lifecycle, materials and energy are utilized for various operations such as raw material extraction, processing, component manufacturing, and the energy needs for heating, cooling, and lighting. Although there are similar tasks between conventional and modular methods, there are also unique opportunities in modular construction to reduce energy and material consumption [79].

Modular construction minimizes waste of material and energy. However, intelligent technologies also contribute much more to this objective. Automated energy control, the use of renewable energy sources, and intelligent materials improve the environmental sustainability of buildings [80]. Smart grids have enabled buildings to communicate flexibly with energy networks where AI and IoT can optimize when and how energy is consumed, stored, or shared. These technologies integrate solar panels, wind turbines, and energy storage for more resilient operating models and lower environmental impact. In circular buildings, energy systems are also consistently designed with changes in a preferred future and modularity in mind [81].

Innovations in passive systems-phase-change materials, dynamic facades, bioclimatic design or natural ventilation modelling-make it dependent on the external environment to regulate the temperatures of buildings and reduce the demand for energy-intensive HVAC systems. AI algorithms can make building planning more effective, reduce energy consumption and operating costs. Meanwhile, smart materials, such as self-healing concrete or adaptive building exteriors, could help extend the lifespan of modular structures, improving their ecological and environmental integration [82].

Nanotechnology is creating materials with enhanced strength-to-weight ratios, self-healing properties, and greater recyclability. For example, nanoparticles in concrete improve durability and lifespan, reducing repair and replacement cycles. This approach decreases resource extraction and supports a more circular material lifecycle [83]. Popular nanoparticles include titanium dioxide, carbon nanotubes, silica, copper, clay, and aluminium oxide used in concrete, asphalt concrete, bricks, timber, and steel [84]. Advanced composites such as fibre-reinforced polymers (FRPs) and graphene-enhanced materials allow long spans and reduce bulk, minimizing embodied carbon, and extending the functional life of the building [85].

Virtual reality and AR technologies are used in design reviews and stakeholder engagement to visualize circularity ideas. Simulating spatial layouts, material flows, and disassembly scenarios enables collaboration and improves design for reuse [79]. VR offers a three-dimensional interactive platform to explore construction scenarios, helping sustainable urban development [86]. Disassembly processes and circular strategies can also be taught via AR/VR training modules, accelerating skill transfer [87].

BIM goes beyond pre-set data and models; real-time data from laser scanning and point-cloud imagery demand robust AI support to interpret and manage. The IoT connects digital models and their physical counterparts, bridging the gap between design and execution. In this context, extended reality (XR), which comprises VR, AR, and MR, offers immersive, interactive visualization for architecture, engineering, and construction (AEC) stakeholders [88].

The integration of emerging technologies into existing frameworks can be complex and may require a steep learning curve for industry professionals. Information about these technologies needs to be effectively disseminated to ensure that stakeholders understand their benefits and applicability. In summary, the convergence of these technological drivers highlights a transformative shift in construction practices, allowing increased resource efficiency, sustainability, and resilience against environmental challenges.

2.2.6 Social and Cultural Drivers

2.2.6.1 Public Awareness and Demand for Sustainability

Public concern about climate change and urban sustainability is transforming the real estate market. Consumers now demand healthier, more adaptable, and more sustainable buildings. This change is reflected in the increased interest in local materials, heritage conservation, and participatory design. The growing awareness of environmental issues among the public is pushing developers and architects to embed CE principles in projects. Building users increasingly demand healthier, adaptable and environmentally responsible spaces [20]. Consumer awareness of sustainability and societal expectations of green practices are pushing companies to adopt CE principles. In particular, in urban contexts, the civic participation in circular food, construction and reuse systems is gaining traction [43].

The most recent consumer behaviour studies are showing a meagre shift, where sustainability will soon stretch only as far as defining a baseline expectation rather than being an actual differentiator. Companies may no longer need to promote environmental responsibility as a unique selling point; sustainability will be assumed as a minimum standard. This shift implies that consumers will purchase products and choose buildings based on deeper principles of performance, adaptation, and long-term sustainability. Increased public awareness leads to proposals for more sustainable buildings and infrastructure. All interested parties from architects, developers, and policymakers are moving to consider eco-friendly designs with sustainable materials and energy-efficient systems. Public participation thus emerges as a powerful driver in sustainable construction [89].

One significant challenge is the gap between awareness and action. Although many people express concern about sustainability, translating this awareness into tangible behaviours or lifestyle changes remains problematic. Factors such as convenience, cost, and accessibility can hinder people from making more sustainable choices.

People willingly change and tend to maintain these changes if they are part of a collective initiative with peers, such as neighbours. They are broader local initiatives that encourage alternative social practices, or adaptations of set or mainstream routines, through citizen participation in local communities and groups of interests. Such interventions are called in theory "social innovation". While they complement top-down applications of CE in cities, these bottom-up efforts often remain small-scale and face significant regulatory, political, and infrastructural barriers [90].

Digital CE has largely focused on technological solutions at the material, product, and system levels. However, little attention has been paid to changes in user behaviour and social practices. Resource efficiency is fundamentally shaped by how consumers produce and consume, not just by innovations in materials or infrastructure. Realizing the full benefits of CE in cities requires a transformation in everyday consumption habits along with changes in technology and infrastructure [91].

2.2.6.2 Stakeholder Collaboration and Education

Educational programs and awareness campaigns are also building cultural acceptance of circularity. Professional training for architects, engineers, and contractors is integrating CE into mainstream practice. A lack of skills remains a barrier, but training programs, professional training, and community awareness initiatives are emerging as key enablers [17, 39]. Organizations with strong internal education and strategic vision tend to be more successful in implementing CE [7]. Incorporating CE concepts into professional training and university curricula empowers industry actors to apply these principles in practice [13]. Awareness and skills are especially critical in the early stages of the design.

Training can increase the efficiency of individuals in applying disaster management and sustainability tools, thus increasing resilience. Circular economy education could help mitigate the social, economic and environmental challenges generated by climate change. Instruction in environmental science and social studies equips students with a broader understanding of sustainability issues [92]. Green technologies, when taught in curricula, enable an Earth-friendly design of processes and products by maximizing resource use, reducing waste, and limiting the demand for new resources. Information drivers support CE implementation by facilitating access to knowledge, training, and learning through structured exchange mechanisms [93].

Multistakeholder engagement, governments, businesses, and academia, is important to advance CE models in the built environment. For CE to work, stakeholders must align interests across sectors: for example, a building component manufacturer may use recyclable waste from unrelated industries (like road asphalt from car tires). This kind of collaboration thrives where mutual trust exists, usually developed over time. Long-term partnerships are essential for coordinated and circular outcomes. Stakeholders, such as architects, engineers, policymakers, and community groups, each play a vital role in making CE operational.

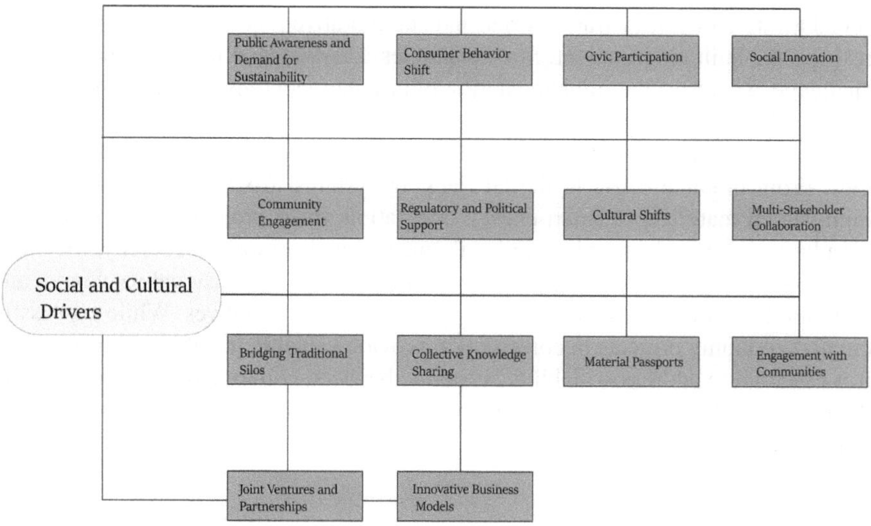

Fig. 2.8 Social and cultural drivers

Public education increases community participation and promotes behaviour changes that support sustainable practices. As social and green infrastructure becomes more prevalent in urban settings, raising public awareness becomes critical for widespread adoption. Public knowledge is also a key determinant in the acceptance of circular strategies and in building resilience to socio-environmental disruptions [94]. These civic efforts contribute to a societal shift where circularity becomes a norm, not an exception.

While the emphasis on stakeholder collaboration and education are commendable, certain gaps remain. Structured frameworks and guidelines are needed to facilitate cross-sector collaboration effectively. Differing priorities and communication barriers between stakeholders can result in inefficiencies or conflict. Furthermore, while there are many educational initiatives, they may lag behind emerging technologies or practices, making some training content outdated quickly. To remain effective, CE education must evolve alongside innovation (see Fig. 2.8).

2.3 Case Studies

2.3.1 Successful Circular Economy Projects in Construction

The principles of circular economy in construction have ceased to be mere conceptual ideals and have turned into practical reality through some innovative projects around the globe. Following are the case studies that represent real-world applications of circular strategy which can show how key drivers—technology-inducing innovations, social awareness, policy frameworks, and collaborative practices—may converge to reshape the built environment. Such examples add evidence in both beneficial and challenge aspects of the implementation of circular economy models and also give valuable lessons to the future.

These case studies represent a varied geography, project types, and circular strategies. From modular reuse to digital tracking systems, from upcycling waste into high-valued materials to smart energy integration, these projects demonstrate how circularity is shaped according to local contexts and priorities. Thereby, they also show the role of stakeholders—municipalities, design professionals, schools, and community members—in ensuring the success of such initiatives. While successful circular economy projects in construction provide valuable insights and inspiration, it is essential to address scalability, systemic barriers, collaborative challenges, and the need for a broader transition framework to fully unlock the potential of circular practices in the industry.

2.3.2 Lessons Learned from Implementation

The realization of implementing the principles of circular economy (CE) in construction worldwide has gained valuable practical insights into a great number of bundled countries. Admittedly, each project is unique in its context, but several common lessons can be learned from these initiatives and presented as a guide for future projects targeting circularity in the built environment.

The finding that appears constant across all case studies is that circular economy strategies need to be integrated into the earliest planning and design stages. Projects that are instead forced to adopt circular solutions after design completion often face logistical and funding constraints. The Doha Stadium 974 stands as an example of circular thinking embedded from the very beginning, with its design entirely relying on modular shipping containers for future disassembly. Cost savings and easy logistics were achieved, ensuring material reuse at the last leg of the building's life.

Public policy and procurement frameworks greatly impact the realization of projects that fall under circular construction. Municipal and national governments that establish clear rules, incentives, and targets for reducing waste, reusing materials, or assessing the life of materials create fertile ground for circularity initiatives. The Circular Materials Bank of the City of Leuven is an important example of local government that promotes circularity through centralized infrastructure and procurement guidelines that aid in the flow of reclaimed materials.

Cooperative collaboration is required for the effective implementation of circularity at all levels, including architect, engineer, contractor, material supplier, policy driver and client. An example is the Tikkurila School Warehouse project in Finland, which reused bricks from the historic Finnish National Theatre. Such a reuse partnership could be created through cultural institutions, demolition firms, and municipal agencies, as showcased by the above example and how partnerships are able to convert an ambitious reuse strategy into actual innovative practices in compliance with regulations, thus generating synergistic opportunities for knowledge sharing.

The role of technology is sometimes underrepresented, while collaborative efforts and public policy are highlighted, the impact of technological advancements should also be considered as a key enabler in overcoming implementation barriers. Incorporating technological solutions could enhance efficiency and contribute to more effective circular transitions.

Technologies such as Building Information Modelling (BIM), Material Passport Systems, and Digital Twins have advanced tremendously to increase transparency in decision making. Digital tracking in the Canada Water Modular Campus project allowed precise recording of reused parts, which further helped project teams manage materials across project stages and, in turn, increased stakeholder confidence in reused materials.

The key factor in the viability of any project is the public perception and understanding of circularity by stakeholders. In many cases, the opposition can be traced to aesthetic, safety, or durability concerns about reused materials. However, projects such as the EcoARK Pavilion in Taipei that combine innovation with public engagement and solid messaging around sustainable practices and transparent visible environmental benefits have shown that strong community support and pride in eco-sustainable practices can be moulded among the public.

Governments can also offer low-interest loans, financial aid, and tax credits to support circular design, business models based on material loops, and strategies involving disassembly and reuse.

In addition, introducing penalties for waste generation or providing financial rewards for circular construction projects can drive innovation in recycling technologies and boost adoption [18].

While circular strategies have the potential to create huge savings in the long run, they often require initial investments, particularly in areas where the reuse market is not so existing or where logistical systems are undeveloped. However, projects such as Golden Lane Redevelopment in London indicated that conserving existing structures and materials not only reduces carbon emissions but also saves costs in construction and external transport. It is increasingly considered important to include lifecycle cost analysis (LCCA) in justification for circular selections (Table 2.1).

2.4 Strategies to Accelerate Adoption

To accelerate the adoption of drivers for a circular economy in the construction sector, several strategic approaches across policy, regulation, global alignment, and partnerships can be implemented [13].

Table 2.1 Case studies

Project name and location	Year	Key CE strategies	Primary drivers	Source	Key enabler/ Highlight
Golden Lane Redevelopment (London, UK)	2024	Retained 96% of existing structure; salvaged steel enabled vertical expansion without new foundations	Policy incentives; material reuse; carbon reduction goals	FOOTPRINT	Emphasis on adaptive reuse of existing infrastructure, showcasing a significant reduction in material consumption
Leuven's Circular Materials Bank (Belgium)	2024	Provides reclaimed building materials to reduce primary raw material consumption; supports local circular economy	Municipal initiative; resource efficiency	BUILD UP	Establishment of a materials bank promoting local circularity and reducing dependence on new raw materials
Taipei EcoARK (Taiwan)	2023	Constructed using over 1.5 million recycled PET bottles; showcases upcycling of waste materials into durable building components	Technological innovation; environmental awareness	Time	Integration of innovative material recycling technology to transform waste into high-performance building components
Modular Campus at Canada Water (London, UK)	2023	Utilized 60–100% reused modular units; designed for disassembly and future reuse	Economic efficiency; design for adaptability	RIBAJ	Focus on future-proof design through modularity and disassembly, reducing long-term environmental impact
Catherine Commons Deconstruction (Cornell University, USA)	2023	Achieved 90% material reuse through careful deconstruction; emphasized environmental benefits	Educational institution initiative; sustainability commitment	BCESG	Institutional commitment to sustainability through educational-driven deconstruction practices

(continued)

Table 2.1 (continued)

Project name and location	Year	Key CE strategies	Primary drivers	Source	Key enabler/ Highlight
Tikkurila School Warehouse (Vantaa, Finland)	2021	Reused bricks from Finnish National Theatre; part of CIRCuIT project promoting material reuse	Municipal leadership; community engagement	Share Your Green Design	Collaboration with local initiatives for material reuse, strengthening the local circular economy model
Stadium 974 (Doha, Qatar)	2021	Constructed using modular steel containers; designed for complete disassembly and reuse	Innovative design; event-driven sustainability	SE2050	Innovative modular design allows for easy disassembly and reuse, minimizing long-term environmental impact
Green Valley Road Bridge (California, USA)	2014	Major renovation with structural component reuse; focused on beams made of steel	Infrastructure sustainability; material conservation	SE2050	Focus on sustainable infrastructure renovation with reused steel beams, highlighting material conservation efforts
Chartwell School (Seaside, California, USA)	2006	Incorporated structural component reuse and design for disassembly; used wood and steel elements	Educational focus; early adoption of CE principles	SE2050	Early adoption of circular design principles, focusing on the potential for future reuse of building components

2.4.1 *Policy Recommendations*

2.4.1.1 Developing Incentives for Circular Practices

Financial incentives play a critical role in encouraging circular economy (CE) practices in the construction sector [10, 13]. One effective approach is the provision of tax breaks for companies that incorporate recycled or reused materials into their projects. These tax incentives help offset the upfront costs often associated with circular processes, thereby improving their financial viability [10, 33].

Another important financial mechanism includes subsidies for retrofitting existing structures. These subsidies promote the extension of building lifespans and reduce the need for new raw materials [8, 10, 13].

2.4.1.2 Strengthening Regulatory Frameworks

To embed circular principles effectively, mandating Life Cycle Assessments (LCA) for new projects is essential. LCAs help quantify the environmental impacts of materials and designs, enabling more informed, sustainable decisions [6, 10]. In parallel, standardizing assessment methods and encouraging their application through clear policies ensures consistency and comparability across projects [10].

Regulations should also set minimum thresholds for material reuse and recycled content, which would create market demand for secondary materials and incentivize suppliers to ensure their quality [8–10]. Incorporating circular economy into building codes is another highly important step. This could include requirements such as submitting a mandatory plan for building disassembly or establishing targets for using salvaged components in new construction.

Broader regulatory reforms are needed to facilitate implementation. Revising antitrust laws to allow for greater collaboration across firms and sectors and enforcing design for disassembly policies will support innovation and enable long-term circular outcomes [9, 13].

2.4.1.3 Aligning with Global Sustainability Goals

Integrating circular economy principles into construction practices supports the achievement of the United Nations Sustainable Development Goals (SDGs), particularly those related to environmental protection, responsible consumption, economic growth and innovation [5, 24, 27]. The circular economy provides a practical framework to operationalize many of these goals by focusing on material efficiency, carbon reduction, and social equity [5].

At a regional level, compliance with the EU Circular Economy Action Plan offers a comprehensive structure for implementing circularity in construction across member states. This framework guides national and local policies toward measurable circular outcomes in material reuse, lifecycle design, and waste reduction [5, 14].

Aligning with these global and regional strategies also ensures a shared vision across countries and sectors, facilitating cross-border collaboration, funding access, and policy coherence [39]. This alignment is essential to scale circular practices from isolated projects to a systemic transformation.

2.4.1.4 Encouraging Public–Private Partnerships

Leadership and collaboration within the construction sector are key enablers of the transformation of the circular economy (CE). Integrating CE into corporate strategies yields not only environmental, but also economic benefits. Companies are embedding CE principles into sustainability reporting and ESG frameworks, while organizational commitment—especially among sustainability-oriented family businesses—has proven pivotal in enabling long-term planning, intergenerational responsibility, and decentralized decision making [7].

Champions of sustainability within firms help foster a culture of innovation and accelerate CE adoption beyond compliance. For example, internal "green teams" and sustainability leaders empower organizations to build circular business models that drive competitive advantage [24].

System-wide circularity demands multi-stakeholder collaboration. Inter-firm partnerships enable material exchanges, joint innovation, and shared infrastructure such as reuse hubs and logistics platforms [36, 40]. Cross-sector collaboration, among developers, demolition contractors, architects, and municipalities, helps deliver circular projects more effectively [20]. Construction companies are also forming coalitions to standardize practices, co-develop digital tools, and influence policy. Public–Private Partnerships (PPPs) play a central role in this ecosystem by allowing experimentation, knowledge sharing, and the demonstration of circular business models. Funding pilot projects for circular construction validates innovative practices and builds stakeholder confidence [9, 10, 30].

Institutional support is essential for scaling circular practices. This includes innovation accelerators, knowledge hubs, and government procurement strategies. For instance, public procurement and demand-side policies have emerged as powerful levers to stimulate reclaimed material markets—especially for structural timber reuse [95]. Regulatory support through streamlined permitting for sustainable designs can reduce bureaucratic inertia and promote circular approaches in the built environment [30]. Establishing certification programs specific to CE practices in construction also helps validate and reward circular business models [10].

At the enterprise level, several drivers are enabling practical CE integration. SMEs in the Netherlands have highlighted intrinsic motivation, leadership commitment, and employee-driven innovation as core enablers [95]. Digital tools such as Building Information Modelling (BIM) and Life Cycle Assessment (LCA) systems are increasingly used to track and optimize circular performance. The infrastructure for reverse logistics, along with marketplaces for reclaimed materials (for example, timber), has helped operationalize CE in practice.

Case Study: Skanska's Circular Construction Model

Skanska, a global construction and development company headquartered in Sweden, has emerged as a prominent actor in the integration of circular economy (CE) principles within the built environment. Although the firm does not present a formalized "Circular Construction Model" under that specific title, its operations and pilot projects reflect a comprehensive approach to circularity that can serve as an effective case study for industry adoption.

One of the central strategies employed by Skanska is the reuse of structural materials, particularly steel, brick, and concrete elements, salvaged from deconstructed buildings. This practice minimizes the extraction of virgin resources and reduces embodied carbon, aligning with broader sustainability and circularity goals. The company also integrates digital tools, notably Building Information Modelling (BIM) and material passports, to trace the lifecycle and condition of building components. These technologies not only facilitate the design for disassembly but also enable systematic reuse across multiple projects.

A significant feature of Skanska's circular approach is its collaborative supply chain involvement. The company develops partnerships with suppliers and waste management entities to co-develop reverse logistics systems and ensure quality standards for secondary materials. These practices contribute to the creation of a robust circular ecosystem capable of supporting multiple construction cycles.

In the UK, Skanska has operationalized its CE strategies through its broader environmental roadmap, the "Journey to Deep Green™". This strategic framework sets targets for zero waste, climate neutrality, and minimal ecological disturbance. The projects under this umbrella aim to demonstrate circular practices through real-world applications, incorporating modular construction techniques, design adaptability, and low-carbon material use.

In addition, Skanska's collaboration with Madaster, a platform for material databases and building material registries, exemplifies its commitment to data-driven circularity. By cataloguing material properties and their potential for reuse, this partnership enhances transparency and supports lifecycle decision-making.

Although Skanska's efforts are distributed across multiple initiatives rather than consolidated into a single model, their integrated circular practice provides a tangible representation of how construction companies can operationalize CE principles. Their experience demonstrates how corporate leadership, digital innovation, and cross-sector collaboration can converge to scale circularity in the built environment.

2.4.2 *Industry Collaboration and Innovation*

2.4.2.1 Fostering Cross-Section Partnerships

The urgency to decarbonize built environments has accelerated the need to adopt circular economy (CE) principles across construction sectors. While policy frameworks and public awareness remain vital, the sector's intrinsic capacity to innovate through collaborative models and new business practices plays a decisive role in enabling this transformation.

The private sector is pivotal in achieving sustainability targets such as the Sustainable Development Goals (SDGs) and the 2030 Agenda. Although traditional government-led interventions are common in the built environment, these alone are insufficient. Successful implementation of circular strategies requires active participation from private actors in tandem with government and civil society. This multi-actor engagement ensures that planning and decision-making processes are participatory and inclusive, allowing circularity to be driven through collaborative arrangements, innovative materials, and smart technologies [96].

Early-stage integration of circularity is critical to overcoming material inefficiencies. When architects, builders, material suppliers, and waste management professionals collaborate from the design phase, they ensure that buildings are not only efficient during use but also disassemble and recyclable at end-of-life. Involving deconstruction experts at the design stage introduces valuable knowledge about how materials behave and what infrastructures are required for their recovery [97].

To address complex sustainability challenges, especially in housing, resilient multi-stakeholder networks offer adaptable and localized solutions. These networks, involving both formal and informal partnerships, foster interdisciplinary collaboration among architects, engineers, policy-makers, community organizations, and private enterprises [98]. Informal industry relationships also play a vital role by facilitating tacit knowledge exchange and fostering a culture of learning that supports CE adoption.

Despite these collaborative potentials, significant barriers persist. Risk aversion, rigid regulations, and the absence of financial or structural incentives often hinder innovation and slow the uptake of circular approaches. Identifying and removing these systemic obstacles is critical for fostering a more enabling environment for CE in the construction sector.

2.4.2.2 Scaling up Circular Business Models

The transition to a circular economy (CE) in the construction sector requires innovative business models and tailored financing mechanisms to overcome systemic barriers and align economic incentives with sustainability goals. The shift of business models toward circularity lies at the core of long-term transformation, enabling a departure from the traditional linear "take-make-waste" paradigm.

One such model that is gaining traction is Product-as-a-Service (PaaS), where manufacturers lease rather than sell building components, such as flooring systems, lighting, or façade panels. This approach preserves the ownership of the manufacturer throughout the product lifecycle, incentivizing the design for durability, disassembly, and recyclability. A notable example is the Dutch company Desso, a pioneer in leased carpets that are entirely recyclable at the end of use.

To complement such business models, the nurturing of markets for secondary materials is essential. Platforms like the Excess Materials Exchange (EME) are developing digital infrastructures to match surplus building components with new construction needs, transforming what would otherwise be waste into high-value inputs [99]. These platforms are central to supporting material loops and lowering barriers to reuse in practice. However, despite their promise, digital platforms face critical infrastructure gaps in many regions. Without adequate digital capabilities, regulatory backing, and logistical frameworks, the potential of these tools to enable circularity may remain underexplored.

In addition to technological readiness, many stakeholders in construction remain anchored to linear models. Shifting to circularity requires fundamental operational and mindset changes. Therefore, it is important to understand and address these behavioural and institutional barriers through education, engagement, and incentives (Table 2.2).

When paired effectively, circular business models and financing mechanisms are mutually reinforcing: innovative business strategies create value from waste, while targeted financial tools help de-risk circular adoption. Projects such as Interface's carpet recycling program and the Buildings as Materials Bank (BAMB) initiative demonstrate both feasibility and value creation.

Finally, while sustainable development is broadly understood as a global societal goal, the circular economy focuses more precisely on changing production and consumption patterns [100]. As such, business and financial innovations in the built environment are not just technical instruments; they are levers for aligning industry practices with a regenerative economic future (Table 2.3).

Key Enablers for Scaling Circular Economy in the Built Environment

The successful implementation and scaling of circular economy (CE) strategies in the construction sector depend not only on innovative business models and financing mechanisms, but also on a set of systemic enablers that create the necessary conditions for transformation. These enablers provide the institutional, technological, and cultural infrastructure required to embed circular practices across project lifecycles and industry value chains:

Policy Alignment: Harmonization of circular economy metrics across national and regional jurisdictions is essential to ensure consistency and comparability. Tools such as material passports, recycled content quotas, and disassembly plans

Table 2.2 Examples of circular business models

Category	Mechanism	Example	Impact
Product-as-a-service (PaaS)	Companies retain ownership and lease products, charging by performance	Philips' 'Lighting as a Service' (the model reduces waste by reclaiming and refurbishing fixtures)	Encourages durability, reduces material use
Material marketplaces and resource banks	Digital platforms trade reclaimed materials and track inventories	Rotterdam's 'Circularity Center' (acts as a hub for storing and redistributing construction waste)	Reduces virgin resource dependency
Performance-based contracts	Payments tied to circular outcomes (e.g., energy savings)	Skanska's 'Green Lease' (agreements with tenants reward reduced energy/water use)	Aligns stakeholder interests, reduces risk
Modular and prefabricated construction	Off-site production allows disassembly and reuse	Project Etopia (UK) (uses modular units designed for multiple lifecycles)	Reduces waste, extends building life
Circular supply chains	Integrates recycled and bio-based materials into production	CIRCULARITI (India) (produces low-carbon cement using industrial byproducts)	Lowers emissions, meets green standards

must be integrated into building regulations and codes. Aligning these regulatory instruments supports transparent reporting, facilitates the trade of secondary materials, and reinforces circular procurement practices.

Collaborative Ecosystems: Interdisciplinary collaboration among architects, developers, material suppliers, recyclers, and financiers is fundamental to the systemic implementation of CE principles. Cross-sectoral partnerships, such as those promoted by initiatives such as Arup's circular building toolkit, demonstrate the potential of integrated approaches to resource efficiency, design innovation, and waste minimization. Such ecosystems also help overcome siloed thinking and foster shared responsibility across the value chain.

Digital tools and infrastructure: Digital technologies are vital for operationalizing circularity. Blockchain systems improve material traceability and accountability, especially in tracking the origin, composition, and future usability of construction components. Simultaneously, Building Information Modelling (BIM) facilitates lifecycle planning by embedding material performance, maintenance, and disassembly information into project workflows. These tools improve decision making and enable circular design from inception to end of life.

Table 2.3 Examples of financing mechanisms

Category	Mechanism	Example	Impact
Green bonds and sustainability-linked loans	Finance tied to sustainability KPIs	Vonovia's €1.5B green bond funds energy-efficient renovations in Europe	Attracts ESG investors, funds CE projects
Blended finance	Combines public and private capital to de-risk projects	EU's Circular Bioeconomy Fund co-invests in circular construction startups	Bridges funding gaps for SMEs and pilots innovative solutions
Pay-for-success models	Returns based on verified CE outcomes	Social Impact Bonds for circular social housing projects in Amsterdam	Links finance to measurable impact
Extended producer responsibility (EPR)	Manufacturers fund recycling via product fees	France's REP framework for construction waste shifts costs to producers	Incentivizes eco-design and closed-loop material flows
Revolving funds for material banks	Public/industry funds to stockpile reclaimed materials	Rotterdam's Urban Mining Fund invests in recovering high-value demolition materials	Stabilizes secondary material markets, reduces landfill use

Education and capacity building: Professional training and curricular integration of CE concepts are critical to equipping stakeholders with the knowledge and skills required to implement circular practices. Educational efforts should focus on circular design principles, procurement strategies, and innovation of CE-adapted business models. Building such capacity within the workforce ensures that circularity is not only technically feasible, but also widely understood and adopted as a new industry norm.

2.4.2.3 Leveraging Technology for Efficiency

By digitizing circular construction planning, implementation, and monitoring, new technologies are reshaping traditional processes in the built environment. Among these, BIM stands out as a methodology that produces digital representations of building components, documenting materials, their properties, and reuse possibilities. It integrates seamlessly with materials passport systems, creating a robust framework for lifecycle planning, waste reduction, and asset management [101].

In the execution phase, the advent of 3D printing has begun to transform traditional construction methods. It reduces labour dependency, material consumption, and waste generation [102]. Once buildings are operational, real-time technologies such as artificial intelligence (AI) and the Internet of Things (IoT) allow dynamic resource monitoring. Embedded sensors can track energy consumption, wear, and maintenance needs, enabling facility managers to anticipate interventions that extend asset life and reduce waste.

While leveraging these technologies promises greater efficiency, realizing their full benefits depends on overcoming several challenges. Addressing high upfront costs, ensuring proper training, and mitigating data security concerns are essential to unlocking their transformative potential.

2.4.2.4 Showcasing Best Practices

For wider inspiration and adoption, circular construction must be demonstrated through practical, tangible examples that prove its feasibility and showcase its advantages (see Fig. 2.9). Recognition platforms such as the Asia Pacific Leadership in Green Building Awards, operated by the World Green Building Council, play a critical role in celebrating these innovations and ensuring they receive visibility [103].

In Taiwan, companies such as Miniwiz and LOTOS C-Slurry are pioneering new concepts for transforming waste into valuable construction materials. Miniwiz, for example, holds more than 1,000 unique patents to convert various waste materials—including rice husks, aluminium, and insect shells—to building products. Their landmark project, the Taipei EcoARK pavilion, constructed using 1.5 million recycled PET bottles, demonstrates the structural potential of recycled waste [104]. On a larger, systemic scale, the Hammarby Sjöstad district in Stockholm exemplifies integrated circular urban planning. Developed as an eco-district, it features closed-loop systems for energy, water, and waste. The heat from wastewater is reused for district heating, food waste is converted to biogas, and stormwater is managed using green infrastructure [105].

To support replication and learning, organizations such as Circle Economy, the Ellen MacArthur Foundation, and the World Green Building Council publish case

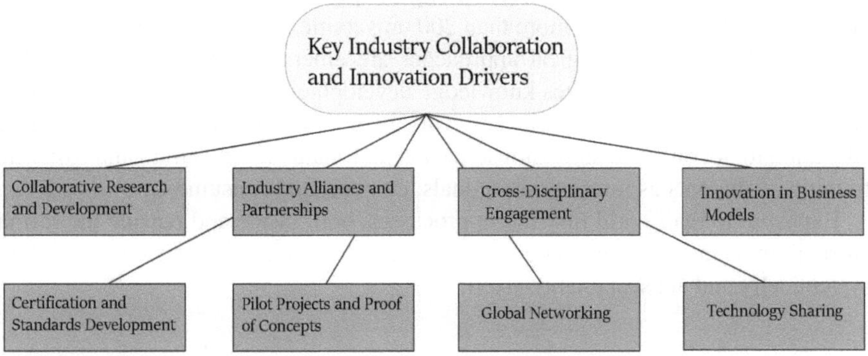

Fig. 2.9 Key industry collaboration drivers

studies, toolkits, and practical frameworks. Resources like the Circular Built Environment Playbook offer actionable guidance including material databases and procurement strategies to help project teams embed circular principles from conception to completion [51].

2.4.3 *Education and Skill Development*

2.4.3.1 Integrating Circular Principles into Curricula

To mainstream circular economy (CE) principles in the built environment, education must play a foundational role. Integrating circular design, material reuse, and lifecycle assessment modules into architectural and engineering curricula is essential to prepare future professionals for sustainable practices. Specialized programs, such as the upcoming Circular Built Environment Specialist certification, are being developed to provide in-depth training on applying CE strategies in real-world construction scenarios.

Equally important is the inclusion of sustainability content in construction trade programs. By updating vocational training to allocate dedicated time for CE concepts, tradespeople can be equipped with the practical knowledge needed to implement circular principles on-site. This includes competencies in responsible material selection, waste minimization, and efficient resource use [106].

2.4.3.2 Raising Public and Professional Awareness

Traditional curricula in universities and educational institutions are increasingly evolving to incorporate concepts of circular economy across a wide range of disciplines (see Fig. 2.10). A notable example is the Circular Economy Club (CEC), which has supported the integration of circularity principles into business, engineering, and other academic programs in more than 200 universities worldwide [107].

In parallel, new pedagogical approaches are emerging. One such approach is co-learning, which emphasizes knowledge development through social interaction, collaboration, and shared experiences. This method aligns with the goals of higher education to not only transmit knowledge but also catalyse innovation and foster the training of students as future professionals, citizens, and consumers.

Exposure to real-world innovation processes, both inside and outside the formal curriculum, is increasingly recognized as a key driver in preparing students to think systemically and act sustainably [108].

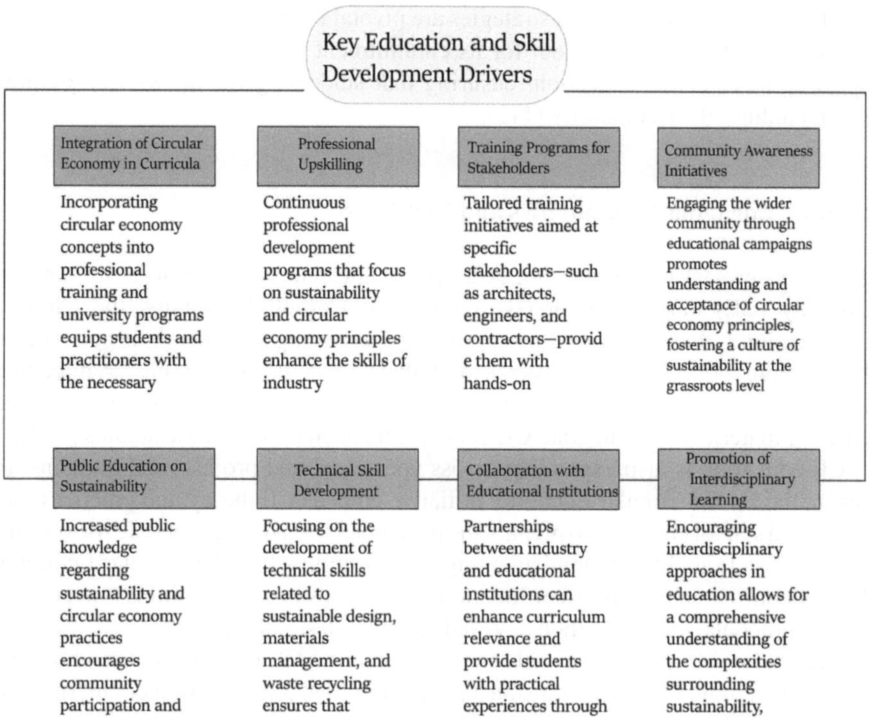

Fig. 2.10 Key education and skill development drivers

2.4.3.3 Building Capacity for Implementation

Hands-on, project-based learning is a cornerstone for translating circular economy (CE) theory into real-world application. At Cornell University, the Circular Construction Lab exemplifies this method by engaging students in design studios that explore building systems and their roles in sustainable architecture. Similarly, the ProSPER.Net Guidebook empowers educators to develop localized teaching materials on circularity in the built environment, encouraging students to apply CE principles through real-world case studies [109].

Beyond academia, vocational education and training programs are also integrating circular principles to equip tradespeople with the skills needed for sustainable construction. Berlin's Fit für Circular Economy initiative embeds CE content into vocational school curricula, preparing the next generation of skilled workers to contribute meaningfully to circular transitions [110]. Professional certifications further support this movement. Programs like the Circular Economy Professional (Foundational) offer comprehensive training on circular business models, design strategies, and practical tools applicable across industries [111].

At the policy level, national strategies are pivotal in embedding CE across education systems. Finland stands out for its commitment to internalizing circular principles at all levels of education, ensuring that students grow up with a systemic understanding of sustainability [112].

2.4.3.4 Engaging Communities and End-Users

Hands-on, project-based learning is a critical mechanism for translating circular economy (CE) theory into practical skills. At Cornell University, the Circular Construction Lab exemplifies this approach by engaging students in design studios that explore both building systems and their roles in sustainable architecture. Similarly, the ProSPER.Net Guidebook supports educators in developing localized teaching materials that integrate CE concepts through real-life case studies [109].

CE education is also expanding across vocational and professional domains. In Berlin, the Fit für Circular Economy initiative embeds CE theory and practices into vocational school curricula to prepare the next generation of tradespeople for sustainable construction [110]. Complementing this, programs like the Circular Economy Professional (Foundational) certification offer industry-wide training on circular business models, material efficiency, and strategic design thinking [111].

National policy plays a key role in mainstreaming CE principles throughout the education system. Finland, for instance, has committed to integrating circularity into all levels of education, ensuring that young learners internalize sustainability values and practices from an early age [112].

A noteworthy limitation is the potential disparity in access to quality education and training resources in different regions and demographics. While some may benefit from comprehensive programs, others, particularly in less developed areas, may find themselves at a distinct disadvantage. Additionally, the focus appears to be primarily on technical skills, possibly neglecting the importance of soft skills and interdisciplinary approaches that are essential for effective collaboration among diverse stakeholders. Lastly, there may be insufficient emphasis on lifelong learning, which is important in a rapidly evolving field where continuous adaptation to new technologies and practices is necessary.

2.4.4 Summary of Key Drivers

Environmental Concerns and Resource Scarcity

Growing environmental pressures, including climate change, biodiversity loss, and pollution, combined with the overexploitation of finite resources, underscore the urgency of rethinking resource consumption patterns. The risk of material scarcity is forcing both policymakers and industries to adopt more resource-efficient and regenerative approaches [7, 39]. This creates a strong impetus for circular models aimed at reducing ecological footprints and promoting long-term sustainability [20].

Economic Opportunities and Benefits

The circular economy offers compelling financial benefits. These include cost reductions through waste minimization and energy efficiency, the creation of value from recovered resources, business growth through innovation, and improved supply chain resilience [26, 51]. Recognizing these economic incentives encourages businesses to adopt circular practices, viewing sustainability not just as a responsibility, but also as a business opportunity [19].

Policy and Regulatory Support

Government intervention remains one of the strongest levers for circular transformation. Policies addressing waste management, green public procurement, and reuse incentives provide essential guidance and motivation for stakeholders [14, 18, 39, 113]. A supportive legal framework that institutionalizes circular practices is necessary to overcome systemic inertia [19, 38].

Technological Advancements and Innovation

Technological innovations, ranging from BIM to AI-driven material marketplaces, play a critical role in enabling circularity [9, 18, 114]. These technologies facilitate more accurate material tracking, enable urban mining and support the design of disassembly, which is essential for lifecycle resource optimization [26].

Increased Awareness, Education, and Collaboration

Rising awareness of sustainability and circular principles is fuelling demand for education and stakeholder collaboration. Educational initiatives and communication campaigns improve understanding, while cross-sector partnerships strengthen the implementation of circular strategies [42, 50, 51]. Collaboration between clients, designers, contractors, and recyclers remains vital to effectively close material loops [4, 18].

Market Demand and Client Requirements

Clients are increasingly demanding sustainable construction solutions, driving suppliers and developers to prioritize circular practices [95, 106]. Supportive clients with clear sustainability goals can catalyse project-level innovations and embed circularity in procurement and design processes.

Social Responsibility and Stakeholder Pressure

Public and investor expectations are shaping corporate behaviour. Organizations are under growing pressure to demonstrate environmental and social responsibility, making circular practices not only a compliance issue, but also a reputational and strategic concern [19, 26, 113].

Together, these drivers allow the built environment to manage the shift to the green transition through several key mechanisms:

- Reduction of Virgin Material Use—Circular strategies reduce reliance on virgin materials through practices such as remanufacturing, refurbishment, reuse, and recycling [13, 20, 113, 114].
- Minimization of Waste Generation: Techniques like design for disassembly and extended product lifespans help reduce construction and demolition waste [51].
- Material Utility Maximization Circular models aim to retain materials at their highest value within closed-loop systems, thereby decoupling economic growth from material consumption [13, 26, 38].
- Systemic Change Enablement—Circular economy drivers influence not only technical practices but also governance structures, business models, and procurement frameworks [50].
- Creation of new business models—Opportunities emerge through innovation in material recovery, reverse logistics, and service-based models, facilitating economic and environmental value generation [18, 26, 27].

2.5 Future Outlook for Circular Economy in the Built Environment

CE principles are gaining traction not only as environmental imperatives, but also as drivers of long-term economic resilience. As the urgency of climate action intensifies, growing recognition of ecological limits, combined with the economic potential of resource efficiency, is creating strong momentum for systemic change [7, 51].

Digital technologies such as Building Information Modelling (BIM), Material Passports, and Resource Tracking Platforms are poised to play a central role in this transition. These tools enable more transparent, efficient material flows and foster the integration of circular strategies across value chains [4, 101]. By improving the accuracy and traceability of data, digitalization supports the lifecycle thinking, from design to deconstruction.

Policy frameworks are also expected to evolve toward greater coherence and ambition. Future regulations will likely provide clearer mandates and financial incentives to guide construction stakeholders toward circular solutions [14, 18]. Coordinated regulatory action across jurisdictions will be essential to address current inconsistencies and encourage broader adoption.

Equally important is the role of collaboration. The construction sector remains highly fragmented, and knowledge silos hinder systemic innovation. Increased cooperation among architects, engineers, contractors, policymakers, industry and clients is critical to fostering a shared understanding of CE principles and embedding them into daily practice [4, 50].

Persistent barriers, such as high upfront costs, lack of standardization, and deeply embedded linear habits, continue to limit progress [7, 19, 39, 106]. However, increasing awareness, technological innovation, and evolving market expectations suggest that a progressive shift is underway. The sector is already moving beyond conventional waste management and recycling, toward holistic strategies that emphasize reuse, remanufacturing, and designing for adaptability and longevity [14, 51].

Although the full realization of a circular built environment may be generational in scope [50], it represents a necessary and transformative path toward sustainability. By aligning technology, policy, market services, and cultural change, the sector can transition to a resource-efficient system that contributes to global climate and development goals [51].

The role of digitalization and smart technologies is essential in facilitating circularity by improving resource tracking, enabling urban mining, and supporting the creation and management of comprehensive material passports. Such digital tools enhance the identification, recovery, and reuse of critical raw materials while providing essential feedback loops for designers and engineers to prioritize materials with greater circular potential. The deployment of multilevel monitoring frameworks, underpinned by digital data collection, supports evidence-based policymaking and ongoing refinement of circular economy practices, aligning resource efficiency efforts with broader sustainability and climate objectives.

The economic benefits derived from circular strategies extend beyond simple cost savings to include the development of resilient industries capable of adapting to resource scarcity and evolving regulatory landscapes. Realizing these benefits requires addressing barriers related to market maturity, quality assurance, and stakeholder acceptance through coordinated policy interventions, industry collaboration, and active social participation. Social and health dimensions are integral to this transition, as circular approaches contribute to the reduction of pollution and healthier communities, although careful management is required to mitigate the potential risks associated with certain recycling processes.

Financial instruments such as green bonds, subsidies, and tax reforms play a decisive role in realigning incentives and mobilizing investments toward circular initiatives. Cultivating a culture that embraces circularity, supported by public education and broad stakeholder participation, is essential to overcome resistance and stimulate demand for circular products and practices. The convergence of urban mining, material passports, and supportive policy frameworks creates a resilient foundation for resource-efficient built environments, addressing both national and international material needs while mitigating supply chain vulnerabilities.

Ultimately, sustained and coordinated efforts that integrate technological advancements, economic strategies, social considerations, and institutional support will drive the transformation of the built environment into resilient and resource-efficient urban systems. This transformation positions the construction sector as a central contributor to the creation of circular value and long-term regenerative development, promoting sustainable urbanization aligned with global environmental and economic goals.

2.6 Conclusion

The diversity and complexity of the drivers influence the transition from a linear to a circular economy in the built environment. These drivers vary from environmental imperatives and economic incentives to regulatory frameworks, technological innovation, and evolving social values. Although challenges in aligning these drivers across stakeholders and lifecycle stages persist, the collective impact influences the way how we design, build, use, and deconstruct our built environments.

Advancing circular economy strategies within the built environment demands an integrated approach that synergizes technological innovation, policy frameworks, economic incentives, and cultural shifts. The substantial potential to reduce embodied carbon emissions underscores the environmental and economic advantages of circularity, but significant obstacles persist, particularly in the establishment of robust markets for reclaimed and secondary materials. Overcoming these challenges requires the development of targeted regulatory measures, financial support mechanisms, and the implementation of standardized circularity indicators to enhance transparency and foster stakeholder trust. These elements are critical for enabling informed decision-making and investment, thus stimulating the growth of secondary material markets and encouraging continuous innovation.

Based on the study's findings, we recommend the development of standardized metrics for evaluating CE practices in the built environment. Future research should focus on longitudinal studies to assess the long-term impacts of CE adoption on environmental sustainability and economic resilience.

References

1. UNEP (2019) Global status report for buildings and construction sector. UNEP—UN Environment Programme. https://www.unep.org/resources/publication/2019-global-status-report-buildings-and-construction-sector
2. Ellen MacArthur Foundation (2013) Towards the circular economy. https://www.ellenmacarthurfoundation.org/towards-the-circular-economy-vol-2-opportunities-for-the-consumer-goods
3. Gillott C, Davison B, Tingley D (2022) Drivers, barriers and enablers: construction sector views on vertical extensions. Build Res Inf 50(8):909–923. https://doi.org/10.1080/09613218.2022.2087173
4. Adams KT, Osmani M, Thorpe T, Thornback J (2017) Circular economy in construction: current awareness, challenges and enablers. Proc Inst Civ Eng Waste Resource Manage 170(1):15–24. https://doi.org/10.1680/jwarm.16.00011
5. Försterling G, Orth R, Gellert B (2023) Transition to a circular economy in Europe through new business models: barriers, drivers, and policy making. Sustainability 15(10). https://doi.org/10.3390/su15108212
6. Marchi L, Luo Z, Gasparini N, Antonini E, Gaspari J (2024) Detecting and understanding barriers and drivers to advance systematic implementation of resource circularity in constructions. Buildings 14(10). https://doi.org/10.3390/buildings14103214

7. Ahrens EE (2023) Circular economy in the construction industry: barriers and drivers for the implementation of circularity in family-owned construction companies in Germany. https://campus-fryslan.studenttheses.ub.rug.nl/333/
8. Mhatre P, Gedam V, Unnikrishnan S, Raut R (2023) Circular economy adoption barriers in built environment- a case of emerging economy. J Clean Prod. https://doi.org/10.1016/j.jclepro.2023.136201
9. Hartwell R, Macmillan S, Overend M (2021) Circular economy of facades: real-world challenges and opportunities. Resour Conserv Recycl. https://doi.org/10.1016/j.resconrec.2021.105827
10. Rao P, Rahman M, Duraman S (2025) Adopting circular economy in construction: a review. Front Built Environ. https://doi.org/10.3389/fbuil.2025.1519219
11. Pomponi F, Moncaster A (2017) Circular economy for the built environment: a research framework. J Clean Prod 143:710–718. https://doi.org/10.1016/j.jclepro.2016.12.055
12. European Commission (2020) 2020 circular economy action plan: international aspects. Publications Office of the European Union. https://doi.org/10.2779/085517
13. Galkina D (2023) Circular economy in the construction industry: barriers and drivers
14. Giorgi S, Lavagna M, Wang K, Osmani M, Liu G, Campioli A (2022) Drivers and barriers towards circular economy in the building sector: stakeholder interviews and analysis of five European countries policies and practices. J Clean Prod. https://doi.org/10.1016/j.jclepro.2022.130395
15. Geldermans B, Tenpierik M, Luscuere P (2019) Circular and flexible infill concepts: integration of the residential user perspective. Sustainability 11(1):261. https://doi.org/10.3390/su11010261
16. Geissdoerfer M, Santa-Maria T, Kirchherr J, Pelzeter C (2023) Drivers and barriers for circular business model innovation. Bus Strateg Environ 32(6):3814–3832. https://doi.org/10.1002/bse.3339
17. Piila N, Sarja M, Onkila T, Mäkelä M (2022) Organisational drivers and challenges in circular economy implementation: an issue life cycle approach. Organ Environ 35(4):523–550. https://doi.org/10.1177/10860266221099658
18. Munaro MR, Tavares SF (2023) A review on barriers, drivers, and stakeholders towards the circular economy: the construction sector perspective. Clean Responsible Consumption. https://doi.org/10.1016/j.clrc.2023.100107
19. Haselsteiner E, Rizvanolli BV, Villoria Sáez P, Kontovourkis O (2021) Drivers and barriers leading to a successful paradigm shift toward regenerative neighborhoods. Sustainability (Switzerland) 13(9). https://doi.org/10.3390/su13095179
20. Hart J, Adams K, Giesekam J, Tingley DD, Pomponi F (2019) Barriers and drivers in a circular economy: the case of the built environment. In: Zhao F, Skerlos SJ, Sutherland JW (eds) Procedia CIRP, 80. Elsevier B.V, pp 619–624. https://doi.org/10.1016/j.procir.2018.12.015
21. Zu Castell-Rüdenhausen M, Wahlström M, Astrup TF, Jensen C, Oberender A, Johansson P, Waerner ER (2021) Policies as drivers for circular economy in the construction sector in the Nordics. Sustainability 13(16):9350. https://doi.org/10.3390/su13169350
22. Environment Act 2021. https://www.legislation.gov.uk/ukpga/2021/30/contents
23. Government of the Netherlands (2016) A circular economy in the Netherlands by 2050 [Report]. Ministerie van Algemene Zaken. https://circulareconomy.europa.eu/platform/sites/default/files/17037circulaireeconomie_en.pdf
24. Pasqualotto C, Callegaro-De-Menezes D, Schutte CSL (2023) An overview and categorization of the drivers and barriers to the adoption of the circular economy: a systematic literature review. Sustainability 15(13):10532. https://doi.org/10.3390/su151310532
25. Truant E, Giordino D, Borlatto E, Bhatia M (2024) Drivers and barriers of smart technologies for circular economy: leveraging smart circular economy implementation to nurture companies' performance. Technol Forecast Soc Chang. https://doi.org/10.1016/j.techfore.2023.122954

26. Tura N, Hanski J, Ahola T, Ståhle M, Piiparinen S, Valkokari P (2019) Unlocking circular business: a framework of barriers and drivers. J Clean Prod 212:90–98. https://doi.org/10.1016/j.jclepro.2018.11.202
27. Barros MV, Salvador R, Do Prado GF, De Francisco AC, Piekarski CM (2021) Circular economy as a driver to sustainable businesses. Clean Environ Syst 2:100006. https://doi.org/10.1016/j.cesys.2020.100006
28. Moghayedi A, Michell K, Hübner D, Le Jeune K, Massyn M (2024) Examine the impact of green methods and technologies on the environmental sustainability of supportive education buildings, perspectives of circular economy and net-zero carbon operation. Facilities 42(3–4):201–222. https://doi.org/10.1108/F-12-2022-0161
29. Peszko G (2023) Drivers for circular economy
30. Ghisellini P, Ulgiati S (2024) Evolution and current development of the circular economy concept with insights from worldwide cities and regions. In: Regions, cities and the circular economy. Edward Elgar Publishing Ltd, pp 14–43. https://doi.org/10.4337/9781035316410.00007
31. Charef R, Morel J-C, Rakhshan K (2021) Barriers to implementing the circular economy in the construction industry: a critical review. Sustainability 13(23):12989. https://doi.org/10.3390/su132312989
32. Rahla K, Mateus R, Bragança L (2021) Implementing circular economy strategies in buildings-from theory to practice. Appl Syst Innov 4(2). https://doi.org/10.3390/asi4020026
33. Adabre M, Chan A, Darko A, Edwards D, Yang Y, Issahaque S (2024) No stakeholder is an island in the drive to this transition: circular economy in the built environment. Sustainability 16(15). https://doi.org/10.3390/su16156422
34. Gustavsson L, Joelsson A, Sathre R (2010) Life cycle primary energy use and carbon emission of an eight-storey wood-framed apartment building. Energy Build 42(2):230–242. https://doi.org/10.1016/j.enbuild.2009.08.018
35. Hossain M, Ng S (2019) Influence of waste materials on buildings' life cycle environmental impacts: adopting resource recovery principle. Resour Conserv Recycl 142:10–23. https://doi.org/10.1016/j.resconrec.2018.11.010
36. Holzer D, Mair-Bauernfeind C, Kriechbaum M, Rauter R, Stern T (2023) Different but the same? Comparing drivers and barriers for circular economy innovation systems in wood- and plastic-based industries. Circ Econ Sustain 3(2):983–1011. https://doi.org/10.1007/s43615-022-00210-9
37. Torres-Guevara LE, Prieto-Sandoval V, Mejia-Villa A (2021) Success drivers for implementing circular economy: a case study from the building sector in Colombia. Sustainability 13(3):1350. https://doi.org/10.3390/su13031350
38. Abdulai SF, Nani G, Taiwo R, Antwi-Afari P, Zayed T, Sojobi AO (2024) Modelling the relationship between circular economy barriers and drivers for sustainable construction industry. Build Environ. https://doi.org/10.1016/j.buildenv.2024.111388
39. SEI (2021) Barriers and drivers for enterprises to transition to circular economy. Stockholm Environment Institute. https://doi.org/10.51414/sei2021.029
40. Van Eijk F (2015) Barriers and drivers towards a circular economy—literature review. Acceleratio (March):1–138
41. Jiang J, Chu C, Song L, Gao X, Huang B, Zhang Y, Zhang Y, Liu Y, Hou L, Ju M, Cao Z (2023) From prospecting to mining: a review of enabling technologies, LCAs, and LCCAs for improved construction and demolition waste management. Waste Manage 159:12–26. https://doi.org/10.1016/j.wasman.2023.01.017
42. Mollaei A, Byers B, Christovan C, Olumo A, De Wolf C, Bachmann C, Haas C (2023) A global perspective on building material recovery incorporating the impact of regional factors. J Clean Prod. https://doi.org/10.1016/j.jclepro.2023.139525
43. Williams J (2022) Challenges to implementing circular development—lessons from London. Int J Urban Sustain Dev 14(1):287–303. https://doi.org/10.1080/19463138.2022.2103822

44. Akbarnezhad A, Xiao J (2017) Estimation and minimization of embodied carbon of buildings: a review. Buildings 7(1), Article 1. https://doi.org/10.3390/buildings7010005
45. Geng Y, Doberstein B (2008) Developing the circular economy in China: challenges and opportunities for achieving "leapfrog development." Int J Sust Dev World 15(3):231–239. https://doi.org/10.3843/SusDev.15.3:6
46. De Wolf C (2022) 4 promising digital technologies for circular construction. World Economic Forum. https://www.weforum.org/stories/2022/09/4-promising-digital-technologies-for-circular-construction/
47. Sánchez-García E, Martínez-Falcó J, Marco-Lajara B, Manresa-Marhuenda E (2024) Revolutionizing the circular economy through new technologies: a new era of sustainable progress. Environ Technol Innov 33:103509. https://doi.org/10.1016/j.eti.2023.103509
48. Malladi RC, Ajayan SA, Chandran G, Selvaraj T (2025) Upcycling of construction and demolition waste: recovery and reuse of binder and fine aggregate in cement applications to achieve circular economy. Clean Eng Technol 24:100864. https://doi.org/10.1016/j.clet.2024.100864
49. Ma W, Hao JL, Zhang C, Di Sarno L, Mannis A (2023) Evaluating carbon emissions of China's waste management strategies for building refurbishment projects: contributing to a circular economy. Environ Sci Pollut Res 30(4):8657–8671. https://doi.org/10.1007/s11356-021-18188-6
50. ARUP (2021) Circular economy in the built environment. https://www.arup.com/insights/circular-economy-in-the-built-environment/
51. World Green Building Council (2023) The circular built environment playbook. https://worldgbc.org/article/circular-built-environment-playbook/
52. Ouda E, Haggag M (2024) Automation in modular construction manufacturing: a comparative analysis of assembly processes. Sustainability 16(21). https://doi.org/10.3390/su16219238
53. Karaca F, Tleuken A, Awan HH, Askar R, Çıdık MS, Iskakova A, Turkyilmaz A, Laudal T, Durdyev S, Varol HA, Salles A, Bajāre D, Tambovceva T, Zsembinszki G, Puma GCC, Kapsalyamova Z, Kripa D, Azhgaliyeva D, Nano X, Bragança L et al (2024) Stakeholder perspectives on the costs and benefits of circular construction. Sci Rep 14(1):30039. https://doi.org/10.1038/s41598-024-81741-z
54. National Institute of Building Sciences (2024) New paper: BIM and Digital Twins coexist to drive sustainability | National Institute of Building Sciences. https://www.nibs.org/blog/new-paper-bim-and-digital-twins-coexist-drive-sustainability
55. Banihashemi S, Golizadeh H, Rahimian FP (2022) Data-driven BIM for energy efficient building design. Routledge
56. Keles C, Cruz Rios F, Hoque S (2025) Digital technologies and circular economy in the construction sector: a review of lifecycle applications, integrations, potential, and limitations. Buildings 15(4):533. https://www.mdpi.com/2075-5309/15/4/553
57. Feng H, Zhao J, Zhang H, Zhu S, Li D, Thurairajah N (2022) Uncertainties in whole-building life cycle assessment: a systematic review. J Build Eng 50:104191. https://doi.org/10.1016/j.jobe.2022.104191
58. Liu Z, Meyendorf N, Blasch E, Tsukada K, Liao M, Mrad N (2025) The role of data fusion in predictive maintenance using digital twins. In Handbook of nondestructive evaluation 4.0. Springer, pp 1–23
59. Na S, Heo S, Han S, Shin Y, Lee M (2022) Development of an artificial intelligence model to recognise construction waste by applying image data augmentation and transfer learning. Buildings 12(2):175
60. Kritzinger W, Karner M, Traar G, Henjes J, Sihn W (2018) Digital Twin in manufacturing: a categorical literature review and classification. Ifac-Papers Online 51(11):1016–1022
61. Uhlemann TH-J, Lehmann C, Steinhilper R (2017) The digital twin: realizing the cyber-physical production system for industry 4.0. Procedia Cirp 61:335–340
62. Fuller A, Fan Z, Day C, Barlow C (2020) Digital twin: Enabling technologies, challenges and open research. IEEE Access 8:108952–108971

63. Islam MM, Alam MM, Sohag K (2025) Recycling through technology diffusion for circular economy in Europe: a decomposed assessment. Technol Forecast Soc Chang 214:124052
64. Xiao B, Chen C, Yin X (2022) Recent advancements of robotics in construction. Autom Constr 144:104591
65. Ozdemir ME, Ali Z, Subeshan B, Asmatulu E (2021) Applying machine learning approach in recycling. J Mater Cycles Waste Manage 23:855–871
66. Tamin O, Moung EG, Dargham JA, Yahya F, Omatu S, Angeline L (2022) A comparison of RGB and RGNIR color spaces for plastic waste detection using the YOLOv5 architecture. In: 2022 IEEE international conference on artificial intelligence in engineering and technology (IICAIET), pp 1–6
67. Moon S, Hou L, Han S (2023) Empirical study of an artificial neural network for a manufacturing production operation. Oper Manag Res 16(1):311–323
68. Lim S, Buswell RA, Le TT, Austin SA, Gibb AGF, Thorpe T (2012) Developments in construction-scale additive manufacturing processes. Autom Constr 21:262–268. https://doi.org/10.1016/j.autcon.2011.06.010
69. Canada Mortgage and Housing Corporation (2024) 3D printing and the construction industry. Canada Mortgage and Housing Corporation. https://www.cmhc-schl.gc.ca/blog/stories/105-canadas-first-3d-printed-multi-unit-homes-part-3-of-3
70. Wakefield J (2016) Tomorrow's buildings: construction industry goes robotic. BBC Technology
71. Gibb AGF (1999) Off-site fabrication: prefabrication, pre-assembly and modularisation. John Wiley and Sons
72. Pandremenos J, Paralikas J, Salonitis K, Chryssolouris G (2009) Modularity concepts for the automotive industry: a critical review. CIRP J Manuf Sci Technol 1(3):148–152. https://doi.org/10.1016/j.cirpj.2008.09.012
73. Signorini M, Frigeni S, Spagnolo SL (2019) Integrating environmental sustainability indicators in BIM-based product datasheets. IOP Conf Series: Earth Environ Sci 296(1):012028. https://doi.org/10.1088/1755-1315/296/1/012028
74. McKinsey and Company (2023) Circularity in the built environment. https://www.mckinsey.com/featured-insights/world-economic-forum/knowledge-collaborations/circularity-in-the-built-environment.
75. Klinge A, Roswag-Klinge E, Radeljic L, Lehmann M (2019) Strategies for circular, prefab buildings from waste wood. IOP Conf Ser.: Earth Environ Sci 225:012052
76. Dokter G, Thuvander L, Rahe U (2021) How circular is current design practice? Investigating perspectives across industrial design and architecture in the transition towards a circular economy. Sustain Prod Consumption 26:692–708. https://doi.org/10.1016/j.spc.2020.12.032
77. Jike N, Xu C, Yang R, Qi Y, Dai Y, Peng Y, Wang J, Zhang M, Zeng Q (2022) Pervious concrete with secondarily recycled low-quality brick-concrete demolition residue: engineering performances, multi-scale/phase structure and sustainability. J Clean Prod 341:130929. https://doi.org/10.1016/j.jclepro.2022.130929
78. Yang Y, Pan M, Pan W (2019) 'Co-evolution through interaction' of innovative building technologies: the case of modular integrated construction and robotics. Autom Constr 107:102932. https://doi.org/10.1016/j.autcon.2019.102932
79. O'Grady T, Minunno R, Chong HY, Morrison GM (2021) Interconnections: an analysis of disassemblable building connection systems towards a circular economy. https://doi.org/10.3390/buildings11030535
80. International Energy Agency (2022a) Energy efficiency in smart buildings and the role of modular construction. https://www.reportlinker.com/market-report/Electric-Power/111787/Smart-Grid?term=smart%20grid%20reportsandmatchtype=pandloc_interest=andloc_physical=1011845andutm_group=standardandutm_term=smart%20grid%20reportsandutm_campaign=ppcandutm_source=google_adsandutm_medium=paid_adsandutm_content=transactionnel-1andgad_source=1andgbraid=0AAAAAD19yGca8hYT9xSERMRyMrwzAiQu0andgclid=Cj0KCQjwiLLABhCEARIsAJYS6unt-yJg9SRI9jrEgv_Nb3EDbAbKJ-AbcC2HEO0EqV4-knKHBdbggF0aAk4xEALw_wcB

81. International Energy Agency (2022b) Smart grids and energy efficiency in the built environment. https://www.reportlinker.com/market-report/Electric-Power/111787/Smart-Grid?term=smart%20grid%20reportsandmatchtype=pandloc_interest=andloc_physical=1011845andutm_group=standardandutm_term=smart%20grid%20reportsandutm_campaign=ppcandutm_source=google_adsandutm_medium=paid_adsandutm_content=transactionnel-1andgad_source=1andgbraid=0AAAAAD19yGca8hYT9xSERMRyMrwzAiQu0andgclid=Cj0KCQjwiLLABhCEARIsAJYS6unt-yJg9SRI9jrEgv_Nb3EDbAbKJ-AbcC2HEO0EqV4-knKHBdbggF0aAk4xEALw_wcB
82. Parracho DFR, Nour El-Din M, Esmaeili I, Freitas SS, Rodrigues L, Poças Martins J, Corvacho H, Delgado JMPQ, Guimarães AS (2025) Modular construction in the digital age: a systematic review on smart and sustainable innovations. Buildings 15(5):765. https://doi.org/10.3390/buildings15050765
83. Ferreira MT, Soldado E, Borsoi G, Mendes MP, Flores-Colen I (2023) Nanomaterials applied in the construction sector: environmental, human health, and economic indicators. Appl Sci 13(23). https://doi.org/10.3390/app132312896
84. Mohajerani A, Burnett L, Smith JV, Kurmus H, Milas J, Arulrajah A, Horpibulsuk S, Abdul Kadir A (2019) Nanoparticles in construction materials and other applications, and implications of nanoparticle use. Materials 12(19). https://doi.org/10.3390/ma12193052
85. Hollaway LC (2013) 18-Advanced fiber-reinforced polymer (FRP) composite materials for sustainable energy technologies. In: Bai J (ed) Advanced fiber-reinforced polymer (FRP) composites for structural applications, 2nd ed. Woodhead Publishing, pp 637–673. https://doi.org/10.1016/B978-0-12-820346-0.00018-6.
86. Shahrokni H, Årman L, Lazarevic D, Nilsson A, Brandt N (2015) Implementing smart urban metabolism in the Stockholm royal seaport: smart city SRS. J Ind Ecol 19(5):917–929. https://doi.org/10.1111/jiec.12308
87. Wang P, Wu P, Wang J, Chi H-L, Wang X (2018) A critical review of the use of virtual reality in construction engineering education and training. Int J Environ Res Public Health 15(6), Article 6. https://doi.org/10.3390/ijerph15061204
88. Alizadehsalehi S, Yitmen I (2021) Digital twin-based progress monitoring management model through reality capture to extended reality technologies (DRX). Smart Sustain Built Environ 12(1):200–236. https://doi.org/10.1108/SASBE-01-2021-0016
89. Goh CS, Ting JN, Bajracharya A (2023) Exploring social sustainability in the built environment. Adv Environ Eng Res 4(1). https://doi.org/10.21926/aeer.2301010
90. Marchesi M, Tweed C (2021) Social innovation for a circular economy in social housing. Sustain Cities Soc 71:102925. https://doi.org/10.1016/j.scs.2021.102925
91. Dodman D, Diep L, Colenbrander S (2017) Resilience and resource efficiency in cities. https://www.unep.org/resources/report/resilience-and-resource-efficiency-cities
92. Raghupathi V, Raghupathi W (2020) The influence of education on health: an empirical assessment of OECD countries for the period 1995–2015. Arch Public Health 78(1):20. https://doi.org/10.1186/s13690-020-00402-5
93. Campbell-Johnston K, ten Cate J, Elfering-Petrovic M, Gupta J (2019) City level circular transitions: barriers and limits in Amsterdam, Utrecht and The Hague. J Clean Prod 235:1232–1239. https://doi.org/10.1016/j.jclepro.2019.06.106
94. Drosou N, Soetanto R, Hermawan F, Chmutina K, Bosher L, Hatmoko JUD (2019) Key factors influencing wider adoption of blue–green infrastructure in developing cities. Water 11(6), Article 6. https://doi.org/10.3390/w11061234
95. Springvloed P (2021) The drivers and barriers for a circular economy in the built environment of the Netherlands [Master thesis]. https://studenttheses.uu.nl/handle/20.500.12932/39916
96. Owojori O, Okoro C (2022) The private sector role as a key supporting stakeholder towards circular economy in the built environment: a scientometric and content analysis. Buildings 12(5). https://doi.org/10.3390/buildings12050695
97. Ding Z, Wang Y, Zou PXW (2016) An agent based environmental impact assessment of building demolition waste management: conventional versus green management. J Clean Prod 133:1136–1153. https://doi.org/10.1016/j.jclepro.2016.06.054

98. Pablo Z, London K (2022) Sustainability through resilient collaborative housing networks: a case study of an Australian pop-up shelter. Sustainability 14(3). https://doi.org/10.3390/su14031271
99. EME (2025) Excess materials exchange. https://excessmaterialsexchange.com
100. Sauvé S, Bernard S, Sloan P (2016) Environmental sciences, sustainable development and circular economy: alternative concepts for trans-disciplinary research. Environ Dev 17:48–56. https://doi.org/10.1016/j.envdev.2015.09.002
101. Behún M, Behúnová A (2023) Advanced innovation technology of BIM in a circular economy. Appl Sci 13(13), Article 13. https://doi.org/10.3390/app13137989
102. Siddika A, Mamun M, Al A, Ferdous W, Saha AK, Alyousef R (2020) 3D-printed concrete: applications, performance, and challenges. J Sustain Cement-Based Mater 9(3):127–164. https://doi.org/10.1080/21650373.2019.1705199
103. World Green Building Council (2024) Winners of the Asia Pacific leadership in green building awards. World Green Building Council. https://worldgbc.org/article/winners-of-the-asia-pacific-leadership-in-green-building-awards/
104. Time (2024) These Taiwanese companies are turning waste into buildings. https://time.com/7172075/waste-construction-taiwan/
105. City of Stockholm (2020) Hammarby Sjöstad—a model for sustainable urban development. https://start.stockholm/en/
106. Kanters J (2020) Circular building design: an analysis of barriers and drivers for a circular building sector. Buildings 10(4). https://doi.org/10.3390/BUILDINGS10040077
107. Giannoccaro I, Ceccarelli G, Fraccascia L (2021) Features of the higher education for the circular economy: the case of Italy. Sustainability 13(20). https://doi.org/10.3390/su132011338
108. Leppänen M, Kuula P (2019) Experiences from collaborative online education for degree and further education on circular economy. In: SEFI 47th annual conference: Varietas Delectat... complexity is the new normality, proceedings, pp 699–709. https://researchportal.tuni.fi/en/publications/experiences-from-collaborative-online-education-for-degree-and-fu
109. United Nations University UNU-IAS (2023) Guidebook for the circular economy. https://unu.edu/ias
110. Circular BERLIN (2024) Fit für circular economy: integrating circularity into schools' curricula. Circular Berlin. https://circular.berlin/portfolio-item/fit-fur-circular-economy-integrating-circular-economy-education-into-vocational-schools-curricula/.
111. Circular Economy Alliance (2025) Circular economy professional (foundational)—circular economy alliance. https://circulareconomyalliance.com/wp-content/uploads/2023/04/SFB-Report-2023.pdf, https://circulareconomyalliance.com/wp-content/uploads/2023/03/Circular-Economy-Alliance-Brochure.pdf, https://circulareconomyalliance.com/product/circular-economy-professional-foundational/
112. Abend L (2022) Inside Finland's plan to end all waste by 2050. Time. https://time.com/6132391/finland-end-waste/
113. Khan SA, Mubarik MS, Paul SK (2022) Analyzing cause and effect relationships among drivers and barriers to circular economy implementation in the context of an emerging economy. J Clean Prod. https://doi.org/10.1016/j.jclepro.2022.132618
114. Ababio BK, Lu W (2023) Barriers and enablers of circular economy in construction: a multi-system perspective towards the development of a practical framework. Constr Manag Econ 41(1):3–21. https://doi.org/10.1080/01446193.2022.2135750

Part II
Policies, Regulations and Standards for Circular Transitions

Chapter 3
Overview of Policy Needs for Circular Construction

Marco Frazão Pedroso and Pedro Frazão Pedroso

Abstract This chapter provides a detailed overview of the evolving policy and regulatory framework that supports the transition to a circular economy within the construction sector, with particular emphasis on both European and national levels. It begins by examining the foundational EU frameworks, such as the Circular Economy Action Plan and the European Green Deal, highlighting their impact on national legislation and sustainable construction practices. The analysis then extends to the national level, emphasising both overarching themes and country-specific strategies that are influenced by local contexts. It also addresses the relationship between European directives and national implementation, highlighting the importance of policy harmonisation, technological readiness, cultural adaptation, and financial capacity. The chapter concludes by outlining future directions for circular construction policy, including digitalisation, green public procurement, and regulatory support for emerging technologies, while advocating for multi-level governance and proactive adaptation to achieve long-term sustainability goals.

Keywords Circular economy · Sustainable construction · European policies · Waste management · Energy efficiency

3.1 Introduction

The nexus of sustainability and circular economy of the built environment is very complex as has been shown in the previous sections, but it is increasingly governed by a system of policies and regulations. With the variety of topics approached in the current CircularB COST Action Deliverable, this section builds upon the foundational understanding established in previous subchapters but at the same time gives its new contributions. Therefore, after the identification and analysis of existing and future standards, this section delves into the specifics of policy and regulatory frameworks at both the European and national levels, providing an overview of the

M. F. Pedroso (✉) · P. F. Pedroso
CERIS, Instituto Superior Técnico, Universidade de Lisboa, Lisboa, Portugal
e-mail: marco-pedroso@sapo.pt

L. Bragança et al. (eds.), *Shaping Circular Transitions in the Built Environment*, Springer Tracts in Civil Engineering, https://doi.org/10.1007/978-3-032-02834-1_3

following sections (2.5 and 2.6), which present a detailed approach to both European and National-level policies and regulations.

At the European Union (EU) level, the policy landscape is shaped by a series of legislative acts and guidelines. The EU's *Circular Economy Action Plan* [7] stands as proof of its commitment to sustainable development, related to a broader scope under the United Nations (UN) *Sustainable Development Goals* (SDG) [26], providing a strategic approach to circular economy transformation. This plan is directly connected to the *European Green Deal* [6], which outlines an ambitious roadmap to transform the EU's economy sustainably by turning climate and environmental challenges into opportunities across all policy areas.

In addition to these broad strategies, specific directives and regulations play a crucial role. For instance, the *Waste Framework Directive* [4] (2008/98/EC) and the *Energy Performance of Buildings Directive* [11] (2010/31/EU) illustrate the EU's approach to integrating sustainability into sector-specific legislation. Such directives not only set requirements and objectives but also drive innovation and development within member states [14].

At the national level, member states interpret and implement these EU directives within their local contexts. The diversity in implementation reflects the unique environmental, economic, and social landscapes of each country. For example, Germany's *Kreislaufwirtschaftsgesetz* [2] (Circular Economy Law) demonstrates how national legislation can align with EU directives while addressing specific national priorities [15]. Similarly, France's *Loi relative à la transition énergétique pour la croissance verte* [13] (Law on Energy Transition for Green Growth) showcases national initiatives driving sustainable development [24]. However, many European countries still lack documents that align with the EU's declared objectives.

As such, this section aims to provide an overview, of the exhaustive work done in the following sections (2.5 and 2.6), of these complex layers of governance. It first executes an analysis of the European policies and regulations and then it analyses how EU countries are implementing and adapting such high-level documents. Moreover, it seeks to highlight the dynamic interplay between EU-wide initiatives and individual member-state actions. This exploration is vital for understanding how policy and regulations at different levels are integrated to shape the trajectory towards a built environment more sustainable and circular.

Therefore, this section aims to bridge the gap between the broader narrative of sustainable development and the detailed, often technical, nature of individual policies and regulations. This holistic approach is crucial for understanding the full spectrum of governance mechanisms that underpin the transition to sustainability and circular economy within the EU.

3.2 European Policies and Regulations: A Detailed Overview

The work that will be presented in Sect. 2.5, shows that the transition from the 20th to the twenty-first century has provided a marked increase in focus on environmental issues, particularly in the European context. This shift has prompted significant efforts to reduce the environmental footprint of buildings, recognizing them as primary sites of human activity and significant contributors to raw materials, energy, and water utilization and GHG (Green House Gases) emissions and waste creation [25]. European policies have evolved to address these concerns, transitioning from a focus on energy consumption and emissions during the use phase of buildings to a more comprehensive lifecycle approach, as highlighted by Sala et al. [22].

The European Commission has been at the forefront of this transition. Key communications can be found from 2014 aimed at reducing waste generation and promoting more efficient resource usage in the building sector (COM/2014/0398 and COM/2014/0445). These policies have laid the groundwork for improving the circularity of buildings and addressing market barriers within the sustainable buildings industry.

Also, the adoption of the *2030 Agenda for Sustainable Development* by the United Nations member states further reinforced the importance of sustainable practices in the building sector. This agenda, encompassing 17 Sustainable Development Goals (SDGs), emphasizes the need for sustainable consumption and production patterns, as outlined in Communication COM/2015/0614. The EU has taken significant steps to ensure the retrieval of valuable resources and efficient waste management in construction and demolition, alongside developing a monitoring framework for the implementation and progress of a Circular Economy.

Another key aspect of the EU strategy has been the consideration of *Green Public Procurement* (GPP) to promote sustainability and circularity aspects in several sectors, including office buildings. Also, the 2016 handbook *Buying Green!* [5] by the Directorate-General for Environment, focused on environmentally friendly procurement practices, including the construction sector.

All these aspects have been of relevance for the EU and for the concerns that it has towards achieving the net-zero GHG goals by 2050. However, the most recent developments and Directives have been showing an increase in documents and regulations in the last few years. In 2018, the EU revised several directives to enhance waste management and promote the efficient and rational use of natural resources in adherence to Circular Economy principles (Directive EU 2018/849-2018/850-2018/051-2018/852). These directives have established significant targets for waste recovery and recycling, particularly in construction and demolition waste since it was aiming for a 70% recovery target by 2020 as per the *Waste Framework Directive* (2008/98/EC).

However, it was with the *European Green Dea*l (COM/2019/640 final) and the subsequent *Circular Economy Action Plan* (COM/2020/98) that all these concerns have further accelerated all efforts towards sustainability and circularity, aiming to make the EU climate-neutral by 2050. These initiatives have led to significant policy activity in the building sector, particularly through the *Renovation Wave* (COM/2020/662) and the *New European Bauhaus Initiative* (COM/2021/573), which focus on decarbonization, renewable energy integration, and circularity in building design and construction [12].

Also, in the *European Green Deal,* there is an effort to decouple economic growth from resource use, two concepts that are commonly intertwined, but the new paradigm shifts with the circular economy aim to create value in what would otherwise be waste and reintroduce it in the economy, minimizing the depletion of virgin resources and production of waste. This led to the development of the *Circular Economy Action Plan*, which became a cornerstone in the *Green Deal*, that applies a holistic life-cycle approach, considering products from design to the end-of-life, and emphasizes the sectors that are resource-intensive and, thus, might have a disproportionate impact when measures are applied, such as the auto industry, textile sector and the construction sector.

With this timeline, and with the knowledge that exists on the significant influence that the construction sector has on the environment (responsible for approximately: 50% of raw materials extraction and energy consumption, 40% of GHG emissions—also associated with energy consumption, 30% of freshwater consumption, and 30% of waste generation) the European Commission's study on *Circular Economy Principles for Building Design* resulted in a publication [8] outlining policy choices to support the implementation of these principles at European, national, and local levels. This document involved different stakeholders with the aim of, among others, extending the lifespan of structures, simplifying the application of recycled materials, and enhancing resource effectiveness throughout the building lifecycle.

The EU's commitment to climate neutrality by 2050, as part of the *European Green Deal*, needs a comprehensive approach across all sectors, including construction. This approach is supported by legislative frameworks such as COM/2021/802, which renovate the *Energy Performance of Buildings Directive* and the *Energy Efficiency Directive*, promoting policies to achieve a highly energy-efficient and, again, trying to achieve a decarbonized building stock by 2050.

To achieve these goals, the EU has been considering the heterogeneity that exists within its territory, countries, and regions, while also acknowledging that environmental issues such as GHG emissions, and resource depletion, among others, are not national but international, and may need several or even all European states to work together to find solutions. Thus, a relevant effort in cooperation between the different countries is needed, to achieve a level of policy harmonization that guarantees the policies designed can be effectively implemented at a European level, being one example of this the adoption of the framework Level(s) as a promising analysis baseline. Adding to that, European policies also need to respect global agreements such as the Paris Agreement that sets emissions targets to contain the rise of temperature to, at most, 2.0 °C.

International cooperation and policy harmonization play a pivotal role in the EU's efforts to foster a green transition and establish a circular economy. While challenges exist, the EU's approach serves as a model for regional cooperation in environmental governance. Continued commitment to collaborative and harmonized strategies is crucial for addressing global environmental challenges. In the circularity in the construction sector front, CEN is currently developing the Technical Committee CEN/TC 350/SC 1—Circular Economy in the Construction Sector, to enable the transition to a circular economy in the construction sector, supporting a climate-neutral and resource-efficient sector.

The *EU taxonomy* is another cornerstone in the European overall sustainability plan. It's designed to help guide investments aligned with the green transition in a transparent way. Its main target is to create a classification system with criteria for economic activities that respect the net-zero targets in 2050. It defines conditions to rate a company or investment according to its environmental merits, helping investors make environmentally friendly decisions [9].

The future directions and challenges associated with European-level policies and regulations specific to the construction sector and the built environment are related to the transition to a circular economy, more specifically in terms of occupational transformation and the adoption of innovative business models. Therefore, this sector must attract a diverse workforce, including women and younger age cohorts with digital skills, to support sustainable practices and circular economy models, while also promoting increased productivity.

This is one of the objectives of the new *Construction Product Regulation* (CPR), proposed in COM/2022/144, which aims to lay the foundation for implementing the circular economy in the construction industry, addressing challenges related to market surveillance, legal complexity, and delivering on green and digital transitions. Additionally, the revision of the GPP manuals to include all public buildings and the recent revision of the *Corporate Sustainability Report Directive* (CSRD) aligned with the *European Taxonomy* and with the Do No Significant Harm (DNSH) principle, show a challenging future for the construction sector.

In conclusion, European policies and regulations have undergone significant evolution to address the challenges of sustainability and circular economy in the construction sector. These developments emphasize the importance of a holistic approach to resource efficiency, waste management, and sustainable design, aligning with global climate goals and fostering a cleaner, more competitive Europe. The ongoing commitment to these principles, as evidenced by the latest directives and action plans, underscores the EU's role as a leader in sustainable development and circular economy practices, while much work still lies ahead.

3.3 National Policies and Regulations: A Detailed Overview

In the evolving landscape of sustainability and circular economy, the built environment plays a crucial role. This section aims to identify and analyse, based on the following document sections, various national-level policies and regulations that govern and influence practices in this sector. Understanding these policies is essential for comprehending the global efforts towards a more sustainable future. By examining different national strategies, it's possible to gain insights into how countries Europe-wide are adapting and innovating to embed circular economy principles into the built environment, showcasing a diverse range of approaches and solutions tailored to specific national contexts and needs.

This overview serves as a starting point to contextualize the detailed work executed in the following sections (2.6). That chapter followed a methodology that allowed to obtain a factsheet for each country, identifying the existing policies, documents, legislation, and regulations for each one of the 16 studied countries. This way, it is possible to understand the multifaceted and dynamic nature of sustainability and circularity governance at the national level, where the European panorama needs harmonization across regions, impacting the development of new policy, and how important is the development of sector-specific plans, tackling unique issues across regions and economic sectors.

3.3.1 Key Themes in National Policies

National policies and regulations play a significant role in shaping the trajectory of sustainability and circular economy in the built environment. These documents are influenced (when not directly transcribed to each language) by the EU-level documents since the objectives are almost always aligned. The analysis that is presented in more detail in the following sections of the current document, however, shows several common key themes:

Governance Structures: Many countries have established specific government bodies and agencies dedicated to overseeing and implementing sustainability and/or circular economy initiatives. These structures often facilitate coordination among various sectors and stakeholders, ensuring a cohesive approach towards environmental performance.

Legislative Frameworks: Laws and regulations form the backbone of national strategies. They include mandates for sustainable practices, building codes, waste management regulations, and incentives for green construction. These frameworks not only enforce compliance but also encourage innovation in sustainable building practices.

Platforms and Networks: To foster a collaborative environment, countries have set up platforms and networks that facilitate knowledge sharing, research, and dissemination of best practices in sustainable development. These platforms often involve public–private partnerships, connecting government entities with businesses, academia, and civil society.

Funding Opportunities: Financial support mechanisms are critical for driving the transition to a circular economy. This includes grants, subsidies, tax incentives, and funding for research and development projects. These financial tools are designed to lower the barriers to adopting sustainable and circular practices in the built environment.

All these themes collectively form the foundation of national strategies, guiding countries towards a more sustainable and circular approach in the built environment, and in the construction sector more specifically. Then, each country adapts its approach based on its unique context and circumstances, reflecting a diversity of strategies and priorities.

3.3.2 Diversity and Specificity of National Approaches

The global approach of integrating sustainability and circularity into the built environment is enriched by the different approaches adopted by various countries, each influenced by unique cultural, economic, and environmental contexts. The examples that are presented in further detail in the next sections show this diversity of scopes on the policy and regulations level.

Some countries focus on urban regeneration and green building practices, leveraging their dense urban landscapes to pilot innovative sustainable building technologies such as Spain and Finland. These initiatives often include actions such as retrofitting existing structures for energy efficiency, promoting green roofs, or encouraging the use of sustainable materials.

In contrast, countries with abundant natural resources might prioritize preserving biodiversity and natural habitats within their urban planning, such as France. This approach often involves implementing strict environmental impact assessments, preserving green spaces, and promoting sustainable land use.

Another area of focus is waste management and recycling, crucial in countries with significant industrial activity, such as the Netherlands. These countries develop robust systems for material recovery, recycling, and waste reduction, setting ambitious targets for waste diversion from landfills.

So, as it can be inferred, each country's approach is tailored to its specific challenges and opportunities, reflecting a deep understanding of the local context and needs. This results in a broad example of strategies that contribute to the global goal of a more sustainable, circular, and resilient built environment. Also, innovation and technological advancements are key themes in national approaches. Some countries, such as Finland, particularly those with strong technological sectors, invest heavily in research and development. This fosters the creation of cutting-edge sustainable building technologies and materials, which can then be shared globally.

Nonetheless, the role of policy and economic incentives cannot be overstated. Incentives such as tax breaks, grants, and subsidies are used to encourage sustainable practices among businesses and consumers, while penalties, such as taxes and fines, are used to make stakeholders avoid less sustainable behaviours. Moreover, policy frameworks that mandate certain standards for energy efficiency, water usage, and sustainable materials in construction play a critical role in driving industry-wide changes.

The diversity in approaches also extends to stakeholder engagement and community involvement. While some nations emphasize top-down governance like Sweden, others promote a more grassroots approach, such as the Netherlands, engaging local communities in the decision-making process and emphasizing the social aspect of sustainability.

By examining the heterogeneous approaches and strategies, it becomes clear that there is no one-size-fits-all solution. The success of sustainability and circularity efforts in the built environment is a function of the ability of each country to develop and implement strategies that resonate with their unique circumstances, many times linked to their culture, leveraging their strengths, and addressing their specific challenges.

3.3.3 Challenges and Opportunities

From the amount of data that resulted from the work presented in the following sections, it can be observed that the journey towards a sustainable and circular built environment is filled with challenges but also offers numerous opportunities for innovation and progress. The main identified challenges that are shared by most of the studied countries are:

Technological and Financial Barriers: implementing sustainable technologies often requires a significant upfront investment, which can be a barrier for many stakeholders. Additionally, the need for skilled labour to deploy these technologies can be a constraint.

Policy and Regulatory Hurdles: Aligning national policies with EU-level, and even UN, sustainable and circularity goals can be challenging, especially in the face of existing political, infrastructural, and economic frameworks that still favour traditional practices.

Cultural and Behavioural Change: Shifting the mindset of consumers, businesses, and policymakers to prioritize sustainability requires significant efforts in education and awareness-raising to change from a business-as-usual mentality to a sustainable-focused approach.

Coordination and Collaboration: Effective implementation of sustainable practices needs seamless collaboration between various stakeholders, which is often challenging due to differing objectives and interests, most times conflicting.

However, among all the difficulties and hurdles, many opportunities can also be identified that are shared by most countries, such as:

Innovation and Economic Growth: the drive towards sustainability can promote innovation, leading to the development of new materials, technologies, and practices, ultimately contributing to economic growth, social development, and environmental impact reduction. This is also linked to European financing and the strong environmental components of such financing, where the EU taxonomy is already showing a high influence [9].

Environmental Benefits: Adopting sustainable and circular practices and approaches significantly reduces the environmental footprint of the built environment, aiding in the preservation of natural resources and biodiversity. Specifically, the adoption of circular economy principles allows for reducing the depletion of natural resources and, at the same time, significantly reducing the production of waste.

Social Well-being: Sustainable buildings and built environments often lead to healthier living environments with, for example, more energy-efficient housing that, not only consumes less energy but also can maintain comfortable temperatures, contributing to the overall well-being of communities. Or even a more enjoyable living experience inside the cities.

International Collaboration: The global nature of sustainability and circularity challenges encourages international cooperation, sharing of best practices, and mutual learning.

These challenges and opportunities highlight the complex but rewarding path towards integrating sustainability and circularity into the built environment, requiring a multifaceted, systemic, and collaborative approach.

3.4 Relations Between National and European-Level Policies and Regulations

As previously seen, the EU has established comprehensive policies to drive sustainability and circularity in the built environment, primarily through the European Green Deal and the Circular Economy Action Plan. Both documents set an ambitious roadmap for making the EU's economy sustainable and circular by transforming climate and environmental challenges into opportunities across all policy areas. Here, some of the main goals are achieving net zero emissions of greenhouse gases by 2050, economic growth decoupled from resource use, and no person or place left behind.

The Circular Economy Action Plan, a part of the Green Deal, focuses on prolonging product lifecycles and minimizing waste. It is instrumental in transforming the EU's economy into a sustainable and circular model, ensuring resources are kept in the EU economy for as long as possible. Additionally, directives related to energy efficiency and waste management are central to the EU's approach. These include the Energy Efficiency Directive, which sets binding measures to help the EU reach its 20% energy efficiency target by 2020 and 32.5% by 2030, and the Waste Framework Directive, which introduces measures to protect the environment and human health by reducing the negative effects of waste generation and mismanagement. These directives are not only about reducing emissions and waste but also about creating sustainable growth and employment opportunities, reflecting the EU's integrated approach to sustainability and circular economy.

When the scope goes from the EU level to the national level, different EU member states have adapted and implemented EU directives in varied ways, reflecting their unique national contexts. For the exemplification of this, three cases can be quickly presented. It must be also stated that these countries have much more preoccupations than the ones presented, as can be seen in more detail in the next sections. However, they show a special focus towards specific themes:

Portugal: Has tried to transform its energy production, being one of the countries in EU-27 with the highest percentage of renewable energy production, also, efforts in the energy efficiency of houses have been passed into law, to comply with EU regulations [19]. It is also one of the few EU countries that published a specific Action Plan for Circularity in the Construction Sector [1].
Spain: Focused on urban regeneration, Spain has implemented policies that encourage the retrofitting of existing buildings, aligning with the European Green Deal's objective of renovating the building stock for energy efficiency [21].
Sweden: A leader in waste management, Sweden's national policies have effectively adopted the EU's Waste Framework Directive [4], focusing on efficient waste management systems and high recycling rates [18].

Each case study reflects the unique challenges and adaptations faced by individual member states in aligning their national policies with EU frameworks. Some countries adopt a more technological approach, others focus on urban regeneration, and others focus on efficient waste management systems, exemplifying the diverse strategies employed across the EU.

Knowing the relation between EU and national-level policies and regulations, it is acceptable that there will exist both synergies and conflicts. A more in-depth analysis shows that the documents can be aggregated into four groups: (i) development of policies and regulations properly integrated, thus supporting the transition through finance, case studies using GPP, and other mechanisms—policies and regulations; (ii) new and innovative digital tools and platforms that allow, for example, to collect large amounts of data and convert it into information, thus supporting the twin transition (green and digital) and the conversion of waste into resources—technological; (iii) education and capacitation of different stakeholders, supporting clarity and empowering them to surpass new challenges and promote their resilience—cultural; and (iv) governance models that support the paradigm shift across the society, based on a continuous flux of information and discussion between government and stakeholders supporting the market shift—market. Therefore, it is possible to cluster such approaches into four main pillars: political, technological, cultural, and market.

In terms of policy and regulatory alignment and disparities, it will be seen that regulatory frameworks also dictate the extent to which EU directives are adopted at the national level. Some countries have seamlessly integrated EU regulations into their national laws, but others encounter legal and bureaucratic challenges, delaying directive implementation. This regulatory mismatch can significantly hinder the collective progress towards the EU's sustainability goals.

When considering technological synergies and conflicts, the integration of EU directives such as the Energy Efficiency Directive is significantly influenced by the technological capabilities of member states. While some countries (such as France) have advanced in energy-efficient technologies, others such as Moldova, lack technological infrastructure, and face difficulties in complying with these standards. This disparity creates a conflict between the EU's uniform objectives and the varied technological readiness of different nations [27].

Integrating EU policies at the national level can sometimes clash with local cultural norms and practices. The one-size-fits-all approach of certain EU directives might not align with the unique cultural contexts of each member state, leading to resistance or difficulty in implementation. For instance, policies regarding agriculture or food production may conflict with traditional farming practices. Similarly, environmental regulations might force industries rooted in national heritage. This can create tensions between the need to comply with EU standards and the desire to preserve cultural identities and practices. Balancing EU-wide objectives with respect for cultural diversity remains a challenge, that needs sensitive and flexible approaches in policy formulation and implementation [20, 23].

Finally, similarly to the previous topics, the same can be seen in terms of market and financial challenges since the financial aspect plays a critical role. While EU funding aids in implementing sustainability projects, economic disparities among member states lead to uneven progress. Nations with stronger economies can promote markets to adopt sustainable and circular initiatives, aligning closely with EU goals, whereas economically weaker countries may struggle, leading to a gap in the implementation of EU policies.

Although those four main pillars are of paramount importance, the socio-economic front (affected by the four pillars) shows that the impact of EU sustainability and circularity policies greatly vary between member states and are multifaceted, affecting the most diverse aspects such as employment, economic growth, and technological innovation. With these relatively new concerns, employment is one of the most important aspects to consider. Here, the so-called green policies have led to job creation, especially in sectors like renewable energy, energy-efficient construction, and recycling. However, this transition has also brought challenges, particularly in regions reliant on traditional industries, leading to a need for re-skilling and job transitioning programs. As for economic growth, the investment in sustainable infrastructure and technologies has promoted economic growth in many regions, opening new markets, and new business models and driving innovation. However, the initial investment required for such transitions can be substantial, posing a challenge for member states with lower financial capacity. Finally, technological innovation is one of the most important drivers and enablers of sustainability and circularity in the built environment. Here, the EU directives have been a catalyst for technological innovation in sustainability. This has included advancements in renewable energy technologies, sustainable building materials, and energy-efficient design [16].

More socioeconomic impacts could be identified, but these three clearly showed their extent across many of the EU member states which are further analysed in the following sections. These also highlight the influence that the individual economic structures, regional priorities, and the degree of alignment with EU policies have. Above all, navigating these challenges, requires a nuanced approach, considering the specific circumstances of each country, while also identifying possible synergies.

The EU's role in offering a high-level analysis, guidance and support is essential to mitigate these disparities and ensure a united front in achieving sustainability and circularity objectives. Therefore, the importance of actions such as the present CircularB COST action is to identify these asymmetries and contribute to levelling out all partners.

3.4.1 Role of Regional and Local Governments

Although the main scope of this work is on the EU and national level, it is also important to reinforce the importance that regional and local governments have in translating EU directives and national policies into concrete actions at the community level. Their proximity to local issues allows for more precise approaches to sustainability and circularity, crucial for the successful implementation of broader objectives [17]. Therefore, many approaches are being used:

Local Adaptation of EU Directives: Local governments often adapt EU directives to fit their specific contexts. This involves customizing strategies for waste management, energy efficiency, and sustainable urban planning to align with both national and EU-level goals, considering local economic and environmental conditions.

Case Studies and Initiatives: Various European cities and regions have launched initiatives that exemplify this local adaptation. For instance, a city might implement a local energy efficiency program that complements the EU's Energy Efficiency Directive, tailored to the city's specific building stock and climate. Similarly, regional recycling initiatives might enhance the implementation of the EU Waste Framework Directive by addressing local waste patterns and recycling capacities [3].

Challenges and Opportunities: While regional and local governments are well-positioned to understand and address local sustainability challenges, they often face resource limitations and regulatory constraints. Balancing local needs with national and EU objectives can be challenging, but it also presents opportunities for innovation and community engagement.

These efforts at the regional and local levels are essential for achieving the overarching sustainability and circularity goals set by the EU and by the respective countries where they are located, demonstrating the importance of multi-level governance in addressing environmental challenges. By bridging the gap between high-level policies and local realities, regional and local governments ensure that sustainability initiatives are effective, relevant, and beneficial for their communities.

3.4.2 Discussion

As can be seen, the interaction between national and European-level policies and regulations in the context of sustainability and circular economy is a dynamic and complex process. Achieving effective synergy between these levels of governance requires a nuanced understanding of the political choices, technological capabilities, cultural involvement, and market shifts of each member state. The role of regional and local governments emerges as a crucial element in this process, acting as the centrepiece that connects high-level EU directives with ground-level implementation.

Not only the environmental aspects are relevant, but also the socio-economic impacts. The impact of such policies and regulations is significant and varied across the EU, underscoring the need for specific approaches that consider the unique contexts of different member states. Employment trends, political and regulatory options, cultural approaches, economic growth trajectories, technological innovations, and market trends in sustainability and circularity are all shaped by this interplay of policies at different governance levels.

When connecting the EU and national levels while looking forward, the continued evolution of EU directives and their national implementations will play a very important role in creating a more sustainable and circular Europe. This will require ongoing dialogue, adaptation, and collaboration across all levels of governance to ensure that the sustainability and circularity objectives and goals are met in a manner that is both effective and equitable. The path forward lies in leveraging the strengths of both national and European frameworks to foster a more sustainable, resilient, and prosperous Europe. Once again, promoting actions such as the current COST action can be of paramount importance due to the enabling of cooperation between countries on different adoption levels.

3.5 Future Directions: Trends and Challenges

After the previous presentation and overview of the work done and presented in the following sections, it is also important to identify future trends and challenges in policy and regulations, particularly in sustainability, circular economy, digitalization, and innovation. As Europe moves towards a more sustainable and circular economy, understanding and trying to identify the future directions of policies and regulations is crucial. Thus, based on the work done for EU and national-level identification and analysis of policies and regulations, it was possible to identify the following important aspects:

Digitalization and Smart Technologies: The integration of Internet-of-Things, Artificial Intelligence, and big data analytics in sustainability initiatives is anticipated to grow, which are two of the 10 KET (Key Enabling Technologies) already identified by the EC [10]. These technologies can enhance resource efficiency, waste management, and energy consumption. Moreover, these technologies are the current drivers and enablers of the twin transition, and the reason why only now this transition is possible. Challenges in this field include addressing cybersecurity risks and ensuring equitable access to these technologies, aspects relevant to be regulated.

Circular Economy Expansion: There will likely be a stronger emphasis on circular principles across various sectors, not just in manufacturing or construction, promoting an urban metabolism approach with synergies and interactions between different industries under a systemic approach. This includes developing policies that encourage product life extension, reuse, and remanufacturing. A key challenge is creating market demand for circular products and services and allowing a more agile transition between waste and resource. Here, both policies and regulations must address the difficulties that exist when considering waste as a resource due to regulations or legislation impositions.

Renewable Energy Transition: Policies are expected to support the transition to renewable energy sources increasingly. This involves not only energy production but also storage and distribution. The challenge lies in managing the transition from fossil fuels, particularly in regions heavily dependent on traditional energy sources and that are many times less capable of investing due to lower economic capabilities.

Sustainable Urban Development: With urbanization on the rise, policies focusing on sustainable urban planning and smart cities are likely to gain prominence. This includes green building practices, sustainable transportation, and urban green spaces. The main challenge lies in creating supportive policies for the retrofitting of existing urban areas and balancing development with environmental preservation.

Regulatory Frameworks for New Technologies: As new technologies emerge, especially in the field of biotechnology and materials science, regulatory frameworks will need to evolve. The challenge is to ensure safety and ethical considerations while not impairing innovation.

Global–Local Policy Alignment: Aligning local, national, and international policies for a cohesive approach to sustainability remains a challenge. There is a need for policies that are globally informed yet locally applicable. To truly achieve a circular economy this is something of great importance to achieve, otherwise, loops will not be able to close, and waste will be more easily discarded than reused, failing to keep the material inside the economy for the longest time possible.

Climate Change Adaptation and Resilience: Policies will need to focus not only on mitigation but also on adaptation to climate change. This includes developing resilient infrastructure and communities, especially in areas most vulnerable to climate impacts. This is greatly associated with risk management and is one of the aspects intrinsically connected to the Environmental, Social and Governance (ESG) aspects that have a great deal of importance to companies.

Economic Models for Sustainability: Developing and implementing economic models that support sustainability and circular economy principles is crucial. This includes revising taxation, subsidies, and financial incentives to support sustainable practices and businesses and discourage unsustainable ones. Here the

European Taxonomy and directives such as the Corporate Sustainability Report Directive (CSRD) clearly show the importance of this aspect.

Stakeholder Engagement and Public Policy: Increasing public awareness and stakeholder engagement in policy development is key. This involves ensuring that sustainability policies are developed with a bottom-up approach, considering local needs and conditions.

Education and Workforce Training: As the demand for green jobs increases, there is a need for policies that support education and training in sustainability-related fields, preparing the workforce for the twin transition and a greener economy.

Market reluctance: There is a perceived market reluctance to accept and/or use recycled resources, which is considered a significant barrier to circularity. The circular economy is also seen as a challenge, indicating a need for more business-oriented support and incentives where regulation and standards can help the decision process. Here policies and regulations can help the reassurance of the quality of such materials and the promotion of their use.

All these trends and challenges indicate a comprehensive approach to policy and regulatory development, encompassing technological innovation, economic restructuring, and societal transformation towards resilience, sustainability, and circularity. There must be an emphasis on the importance of proactive and forward-looking approaches to address the emerging trends and challenges in sustainability and circularity, that should promote the creation of new policies and regulations.

3.6 Conclusion

The European Union (EU) and its member states face the complex task of aligning national and EU-level policies and regulations to promote sustainability and circular economy in the built environment. This alignment is crucial for achieving the EU's ambitious sustainability goals, including the *European Green Deal* and the *Circular Economy Action Plan*. The interplay between these levels of governance is dynamic, involving a nuanced and diverse understanding of political, technological, cultural, and market shifts in each member state.

At the EU level, key policies and regulations have been developed to address sustainability and circular economy adoption challenges in the built environment, which will be discussed in detail in the subsequent Sect. 2.5. These include, among others, the *Waste Framework Directive*, the *Energy Performance of Buildings Directive*, and the *European Green Deal*, which collectively set the stage for a sustainable, circular EU economy. These policies are designed, above all, to transform climate and environmental challenges into opportunities and aim for net zero emissions of greenhouse gases by 2050.

When considering national-level policies in EU member states, it was identified a heterogeneous implementation of EU directives, adapted to unique national contexts, discussed next in Sect. 2.6. Most of the time, the national documents showed alignment with, at least, a portion of EU directives while addressing specific national priorities. However, there is variation in the degree of policy and regulatory alignment among EU countries, influenced by their economic structures, technological capabilities, and cultural approaches.

This overview also highlights the role of regional and local governments in translating EU directives and national policies into concrete actions at the community level. It emphasizes the importance of local adaptation of EU directives, considering specific contexts and needs. The role of local and regional governments is central in effectively translating national and European policies on sustainability and circular economy into practical, localized actions. These local governments act as critical intermediaries, adapting broad directives to fit the unique environmental, economic, and social contexts of their regions. Their proximity to the impacted communities enables a more adapted, responsive, and effective policy implementation, which is crucial for achieving the main goals and objectives of these initiatives.

When analysing all documents, and while also looking into the future, this overview identifies key trends and challenges in policy and regulation, including digitalization, the expansion of the circular economy, renewable energy transition, sustainable urban development, and the need for regulatory frameworks for emerging technologies. These trends underscore the importance of a comprehensive, multi-level and systemic approach to policy development, encompassing technological innovation, economic restructuring, and societal transformation towards sustainability and circularity.

In conclusion, the effective synergy between national and European-level policies and regulations in the context of sustainability and circular economy is a dynamic and complex process. Achieving this synergy requires ongoing dialogue, adaptation, and collaboration across all levels of governance. The path forward lies in leveraging the strengths of both national and European frameworks to stimulate a more sustainable, resilient, and prosperous Europe. Moreover, in the following sections, the detailed results for the EU (Sect. 2.5) and national-level analysis (Sect. 2.6) can be found, showing, and reinforcing the ideas here discussed.

References

1. BuiltCOLAB (2022) Portuguese action plan for circularity in the construction sector. https://circularidade.builtcolab.pt/#inicio
2. Bundestag (2020) Circular economy law—Germany
3. Colocci A, Gioia E, Casareale C, Marchetti N, Marincioni F (2023) The role of sustainable energy and climate action plans: synergies with regional sustainable development strategies for a local 2030 agenda. Environ Dev 47:100894. https://doi.org/10.1016/J.ENVDEV.2023.100894
4. European Commission (2008) Directive 2008/98/EC: waste framework directive
5. European Commission (2016) Buying green! A handbook on green public procurement. https://doi.org/10.2779/837689
6. European Commission (2019) European green deal

7. European Commission (2020a) A new circular economy action plan for a cleaner and more sustainable Europe. https://www.un.org/sustainabledevelopment/sustainable-consumption-production/
8. European Commission (2020b) Circular economy: principles for buildings design
9. European Commission (2020c) EU taxonomy. https://ec.europa.eu/sustainable-finance-taxonomy/
10. European Commission (2021) Digitalisation in the construction sector
11. European Union (2010) Directive 2010/31/EU of the European Parliament and of the Council of 19 May 2010 on the energy performance of buildings
12. European Union (2023) EU's new Bauhaus. https://new-european-bauhaus.europa.eu/about/about-initiative_en
13. French Government (2015) Law on energy transition for green growth—France. https://www.ecologie.gouv.fr/loi-relative-transition-energetique-croissance-verte-tepcv
14. Geissdoerfer M, Savaget P, Bocken NMP, Hultink EJ (2017) The circular economy—a new sustainability paradigm? J Clean Prod 143:757–768. https://doi.org/10.1016/J.JCLEPRO.2016.12.048
15. Kirchherr J, Reike D, Hekkert M (2017) Conceptualizing the circular economy: An analysis of 114 definitions. Resour Conserv Recycl 127:221–232. https://doi.org/10.1016/J.RESCONREC.2017.09.005
16. Lamnatou C, Cristofari C, Chemisana D (2023) Renewable energy sources as a catalyst for energy transition: technological innovations and an example of the energy transition in France. Renew Energy. https://doi.org/10.1016/J.RENENE.2023.119600
17. Masuda H, Kawakubo S, Okitasari M, Morita K (2022) Exploring the role of local governments as intermediaries to facilitate partnerships for the Sustainable Development Goals. Sustain Cities Soc 82:103883. https://doi.org/10.1016/J.SCS.2022.103883
18. Moalem RM, Schmidt K (2023) Municipal solid waste management in the interface between commercial and non-commercial repair: lessons from Denmark and Sweden. Clean Waste Syst 5:100095. https://doi.org/10.1016/J.CLWAS.2023.100095
19. OE, DGEG, & ADENE (2022) Energy in numbers
20. Pedroso MF, Tavares V (2024) Circular economy supporting policies and regulations: the Portuguese case, pp 277–290. https://doi.org/10.1007/978-3-031-45980-1_23
21. Rodríguez Ochoa D, Arranz N, de Arroyabe JCF (2023) The role of Horizon (2020) in achieving climate-neutral urban objectives: a study of 14 Spanish cities innovation networks. J Clean Prod 432:139820. https://doi.org/10.1016/J.JCLEPRO.2023.139820
22. Sala S, Amadei AM, Beylot A, Ardente F (2021) The evolution of life cycle assessment in European policies over three decades. Int J Life Cycle Assess 26(12):2295–2314. https://doi.org/10.1007/s11367-021-01893-2
23. Smith RC (2004) Culture and European Union Law. Culture Eur Union Law, 1–460. https://doi.org/10.1093/ACPROF:OSO/9780199275472.001.0001
24. Stavins RN, Stowe RC (2016) The Paris agreement and beyond: international climate change policy post-2020. http://belfercenter.ksg.harvard.edu/publication/26833
25. United Nations (2009) Buildings and climate change: summary for decision-makers. http://www.tge.ca
26. United Nations (2023) United Nations 17 sustainable development goals. https://www.undp.org/sustainable-development-goals/climate-action?gad_source=1&gclid=CjwKCAiAmsurBhBvEiwA6e-WPOhCx6VL1Z3Sm6wzfdDWRxGXsXhA2lXn8hJGFDfzgVBuvs4T1ZAKQhoCo7YQAvD_BwE
27. Wang R, Usman M, Radulescu M, Cifuentes-Faura J, Balsalobre-Lorente D (2023) Achieving ecological sustainability through technological innovations, financial development, foreign direct investment, and energy consumption in developing European countries. Gondwana Res 119:138–152. https://doi.org/10.1016/J.GR.2023.02.023

Chapter 4
Circular Economy Policies and Regulations at the European Level

Marco Lamperti Tornaghi

Abstract This chapter examines the evolution and implementation of circular economy policies and regulations at the European level, with a particular focus on the construction sector. It highlights the shift from conventional linear approaches to more sustainable and circular practices, driven by key EU initiatives, including the Circular Economy Action Plan, the European Green Deal, and the Renovation Wave. The chapter outlines various policy instruments, including directives, communications, and frameworks, that have been developed to promote waste reduction, energy efficiency, and life cycle thinking within the built environment. Additionally, it addresses tools such as Level(s), Green Public Procurement, and the proposed Construction Product Regulation, which aim to support the integration of circular principles. Significant challenges are discussed, including market fragmentation, regulatory complexity, and the need for occupational transformation. Ultimately, the chapter emphasises the need for harmonised EU strategies to promote innovation, improve resource efficiency, and facilitate the green transition in the construction industry.

Keywords Circular economy · Construction sector · European Union policies · Green public procurement · Sustainable buildings

4.1 Introduction

During the transition from the twentieth to the twenty-first century, increased focus on environmental issues has prompted efforts to reduce the ecological footprint of buildings. Buildings are primary sites for human activity and significant contributors to energy utilisation and CO_2 emissions. Initial efforts have prioritised the reduction of energy consumption and emissions, particularly during the use-phase of buildings. More recently, the entire lifecycle of construction has also been taken into account for buildings, as was already the case for other industrial sectors [1]. The next step

M. Lamperti Tornaghi (✉)
European Commission - Joint Research Centre (JRC), Ispra, Italy
e-mail: Marco.LAMPERTI-TORNAGHI@ec.europa.eu

L. Bragança et al. (eds.), *Shaping Circular Transitions in the Built Environment*, Springer Tracts in Civil Engineering, https://doi.org/10.1007/978-3-032-02834-1_4

was therefore to consider each cycle repeatable, thus envisaging the transition from a linear economy to a circular economy. Considering that building products and materials usually have a significantly longer lifespan than other consumer goods (at least 20–50 years), different strategies are needed to implement circularity for constructions and building products.

4.1.1 Review of Policies and Regulations at European Level

In this context, the European Commission published two communications in 2014. The first policy has a wider goal of reducing waste generation (COM/2014/0398), while the second specifically targets the building sector by promoting more efficient resource usage (COM/2014/0445). In the latter, the assessment of the environmental performance of buildings was considered as a practical, achievable and viable avenue to pursue that would both facilitate the improvement of building sustainability and address existing market barriers to enhance business opportunity within the sustainable buildings industry.

Later in 2015, member states of the United Nations adopted the 2030 Agenda for Sustainable Development. At the heart of this plan are the 17 Sustainable Development Goals (SDGs), which urgently call for action by all nations, both developed and developing, in a global partnership. Ending poverty and other forms of deprivation must be accompanied by measures that enhance health and education, minimize inequality, stimulate economic development, address climate change, and safeguard our oceans and forests. The Agenda confirms the relevance of buildings in at least five of the seventeen goals:

1. Ensure access to affordable, reliable, sustainable and modern energy for all;
2. Promote sustained, inclusive and sustainable economic growth, full and productive employment and decent work for all;
3. Build resilient infrastructure, promote inclusive and sustainable industrialization and foster innovation;
4. Make cities and human settlements inclusive, safe, resilient and sustainable;
5. Ensure sustainable consumption and production patterns.

In order to achieve the Sustainable Development Goals (SDG) by 2030, in particular Goal 12—ensure sustainable consumption and production patterns the Commission released a new Communication (COM/2015/0614). It identifies a range of measures to guarantee the retrieval of valuable resources and efficient waste management in the construction and demolition industry and to facilitate assessment of the environmental performance of buildings. Furthermore, it highlights the need to develop a monitoring framework to assess the progress of the Circular Economy.

Green public procurement could encourage public authorities to purchase environmentally friendly goods and services. At the same time, it can also promote the Circular Economy and stimulate the market in this direction, resulting in benefits in terms of sustainability. This is the aim of *Buying green!*—A handbook on green public procurement published by the Directorate-General for Environment in 2016 [5]. The Handbook identifies buildings as one of the four main procurement categories for GPP, with food and catering services, vehicles and energy-using products. It looks at environmental impact, budgetary importance, potential to influence the market and the availability of green alternatives. From an operational point of view, the Handbook is still much centred on reducing the impact of the use phase (energy and water consumption) and lacks a thorough approach to the circular economy (cradle-to-cradle), which is crucial for buildings, given that they are highly complex systems, consisting of numerous component parts, which all influence the overall performance of the structure.

In 2018, extensive action to update previous directives covered several sectors (Directive EU 2018/849-2018/850-2018/051-2018/852). With regard to building, these documents establish guidelines to enhance waste management in the EU with the aim to protect, preserve and improve the environment, as well as safeguard human health, while promoting efficient, rational use of natural resources, adhering to the principles of the Circular Economy. Additionally, the directives would increase energy efficiency and decrease dependency on imported resources; establish a sorting system for construction and demolition waste, with a focus on wood, mineral fractions (concrete, bricks, plaster, tiles and ceramics, stones), metal, glass and plastic. Significant targets have been established for the coming years. By 2030, no waste suitable for recycling or alternative recovery, particularly in municipal waste, will be allowed in landfills, unless landfilling is the most ecologically sound option for that waste. Moreover, from 2024, the Commission evaluates the establishment of objectives for the preparation of construction and demolition waste and related materials for re-use and recycling.

The Communication COM/2018/029 establishes a framework for monitoring the circular economy to measure its progress and a Commission's Staff Working Document (SWD/2018/017) that sets key indicators for the monitoring framework accompanies the Communication. The recovery rate of construction and demolition waste is set to a 70% recovery target, to be achieved by 2020 as per the Waste Framework Directive (2008/98/EC). The monitoring framework resulted in the EUROSTAT dataset to keep track of progress towards a circular economy. This framework provides a holistic view as it:

- measures direct and indirect benefits of 'becoming circular' and
- values the contribution of a circular economy in living well within the limits of the planet
- addresses energy and material supply risks.

It consists of five thematic sections with eleven statistical indicators, some of which have additional sub-indicators. In some cases, policy targets exist which should be achieved in the future, and the indicators monitor progress towards these targets. The indicators were selected in order to capture the main elements of a circular economy. The list of the indicators chosen is short and concise. It makes use of already available data, while also pointing out areas for which the development of new indicators is underway, for example for green public procurement. Following the launch of the new circular economy action plan for a cleaner and more competitive Europe, a revised framework has just been adopted to capture the circular economy focus areas and the interlinkages between circularity, climate neutrality and the 'zero pollution' ambition. The current monitoring framework is a revision of the original framework set up in 2018.

With the new Commission taking office in 2019, the Circular Economy has become one of the Commission's priorities and thus regulatory activity in this field has gained further momentum. The central element of this ambitious programme is the European Green Deal (COM/2019/640 final), approved in 2020, it is a set of policy initiatives by the European Commission with the overarching aim of making the European Union climate neutral in 2050. The Circular Economy Action Plan (COM/2020/98) presents a forward-thinking strategy to accomplish a cleaner and more competitive Europe in collaboration with economic entities, consumers, citizens, and civil society organisations. It aims to accelerate the transformation required by the European Green Deal, while building on circular economy policies. The plan guarantees that the regulatory framework is simplified and adjusted to meet the requirements for a sustainable future.

The Circular Economy Action Plan led to a policy activity aimed at buildings in particular. The Renovation Wave (COM/2020/662) significantly influences the construction industry, particularly concerning decarbonisation and the integration of renewables, as well as Life-cycle thinking and circularity. It strives for elevated health and environmental standards, addressing the two-fold obstacles of both eco-friendliness and technological advancement. Considering the principle of prioritising energy efficiency 'Energy Efficiency first', alongside a consideration for aesthetics, architectural quality and affordability. The New European Bauhaus Initiative (COM/2021/573) shares similar goals of combining sustainability, inclusion and aesthetics.

On the operational side, Level(s) provides a common language for assessing and reporting on the sustainability performance of buildings. It is a simple entry point for applying circular economy principles in our built environment. Level(s) offers a system for measuring and supporting improvements, from design to end of life for residential buildings or offices [16]. DG GROW and EISMEA, the Executive Agency for Small and Medium-sized Enterprises, published a study on circular economy principles for building design [17]. It outlines policy choices to back the implementation of circular economy principles in building design at European, national, and local levels. The objective is to extend the lifespan of structures, simplify the application of recycled materials, and enhance resource effectiveness throughout the building lifecycle.

4.1.2 *Identification of Challenges at European and Regional Levels*

The EU aims to be climate-neutral by 2050—an economy with net-zero greenhouse gas emissions. This objective is at the heart of the European Green Deal and in line with the EU's commitment to global climate action under the Paris Agreement. The shift towards a carbon–neutral society requires all sectors of society and the economy to play their part—from the energy sector to industry, mobility, buildings, agriculture and forestry. The EU faces this challenge with the opportunity to build a better future for all by taking the lead in investing in realistic technological solutions, empowering citizens and coordinating action in key areas such as industrial policy, finance and research, while ensuring social fairness for a just transition.

The communication COM/2020/102 address the sustainability of construction products and improve the energy efficiency and environmental performance of built assets. It also expresses the objective of creating 700,000 green new jobs throughout the EU by 2030.

To boost energy performance of buildings, the EC proposed a legislative framework (COM/2021/802) that renovates the Energy Performance of Buildings Directive 2010/31/EU and the Energy Efficiency Directive 2012/27/EU. The directives promote policies that will help:

- achieve a highly energy efficient and decarbonised building stock by 2050;
- create a stable environment for investment decisions;
- enable consumers and businesses to make more informed choices to save energy and money.

Following the introduction of energy performance rules in national building codes, buildings consume only half as much today, compared to typical buildings from the 1980s. While the objective of achieving a zero-emission building stock by 2050, the EC's proposal emphasizes the importance of minimizing the entire life cycle greenhouse gas emissions of buildings through resource efficiency and circularity. Calculating the life-cycle greenhouse gas emissions of new buildings is a crucial initial step towards improving the overall performance of buildings throughout their life cycle, and accomplishing a Circular Economy. Furthermore, the prevention and advanced management of construction & demolition waste (CDW) is critical to the achievement of circular economy goals.

Further initiatives are currently in the planning stage, with a particular focus on the construction sector.

The Green Public Procurement revision process encompasses reviewing the potential of extending the scope to other types of buildings procured or upheld by public authorities, notably schools and social housing. Moreover, the proposed criteria will align with current policy advancements regarding the Renovation Wave, the Level(s) common framework, and the EU Taxonomy [20].

The new Construction Product Regulation (CPR) will bring the foundation for implementing the circular economy in the construction industry, enabling the utilization of sustainable materials with comparable safety and reliability to conventional products (COM/2024/3110).

The Regulation (EU) 2024/3110 of the European Parliament and of the Council of 27 November 2024 laying down harmonised rules for the marketing of construction products and repealing Regulation (EU) No 305/2011, it includes the following main changes compared to the previous CPR:

- reused and 3D-printed construction products and prefabricated one-family-houses are included;
- the Commission would be allowed to adopt technical specifications in cases where the standardisation system is not delivering as expected, and to set product requirements;
- it sets out new environmental, functional and safety product requirements for construction products;
- it establishes a 'harmonised zone' (as opposed to the areas under the responsibility of Member States) and a mechanism to deal with Member States 'imperative regulatory needs' on health, safety or protection of the environment, including climate;
- it creates a new obligation for manufacturers to provide a declaration of conformity (concerning compliance with product requirements), on top of a declaration of performance, and introduces the possibility to give information via electronic means;
- it defines a list of general sustainability requirements (to be further defined in Commission acts/harmonised standards);
- it introduces simplification and exemption provisions for micro-enterprises;
- it strengthens enforcement powers of market surveillance authorities;
- it reinforces the product contact points for construction, to better support economic operators;
- it sets up a new Commission system allowing any natural or legal person to share complaints or reports related to possible breaches of the Regulation;
- it is aligned with the proposed regulation on ecodesign requirements for sustainable products on climate and environmental sustainability and on the digital product passport.

4.1.3 Potential Improvements

Construction is a key sector for the EU economy and an important source of employment, with more than 13 million direct jobs (9% of EU employment). In terms of number of enterprises, 99.9% of the sector consists of SMEs with less than 250 employees and 94% have less than 10 employees. One of the key sensitive points for the wealthy EU economies, and more generally in the OECD, relates to the distributional effects of the green economy, and in particular to occupational transformation.

The transition from linear to circular economy models needs to combine the peculiar aspects of the construction sector with the transformative processes that are affecting the labour market and reshaping the world, including action against climate change, demographic ageing, digitalisation and globalisation.

The paradigm shifts from traditional selling/buying to innovative business models can support the promotion of circular economy. Business models such as product service, pay-per-use, buy-back, lease or rent/own systems could extend producers' responsibility, improving product maintenance, reuse during the use phase and recycling at the end of the life cycle.

In this ever-changing employment landscape, the building sector faces the challenge of attracting and retaining a sufficiently large pool of talent with the right skills. The sector is traditionally male-dominated and characterised by an ageing workforce. In the green transition, it will be essential to attract more people into the construction workforce by expanding to different talent profiles, including women and younger age cohorts with increasingly digital skills. Sustainable industries have a more positive image, especially among women and young people, compared with traditional ones. Occupational transformation must be ensured in order to avoid that the shift towards circularity models in the construction sector results in unemployment issues, particularly at the local level (Pacheco et al.).

References

1. Sala S, Amadei AM, Beylot A, Ardente F (2021) The evolution of life cycle assessment in European policies over three decades. Int J Life Cycle Assess 26:2295–2314. https://doi.org/10.1007/s11367-021-01893-2
2. European Commission (2014) Communication COM/2014/0398 final, Towards a circular economy: a zero waste programme for Europe. https://eur-lex.europa.eu/legal-content/EN/TXT/?uri=celex:52014DC0398
3. European Commission (2014) Communication COM/2014/0445 final, Resource efficiency opportunities in the building sector. https://eur-lex.europa.eu/legal-content/EN/TXT/?uri=celex:52014DC0445
4. European Commission (2015) Communication COM/2015/0614 final, Closing the loop—An EU action plan for the Circular Economy. https://eur-lex.europa.eu/legal-content/EN/TXT/?uri=CELEX:52015DC0614
5. European Commission, Directorate-General for Environment (2016) Buying green!—A handbook on green public procurement. Publications Office. https://data.europa.eu/doi/10.2779/837689
6. European Parliament and Council (2018) Directive (EU) 2018/849 amending Directives 2000/53/EC, 2006/66/EC and 2012/19/EU. http://data.europa.eu/eli/dir/2018/849/oj
7. European Parliament and Council (2018) Directive (EU) 2018/850 amending Directive 1999/31/EC on the landfill of waste. http://data.europa.eu/eli/dir/2018/850/oj
8. European Parliament and Council (2018) Directive (EU) 2018/851 amending Directive 2008/98/EC on waste. http://data.europa.eu/eli/dir/2018/851/oj
9. European Parliament and Council (2018) Directive (EU) 2018/852 amending Directive 94/62/EC on packaging and packaging waste. http://data.europa.eu/eli/dir/2018/852/oj
10. European Commission (2018) Communication COM/2018/029 final, On a monitoring framework for the circular economy. https://eur-lex.europa.eu/legal-content/EN/TXT/?uri=COM:2018:29:FIN

11. European Commission (2018) Commission Staff Working Document SWD/2018/017 final, Measuring progress towards circular economy in the European Union—Key indicators for a monitoring framework. https://eur-lex.europa.eu/legal-content/EN/TXT/?uri=SWD%3A2018%3A17%3AFIN
12. European Commission (2019) Communication COM/2019/640 final, The European Green Deal. https://eur-lex.europa.eu/legal-content/EN/TXT/?uri=COM:2020:98:FIN
13. European Commission (2020) Communication COM/2020/98 final, A new Circular Economy Action Plan for a cleaner and more competitive Europe. https://eur-lex.europa.eu/legal-content/EN/TXT/?uri=COM:2020:98:FIN
14. European Commission (2020) Communication COM/2020/662 final, A renovation wave for Europe—greening our buildings, creating jobs, improving lives. https://eur-lex.europa.eu/legal-content/EN/TXT/?uri=CELEX:52020DC0662
15. European Commission (2021) Communication COM/2021/573 final, New European Bauhaus Beautiful, Sustainable, Together. https://eur-lex.europa.eu/legal-content/EN/TXT/?uri=COM:2021:573:FIN
16. European Commission, Directorate-General for Environment (2021) Level(s), Putting circularity into practice. Publications Office of the European Union. https://data.europa.eu/doi/10.2779/19010
17. European Commission, Executive Agency for Small and Medium-sized Enterprises (2021) Study on circular economy principles for buildings' design—final report. Publications Office. https://data.europa.eu/doi/10.2826/3602
18. European Commission (2020) Communication COM/2020/102 final, A new industrial strategy for Europe. https://eur-lex.europa.eu/legal-content/EN/TXT/?uri=CELEX:52020DC0102
19. European Commission (2021) Communication COM/2021/802 final, Proposal for a Directive of the European Parliament and of the Council on the energy performance of buildings (recast). https://eur-lex.europa.eu/legal-content/EN/TXT/?uri=celex:52021PC0802
20. Pérez Arribas L et al (2023) EU Green Public Procurement (GPP) criteria for the design, construction, renovation, deconstruction and management of buildings—Draft Technical Report v.1.1. https://susproc.jrc.ec.europa.eu/product-bureau/sites/default/files/2023-06/GPP_Buildings_TR2_v1.1_0.pdf. Accessed 30 Sep 2023
21. Regulation (EU) 2024/3110 of the European Parliament and of the Council of 27 November 2024 laying down harmonised rules for the marketing of construction products and repealing Regulation (EU) No 305/2011. https://eur-lex.europa.eu/eli/reg/2024/3110/oj
22. Pacheco J, de Brito J, Lamperti Tornaghi M (2023) Use of recycled aggregates in concrete: opportunities for upscaling in Europe. Publications Office of the European Union, Luxembourg. https://doi.org/10.2760/144802

Chapter 5
National Policies and the Challenge of Coherence

Ana Andabaka, Martina Basarac Sertić, Stylianos Yiatros, Olympia Nisiforou, Marilena De Simone, Tatjana Tambovceva, Dzintra Atstāja, Natālija Cudečka-Puriņa, Ruben Paul Borg, Victoria Iordachi, Neli Muntean, Ana Trombeva Gavriloska, Teodora Mihajlovska, Vanessa Tavares, Ana Nadaždi, Mirjana Laban, Emília Huttmanová, Jana Chovancová, Cristina Jiménez-Pulido, Mustafa Sahmaran, Philip Griffiths, Raluca Buzatu, Jan Pešta, Katerina Tsikaloudaki, and Milen Baltov

Abstract This chapter explores the complexities of aligning national policies with the European Union's objectives in transitioning to a circular economy within the construction sector. While EU directives establish general frameworks, national

A. Andabaka (✉)
Faculty of Economics and Business, University of Zagreb, Zagreb, Croatia
e-mail: ana.andabaka@efzg.unizg.hr

M. Basarac Sertić
Economic Research Division, Croatian Academy of Sciences and Arts, Zagreb, Croatia

S. Yiatros
Department of Civil Engineering and Geomatics, Cyprus University of Technology, Limassol, Cyprus

O. Nisiforou
Department of Shipping, Faculty of Management and Economics, Cyprus University of Technology, Limassol, Cyprus

M. De Simone
Department of Environmental Engineering, University of Calabria, Rende, Italy

T. Tambovceva
Faculty of Engineering Economics and Management, Riga Technical University, Riga, Latvia

D. Atstāja
Faculty of Social Sciences, Riga Stradins University, Riga, Latvia

N. Cudečka-Puriņa
Management Department, BA School of Business and Finance, Riga, Latvia

R. P. Borg
Faculty for the Built Environment, University of Malta, Msida, Malta

L. Bragança et al. (eds.), *Shaping Circular Transitions in the Built Environment*, Springer Tracts in Civil Engineering, https://doi.org/10.1007/978-3-032-02834-1_5

governments encounter challenges in adapting these directives into coherent and context-specific regulations. The chapter identifies key obstacles to effective policy implementation, including fragmentation, institutional overlap, and inconsistent terminology. Additionally, it examines how governance structures, planning traditions, and institutional cultures influence policy coherence. The analysis is based on survey results collected in 2023. Although subsequent policy and regulatory changes have occurred in the past two years, the data remain relevant, as they capture the main trends and challenges that continue to shape national approaches to circular construction. By analyzing case studies and examples from various member states, the chapter

V. Iordachi
National Institute for Economic Research, Academy of Economic Studies of Moldova, Chisinau, Moldova

N. Muntean
Technical University of Moldova, Chisinau, Moldova

A. Trombeva Gavriloska · T. Mihajlovska
Faculty of Architecture, Ss. Cyril and Methodium University, Skopje, North Macedonia

V. Tavares
BUILT CoLAB Porto, Porto, Portugal

A. Nadaždi
Faculty of Civil Engineering, University of Belgrade, Belgrade, Serbia

M. Laban
Faculty of Technical Sciences, University of Novi Sad, Novi Sad, Serbia

E. Huttmanová · J. Chovancová
Faculty of Management and Business, University of Prešov in Prešov, Prešov, Slovakia

C. Jiménez-Pulido
Research Group Sustainability in Construction and Industry (giSCI-UPM), Universidad Politécnica de Madrid, Madrid, Spain

M. Sahmaran
Department of Civil Engineering, Hacettepe University, Ankara, Turkey

P. Griffiths
Belfast School of Architecture and the Built Environment, Ulster University, Belfast, UK

R. Buzatu
Department of Steel Structures and Structural Mechanics, Politehnica University Timisoara, Timisoara, Romania

J. Pešta
University Centre for Energy Efficient Buildings, Czech Technical University in Prague, Buštěhrad, Czech Republic

K. Tsikaloudaki
School of Civil Engineering, Aristotle University of Thessaloniki, Thessaloniki, Greece

M. Baltov
Burgas Free University, Burgas, Bulgaria

emphasizes the necessity for enhanced multilevel coordination, harmonized regulatory frameworks, and knowledge-sharing platforms. Bolstering policy coherence is crucial to ensure that national initiatives make meaningful contributions to the EU's circular economy goals and the broader sustainability agenda.

Keywords Circular economy · Policy coherence · National implementation · Construction sector governance · Multilevel coordination

5.1 Introduction

The shift towards a circular economy in the construction sector presents not only technical and operational challenges but also significant policy and governance obstacles. Although the European Union has established ambitious strategies and directives, such as the Circular Economy Action Plan and the European Green Deal, the successful implementation of these initiatives relies heavily on national-level policies and regulatory coherence. However, variations in legal frameworks, institutional arrangements, and policy-making traditions among EU member states often result in fragmented approaches. This misalignment can hinder the achievement of circular economy objectives, particularly in complex sectors such as construction, where responsibilities are dispersed across multiple levels of governance.

To gain a clearer understanding of these challenges and opportunities at the national level, the following sections provide factsheets from more than 20 EU countries. Each factsheet follows a standardized structure, which includes an overview of the governance framework involved in implementing circular economy practices, a summary of policy instruments that support circular construction, and a description of implementation mechanisms such as incentives, public procurement, and certification systems. Furthermore, each country analysis identifies existing barriers and regulatory gaps, highlights successful practices that can serve as models, and concludes with recommendations for enhancing alignment with EU objectives. Through these case-based insights, this chapter offers a comparative perspective on how national frameworks contribute to, or hinder, the broader European transition towards a circular built environment.

5.1.1 Bulgaria Factsheet

5.1.1.1 Policy and Regulatory Framework

Governance

Bulgaria has significantly strengthened its governance framework for the circular economy through institutional structures, strategic planning, and alignment with EU standards.

Circular Economy Council was established in April 2024 as a permanent consultative body under the Council of Ministers. This body has mandate to: coordinate across ministries, local governments, and NGOs; issue guidelines; assess progress; and propose national legislation updates. It is supported by three specialized working groups focused on:

1. Production and Consumption
2. Competitiveness and Innovation
3. Waste Management and Secondary Raw Materials.

Legislation

The main document framing the future legislation is the Strategy for the Transition to a Circular Economy in Bulgaria 2022–2027. It also includes a plan for monitoring the implementation and updating of the Strategy and Action Plan. On October 26, 2022, the Council of Ministers approved a strategy and an action plan for transition to circular economy in the period 2022–2027. The aim of the two documents is to achieve resource efficiency through a hierarchic approach in waste management, eliminating waste generation, promoting reuse, recycling and recovery of materials and products, reduction of landfilled waste and limiting the harmful impact of waste on the environment and human health.

The strategy formulates three strategic goals: green competitive economy; less waste and more resources; economy to the benefit of consumers. Concrete measures are planned to achieve the goals formulated as concrete activities in the action plan accompanying the strategy. The implementation of the plan seeks to redress the imbalances and overcome the obstacles in achieving the goals of the strategy for circular economy.

Circular Economy Strategy, Roadmap or Action Plan

Besides the Strategy for the Transition to a Circular Economy in Bulgaria 2022–2027, and according to the ETC CE Report 2022/5—Bulgaria the circular economy policy elements included in other policies are focused on four main areas, i.e.—the material efficiency, the product design, the resource-efficient production processes, and on the waste management [86].

On the material efficiency the planned measures (an action plan is currently in preparation) will promote the sustainable and reduced use of raw materials in production, stimulating the use of alternative raw materials and increasing the use of recyclable materials. A particular focus will be on the improvement of the knowledge base on the CE and the monitoring of waste and material flows, as well as the introduction of new forms of interaction between producers and consumers to support the CE [46].

On the issues of the product design the planned activities are, for example, improving the possibilities for product recycling, including the recycling of individual materials contained in the products by, amongst other things, substituting or reducing the use of products and materials which are not subject to recycling; reducing the content of hazardous substances in materials and products throughout their lifecycles, including by replacing them with safer alternatives; and increasing the durability, repairability, updateability or reuse of the products (funding program "Competitiveness and Innovation in Enterprises" (CIE) [20].

Concerning the resource-efficient production processes the measures under consideration are the more efficient use of natural resources in production, including reducing the use of primary raw materials or increasing the use of byproducts and secondary raw materials; supporting partnerships between enterprises to achieve industrial symbiosis, sharing of resources, services and secondary products, creating a link between production through which waste from one process is recognised as raw material for another; creation of industry platforms to exchange good practices for, for example, material flows of plastics, wood and glass.

In the waste management the activities planned to foster the prolonged use of products, including reuse, design for durability, modification of the product's purpose, disassembly, remanufacturing, updating and repair, and sharing of products; support for improving waste management in enterprises, including the prevention and reduction of waste generation, preparation for reuse and recycling. Support will be provided for technologies for the treatment and reduction of pollution of industrial wastewater.

Construction and Demolition Waste Management Regulatory Framework

Bulgaria's regulatory framework for Construction & Demolition (C&D) Waste Management is comprehensive and aligned with EU standards, incorporating detailed governance, technical requirements, and enforcement mechanisms.

Waste Management Act (2012, amended) transposes EU's Waste Framework Directive into national legislation, including specific provisions for C&D waste.

Ordinance on Management of C&D Waste and Insertion of Recycled Materials, was issued in 2012 and updated in 2017 under Art. 43(4) of the Waste Management Act. This document sets binding recycling targets for public construction projects: e.g., $\geq 2\%$ in buildings, $\geq 10\%$ for roads, $\geq 12\%$ for backfill by 2020.

Construction Waste Management Plan (CWMP) is mandatory for major C&D projects and has to be part of investment documentation. Since 2017, CWMP must be submitted before site opening or demolition and municipal approval is required (except for minor works or emergency removals).

5.1.1.2 Platforms and Networks Facilitating the Transition to a Circular Economy

Circular Economy Platforms

Key for Bulgaria for the last year was the adoption of a new Innovation Strategy for Smart Specialization. The projects eligible for funding under the procedures of the European Regional Development Fund for Bulgaria are those that fall within the specific thematic areas defined by the Innovation Strategy for Smart Specialization 2021–2027 (ISIS 2021–2027). These areas include "Informatics and ICT", "Mechatronics and Microelectronics", "Healthy Living, Bioeconomy and Biotechnology Industries" and "New Technologies in Creative and Recreational Industries". It also includes "Clean Technology, Circular and Low Carbon Economy", which shows the programme's commitment to sustainable development and environmental innovation. Examining the current policy context and frame S3 in the broader EU agenda. We stress the importance of aligning S3 to the EU New Industrial Strategy and the European Green Deal, mustering momentum to address social and environmental challenges.

Introducing conceptual and policy frameworks which can support the development of a transformative EDP, namely Sustainability Transitions, Transformative Innovation Policy, and Responsible Research and Innovation. We then report on two pioneering policy experiences that -building on the above conceptual foundations and approaches—have attempted to imbue the S3 and EDP with transformative elements, namely the POINT Reviews [177] by the Joint Research Centre and the Shared Agendas in Catalonia.

Circular Buildings Platform

For Bulgaria there are no circular buildings platform has been established to date.

Building Materials Passport Platform

No building materials platform has been established to date.

But in parallel for the mining industry, the situation is slightly different from the traditional one—material flows are usually large and concentrated and therefore the useful use of waste occurs much more often [141]. Given the specifics of the activity, the business models that have been proven to be applicable are "Resource efficiency and recycling" and "Waste resource recovery".

An example in the circular economy is one of the best underground mines in the world—Dundee Precious Metals Chelopech. Still in 2005, the so-called chamber system was introduced with the subsequent filling of the seized mining spaces with hardening material, which includes all sterile rock masses and a large part of the flotation waste mixed with cement [268].

In Assarel Medet an oxide embankment (a hill formed by the disposal of oxide copper ores, once treated as waste) is simultaneously reclaimed and operated through a drainage system for catching and draining water to be directed to the production of cathode copper or treatment plants. Lavender is grown on the reclaimed land for oil production and honey is also extracted at the same time.

Public Procurement Platform

Under the Public Procurement Portal the Bulgarian Public Procurement Agency published a Practical Handbook for Green Public Procurement [283]. The practical tool was developed by PricewaterhouseCoopers Bulgaria EOOD under a public procurement contract of May 15, 2018 with the subject: "Study and forecasting of the potential of the national market for green public procurement in Bulgaria and preparation of a practical tool for their awarding", part of the project "Methodological Support for the Development of Green Public Procurement in Bulgaria", funded under Priority Axis 3 "Private Sector" of the Bulgarian-Swiss Cooperation Program (BCSP).

The structure is made up of three modules:

- Strategic module—presents the potential for development of the market of green public procurement in Bulgaria and the product groups included in the Operational module;
- Legal module—outlines the applicable legal framework for incorporating green criteria in public procurement throughout the life cycle of the contract;
- Operational Module—provides exemplary environmental criteria for 12 product groups: "Wastewater infrastructure", "Food preparation and delivery services", "Computers and monitors", "Imaging devices", "Copier and Graphic Paper, Furniture, Cleaning Products and Cleaning Services, Public Space Maintenance, Design, Construction and Management of Buildings, Design, Construction and Maintenance of Roads, Transportation and Street lighting." For the Lighting and Computers and Monitors product groups, models (calculators) have been applied to calculate Lifetime Costs (RLC). The examples given are illustrative and advisory in nature and are intended to serve as a reference for stakeholders.

At the Public Procurement Portal there is also a special Guidance for public authorities on Public Procurement of Innovation https://www2.aop.bg/wp-content/uploads/2019/04/PPI-Platform_Guide_new-final_download.pdf. Part of it are the recommendations for the planning of organisation's PPI strategy. One of the recommendations is on evaluating in the tender the "Whole-life and Life-cycle Costing", e.g. How will the costs of new products and services be assessed across their life-cycle? Have relevant tools for this been developed within the sector(s) you are targeting?

5.1.1.3 Funding Opportunities

Besides the Horizon Europe that is the EU's flagship programme for funding research and innovation, in Bulgaria the circular economy promotion stands with the businesses through local financial institutions and the funding access portal, as well as instruments such as the European Agenda for Employment and Social Innovation, the European Social Fund Plus (ESF+) and the European Maritime, Fisheries and Aquaculture Fund (EMFF). Mainly the funds are fueled through the Competitiveness and Innovation in Enterprises Programme 2021–2027 supporting the Bulgarian enterprises in developing innovative products and business processes. The main objective of the programme is to strengthen the country's innovation capacity and support the transition to a more sustainable and efficient economy. It is divided into two priorities: "Innovation and Growth" and "Circular Economy".

The programme provides financial support through various mechanisms, including grants, loans and guarantees. It also offers technical assistance and advice to SMEs looking for development and innovation opportunities. Funds are allocated in accordance with the specific needs and project proposals of enterprises. The procedure provides for the financing of research and innovation activities, as well as networking between enterprises and scientific institutes. This includes the development of industrial research, experimental development, and studies to establish the feasibility of new ideas and technologies. Thus, the procedure aims to strengthen the links between the scientific sector and business and to help translate scientific discoveries into practical innovations.

For Bulgarian enterprises, procedure "Circular Economy" represents a valuable opportunity for obtaining support in the process of innovative development. It provides financial resources and expert assistance necessary for the implementation of innovative projects that can significantly contribute to the development of the Bulgarian economy and to increase the competitiveness of local enterprises on the European and global markets.

Under the priority 2 of the programme "Circular Economy" are included two specific objectives aimed at preparing businesses for the challenges of the green transition and achieving climate neutrality. The interventions under the first specific objective to promote energy efficiency and reduce greenhouse gas emissions, to which 6.87% of the CIE budget, around EUR 1493 billion, is allocated, aims to improve energy efficiency in enterprises through targeted action including the implementation and certification of energy management systems, and energy consumption monitoring and control systems.

International Cooperation and Support from International Organizations

Bulgaria's progress in the transition to a CE and resource efficiency has been analysed by the European Commission, in the context of the European Semester, in the chapter Green transition and resource efficiency and Annex 7 Resource Efficiency and Productivity of a document entitled 2022 Country Report—Bulgaria [81]. This report states that Bulgaria is "among the Member States lagging most behind in the implementation of circular economy policies" and that "the Bulgarian economy is among the most resource intensive in the EU and lags behind Member States in the implementation of the circular economy principle and eco-innovation". These conclusions are based on statistical data—the country's circular material use rate was 2.6% in 2020, almost five times lower than the EU average of 12.8%, and resource productivity was 0.82 purchasing power standards (PPS) per kilogram in 2020, compared to 2.23 PPS per kilogram in the EU. Some possible reasons for this are outlined below.

The share of SMEs offering environmentally-friendly products or services is one of the lowest in the EU, due to the low demand for environmentally-friendly products, which tend to be more expensive. Promoting the higher quality of ecolabelled products would boost demand and production. At the same time. the low interest in ecolabelling and the EU's Eco-Management and Audit Scheme (EMAS) is due to the fact that manufacturers do not take full advantage of the opportunities offered by these voluntary instruments.

In November 2023 it was decided the Bulgarian micro-, small and medium enterprises (MSMEs) along the border with Turkey will be able to apply for funding of over 6.6 million euro ($7 million) in total for projects focused on energy efficiency and the transition to a circular economy, the regional development ministry said. Companies can receive up to 200,000 euro each in financing under the cross-border cooperation programme between Bulgaria and Turkey, the ministry said in a statement. The proposed projects must target energy efficiency, greenhouse gas reduction and shifting to a circular and resource-efficient economy. The initiative will prioritise the promotion of circular production processes through the purchase of technological equipment and the development of new business models. The funds can be used to buy energy-saving equipment, upgrade production facilities for energy efficiency and implement automatic energy consumption monitoring systems. Project proposals were accepted starting mid-November 2023.

5.1.1.4 Challenges, Barriers and Potential Improvements

The beginning of 2020, in connection with INTERREG Europe project "REDUCES", a survey was conducted among the companies from the Bulgarian mining industry, in order to establish the understanding and application of the models for circular economy. The survey was provided to 65 companies in the business, with only 13 of them participating in its completion.

The analysis of the results shows that in most of the surveyed companies circular models for modernisation and innovation are applied, namely:

- Opportunity for efficient processing of poor ores;
- Industrial water supply in a closed cycle to supply the production process;
- Mounting of an installation for extraction and electrowinning of cathode copper from mine water, through which an additional by-product is obtained—pure copper;
- Technology for extraction of accompanying gold in copper concentrate, so that precious components are not lost in the waste;
- Application of a method of extraction, which includes backfilling, i.e. the ground sterile rock mass is returned to the underground mine in the form of a pasty filling, is an example of a circular economy, which not only reduces the amount of waste deposited on the surface, but is also used the waste as a raw material, which is mixed with cement and returned to the seized chambers, ensuring the stability of the rock massif;
- Use of part of the mine ballast for filling the used spaces underground, for paving during the construction of the underground roads after crushing and for upgrading the walls of the tailings pond;
- In the extraction of limestone, necessary for the production of lime, many additional fractions of material are released, which are not discarded, but sifted and placed on the market for aggregates;
- Use of earth masses and sludge for reclamation of disturbed terrains;
- Delivery to specialised companies of used oils for recycling and use in other industries;
- Recycling of metals used for drilling tool crowns.

5.1.1.5 Examples of Successful Implementation

Public Policy Initiatives

Supposing the best benchmark for public policy, and with elements of PPP is the case of the Cleantech Bulgaria (Bulgaria, project coordinator: Hamanova Mariyana) and ATHENA (Greece, project coordinator: Prof. Phoebe Koundouri, School of Economics and ReSEES Laboratory, AUEB, Director EIT Climate KIC Greece) with the support of EIT Climate-KIC will pilot the adoption of CE in their respective S3s, working together with the responsible authorities. Country will identify challenges, priorities, the potential for synergies and support instruments for CE. Participatory workshops with experts from EIT Climate-KIC will be run. Specific approaches to implementation and supporting funding instruments will be discussed and shared with participants. The end goal of the project is to stimulate the timely and systemic adoption of the CE in S3s for the 2020–2027 programming period in the EU Member States [58].

Together with the support of the Ministries of Economy of Bulgaria and the Hellenic Ministry of Energy and Environment of Greece they focus on mapping of the key aspects of the National Strategy on Circular Economy in relation with the Research and Innovation Strategies for Smart Specialisation (RIS3) and the discussion on the implementation of these strategies identifying the needs, barriers and strengths in Greece. Mutual learning between Greece and Bulgaria was guaranteed by providing exchange visits between members of CTBG and ATHENA.

Stepping on the outputs provided, a final event in Brussels with the participation of external EU experts and other RIS countries was held during the European Week of Regions and Cities in 2019, an executive meeting for the CE in S3 project from took place at EIT House, Brussels. Both Greece's and Bulgaria's work and upgrade the project to a national based deep demonstration of the circular economy project, which will be led by ATHENA RC and the Ministry of Environment and Economy supposed. Potential synergies with existing projects of EIT Climate-KIC Greece with focused areas the ports and the plastic are discussed, while the Ministry of Environment and Energy stimulated the discussions by being represented at the meetings in Brussels with Greek delegates from the Ministry of Environment and Energy.

Another case is a project supported by the Bulgarian "Research and Education for Smart Growth" Operational programme, that just finished on December 31, 2023. The project title is MIKS-IP "Modernising Educational Solutions for Achieving Circular Economy in the Strategic Infrastructures and Industries" https://www.bfu.bg/bg/proekt-modernizatsiya-na-obrazovatelnite-resheniya-za-kragova-ikonomika-strategicheski-infrastrukturi-i-proizvodstva-miks-ip. It was run by a consortium of the Burgas Free University, the University of Chemical Technologies and metallurgy, and the International Business School, with associated partners in Poland, Hungary and Romania, and coordinated by prof. Milen Baltov. One of the achievements was introducing a Masters degree joint programme on "Circular Economy".

Private Policy Initiatives

There are more and more companies involved, but especially the three part of the European Circular Economy Stakeholder Platform https://circulareconomy.europa.eu/platform/en/main-language/bulgarian.

Three of the benchmark example are mentioned below.

The Remix, with its CE project slogan Reuse, reduce, remix. The Remix Shop, as a type of organisation or company: Private company; Key Area: Consumption; Sector: Clothing and Fashion industry, Reuse, Textiles, apparel and leather. With this company the scope is national, and EU. Remix is a shop based in Bulgaria, which sells second-hand and outlet clothing online to nine countries.

LanaTerm, with its CE project slogan: wool can keep buildings warm as well as sheep. Lana Term as a type of organisation or company is a private company. It operates both in Bulgaria and in Romania. The languages for its original content are Bulgarian, English, German, and Romanian. The key area is production, and its sectors are in construction, buildings and infrastructure. There are not many eco-friendly products on the Romanian construction market, so there was definitely a niche in the thermal insulation market. LanaTerm uses sheep's wool to create thermal insulation for buildings.

ETV4, with its CE project slogan: Innovation for Environmental technology verification. ETV4INNOVATION is rather an organization, than a company and it's focused on Environmental Technology Verification (ETV). Besides Bulgaria it operates mainly in Ireland. The languages for original content are Bulgarian, Danish, English, French, Polish, and Spanish. Its key areas are in the waste management; the secondary raw materials, and the innovation and investments. The sector is the B2B services, Education and skills, and Environmental technology verification. Its scope is international, mostly in the EU. ETV4INNOVATION is a two-and-a-half-year long European Commission funded vocational education and training Strategic Partnership project under Erasmus + programme. It has been designed with the aim of supporting the development and the implementation of an innovative practice and a new training path in the field of Environmental Technology Verification (ETV) [75].

5.1.2 Croatia Factsheet

5.1.2.1 Policy and Regulatory Framework

Governance

According to the Implementation program of the Ministry of Economy and Sustainable Development for the period 2021–2024 [226], which entered into force on December 15, 2022, the Ministry of Economy and Sustainable Development (MoESD) determined strategic directions, developed programmes and plans for sustainable development of society based on the principle of green and circular economy in order to transform it into a just and prosperous society with a resource-efficient and competitive economy, ensuring climate neutrality, preservation and sustainable use of natural resources. The MoESD's vision was: "Sustainable growth and development based on efficient use of resources, circular economy, clean industries and new technologies with a strong reduction in greenhouse gas emissions."

The Ministry of Physical Planning, Construction and State Assets is in charge of implementing the Programme for the development of circular spatial and building management for the period 2021–2030 [227], which entered into force on December 23, 2021. According to the Programme's medium-term vision, "In the year 2030, Croatia is a country where urban areas are successfully implementing the model of circular management of space and buildings, in which unused existing spaces and buildings are renovated, the amount of total produced construction waste is reduced, innovations in the development of circular materials and products are promoted, and the principles of eco-design are applied in the design and construction of buildings. Croatia in 2030 is a country that recognizes, revitalizes, improves and values its built environment as a fundamental resource, and its citizens are provided with a healthy, socially functional, safe living and working environment."

Legislation

The development of the Programme for the development of circular spatial and building management for the period 2021–2030 is required by the Building Act (Official Gazette 153/13, 20/17, 39/19, 125/19, 145/24) [109], which also incorporates the circular economy principles. Article 14 states that construction works and their heating, cooling, lighting and ventilation installations must be designed and built in such a way that the amount of energy they require in use shall be low, when account is taken of the occupants and of the climatic conditions of the location of the construction work. Construction works must also be energy-efficient, using as little energy as possible during their construction and dismantling. According to Article 15, construction works must be designed, built and demolished in such a way that the use of natural resources is sustainable and in particular ensure the following: 1. reuse or recyclability of the construction work, its materials and parts after demolition; 2. durability of the construction work; 3. use of environmentally compatible raw and secondary materials in the construction works.

Circular Economy Strategy, Roadmap or Action Plan

No specific political communications has been adopted in the field of circular economy on national, regional and local level to date. The Circular Economy Action Plan on Construction and Demolition Waste was proposed by the World Bank [362] as a deliverable of the project "Technical assistance to the Ministry of Economy and Sustainable Development for sustainable waste management—transition to a circular economy".

Construction and Demolition Waste Management Regulatory Framework

The Waste Management Act (Official Gazette 84/21, 142/23) [110] defines construction waste as waste generated by construction and demolition activities, and recycling centre for construction waste as any facility intended for sorting, mechanical treatment and interim storage of construction waste. In order to contribute to the circular economy of the European Union, the Republic of Croatia should achieve the following objective set in Article 54: "At least 70% of the mass of non-hazardous construction waste, other than naturally occurring materials as determined by the waste code 17 05 04—soil and rocks not listed under 17 05 03, shall be recovered by means of recycling, preparing for re-use and other material recovery operations, including backfilling operations, whereby the waste is being used as a replacement for other materials." In 2020, the recovery rate of the construction and demolition waste was 60% (driven by two devastating earthquakes in 2020 and the large amounts of construction debris) [85] and 64% the following year. The Ministry of Economy and Sustainable Development and the Environmental Protection and Energy Efficiency Fund shall, pursuant regulations governing state funded subsidies, encourage the prevention of waste generation and waste management for the following activities related to the implementation of circular economy in the built environment: (1) prevention of waste generation; (2) the application of the waste management priority order; (3) the designing of products and parts of products that reduce environmental impact and waste quantity during their production and use, as well as ensure the recovery and disposal of products which have become waste, in accordance with the waste management priority order and the fundamental requirements of waste management, including the development, production and promotion of products and parts of products which are suitable for repeated use, contain recycled materials, are technically permanent and easy to repair, and which are suitable for preparing for re-use and recycling, when they become waste; (4) the promotion of high quality recycling based on the separate collection of waste; (5) the development of secondary raw materials markets; (6) the promotion of targeted deconstruction (disassembly) of buildings in order to extract hazardous substances and to facilitate re-use and the high-quality recycling of targeted materials that were excluded so as to ensure that a construction waste sorting system exists, at least for wood, mineral fractions (concrete, brick, ceramic tiles and other ceramics, stone), metals, glass and gypsum, as well as the establishment of recycling centres for construction waste, etc.

In addition to the Waste Management Act, the conditions for the construction waste management are regulated by the Ordinance on Waste Management (Official Gazette 106/22) [228] and the Ordinance on Construction Waste and Waste Containing Asbestos (Official Gazette 69/16) [229]. The concept of circular economy is incorporated into the Waste Management Plan of the Republic of Croatia for the period 2023–2028 [111], which was published in June 2023. In accordance with the concept of a circular economy, it is envisaged to create instructions for the selection and removal of hazardous construction waste during the removal of the building.

5.1.2.2 Platforms and Networks Facilitating the Transition to a Circular Economy

Circular Economy Platforms

Within the project "Technical assistance to the Ministry of Economy and Sustainable Development for sustainable waste management—transition to a circular economy", the creation of a national digital platform for the circular economy was planned, which would also contain a user portal for monitoring progress in achieving the goals of the circular economy set by European and national legislation. The platform would also serve as a tool for joint work and cooperation and a place for exchanging knowledge about the circular economy for citizens, businesses and local government, as defined in the National Reform Programme of the Republic of Croatia for 2020 [112].

The Circular Economy Hub as a physical location for demonstrating circular technologies and sharing knowledge, innovations, and best practices among all stakeholders is announced to be established [221].

Circular Buildings Platform

No circular buildings platform has been established to date.

Building Materials Passport Platform

No building materials platform has been established to date.

Public Procurement Platform

The national green public procurement platform was established in 2015 and was operated by the MoESD in 2023. It communicates green public procurement (GPP) criteria, publishes educational materials, runs webinars and free seminars, lists eco-labels, and contains life-cycle costing tools and good practice examples [85]. Among other, it contains EU green public procurement criteria for office building design, construction and management. According to the Croatian Government's Decision on Green Public Procurement in the Central Public Procurement Procedures (Official Gazette 49/2021) [113], the Central State Office for Central Public Procurement is obliged to apply GPP criteria in central public procurement procedures and the MoESD is obliged to measure and publish details of annual savings in carbon-dioxide emissions from GPP [85].

5.1.2.3 Funding Opportunities

National Recovery and Resilience Plan

Croatia's recovery and resilience plan will finance the post-earthquake reconstruction and energy upgrade of buildings damaged in devastating earthquakes that hit Croatia in 2020 and transitioning to circular economy. Besides increasing the seismic resilience of buildings (build back better principle), all buildings should achieve at least 30% energy savings compared to pre-renovation state and therefore contribute to a cleaner and safer environment. In the period of 1 June, 2023 to 1 September, 2023 there was an open call for the allocation of grants for the implementation of the Pilot project for the development of green infrastructure and/or circular management of space and buildings at the local level intended for the local self-government units. The Plan also foresees a programme of support for companies for the transition to an energy- and resource-efficient economy that is aimed at companies with up to 3000 employees [85].

Cohesion Fund

The technical assistance program provided by the World Bank was financed from the European Union's Cohesion Fund. The aim of the project "Technical assistance to the Ministry of Economy and Sustainable Development for sustainable waste management—transition to a circular economy" was to integrate the circular economy (CE) into the waste management sector. This involved aligning the Waste Management Plan of the Republic of Croatia 2017–2022 with the national legislation due to amendments to the EU Waste Management Directive, preparing the starting point and elements of the CE to be integrated into the future Croatian Waste Management Plan after 2022 in order to transition towards the CE 2030 targets, and strengthening the capacity of relevant stakeholders for CE in the waste management sector. The project was implemented within the Operational Programme Competitiveness and Cohesion 2014–2020 Priority Axis 6 Environmental Protection and Sustainability of Resources, Specific Objective 6i1 Reduced amount of waste landfilled. The applicant and beneficiary of the project was the Ministry of Economy and Sustainable Development.

WeBSEFF

Western Balkans Sustainable Energy Financing Facility (WeBSEFF) is a financing facility under which the EBRD provides credit lines to partner banks (Zagrebačka banka, Privredna banka Zagreb and Erste banka) in the Western Balkans to lend to businesses and municipalities wanting to invest in energy efficiency and small-scale renewable energy projects [76].

International Cooperation and Support from International Organizations

World Bank provided technical assistance in advancing circular economy principles in the Croatian waste management sector and transitioning towards a circular economy. World Bank supported Ministry of Economy and Sustainable Development in the preparation of the new National Waste Management Plan 2023–2028, integrating the circular economy aspects in waste management, in line with the European Green Deal and Circular Economy Action Plan. During the course of this project, the World Bank prepared relevant resources including documents, presentations from workshops and webinars, and reports from study tours. The study tours included visiting Slovenia and the Netherlands as good practice examples. The Croatian Circular Economy Committee (CEC) engaged with stakeholders from Slovenia and learnt about Government's strategy to accelerate circularity by maximizing the use of EU funds towards CE, enabling national policy framework for CE and ensuring strong stakeholder engagement, while providing business opportunities with good returns on investments [363]. The Croatian Circular Economy Committee visited Netherlands in July 2020 seeking to learn from Amsterdam's and Almere's approach to transitioning to the circular economy and managing solid waste [221]. Participants visited facilities with best practices in recycling and using waste as a resource for new products, including the construction of new buildings [363].

5.1.2.4 Challenges, Barriers and Potential Improvements

There are several challenges and barriers related to different areas of implementation of circular economy in the built environment. The World Bank [362] identified challenges and barriers in the waste management sector and proposed several measures and actions for potential improvements (Table 5.1).

5.1.2.5 Examples of Successful Implementation

Public Policy Initiatives

The new Zagreb International Airport terminal building has been designed for an annual passenger throughput of 5.5 million passengers per annum, with added resilience to allow further extension to accommodate 8 million passengers. The terminal was constructed by MZLZ (a consortium of Bouygues Bâtiment International, Aeroports de Paris Management and Viadukt) who were awarded the Airport Concession by the Croatian government in December 2013 to operate the airport for a 30-year period, including the Design and Build of a new terminal building. Arup was appointed to act as the Independent Engineer for the New Passenger Terminal at Zračna Luka Airport in Zagreb. They were responsible for tracking performance, quality and compliance of the construction works over the lifetime of the project and also supervised and witnessed the testing and completion of elements of the works as they developed. Arup reported jointly to the Concessionaire (MZLZ) and also the Grantor (Croatian Ministry of the Sea, Transport and Infrastructure), providing monthly reports of compliance and performance tracking [14]. This was the fourth new airport terminal to be built by Bouygues Bâtiment International as a Public–Private Partnership (PPP), within the framework of a project company in which it is one of the principal shareholders. As the project aimed to secure LEED certification, particular attention was paid to sustainable construction [187].

The environmental sustainability of the Airport design solution was based on [162: 477]:

- ventilation of the facade and roof, principled on the double membrane envelope
- large area of the photo-voltage cells (8500 m^2) for the environmentally friendly production of electricity

Table 5.1 Overview of challenges and potential improvements for the implementation of the circular economy in CDW sector in Croatia

Challenges	Recommended measures	Recommended actions
Absence of a shared vision among stakeholders	1. Enhance dialogue and cooperation among CDW stakeholders	1. Creation of CE platform for CDW management to discuss regulatory provisions, new opportunities and share information
	2. Enhance education and support for research, innovation, and development	1. Organization of specific annual educational programs including meetings with local administrations and the drafting of a handbook related to innovative recovered materials, reuse, and recycling within the framework of CE and CDW regulatory requirements 2. Development of R&D program for CE in CDW sector through the creation of nation or EU funds
Limited information about CDW generation and management	1. Implement educational and informative activities related to CDW reporting	1. Educational activities, including training, guidelines, and handbook, on the topic of CDW management reporting data within the framework of the CE

(continued)

Table 5.1 (continued)

Challenges	Recommended measures	Recommended actions
	2. Improve CDW management information system applications	1. Enhanced interoperability between waste registers EONTO and ROO (Environmental Pollution Register of the MoESD) for data collection regarding CDW quantities and treatment 2. Enhanced the usability of EONTO through the creation of a mobile application 3. Integration of recovered material data generated on-site in the construction permit database for building sites with an expected CDW generation greater than 200 t 4. Integration of digital applications related to CDW on the joint IT platform of waste management in the Republic of Croatia, as part of the national environmental protection and circular economy platform 5. Creation of a Materials Passport and a related database to register materials used in construction projects in order to ease the recognition of hazardous material, future dismantling phase and recovery of resources

(continued)

Table 5.1 (continued)

Challenges	Recommended measures	Recommended actions
	3. Improve availability and quality of CDW management data before and during operating sites	1. Introduction of obligations to prepare CDW Management Report as part of Main Project before issuing Construction Permit for sites with an expected CDW generation greater than 200 t 2. Drafting of standard forms and an on-line tool for compiling and submitting the CDW Management report 3. Approval of a bylaw for the enforcement of a mandatory financial bond or bank surety for construction projects to cover the cost in case of mismanagement of CDW on the building site
	4. Implement pre-demolition audit for construction sites	1. Amendment to Law on Construction to include pre-demolition audit based on EC guidelines for construction sites with an expected CDW generation greater than 200 t 2. Definition of guidelines on Selective and Removal of Hazardous Components based on the European Guideline 3. Training for the enforcement of guidelines on Selective and Removal of Hazardous Components
Uncontrolled streams of CDW	1. Increase education of relevant bodies involved in CDW management inspection processes	1. Training programs for CDW management supervision and inspection bodies and parties

(continued)

Table 5.1 (continued)

Challenges	Recommended measures	Recommended actions
	2. Strengthen regulation and controls to reduce illegal dumping of CDW	1. Enforcement of waste management regulation to strengthen the control powers of local authorities and state inspectorate on environmental crimes related to illegal waste management 2. Enforcement of construction law and introduction of mandatory cost breakdown for CDW management in technical project proposals to support companies that propose proper CDW management. 3. Investment in technological equipment such as capturing cameras and remote cameras to increase and automatize controls against illegal dumping 4. Improvement of ELOO (record of the location of discarded waste) mobile application for citizens to ease the anonymous reporting of CDW illegal dumping 5. Strengthening of sanctions and fines against illegal dumping
	3. Remove CDW from illegal dumpsites	1. Removal of CDW (or mixed waste) from locations contaminated by waste dumped into the environment

(continued)

Table 5.1 (continued)

Challenges	Recommended measures	Recommended actions
Limited capacity for CDW recovery and recycling	1. Improve End of Waste (EoW) status legislation for CDW	2. Feasibility study for the identification of potential "by-product for non-contaminated excavated soil" and "EoW status for specific CDW streams." 3. Definition of a guideline on the enforcement of EoW and relevant Ordinance No. 117/14 4. Training sessions on the guideline on the enforcement of EoW
	2. Improve hazardous waste removal and asbestos disposal system	1. Analysis and assessment (report) of the existing and required number and capacity of areas for the disposal of CDW containing asbestos and treatment of hazardous CDW 2. Grants for the development of installation for the recycling of contaminated soils (EWC 170,503*) based on BAT 3. Financial support for the collection and disposal of asbestos-containing material from existing buildings
	3. Construct and equip recycling yards for construction and demolition waste "Type A"	1. Construction and equipping of municipal or regional "Type A" recycling yards for CDW on the mainland 2. Construction and equipping of municipal or regional "Type A" recycling yards for CDW on the most populated island

(continued)

Table 5.1 (continued)

Challenges	Recommended measures	Recommended actions
	4. Construct and equip recycling yards for construction and demolition waste "Type B" for processing CDW	1. Construction and equipping of "Type B" recycling yards for processing CDW, equipped with mobile equipment to service municipal and regional requirements 2. Construction and equipping of "Type B" recycling yards for processing CDW, as part of waste management centers equipped with mobile equipment to service municipal and regional requirements
	5. Build and equip facilities for CDW analysis and quality control	1. Financial funds for the development of authorized quality control infrastructure for CDS SRM
Low market share on CDW secondary raw materials (SRM)	1. Improve the existing application to develop a CDW Exchange	1. Improvement, development, and implementation of a web-based tool such as CDW Exchange or marketplace for materials recovery and secondary materials exchange
	2. Create guidelines on GPP	1. Creation of guidelines for GPP of the office building design, construction and management 2. Implementation of GPP pilot project in accordance with guidelines for office building design, construction and management 3. Creation of guidelines for GPP of road design, construction and maintenance 4. Implementation of GPP pilot project in accordance with guidelines for road design, construction and maintenance 5. Feasibility study for the introduction of a minimum quantity of recycled/reused materials in public constructions and infrastructures in the construction law

(continued)

Table 5.1 (continued)

Challenges	Recommended measures	Recommended actions
Increasing generation of CDW	1. Develop guidelines for the reuse of building components, conversion, and renovation of buildings	1. Drafting of a guideline on identification of reusable component including a positive list of elements such as doors, windows, and other non-structural parts 2. Development of guidelines for sharing, conversion and renovation of buildings including recommendation for amendments and supplements to spatial plans 3. Dedicated grants to support the local administration and private companies on restoring buildings instead of demolition in accordance with guidelines and to promote sharing, conversion, and multifunctional uses of the public buildings
	2. Enhance take-back systems in CDW to increase reused and recycled content	1. Feasibility study for the introduction of "take back system" on specific CDW scraps and unused materials as: 17 01 concrete, bricks, tiles and ceramics; 17 03 bitumen mixtures, coal tar and products containing tar; 17 08 building material based on gypsum 2. Pilot project to enhance "take back system" on select CDW stream based on result of the aforementioned study

(continued)

Table 5.1 (continued)

Challenges	Recommended measures	Recommended actions
	3. Implement educational and informative activities on eco-design in public and private buildings	1. Implementation of educational and informative activities on the topic of eco-design in public and private buildings to raise technical knowledge and education for engineers, architects, technicians, contractors and public administration 2. Design and construction of an eco-design pilot project in a public building 3. Reduction of planning fees for construction projects that use recycled or reused materials or adopt eco-design criteria focused on future reduction of CDW generation
	4. Develop guidelines for sharing, conversion and renovation of buildings	1. Development of guidelines for sharing, conversion and renovation of buildings including recommendation for amendments and supplements to spatial plans 2. Dedicated grants to support the local administration and private companies on restoring buildings instead of demolition in accordance with guidelines and to promote sharing, conversion, and multifunctional uses of the public buildings

Source World Bank [362]

- trigeneration plants for synergetic production of electricity and preparation of warm and cold water
- collection, processing, purifying and managing of water from all parts of the complex, such as roofs, aprons, runways, sanitation facilities, etc.
- centralized control and management of all power and utility resources by means of the efficient management systems (EMS)
- selection of best technological solutions and usage of materials which contribute to quality and ecological sustainability.

The roof structure is made up of 45,000 elements, each different from the others. The works switched to post-tensioned concrete for the floors and connected the drainage system to the city's water treatment plan. To avoid any risk of flooding, 40 km of pipe were installed under the 66,000 m^2 of the terminal, the 100,000 m^2 of taxiways, and the 40,000 m^2 of aircraft parking stands. LEED silver certification acknowledges measures taken to increase its environmental performance, including a rainwater recovery system to provide water for the toilets and the use of concrete with titanium content to capture CO_2 [18]. Zagreb Airport received an Airport Carbon Accreditation from the ACI (Airport Council International) in 2021, which confirms the level 3 of management and reduction of CO_2 emissions in everyday airport activities. According to the classification of the global ACI ACA program (Airport Carbon Accreditation), this level is called "Optimization" and represents a significant step towards fulfilling the commitment of the Zagreb Airport to achieve zero CO2 emissions by 2050. In recent years, Zagreb Airport has introduced a wide range of energy management measures that have enabled it to monitor and reduce overall energy consumption, such as the installation of efficient LED lighting, low voltage reconstruction in transformer station, boiler room reconstruction, reconstruction and modernization of the heating/cooling substation and hot water using solar collectors. Thanks to all these activities, the carbon emissions at the airport were reduced by 4% between 2017 and 2019, despite a significant increase in the number of passengers in that period. Plans for the airport include the construction of a photovoltaic power plant that will enable the direct conversion of solar energy into electricity and the replacement of existing vehicles and equipment in an environmentally friendly way [368].

Private Policy Initiatives

Valamar Amicor Green Resort is the first family resort on the Croatian Adriatic to be built according to the principles of sustainable construction and designed to exist in harmony with nature. Valamar Amicor Green Resort was built in 2022 in partnership with the German Sustainable Building Council (*Deutsche Gesellschaft für Nachhaltiges Bauen*—DGNB). The interior design used natural and recycled materials from local sources, reinterpreted by local artists. Wood, as one of the most sustainable building materials, has been used to build the villas, and the paint on the walls does not contain harmful particles, so maximum air quality is achieved inside the accommodation units. With regard to the new accommodation units, the prefabricated construction was used, which saves energy, speeds up the construction process, creates controlled production conditions, and allows easy disassembly of parts of the building that can be reused later [352].

5.1.3 Cyprus Factsheet

5.1.3.1 Policy and Regulatory Framework

Governance

Circular Economy policy in the Republic of Cyprus is under the jurisdiction of the Ministry of Energy, Commerce and Industry (MECI), which in 2021 launched the Cyprus Action Plan for Circular Economy [301], in collaboration with the Ministry of Agriculture, Rural Development and Environment (MoA) and the Deputy Ministry of Research, Innovation and Digital Policy (DMRID). The development of the national action plan was a follow up of the development of the European Action Plan [83] providing member states to take the lead in circular economy transition according to their Smart Specialization Strategies and national targets. At the launch of the National Action Plan, the then Minister of Energy, Industry and Commerce, Ms Natassa Pilidou, mentioned that: "..beyond the fact that Circular Economy contributes directly in the reversal of climate change effects, due to the decrease of greenhouse gas emissions, serving towards our national targets for net zero emissions by 2025, Circular Economy Best Practices, create opportunities for the transformation of our businesses and industry, in order for them to acquire more competitive advantages and become more competitive and sustainable locally, and more importantly internationally. Circular Economy, just as any green investment, creates new jobs and new novel products" (DMRID Press release, 27/7/2021). The collaboration of the two ministries and the deputy ministry is key in the implementation of Cyprus Circular Economy goals through incentives, waste management regulation and digitilisation which can play a significant role in the success of the action plan. Actions from the action plan are part of the Cyprus National Recovery and Resilience plan [82].

Legislation

Circular Economy appears in the Republic of Cyprus Legislation under the Waste Management Law (Cylaw, 185(I)/2011 and its latest amendment 48(I)/2022). The latest amendment to contains an act for Construction and Demolition Waste (CDW) which is segregated from the municipal solid waste (under the jurisdiction of Ministry of Interior) and is now under the jurisdiction of the Department for the Environment (DfE) of MoA. In 2023 the DfE issued new regulations for CDW management (Κ.Δ.Π. 112/2023) which aim to protect the environment and human health by adopting new measures for the rational management of CDW, targeting the reuse and exploitation of at least 70% of the volume generated.

Circular Economy Strategy, Roadmap or Action Plan

In June 2021, MECI in collaboration with MoA and DMRID issued the National Plan for supporting the implementation of Circular Economy in Cyprus 2021–2027. The plan includes prospects for circular economy in Cyprus, the priorities and special goals and the governance of the plan which includes participation, implementation and monitoring.

The national plan implementation budget is set to 90 m Euros, elements of which are included in the Cyprus Recovery and Resilience Plan. The four pillars of the plan are as follows:

- Development of Culture (mindset) for Circular Economy
  - Developing awareness of the business community and consumers for the prospects and opportunities of circular economy, public campaigns for sustainable waste management strategies and training in circularity topics.
- Provision of incentives
  - Creation of a plan for the provision of consulting services and guiding grants (1 m Euro total, up to 15 k per organization),
  - Creation of grant schemes from the Cyprus Research & Innovation Foundation (RIF) "Go Circular" for new circular products and services (1 m Euro).
  - Utilization of courses certified by the Authority for the Development of Human Resources (ADHR) for circular economy training.
- Infrastructure
  - Investigation for the identification of waste categories fit for decharacterisation,
  - Creation of online platform which will allow organisations to share infrastructure, services, waste management etc.
  - Development of an integrated system for monitoring waste management projects (2 m Euros)
- Municipal Solid Waste (MSW) Management
  - Development of "pay as you waste" programme (25 m Euros)
  - Segregated collection systems for mountainous areas and operation of "Green Points" (Recycling centres) for remote areas (3.5 m Euros)
  - Domestic composting of organic waste (7 m Euros)
  - Creation of Product Repair and Reuse centres (4 m Euros)
  - Reduction and segregated collection of MSW (5 m Euros)
  - Plan for utilization of MSW through segregated collection (15 m Euros)

The plan has targeted four priority areas

- Primary Sector
- Food and Beverage Sector
- Construction
- Hotels, Restaurant, Café (HoReCa)
- Other individual industries

The mechanism to monitor the implementation of the plan is through a committee under the realm of the two ministries (MECI and MoA) and deputy ministry (DMRID) and it will also include representatives from EPSA (DG Growth RoC), industrial stakeholders and experts. The committee will be convening at regular intervals to:

- Ensure the achievement of the action plan targets.
- Keeping all stakeholders committed to the plan.
- Monitoring of actions included in the plan.

Construction and Demolition Waste Management Regulatory Framework

In 2023 the DfE issued new regulations for CDW management (Κ.Δ.Π. 112/2023) which aim to protect the environment and human health by adopting new measures for the rational management of CDW, targeting the reuse and exploitation of at least 70% of the volume generated. The regulations apply to the construction project owners, the producers of CDW and any entities involved in CDW management as long for projects that generate a minimum of $12m^3$ of CDW. For the first time these regulations include CDW management as part of the owner obligations and it is incorporated in the technical specifications of the project. CDW producers (i.e. construction companies, etc.) have the obligation to organize or participate in a collective CDW Management System or to run their own. The regulations include templates for all the CDW management actions to be fully documented and monitored in order to achieve the targets at hand. The CDW Management plan includes:

- Name and address of owner in case of a natural person or legal entity
- Description of the project including timeline and land registry/town planning details of the plot on which the construction activity will take place
- Estimated quantities per type of CDW (either by volume or weight) expected to be produced during the project, as well as the expected quantities to be reused or recovered, taking into account the opinion of the project consultant.
- The CDW management methods to be implemented, the necessary equipment which could be used for processing CDW.
- The agreement with a licensed CDW Management unit for collecting and processing CDW.
- The location within the construction site for temporary storage and processing of CDW.

The regulations further describe the obligation of single and collective CDW systems in order to achieve the 70% target as well as the necessary licenses and certifications required according to the legal framework.

5.1.3.2 Platforms and Networks Facilitating the Transition to a Circular Economy

Circular Economy Platforms

Following the European and National action plans a number of platforms and networks have begun appearing in Cyprus pertinent to Circular Economy. Most of the initiatives are still in their infancy, but in all cases the composition of this initiatives engages a diverse spectrum of stakeholders including academia, policymakers and industry amongst others. A few of these initiatives are presented below:

- Cyprus Circular Economy Network (CCEN)
  - The first pioneering network for Circular economy in Cyprus founded by the Cyprus Federation of Industrialists and Employers (OEB), the Cyprus Institute of Management (CIIM), the Cyprus Institute (CyI), the Cyprus University of Technology (CUT) and the Cyprus Organisation for Standardisation (CYS)
  - The ultimate goal of the CCEN is to enable and accelerate the transition of Cyprus economy to a circular and green economy, especially after the COVID-19 pandemic, offering its services in a multilevel stakeholder approach; businesses, academia and public sector, contributing to the achievement of the economic and social resilience of Cyprus, for a sustainable future.
- The mission of CCEN is
  - Promote the implementation of the European and Cyprus strategy, policy and contribution for achieving the national goals for the circular economy.
  - Enhance the multilevel development of the circular economy, resource efficiency and sustainable production and consumption in Cyprus.
  - Ensure the application of the circular economy principles to reduce value loss by recovering resources via reusing, repairing, refurbishing, remanufacturing, repurposing, and/or recycling.
  - Promote the implementation of the circular economy principles in the design of products, services, infrastructure and processes.
  - Support the utilization of green procurement and ensure circular supply chains and green contracts and supplies.
  - Facilitate the development of smart and circular cities, smart and circular industrial zones/areas with zero waste.
- KyklOIKOdromio (KYK)

- kyklOIKOdromio is a non-profit organization, which aims to implement educational and research programs, conduct studies, as well as organize events, workshops and other activities in the fields of energy, environmental protection, climate change and Circular economy. It also aims to develop information technologies on related topics.

- Circular Economy Alliance (CEA)
  - The Circular Economy Alliance is a training and consulting organisation for the transition to circular economy offering courses and certifications to several aspects of CE. The mission of the organisation is to strengthen the global workforce by giving professionals and organisations the complete, accredited, reliable, actionable skills and knowledge they need to become change agents while deploying Circular Economy best practices.

Circular Buildings Platform

No circular buildings platform has been established to date.

Building Materials Passport Platform

No building materials platform has been established to date.

Public Procurement Platform

The national action plan for Green Public Procurements was established in 2014 and is operated by the MoA. It communicates green public procurement (GPP) criteria, the categories and national targets. It has been approved by the Council of Ministers of the Republic. Organisations like the Cyprus Energy Agency are GPP Supporter organisations and provide supporting materials and training for achieving the GPP targets. The revised action plan has 4 annexes:

- Annex A: Summarises 12 groups of products/services, n 30 categories. This annex further includes measures and application targets, alongside some guidelines to be included in the procurement documents.
- Annex B: Refers to environmental management pertinent to:
  - Paper consumption
  - Ink cartridges
  - Batteries
  - Electric appliances
  - Hazardous waste
  - Packaging recycling
  - Energy savings

- Food purchases
- Noise

- Annex C: EU and National Criteria for GPP
- Annex D: Includes lists of contracting authorities and bodies

The target for Cyprus for GPP is 50% of green public procurement contracts, subject to exceptions. The target for each category is cited in annex A.

5.1.3.3 Funding Opportunities

National Recovery and Resilience Plan

Circular Economy features in the Cyprus National Recovery and Resilience Plan funding schemes and specifically in Measure 3.1 "New growth model and diversification of the economy" and particularly intervention 3.14.—Strengthening Circular Economy in industry which includes the following actions:

- Promotional campaign for building awareness in consumers regarding the advantages of circular economy.
- Operation of a Grant scheme to support the transition of business organisations in a circular economy model.
- Development of a Sharing Marketplace platform for Circular Economy.

Cyprus Research and Innovation Foundation (RIF)

The Circular economy action plan includes provisions for research and innovation grants from RIF which are in the pipeline ("GoCircular grants"). In the meantime, Circular Economy is considered as one of the priorities in horizontal research and innovation calls. A successful integrated project includes the Development of an innovative insulation fire resistant façade from CDW, [311] funded from the EU Regional Fund, the RoC and the Structural Funds through RIF.

International Cooperation and Support from International Organizations

Cyprus through the Cyprus Organisation for Standardisation (CYS) participates in Technical Committees for standards on Circular Economy, including TC350/SC1 Circular Economy in the Construction sector [24].

Challenges, Barriers and Potential Improvements

There are several challenges and barriers related to different areas of implementation of circular economy in the built environment. A pathfinder study [186] for the development of a roadmap for the implementation of Circular Economy Thinking for the Construction sector in Cyprus has identified the following challenges and barriers in 2018.

- Regulations were until very recently in transition. New regulations just published.
- No strict control and enforcement of environmental policy on the precautionary principle, the "polluter pays" principle and the principle of collaboration between industry, regulators and academia.
- Manufacturers and distributors in their majority, are most likely not designing their products in such a way as to minimize the amount of waste produced during manufacturing and to facilitate an ecological removal of those components of the waste which can no longer be reused.
- The absence of a policy requiring designers, builders, and contractors to use an appropriate percentage of recycled CDW materials.
- Lack of environmental inspectors and low organizational capacity for implementation and/or enforcement of the law, causing delays in the administration of fines or non-conviction of CDW management rules violators.
- The general mentality in the construction sector and of the general public in Cyprus is based on the misperception that CDW can be disposed somewhere and left there, since its inert nature makes it harmless for human health and the environment.
- There is a preference to avoid the cost of CDW management (illegal fly tipping or re-use on site) even if it means paying the fine for not complying with the regulations due to the high cost of transport and disposal in collection areas, and general lack of skills and knowledge to organize effective systems of CDW management.
- No perceived value in recycled CDW due to virtually no market demand natural materials are always preferred over recycled materials in the construction works
- Lack of knowledge from construction industry professionals regarding the implementation of CDW legislation, including on knowledge on how to prevent and minimise CDW.
- No market demand for recycled CDW, virgin materials are usually preferred
- There is no implementation of secondary raw material regulation and standards in Cyprus's waste framework policy that would specify: the nature of the waste which could be used as a secondary raw material in construction and/or the minimum concentrations of heavy metals and aromatic hydrocarbons.
- Lack of confidence in the quality of construction and demolition recycled materials restricts the demand for CD recycled materials, which inhibits the development of CDW management and recycling infrastructures in the EU.

A lot of these barriers and issues have been addressed with the recent regulations and their implementation will accelerated the implementation of circular economy and achievement of the action plan targets. On the other hand, the small scale and isolation (island state) of the Republic of Cyprus, amplifies a number of challenges that can make Circular Economy principles seem unachievable.

5.1.3.4 Examples of Successful Implementation

Public Policy Initiatives

No data to this point.

Private Policy Initiatives

- Nicolaides & Kountouris Metal Company.
  - Circular products: THERMINK and PS THERMINK
  - The company is manufacturing metal panels with polyurethane core. They are using their polyurethane waste stream to create a circular thermal insulation mortar (THERMINK), reducing their polyurethane trimmings waste to zero and managing polyurethane waste from others too. A few years later they developed a similar (lower quality) product using styrofoam waste.
- Elysee plastics
  - Elysee strategically approaches circular economy model, moving on to a sustainable world. GreenDrip project aims to recycle post-industrial polypropylene (PP) material (PPM) generated through the manufacturing processes of piping fittings, producing high-quality recycled PP thermoplastic material—green PP—to be used as secondary material in co-injection molding. GreenDrip project runs for 9 months, a sufficient timeframe for successfully fulfilling all industrial research and development activities/tasks and falls within (TRL) 3–4 (TRL 3—experimental proof of concept, TRL 4—Technology validated in lab) Technology Readiness Levels. The project's implementation plan consists of the creation and lab validation of GreenDrip prototypes.
- KAMERIS GALLERY (no website)
  - Kameris group is a leading Demolition Company. Over the years they began salvaging building material and household components and appliances and created a depot on the outskirts of Nicosia.

5.1.4 *Czech Republic Factsheet*

5.1.4.1 Policy and Regulatory Framework

Governance

The responsibilities in relation to the implementation of circular principles in the building industry are divided between three ministries and their departments.

The key actor in the implementation of a circular economy in the Czech Republic is the Ministry of the Environment. On its behalf, the strategy "Cirkulární Česko" was issued in 2021 [199]. This ministry has several competencies related to circularity such as waste legislation and protection of nature. On the other hand, the competencies concerning the production of materials and products (even secondary raw materials) are covered by the Ministry of Industry and Trade. Moreover, this ministry also organizes national funding schemes to support business development and circularity of SMEs.

Concerning buildings, another key actor is the Ministry of Regional Development, which has competence over the Building Act and related documentation. Therefore, the parameters of buildings and their limits for building permits and permits for building demolition are defined by departments of this Ministry. As the building industry is an important part of the economy, the parameters of buildings are also developed in coordination with the Ministry of Industry and Trade.

There are also national agencies, which are responsible for the development of knowledge for ministries and the preparation of background data for preparation of strategies. One of the key agencies is the Czech Standardization Agency, which is currently supporting the development of BIM and digitalization standards and standards for demolition audits. Another agency is the Czech Environmental Information Agency, which collects for example data about waste management. On the other hand, the collection of data about raw materials is the responsibility of the Czech Geological Survey.

Legislation

The main act, which regulates the circularity of materials in the Czech building industry, is the Waste Management Law and linked regulations [198]. This act defines the obligation of a producer of waste to manage it properly in compliance with the waste hierarchy. However, recycling of waste is limited by regulations to disable the possibility for recycling of hazardous waste. The key regulation for recycling of aggregates, as a main part of construction and demolition waste, is regulation about waste management [197], which defines the limits of hazardous compounds, methods of sampling and possible final use. Furthermore, the recycling of some materials is then subject to other regulations such as regulation about asphalt [200], which defines the limits of polyaromatic hydrocarbons concerning the final use of asphalt. A similar regulation considering also final use and function of the recycled material is also being prepared for concrete and other materials.

The Building Act in the Czech Republic supports circularity only in a general way. At first, the act declares that the building must be used only as a product, which supports the proper and safe function of the building, which can often be used as an excuse to not use recycled materials. On the other hand, the act refers to the regulation of building documentation. A new version of this regulation, which is being prepared will define in more detail the documentation needed for pre-demolition audits, which should support higher use of recycled materials.

Circular Economy Strategy, Roadmap or Action Plan

The main strategic document related to circular building industry in the Czech Republic is "Cirkulární Česko 2040" [199]. This document introduces the principles of circular economy in the main areas of industry and economy in the Czech Republic. One of these areas is the building industry as a part of the chapter about resources consumption, the building industry and energy consumption. In this part, the state of the art is described in eleven paragraphs pointing out the significance of European initiatives such as the renovation wave, waste management goal (70% rate of recycling), energy refurbishment of the old housing etc. Based on this state-of-the-art description, main goals are set with the support of principles and measures. Concerning the building industry, the main measures cover support of digitalization in the building industry, increased rate of recycling materials, and support of secondary materials in green public procurement. Unfortunately, no specific indicators are defined for these measures.

Construction and Demolition Waste Management Regulatory Framework

The main document affecting waste management is the Waste Management Plan of the Czech Republic. This document declares the goal of a 70% recycling rate and defines measurable indicators and measures. Following this document, regions and cities have the same goal stated in their local waste management plans.

The reaching of the above-mentioned goal is regulated mainly using the Waste Management Act. Also, the circular strategy for the Czech Republic was published to indicate key activities to increase circularity of the Czech Republic.

5.1.4.2 Platforms and Networks Facilitating the Transition to a Circular Economy

Circular Economy Platforms

One of the key platforms is a working group on circular economy and sustainable materials organized by the Czech Green Building Council [28]. The main aim of this working group is to support their members in the adoption of circular measures and communicate the needs of the building industry to the state representatives.

The Circular Economy Hotspot organized by the Institute for circular economy covers different areas of industry and economy and creates a space for networking and raising awareness about topics for circular building economy [153].

The demolition companies and recycling companies are grouped in ARSM., which is the association for the development of construction materials recycling [2].

Circular Buildings Platform

No circular building platform has been established to date.

Building Materials Passport Platform

No building materials platform has been established to date.

Public Procurement Platform

The national green public procurements platform was founded with the support of the Ministry of Labour and Social Affairs in 2021 [196]. The platform aims to raise awareness about public procurement and implementing other facilities concerning social and environmental perspectives of sustainable development. The platform creates space for networking and sharing of good practices among cities, state and regional authorities.

5.1.4.3 Funding Opportunities

The National Recovery Plan of the Czech Republic is organized by the Ministry of Industry and Trade and it aims to support tackling of challenges of green transition and innovation [176]. It covers multiple activities and supports research via the Technology Agency of the Czech Republic.

International Cooperation and Support from International Organizations

No data to this point.

5.1.4.4 Challenges, Barriers and Potential Improvements

Based on the discussion with stakeholders grouped in professional organization for building industry—Czech Green Building Council [28], the main barriers were summarized in the Table 5.2 below with recommended actions.

5.1.4.5 Examples of Successful Implementation

Public Policy Initiatives

No data to this point.

Private Policy Initiatives

No data to this point.

5.1.5 Finland Factsheet

5.1.5.1 Policy and Regulatory Framework

Governance

According to the Government Resolution on the Strategic Programme for Circular Economy [230], which entered into force on April 8, 2021, the Ministry of the Environment is the official body of Finland that shapes circular economy policy. The aim of the Ministry of the Environment [231] is a circular economy that provides a foundation for all economic activities so that the production and consumption are within the Earth's carrying capacity. A circular economy offers solutions to mitigate climate change and halt the decline in biodiversity. In a circular economy natural resources are used more sustainably and they are kept in circulation safely and for a longer time. At the Ministry of the Environment, work on a circular economy is being done at the Department of the Built Environment, Environmental Protection Department and Department of the Natural Environment.

According to the Government Programme, "Finland's role as a pioneer of the circular economy will be strengthened and a cross-administrative, Strategic programme to promote a circular economy and its indicators will be implemented.

Table 5.2 The main barriers

Barrier	Recommend action
Removing legislative barriers	Definition of waste and related terms Proper waste transfer approach Monitoring of national policies and strategies (ask someone from MIT's industrial ecology department to join the group) Sharing good practice on donation contracts (waste donation, valuation) Definition of guarantees for reused elements Mapping the implementation of mandatory pre-demolition audits
Increasing the demand from contracting authorities for environmentally friendly materials	Sharing examples of how to make selections to favour environmentally friendly materials
Adjustment of standards	Support for the extension of standards for individual product categories to take into account (and therefore allow) the recycled content Definition of the content of the pre-demolition audit documentation Support for the preparation of a standard for pre-demolition audits
EPD support	EPD mandatory only from 2028 (CPR)—map what is covered, help/support/require to implement earlier
Increase the availability of information on materials contained in buildings	Obligatory statements of works for refurbishments, BIM standards, passporting, how to store and submit data and who should do it (for new and existing buildings)
Education—circular and green products and building design	Architects, designers—circular buildings: recycling myths, design for deconstruction Contractors and suppliers—concerning circularity, wood and other natural materials Good circular practice in building management—fit-outs
Increase the take-back capacity of manufacturers and other processors	Mapping the readiness of manufacturers and other processors to take back used products (list of materials and entities able to take it back)
Information support for investors and developers	Mapping the readiness of producers to decarbonise production Sharing of good examples for reaching 70% recycling for taxonomy (DNSH)—technical solutions for realistic achievement and reporting (for funding)

The programme sets objectives, defines the necessary measures and allocates the necessary resources. The transition to a circular economy is an important opportunity for Finland. The circular economy offers a way to strengthen Finland's export-driven economy and employment. At the same time, it will reduce the consumption of natural resources and the resulting CO_2 emissions and other environmental impacts. The circular economy is not just about the economy and technological solutions. The transition to a carbon–neutral circular economy will require a comprehensive change both in decision-making and planning by society and in the attitudes and behaviour of companies, households and consumers." [224].

Legislation

The Government Resolution on the Strategic Programme for Circular Economy [230] has been prepared in cooperation between key ministries (Ministry of the Environment, Ministry of Economic Affairs and Employment, Ministry of Agriculture and Forestry, Ministry of Finance, Ministry of Transport and Communications, Ministry of Education and Culture and the Ministry of Foreign Affairs) and research institutes (Natural Resources Institute Finland, Finnish Environment Institute, VATT Institute for Economic Research, VTT Technical Research Centre of Finland) as well as SITRA and Business Finland (BF). The steering group has been led by Professor of Working Life Reijo Karhinen, and the Ministry of the Environment and the Ministry of Economic Affairs and Employment have been responsible for the practical preparation. This resolution outlines the key measures that the ministries have committed to implementing in their areas of responsibility in 2021–2024 to the extent that this is possible within the framework of their resources.

According to the Government Resolution (2021), the vision of the programme is that in 2035, a carbon–neutral circular economy society will be the foundation of successful economy in Finland, in which:

- Sustainable products and services will be part of the mainstream economy and the sharing economy will a normal part of daily life.
- Our choices will be future-proof and strengthen a fair welfare society.
- More for less: The use of natural resources will be sustainable and materials will remain in circulation for longer and safely.
- The circular economy breakthrough has been made with the help of innovations, digital solutions, smart regulation and responsible investors, companies and consumers.
- Circular Economy Finland will influence the world and provide sustainable solutions in international markets.

The implementation of the vision will require the sustainable and efficient use of natural resources. This is outlined by the following steps and objectives:

- The consumption of non-renewable natural resources will decrease, and the sustainable use of renewable natural resources may increase to the extent that

the total consumption of primary raw materials in Finland in 2035 will not exceed what it was in 2015. The natural resources used to manufacture exported products are not covered by the objective.
- The profitability of resources will double by 2035 from what it was in 2015.
- The circular economy rate of materials will double by 2035.

Circular Economy Strategy, Roadmap or Action Plan

National. Leading the cycle—Finnish road map to a circular economy 2016–2025 [324] is a Finnish circular economy roadmap, published in 2016 by the Finnish Innovation Fund SITRA [325]. Finland's circular economy road map describes the concrete actions that can accelerate the transfer to a competitive circular economy in Finland. The road map highlights best practices and pilots that can be easily replicated and provide added value on a national scale. According to the SITRA [314], the target of the Finnish government and the road map is to make Finland a global leader in the circular economy by 2025. The goal and mindset are the starting points for the road map according to which the road map will make Finland a leading circular economy country by 2025. This change will emphasise the state's role as a facilitator and supporter, research, development and innovation activities, and a strong company, export and technology orientation combined with the search for comprehensive solutions and co-operation covering the entire value chain.

Finland will seek a pioneering role by focusing on five interlinked focus areas: (1) a sustainable food system, (2) forest-based loops, (3) technical loops, (4) transport and logistics, and (5) joint actions. Synergies between these areas will also be taken into account. The actions in the different focus areas of the road map are divided into three levels: policy actions, key projects and pilots. Out of more than 100 ideas, the pilots with the greatest opportunities for expansion and best supported the target were included in the project portfolio. The process of compiling the road map also identified areas in which Finland should be active but where the necessary initiatives haven't been clarified or responsible parties still have to be found [314].

The Roadmap was revised in 2019—Finland 2016–2025 Circular Economy Roadmap 2.0. [326]—and includes the following strategic objectives [315]:

- Renewal of the foundations of our competitiveness and vitality, putting the focus of circular economy solutions on competitiveness and an economic growth strategy.
- Making a shift to low-carbon energy; in addition, we need to promote the efficient use of energy and set more ambitious goals in our climate and energy policies.
- Natural resources are regarded as scarcities, because to reach the climate targets of the Paris Agreement consumption and production in Finland can no longer be based on the limitless use of natural resources.
- Everyday decisions act as catalysts for change. Cutting our carbon footprint in half by 2030 requires the adoption of a new kind of approach to ownership, in terms of culture, taxation and income distribution.

Regional. Ostrobothnia in Transition—Roadmap for sustainable development and circular economy [163] is based on the UN's 17 Sustainable Development Goals. The report has been prepared by Ostrobothnian development companies in co-operation with the Regional Council of Ostrobothnia and the EU Regional Development Fund. The report is published as part of the CERM project, Circular Economy Road Map and Action Plan for Ostrobothnia Region in Finland. The CERM project creates a roadmap including a strategy and an action plan for the three regions in Ostrobothnia, Finland to develop a society that is based on circular economy and sustainable principles. The goal is to identify the most important branches in each region that have the biggest volumes and where circular economy can be implemented in the easiest way. The strategy and action plan are based on statistics, structure of the society, inhabitants and business life.

Local. The City of Helsinki's Roadmap for Circular and Sharing Economy [38] was laid out in 2020. The City of Helsinki's Roadmap for Circular and Sharing Economy is one of the 147 actions in the Carbon–neutral Helsinki 2035 Action Plan. The roadmap includes the following four focuses: construction, procurements, green waste, and sharing economy and new business opportunities in circular economy. One topic in the roadmap is the reduction of the consumption of plastics and increased utilization of recycled plastics. The roadmap has been processed in workshops in collaboration with experts from both inside and outside the City. A team consisting of representatives of the City's Environmental Services has coordinated the work. The goals for each focus are set until 2035. At this point, more detailed reviews cover only the council terms of 2020–2021 and 2021–2025. Interim goals and supporting practical actions have been set for each focus and for both of these terms. The intention is to update the interim goals and actions in the roadmap during each council term starting from 2025 and until 2035.

Construction and Demolition Waste Management Regulatory Framework

The National Waste Plan to 2027 [331] has been approved is 2022 (in Finnish, only abstract in English). The plan presents the objectives for waste management and for preventing the creation of waste and the measures to reach the objectives. In accordance with the Government Programme, the Waste Plan sets out a vision extending to the 2030s that supports the recycling and circular economy objectives. The aim is to increase the recycling rate at least to the level of the recycling targets of the EU.

The National Waste Plan to 2027 includes both a plan to reduce the volume and harmfulness of waste and a waste management plan. In terms of the geographical area, the plan covers the whole of Finland except for the Åland Islands. Vision of the National Waste Plan to 2030:

- Material-efficient production and consumption save natural resources and mitigate climate change.

- The volume of waste has decreased from the present. Recycling and reuse have risen to a new level.
- High-quality waste management is part of a sustainable circular economy.
- Circular economy markets function well. Recycling and reuse create new jobs.
- Valuable raw materials present in recycled materials even in small concentrations can be recovered.
- Material cycles cause no harm and hazardous substances are used less and less in production.
- Cooperation between operators in the sector promotes high-quality material cycles.
- Reliable and comprehensive data supports the circular economy. Information is available for use in digital form.
- There is high-quality research and experimentation in the waste sector and expertise on waste management is of a high standard.
- Legislation supports circular economy innovations and the conditions where it operates.

The Waste Plan proposes measures to reach the vision and objectives. The implementation and impact of the Waste Plan will be monitored annually.

5.1.5.2 Platforms and Networks Facilitating the Transition to a Circular Economy

Circular Economy Platforms

Circular Economy Finland (KiSu) [35] is a hub for circular economy know-how. Circular Economy Finland (KiSu) is a hub of skills and knowledge which unites those who seek solutions with those who offer them. The network supports different actors in finding their own circular economy paths. KiSu is one of the actions launched by the national circular economy programme of Finland. The programme is coordinated by the Ministry of the Environment and the Ministry of Economic Affairs and Employment.

KiSu creates a shared image of the current situation of circular economy, the direction for the future, and state of mind. KiSu:

- Unites networks and actors in developing solutions and promoting the spread of the best circular economy practices
- Compiles and shares information on concrete actions, benefits, solutions, and possibilities related to circular economy
- Helps in finding the right partners, channels, and sources of funding, and supports the search for solutions to bottlenecks
- Works together with municipalities to build ways of committing to the long-term advancement of circular economy in the different sectors where the municipality operates
- Supports companies in developing circular economy-based business activities.

Circular Buildings Platform

No circular buildings platform has been established to date.

Building Materials Passport Platform

The Green Building Council Finland (FIGBC) [134] is a non-profit association founded in 2010 that gathers and refines sustainability know-how in the building and construction sector. FIGBC's network consists of around 300 member organizations and other active partners.

FIGBC's core functions regarding the built environment are advancing sustainable development practices, providing knowledge and know-how, activating dialogue and conversation, and connecting Finland to the international Green Building Council network.

Within that, the MADASTER materials passport [179] is set to become the new standard in the real estate sector. Generating a digital materials passport for every building opens up the possibility of creating a circular economy within the construction industry.

FIGBC's vision is that by 2035 Finnish built environment is a fundamental part of the solution to climate change and functions according to the principals of circular economy. The association's strategic goals for the years 2021–2023 include urging over 50% of the Finnish professional real-estate owners to set a goal for carbon–neutral energy use, produce in cooperation with the built environment sector a carbon neutrality action plan for the building and construction sector and ensuring that in all major built environment projects low-carbon solutions and circularity are considered.

Public Procurement Platform

KEINO [142] is a network-based Competence center for Sustainable and Innovative public procurement in Finland. KEINO started its operations on March 2018. KEINO started its operations on March 2018. KEINO is part of the Government Programme's implementation. It is funded and steered by the Ministry of Economic Affairs and Employment. The strategic management of KEINO is supported by a secretariat appointed separately from the representatives of the ministries and a broad-based cooperation group.

The priorities of KEINO's activities in 2022–2023 are:

- procurement of new solutions,
- implementation of ecosystem agreements with cities and
- refining lessons learned into practical tools and operating models for procurement units.

The value of the Finnish public sector's procurements is approximately EUR 47 billion annually, or on average 20% of the country's GDP.

5.1.5.3 Funding Opportunities

According to Finnish Ministry of Environment, the EU regional and structural policy programme Innovation and Skills in Finland (2021–2027) [232] to an external website aims to promote energy efficiency and the circular economy and to reduce greenhouse gas emissions. Funding will also be targeted to actions to prepare for climate change. Responsible organisations: Ministry of Economic Affairs and Employment and Ministry of the Environment.

SITRA [327] launches and implements projects together with the private, public and third sector, all aimed at increasing sustainable well-being in Finland. There is no specific funding call for applications or application form, and interested stakeholders should contact SITRA directly.

International Cooperation and Support from International Organizations

Nordic Circular Hotspot [248] is a network and facilitator that aims to accelerate the transition to a sustainable and circular economy in the Nordic countries. Circular economy is an economic system that designs out waste and pollution, keeps products and materials in use as long as possible, and regenerates natural systems.

Vision—The Nordic Circular Hotspot envisions a vibrant, prosperous and regenerative region that will be fully circular and a role model for others in the world to follow. The old ways of achieving success are no longer working and, therefore, the Nordic Circular Hotspot envisions a new status quo in the Nordics. Through its strategic and systemic approach, the Nordic Circular Hotspot aims to accelerate the transition to an inclusive, circular and sustainable economy in the Nordics.

Mission—Through propelling collaboration, knowledge, capacity building and investments in circular economy solutions, the Nordic Circular Hotspot's mission is to accelerate the transition to an inclusive, resource-efficient, regenerative and circular market in the Nordic region and contribute in a meaningful way to:

- reinventing how the Nordics design, produce and market products;
- rethinking how the Nordics use and consume goods and services; and
- redefining growth in the Nordics and what is possible through reuse, reduction, repairing, regeneration and, most importantly, systems change.

The World Circular Economy Forum 2023 [365] gathered together nearly 1900 forward-looking thinkers and doers and presented the game-changers in the circular economy. The Forum brought together representatives from 155 countries, and the main event had more than 12,500 online views. The annual World Circular Economy Forum (WCEF) presents the world's leading circular economy solutions with business leaders, policymakers and experts participating from around the world. Circular economy approaches can help businesses seize new opportunities and gain a competitive advantage, as well as contribute to achieving the United Nations Sustainable Development Goals. WCEF is a global initiative of Finland and SITRA, the Finnish Innovation Fund. The first Forum, WCEF2017, identified key elements of a circular economy and showcased solutions and learnings from around the world, bringing together 1500 people from more than 100 countries.

5.1.5.4 Challenges, Barriers and Potential Improvements

Despite its leadership in thought, Finland—like most other countries in the world—is, in practice still far from circular economy objectives. In an assessment of Finland's sustainable development policy, climate change, the state of the environment and overconsumption as well as the increasing inequality of society were identified as the greatest challenges in Finland. Finland is a European leader in competitiveness and in the number of eco-innovations that save the environment in new ways, which their demanding environmental legislation acts as an incentive for. They have also traditionally been strong in research, development and innovation funding. In the 2010s, Finland lost its position as a pioneer in innovation activities [224]. Bottlenecks in the progress of the circular economy include attitudes, a shortage of experts, investments in established operating models and a high threshold for the introduction of new business models, as well as the poor steering effect of prices, as harmful external impacts are not sufficiently priced [224].

5.1.5.5 Examples of Successful Implementation

Public Policy Initiatives

CIRCWASTE [96] is a seven-year LIFE IP project that promotes efficient use of material flows, waste prevention and new waste and resource management concepts. It is a regional cooperation network—regions and forerunner municipalities. The project is implemented during the years 2016–2023. All actions contribute to implementing the national waste management plan and directing Finland towards a circular economy. CIRCWASTE is a creation of 20 partners and 10 funding organisations. The project is coordinated by the Finnish Environment Institute. CIRCWASTE is funded in large part by the EU LIFE programme.

Fisu [97] (Finnish Sustainable Communities) is a network of Finnish municipalities committed to working towards becoming carbon neutral and waste-free and curbing overconsumption by 2050. Today, the network consists of 11 municipalities: Forssa, Hyvinkää, Ii, Joensuu, Jyväskylä, Kuopio, Lahti, Lappeenranta, Riihimäki, Turku and Vaasa. The Fisu network is a competence community of resource-smart pioneer cities and municipalities. They help their members to develop new low-emission operating methods and solutions that create sustainable wellbeing and growth based on local strengths.

Private Policy Initiatives

EcoUp [69] is a circular economy group that manufactures carbon–neutral thermal insulation and reuses the side streams from construction. They promote green construction by developing and manufacturing construction products and raw materials from discarded materials. To solve the waste problem, they develop equipment and methods to give new life to construction and demolition waste as ecological and competitive construction end products. They turn waste fractions that traditionally have been difficult to utilize into resources for new building and infrastructure products. By licensing the technology to their customers and partners, they multiply the benefits for the environment.

Environmental management company Remeo [299] invested 35 million euros in a recycling plant in Vantaa that uses artificial intelligence to recycle waste from the construction industry and to reuse it much more efficiently than before. Thanks to artificial intelligence, they can sift out pure raw material from waste and recover it to be reused by companies and industry. Domestically generated secondary raw materials raise the recycling rate in Finland and reduce the carbon dioxide emissions by removing the need to transport waste abroad [316].

5.1.6 *France Factsheet*

5.1.6.1 Policy and Regulatory Framework

Governance

The Ministry for an Ecological and Solidary Transition [235] is the official body for shaping circular economy policy. According to the competent Ministry, in France, the transition to a circular economy is officially recognized as one of the objectives of the energy and ecological transition and as one of the commitments of sustainable development.

It requires progress in several areas.

- Sustainable sourcing: taking into account the environmental and social impacts of the resources used, in particular those associated with their extraction and exploitation.
- Eco-design: taking into account environmental impacts over the entire life cycle of a product and integrating them from the design stage.
- Industrial and territorial ecology: synergizing and pooling between several economic actors the flow of materials, energy, water, infrastructure, goods or services in order to optimize the use of resources in a territory.
- Functionality economy: favoring use over possession, selling a service rather than a good.

- Responsible consumption: taking into account the environmental and social impacts at all stages of the product's life cycle in purchasing choices, whether the buyer is public or private.
- Extending the useful life of products through the use of repair, sale or second-hand purchase, through donation, within the framework of re-use and re-use.
- Improving the prevention, management and recycling of waste, including reinjecting and reusing materials from waste in the economic cycle.

Further, the National Circular Economy Council (CNEC) [236] is a consultative body created in 2021 to replace the National Waste Council (CND), set up in 2001 and dealing solely with the field of waste. This body makes it possible to ensure regular dialogue and close association of the main stakeholders in the areas of action of the circular economy in the construction of Government policy. CNEC is the forum for dialogue on the circular economy. It succeeds the National Waste Council, a consultative body created in 2001 and informally replaced in 2017 by the Circular Economy Roadmap Steering Committee. This committee included the members of the National Waste Council but was also open to all the themes of the circular economy beyond waste management. It had also associated with its work "*circular economy ambassadors*" whose mission is to contribute to the Government's roadmap and to the dissemination of best practices in this area to citizens, businesses and communities. The CNEC's field of action goes beyond waste management to encompass areas of action for sustainable consumption and production such as the eco-design of products, the functional economy, repair, reuse, etc.

The National Council for the Circular Economy *is an advisory administrative body, attached to the Ministry* responsible for the environment and which, as provided for in the Environment Code (article D. 541-1), has the following missions:

- The ministers responsible for the environment and the economy can call on it for advice on issues relating to the circular economy, from the extraction of materials to waste management, etc. Opinions are submitted to the Government.
- The National Council for the Circular Economy can be consulted on draft legislative and regulatory texts having an impact in this area.
- The National Council for the Circular Economy can take up any question of national interest concerning the circular economy.
- The National Circular Economy Council participates in the development and monitoring of national circular economy strategies.
- It monitors the implementation of circular economy policy guidelines, in particular as defined by European directives and related laws.
- The National Council for the Circular Economy is kept informed of the orientations of the strategic committees of the industrial sectors of the National Council for Industry in terms of the circular economy and in particular the extension of the lifespan of products, eco-design, recycling and waste recovery.

Legislation

France adopted in 2020 an ambitious law to shape a system-wide transition towards a circular economy—Loi relative à la lutte contre le gaspillage et à l'économie circulaire [98] (the anti-waste law for a circular economy) referred to in short as the Anti-waste Law or AGEC. The French law on the fight against waste and for the circular economy encourages businesses across various sectors, municipalities, and citizens to eliminate waste and adopt more circular practices. The law also aims to promote a societal transformation and support the solidarity economy. It introduced several measures that are a world first, such as the ban on the destruction of unsold goods and the repairability index. The law originates from France's climate change reduction commitments and successive national policies, such as the 2015 Energy Transition Law for Green Growth and the 2017 French Climate Plan, which have helped to build the momentum towards its adoption [73].

According to the Ministry's official website, the anti-waste law for a circular economy aims to transform the French system in depth, with its 130 articles addressing all forms of waste. The law aims to transform the linear economy (produce, consume, throw away) into a circular economy.

It is divided into five main axes:

- phasing out of disposable plastic;
- better inform consumers;
- fight against waste and for solidarity reuse;
- act against planned obsolescence;
- produce better.

It sets new objectives for the years to come.

France's new regulation, known as RE2020 law [237], significantly tightens the existing energy efficiency requirements on the construction industry, aiming for a 50% reduction in embodied carbon emissions of buildings by 2030 and full decarbonization by 2050. Under this law, a project's carbon footprint will be calculated based on the emissions created during its whole life cycle, including the construction phase and even the manufacture and transportation of the building materials, which puts emphasis on the use of environmentally sustainable materials and manufacturing. Developers are required to create more comfortable spaces in extreme temperatures without having to rely on air conditioning, and users should be provided with buildings which consume less energy and generate a lower carbon footprint [160]. The law forcing the construction industry to reduce its carbon footprint took effect on 1 January, 2022, and becomes more stringent in 2025, 2028 and 2031 to ensure emissions go down. It currently applies to new offices, residential buildings, and primary and secondary education facilities, and will also apply to specific tertiary buildings (hotels, restaurants, retail, and gyms) before the end of 2023 [330].

Circular Economy Strategy, Roadmap or Action Plan

The roadmap for the circular economy (FREC) operationally describes the transition to be made to move from a linear economic model "manufacture, consume, throw away" to a circular model that will integrate the entire life cycle of products, from their eco-design to waste management.

In April 2018, the Ministry for an Ecological and Solidary Transition issued the Circular Economy Roadmap: 50 measures for a 100% circular economy [233]. The document charts a path for rapid progress towards the ambitious goals of the Climate Plan in matters relating to the circular economy. It is the results of 4 workshops between stakeholders around 4 topics: territories, plastics, sustainable consumption/ production and economic instruments. At the same time, an online platform was opened to collect citizens' opinions. The French roadmap includes 50 measures for a 100% circular economy, divided into four key priority areas:

1. A Roadmap for Better Production;
2. A Roadmap to better consumption;
3. A Roadmap for Better Managing Our Waste;
4. A Roadmap for Mobilizing all Actors.

Objectives:

- Reduce natural resource use related to French consumption: 30% reduction in resourceconsumption in relation to GDP between 2010 and 2030.
- A 50% reduction in the amount of non-hazardous waste landfilled by 2025, compared to 2010.
- Aim towards 100% of plastics recycled by 2025.
- Reduce greenhouse gas emissions: avoid the emission of 8 million additional tonnes of CO_2 each year thanks to plastic recycling.
- Create up to 300,000 additional jobs, including in new professions.

The Roadmap development and the implementation of the AGEC law is the joint responsibility of the Ministry for the Ecological Transition and the Ministry of the Economy, Finance and the Recovery. The strong coordination between these two key ministries enhances the coherence and impact of the law, ensuring that it will significantly influence the economic growth path of the country.

Construction and Demolition Waste Management Regulatory Framework

The 2018 revision of the 2008 Waste Framework Directive, which follows the Grenelle French Environmental Act, is the new reference text for waste management policy in the European Union. The Grenelle Act is an example of a national legislative framework that has set the tune for all the European Member States until they adopted the CE Principles for building design by 2020. In addition, consultation work was carried out in 2012 and 2013 within the framework of the National Waste Council's prevention working group and resulted in the drafting of a National Waste

Prevention Programme for the period 2014– 2020. The programme has provided a reduction of 7% in the production of household and similar waste per inhabitant in 2020 compared to the 2010 level, as well as a stabilization of waste production from economic activities (DAE) and construction [53].

The plan stands for a gradual implementation of 54 concrete actions, divided into 13 strategic areas, which will contribute to achieving 4 actions in the B&C sector, the first of which has already been completed [53]:

1. Sector-wide voluntary recycling protocol for construction and demolition waste;
2. Whitepaper of possible uses of the environmental footprint of products, including certain materials in the construction sector, to measure and indicate environmental information (PEF) (in progress);
3. Guidelines on pre-demolition analysis for the construction sector, the French version of which is a diagnosis (currently being drawn up);
4. Guidelines to develop indicators for assessing the environmental performance of buildings throughout their life cycle and measures for their use, also known as Levels.

According to the information [238] about products and construction materials of the building sector (PMCB) on the website of the competent ministry, the law of February 10, 2020 relating to the fight against waste and the circular economy (AGEC) provided for the establishment of an EPR sector for construction products and materials in the building sector from 2022 with a view to:

- Reduce illegal dumping by improving collection by free collection of waste, increasing the density of collection points, and improving traceability;
- Prevent the saturation of landfills by developing material recycling as well as reuse and reuse.

EPR requires companies that sell or import waste-generating products or components and materials used in their manufacture to ensure the prevention and management of this waste [353].

Legislative and regulatory principles. Article L. 541-10-1 (4°) of the Environment Code in its wording resulting from Article 62 of the "AGEC" law provides that waste from construction products and materials in the building sector is taken back free of charge when it is subject to separate collection and when traceability of this waste must be ensured. It provides for a Conseil d'Etat decree to define the minimum conditions for the territorial network [239].

In addition, Article L. 541-10-23 of the Environment Code in its wording resulting from Article 72 of the "AGEC" law sets the obligations incumbent on eco-organizations and distributors affected by this new REP (Extended Producer Responsibility) sector. In particular, it provides for a mixed financial and operational operation for the eco-organizations which are required both to cover the costs of any person who ensures the recovery of waste from the building collected separately and to provide for this recovery when this is necessary to ensure the territorial network.

5.1.6.2 Platforms and Networks Facilitating the Transition to a Circular Economy

Circular Economy Platforms

The economiescirculaire.org [70] brings a number of free services:

- Follow circular economy news and identify opportunities;
- Promote your company locally and internationally;
- Join the cast and participate in the network
- Create partnerships and integrate project communities;
- Access implementation tools and methodologies;
- Use feedback and highlight your achievements.

Circul'R [32] is an international network of circular economy startups. Its mission is to unlock the circular economy's potential by connecting innovative startups with companies so that they can co-create solutions to accelerate their transition towards the circular economy. Its main services are [77]:

- raising awareness (conferences to explain circular economy by presenting the most innovative circular solutions),
- learning expeditions (bringing people on the field to meet with circular economy entrepreneurs and their projects),
- consultancy (business opportunities, funding, etc.),
- Circul'R Club (bringing together large companies and startups with the objective of co-creating concrete projects in the field of circular economy: waste management, eco-design, new business models, etc.).

Founded in 2013, the National Institute for the Circular Economy (INEC)—l'Institut National de l'Économie Circulaire [155] aims to promote the circular economy and accelerate its development through the dynamics of cooperation. A multi-stakeholder organization, it consists of more than 200 members, public and private organizations: companies, federations, communities, institutions, associations, non-governmental organizations and universities. The diversity of these members enables nurturing a holistic vision of the circular economy, taking into account all economic, social and environmental issues. The activity of the Institute mainly revolves around 3 axes:

- animation of reflection,
- promotion of circular economy,
- implementation of circular economy.

Eclaira [68], is the regional collaborative platform dedicated to the circular economy in the French region Auvergne-Rhône-Alpes. It aims to:

- Identify and highlight initiatives carried out in the region;
- Encourage collaborative projects in the circular economy;
- Promote the implementation of the circular economy in the region.

Its main objectives: to identify initiatives, encourage projects and thus promote the implementation of the circular economy in the territory by all the actors involved.

Circular Buildings Platform

No circular buildings platform has been established to date.

Building Materials Passport Platform

No building materials platform has been established to date.

Public Procurement Platform

The action-program "Circular procurement" [78] was a one-year experiment, inspired by the Dutch and Belgian "Green Deals" models and launched in 2018. It aimed at bringing together actors from the public and private sectors to collaborate on a project to foster the circular economy as an essential factor in their purchasing decision-making.

The objectives of this program were:

- to create and share knowledge by promoting excellent approaches
- to have real impacts on the use of resources and social and economic realities
- to raise awareness and convince on the benefits of integrating the circular economy in the purchasing process for the economy, social and environment.

5.1.6.3 Funding Opportunities

The ambitious objectives set for 2020–2025 to initiate the transition to a circular economy require a substantial volume of investment [240], as well as aid for innovation to accelerate the dynamics of transition. Certain investment projects may be the subject of public aid, in particular from ADEME [1]—the French Agency for Ecological Transition. Their mission is to accelerate the transition to a more sober and supportive, job-generating, humane and harmonious society. They support innovation and research up to the application and sharing of solutions. As such, they advise, facilitate and help finance many projects of companies, local authorities, associations and doctoral students related to the ecological transition. ADEME is an important vector for the massification of solutions and the mobilisation of stakeholders. As such, we have several financing levers. Through the Renewable Heat Fund, the Circular Economy Fund and the Air-Mobility Fund, we finance all stages, from research to development to industrialisation, including demonstrators.

ADEME also relies on the State's major investment plans, such as the Investments for the Future Pogramme, the Recovery Plan and today the France 2030 plan. The

France 2030 plan, launched in October 2021, offers unprecedented ways to meet ecological, economic, industrial and social challenges. This innovation and industry plan aims to sustainably transform key sectors of our economy (energy, automotive, aeronautics, digital and space) and to position France as a leader by bringing out future technological champions. France 2030 builds on the achievements of the Investments for the Future Programmes (Programmes d'investissements d'avenir—PIA), and in particular PIA 4 endowed with €20 billion that it integrates and exceeds in ambitions and means. In total, France 2030 mobilises €54 billion to ensure that companies, universities and research organisations fully succeed in their transitions in strategic sectors.

Their financial aid to companies, local authorities, research organisations and advocacy structures covers four areas of intervention:

- Reducing energy consumption and greenhouse gas (GHG) emissions,
- Developing renewable energies and the circular economy,
- Converting brownfields and polluted sites,
- Improving indoor and outdoor air quality.

International Cooperation and Support from International Organizations

A multi-stakeholder association created in 1992, ORÉE [254] has for 30 years brought together and led a network of stakeholders committed to exchanging and implementing an environmental dynamic serving the territories. Today, ORÉE brings together, supports and equips a network of more than 200 members, companies, communities, professional and environmental associations, academic and institutional organizations, etc.

The association's action focuses on three priorities:

- Biodiversity and economy, or how to integrate biodiversity into organizational strategy
- Circular economy covering approaches focused on both products/services/equipment (functionality economy, eco-design), sectors (recycling/recovery) and territories (industrial and territorial ecology)
- CSR/ESG reporting in line with French and European regulations on the publication of extra-financial information.

ORÉE offers a global, systemic and integrated approach to all physical flows in order to benefit from the potential of the circular economy and to propose the bases and trajectory of a new economic model. In this sense, all the operational pillars for a systemic and integrated global economy invested by ORÉE for many years are grouped into a "Circular Economy" priority. In line with its two other priorities—CSR/ESG Reporting and Biodiversity and Economy—ORÉE promotes a vision of the circular economy that is responsible and supportive of societies, as well as respectful of biodiversity and ecosystem services.

5.1.6.4 Challenges, Barriers and Potential Improvements

According to French Roadmap 50 measures for a 100% circular economy [241] the main challenge is mass mobilization—citizens, communities, businesses, associations, government agencies, researchers and development actors in all regions—for real action and upscaling. To meet this challenge, the Roadmap aims to shake up habits, showcase good practices, convince people and inspire action.

5.1.6.5 Examples of Successful Implementation

Public Policy Initiatives

In mid-May 2021, Toulouse Métropole won the European call for projects from the Life program by proposing its Waste2Build project. It is about building a sustainable economy by optimizing resources and recovering building and public works waste on a local and then regional scale. The project follows the experimental deconstruction of the Exhibition Center. Life Waste2Build relies on public procurement (works contracts) to develop the use of second-life materials in construction, renovation or landscaping.

The operational objectives of the project are the following:

- reduce by 20% the impact of construction in resource consumption and waste production
- structure the circular construction sector
- implement more circular purchasing policies
- support the rise in skills of the circular construction sector.

This project, coordinated by Toulouse Métropole, is carried out in collaboration with 6 beneficiary partners: the National Institute of the Circular Economy (INEC), Synéthic, the Center Scientifique et Technique du Bâtiment (CSTB), Envirobat Occitanie, Toulouse Business School (TBS), the French Federation of Building and Public Works of Haute-Garonne (FFB 31) [156].

Private Policy Initiatives

VALOBAT [354] is the eco-organization by and for the building industry born from an initiative carried by 48 leading companies in the manufacture of Building Products and Materials (BPM). They offer a multi-material solution, based on a representative governance of all products and materials of the building industry. Given the implementation of an Extended Producer Responsibility channel for construction waste required by APEC law, they wish to propose a solution for each construction waste and for each stakeholder of the ecosystem of the sector. Their mission is to set up the entire collection, sorting, recycling and recovery chain of the PMCB (Construction and Building Products and Materials) waste in France, by an action at 4 levels:

1. Support and services: To support its members in their regulatory compliance as producers and to ensure the management of the end-of-life of PMCB
2. Collection and processing: Collect all separately collected waste streams free of charge, thanks to a network of adapted collection points
3. Value chain: Accelerate the development of reuse, recycling and recovery of building waste to create value
4. Information and education: Supporting the evolution of practices of all the actors of the building sector.

5.1.7 Greece Factsheet

5.1.7.1 Policy and Regulatory Framework

Governance

The government institution in charge of implementing the circular economy in Greece is the Ministry of Environment and Energy (https://ypen.gov.gr/). In 2018 it published the National Strategy for circular economy [144], which identified the sustainable management of resources, the circular entrepreneurship and the circular consumption as the main pillars for the transition to new, more sustainable and circular development models. In the first National Strategy for circular economy the main longtherm (2030) targets were formulated and included the application of ecological design criteria and life cycle analysis of products, waste management, energy improvement of the production processes, innovative consumption models, transition from the linear to the circular economy model, etc.

In line with the National Strategy for circular economy, the first National Operational Plan for 2018–2019 was published, which included specific interventions and actions towards regulatory and legislative provisions to promote circular economy and remove bureaucratic barriers; financing and incentives; improvement of knowledge and linking entrepreneurship and social economy with technological innovation; and support of the Circular economy plan through governmental actions and process acceleration.

In 2021, the operational plan was revised and the new Action Plan for Circular Economy for the period 2021–2025 [145] was entered in force, which: (i) is fully in line with the objectives and commitments of the new European Circular Economy Action Plan; (ii) is compatible with new European directives and institutional initiatives; (iii) takes into account all new legislative developments in Greece such as the new legislative framework on waste, alternative management, single-use plastics and related development policies at sectoral and national level, as well as the National Action Plans for waste and the National Plan for waste prevention. The actions that are foreseen in the new Action Plan for Circular Economy address the following axes: (i) sustainable production and industrial policy, e.g. ecodesign, eco-certification, industrial symbiosis, tax exemptions; (ii) sustainable consumption, e.g. promotion of green public procurement, repair, reuse services; (iii) less waste with greater value, e.g. funding programmes for prevention, institutional framework for prevention, (iv) horizontal actions, e.g. national observatory, voluntary agreements, coordination body, indicators, and (v) specific product categories to be addressed as a matter of priority e.g. plastic products, batteries and vehicles.

Legislation

The national policy on circular economy, which is expressed through the relevant action plan described above, focuses on the promotion of waste hierarchy, the handling of management as a resource and the transition towards a circular production and consumption model.

The waste sector is a key pillar for the implementation of circular economy and is included in the EU Action Plan for Circular Economy. The implementation of the national waste policy is applied through the National Waste Management Plan, which was ratified by a Cabinet Act on August 31, 2020, and the National Waste Prevention Program, which was approved by the Cabinet in June 2021 and ratified on April 24 2022. These plans set out the strategic priorities and targets until 2030 for the implementation of the new EU waste legislation based on circular economy principles.

Other national plans of high relevance to the implementation of circular economy are the National Green Public Procurement Action Plan, which was announced in February, 2021, and the National Plan for Energy and Climate, which was ratified at the end of 2019.

The legislative framework includes the Framework Law on Waste (Law 4819/2021), the Law on single-use plastics (Law 4736/2020) and the Joint Ministerial Decision on landfill (90439/1846/2021).

The New Building Regulation (official gazette 79/2012) foresees in article 17 that for every activity concerning the building construction and the configuration of the surrounding open space, the measures for the alternative management of all waste should be implemented.

Circular Economy Strategy, Roadmap or Action Plan

On a regional level, the Circular Economy is among the targets that the 13 Regions of Greece have set in their operational roadmaps for 2021–2027. More specifically, the Regions of Greece have integrated specific targets, actions and indicators for the transition towards circular and resource efficient economy and pay particular attention to the management of solid urban waste. Examples of such regional strategic plans can be found here.

In 2020, the Greek Association for Local Development and Government Administration, which is the technical and scientific advisor of the Municipalities and Regions published in 2020 a guide on Circular Economy and local government administration [133], in order to provide the necessary background information and present tools and best practices to Municipalities.

On a local level, many municipalities have integrated circular economy in their strategic plans, i.e. Athens, in its Resilience Strategy for 2030 [37].

All national strategies and action plans for circular economy make reference to the construction sector. In fact, the National plan for circular economy includes the building sector as one of the 7 sectors which should have priority for implementing actions for circular economy. The targets are multiple: (1) setting a holistic national strategy for sustainable built environment, and introducing the CE principles during the whole life cycle of buildings, (2) Improvement of the performance of alternative management of excavation, construction and demolition waste, (3) development of the market for the secondary materials produced from the management of the excavatio, construction and demolition waste, (4) integration of circularity criteria on the programs for national and private buildings, (5) Creation of a financial program for the refurbishment and reuse of abandoned property and rehabilitation of degraded lands and areas. The specific actions for reaching these targets, along with the timeframe and the responsible bodies for implementing the actions are mentioned in the strategy plan.

Construction and Demolition Waste Management Regulatory Framework

The National Waste Management Plan (2020–2030, issued in 2020) [147] defines as CDW every material or product derived by excavation, construction and demolition procedures. It also presents the current status and its assessment, the predictions for the production of CDW till 2030, the targets for their management and the proposed measures and actions for the plan implementation, accompanied with a relevant progress monitoring indicators.

The targets concerning Excavation, Construction and Demolition Waste management (ECDW) are:

- 70% w/w preparation for reuse, recovery and ecycling (excluding however natural aggregates from the excavation).
- Increased geographic coverage of the country by Alternative Waste Management Systems (was at 63% in 2018).
- Impose the obligatory management of ECDW of public and private projects (already implemented through Law 4819/2021).
- Streamlining the Alternative Waste Management functions and fees.
- Establish separate collection of excavation waste, which is excluded from ECDW targets, as well as for concrete residue.
- Develop the secondary ECDW materials market.

Accordingly, the National Waste Prevention Program (2021–2030) [146] presents and evaluates the current situation in waste prevention, as well the national and EU legislation and strategies, defines critical material flows, sets targets and defines measures and actions to achieve them, taking into consideration goals provisioned for in National and EU legislation. Specifically concerning CDW, it proposes the following actions:

- Development of sustainable building design and renovation (through promoting correct practices, education, incorporating sustainability elements in building energy renovation programs, standardizing sustainable building design processes and developing certification systems, incentivizing building renovation rather than demolition).
- Development of CDW reuse practices (through encouraging construction materials reuse by developing selective demolition standards, promoting construction elements reuse research and education).
- Prioritization of excavation waste reuse on site (through minimizing them and balancing them with filling material).
- Creation of CDW reuse centers or spaces inside CDW processing units.
- Development of new circular construction practices (through pilot projects that focus on CDW minimization, dissemination of knowledge concerning building life cycle extension practices and structural elements recyclability).

Additionally, a joint ministerial decision, issued in 2010, addresses the management of excavation, constrution and demolition waste and foresees their alternative management within the scope of circular economy.

5.1.7.2 Platforms and Networks Facilitating the Transition to a Circular Economy

Circular Economy Platforms

The main platforms that are associated with circular economy in Greece are:

- The National Producers' Registry (for packaging, vehicles, tyres, mineral oils, batteries and accumulators, electrical and electronic equipment). Its aim is to record all domestically produced or imported quantities of the aforementioned products. It should be mentioned though that this does not currently apply to ECDW producers.
- The Digital Waste Registry (CDW included). It is used to collect waste production and management data for all materials of the European List of Waste, from waste producers and organizations/bodies/private companies active in the collection, transportation and processing of waste, in order to facilitate monitoring the waste management supply chain. It also allows searching for appropriate management companies by material code.

Other platforms that accelerate the implementation of circular economy in Greece are:

- The Hellenic Recycling Agency, a public organization aiming to develop and apply national policies for the sustainable—alternative management of waste, including ECDW (www.eoan.gr).
- The Hellenic Solid Waste Management Association, a non-profit, non-governmental organization that promotes the adoption of socially and environmentally acceptable methods for Solid Waste Management, including ECDW (ΕΕΔΣΑ—Hellenic Solid Waste Management Association (eedsa.gr)) The objectives of HSWMA are to promote the adoption of socially and environmentally acceptable methods for Solid Waste Management, the development and dissemination of relevant scientific methods and applications, the protection of the quality of our environment, and the preservation of natural resources and energy.
- Circulargreece, the website of the integrated LIFE project LIFE-IP CEI-Greece (https://circulargreece.gr/), which aims to contribute towards the implementation of the National Waste Management Plan, the National Waste Prevention Plan and the National Strategy for Circular Economy. According to the project's website, a new concept for waste management emerges with-in this project; based on circular economy principles, promoting behavioral changes and supporting actions to increase product life cycle, convert waste into valuable resources and implement successfully the legislative framework for waste.

Circular Buildings Platform

No circular buildings platform has been established to date.

Building Materials Passport Platform

No building materials passport platform has been established to date.

There is the national Product contact point, https://pcp.ggb.gr/?lang=en, which provide information on goods and products, in order to enable their free movement across the member states.

Public Procurement Platform

The Green Public Procurement (GPP) is defined by the European Commission as "a process whereby public authorities seek to procure goods, services and works with a reduced environmental impact throughout their life cycle when compared to goods, services and works with the same primary function that would otherwise be procured". Contracting authorities and contracting entities are major buyers of goods, services and works, therefore public procurements are a great way for implementing environmental, social and economic policies. By using their purchasing power to choose environmentally friendly goods, services and works, they contribute locally, regionally and nationally to the achievement of national and international sustainability and environmental policy objectives. Green public procurements (GPPs) are related to the principles of Circular Economy. According to European Commission, Circular Public Procurements are defined as "a process whereby public authorities seek to procure goods, services and works with a reduced environmental impact throughout their life cycle when compared to goods, services and works with the same primary function that would otherwise be procured". GPPs play a main role in circular economy, that's why European Commission uses this in its own supplies and emphasizes on aspects such as durability and ability to repair.

In February 2021 the National Action Plan for the Green Public Procurement (2021–2023) [143] was approved. Its main objectives are:

- Establishment and implementation of an elementary level of adoption of green criteria in public procurement of products, services and projects.
- Gradual increase of GPP during the next three years in certain sectors of goods, services and projects.
- Wider integration of life cycle cost estimation in public procurement.
- Dissemination of environmental and economic benefits of GPP.
- Raising awareness, building capacity and active participation of stakeholders (contracting authorities and economic operators) in GPP.
- Monitoring the achievement and updating the objectives.

The Green Public Procurement National Action Plan has set obligatory targets for certain service/product categories (copying and graphic paper, imaging equipment, consumables, and print services, computers, monitors, internal lighting—LED lamps, road lighting and traffic signals, road transport, reclaimed or biodegradable lubricants). According to the plan, services and products associated with road design, construction and maintenance and office building design, construction and management are sectors where the green public procurement can be applied on a voluntary basis. One significant aims mentioned in this specific plan is the use of recycled aggregates at a minimum rate of 15% for the construction of public works and public buildings.

5.1.7.3 Funding Opportunities

At the moment, there are no specific instruments, private–public partnerships or private institutions which provide financial support explicitly to the implementation of circular economy at a national level.

However, the Economic and Financial Affairs Council of the European Union has approved the National Recovery and Resilience Plan Greece 2.0 on 13 July 2021, which ncludes an integrated and coherent set of reforms and investments structured in four Pillars that make up eighteen components (Greece 2.0—National Recovery and Resilience Plan (greece20.gov.gr)). Among them, the 1st Pillar "Green Transition" includes the component "Sustainable use of resources, climate resilience and environmental protection", whose main objective is the promotion of the green transition through alignment with the principles of circular economy, efficient use of natural resources, climate change adaptation and mitigation via protection of the natural environment. Among others, the first set of actions addresses waste management, following the principles of waste hierarchy and circular economy.

Additionally, the Cohesion Fund supports investments in the field of environment to member states with a gross national income (GNI) per capita below 90% EU-27 average. Within this framework, cross border, transnational and interregional programmes that address circular economy, such as Interreg VI-A Greece-Bulgaria, Interreg VI-B Adriatic Ionian and NEXT Mediterannean Sea Basin, as well as the national regional programme Peloponisos, are being funded for the period 2021–2027.

On a national basis, there are some funding opportunities that address circular economy and the building sector:

- The "Green Transition" action, funded by the National Strategic Reference Framework (https://www.espa.gr/el/Pages/ProclamationsFS.aspx?item=5666).
- The National Development Plan (2021–2025, epa.gov.gr). The pillar "Green Developmen" includes targets for the circular economy transition, environmental protection and climate change mitigation.
- The Green Fund (Prasino Tameio—| Official Website), which funds projects and programs set by the Ministry of Environment, Energy and Climate Change or other Ministries that promote the national environmental strategy.
- Through the Public–Private Partnership Scheme (Implementation of PPP development projects—ependyseis.mindev.gov.gr), monitored by the Ministry of Development, which promotes large scale projects between the private and the public sector resulting to to significant positive impact in favour of the local societies and the Greek economy.
- The Joint European Support for Sustainable Investment in City Areas—JESSICA and the European Investment Bank (Infrastructure Fund of Funds—Greece (eib.org)) invest in local and regional projects that contribute to job creation and a sustainable and healthy European economy.

Moreover, the European Commision has established programmes and funding opportunities, such as the Urban Initiative Action initiative (UIA—Urban Innovative Actions (uia-initiative.eu)), which provides urban areas throughout Europe with resources to test new and unproven solutions to address urban challenges, etc. For more information: Circular economy in cities (europa.eu).

International Cooperation and Support from International Organizations

The European Circular Economy Stakeholder Platform is a joint initiative by the European Commission and the European Economic and Social Committee (EESC), launched in March 2017. The objective of the platform is to serve as a space for the exchange of ideas and information, addressing all stakeholders, i.e. public authorities, businesses, trade unions and civil society (Homepage | European Circular Economy Stakeholder Platform (europa.eu)).

5.1.7.4 Challenges, Barriers and Potential Improvements

The implementation of circular economy principles in the built environment is at its initial stage in Greece. This is due to several challenges and barriers, which can be identified as regulatory, market, technological and cultural-societal. More specifically:

- Regulatory and legislative barriers: Currently the regulatory framework for implementing the principles of circular economy is mainly associated with the waste prevention and management. Although provisions for CDW have been made, there are no specific regulations and legislative contexts for applying the circular economy principles in the built environment across the whole lifespan of the constructions. Additionally, there is no specific framework for assessing circularity in buildings and constructions. The application of international standards is also
- Market barriers: In Greece, selective demolition, source separation and collection of wood, concrete, bricks, tiles and ceramics, stone, metals, glass, plastics and plaster for all types of construction works are established in Law 4819/2021. For implementing the management of the CDW, several companies have been established, which mainly focus on recycling the waste. The lack of business models limits the potential of reusing and repairing the waste in order to create more closed loops of materials. Furthermore, it must be highlighted that not all of the Greek regions have licenced units for CDW management. Additionally, high investment costs are needed for implementing CE in the building industry, not only for the design of new, circular materials, but also for the whole value-chain. The lower prices of virgin materials create an additional obstacle for the wider use of circular ones, while landfilling is often cheaper than other waste management processes. In case of products/materials reuse, the logistics (transportation, storage and resell) are not always supported. In general, it is very difficult for all building industry players to make the transition from a linear to a circular construction model. Many of the above barriers were identified as challenges for the Greek Enterprises in the study on the Circular Economy in Greece, published in 2016 from the Hellenic Federation of Enterprises. The study concluded that, among others: (i) there is a fragmented adoption of circular models, in the context of inefficient systemic approach; (ii) Emphasis is placed on the 'end-of-life' stage, rather than on the design stage; recycling and recovery of construction and demolition waste requires considerable improvement, (iii) the waste market in Greece faces several distortions most important of which is that the established price mechanism does not reflect the environmental benefits of recycled materials, hindering their competitiveness and resulting in low demand.

- Technological barriers: The implementation of circular economy principles in the building industry requires the transition to new design and construction techniques. For Greece, it is particularly difficult to change the conventional structural system of the buildings, the vast majority of which are made of reinforced concrete and clay bricks, to other building types, such as prefabricated or modular ones, made of wood or steel. Accordingly, the use of reclaimed, reused, or repaired materials/products is very limited, namely because their quality and performance is not yet certified. The lack of protocols, data and information (e.g. material passports) create insecurity to the stakeholders, including architects and engineers. At the moment, there are not many materials and products in the market that promote circularity. Moreover, there is a lack in necessary equipment and infrastructure to implement circular economy principles.
- Cultural-societal barriers: Knowledge and awareness on the benefits and practices of circular economy in the built environment are very limited among the stakeholders in Greece, especially the civil society and the members of the technical pyramid (architects, engineers, workers). Therefore, there is no particular interest and requests from building owners, users and investors for circular buildings. The inadequate presentation of good building practices and successful implementations limit the acceleration of circular buildings in Greece. Cultural barriers, lack of consumer interest and awareness and hesitance or reluctance due to company culture are the main barriers for businesses and policy makers.

5.1.7.5 Examples of Successful Implementation

Private Policy Initiatives

There are some good practices of implementing the circular economy principles in the built environment in Greece, initiated by private companies. Indicatively:

- Titan Greece (Home Page—Titan Greece) produces building materials. It plays an active role in implementing a circular economy model at various stages of the production process and has created synergies (industrial symbiosis) with other industries. According to the company, the main contributions to circular economy include the full inclusion of concrete in a circular economy; the promotion of sustainable construction; durable construction to adapt to climate change; recovery of materials and energy through the use of alternative raw materials and fuels.
- The ReWeee Project (ReWeee—Home) entitled Development and Demonstration of Waste Electrical and Electronic Equipment (WEEE) Prevention and Reuse Paradigms aims to reduce WEEE in Greece by preventing the production of WEEE itself, and also by demonstrating, via appropriate paradigms, that WEEE can be efficiently sorted and prepared for re-use. The goal is also to make re-used electrical and electronic equipment (REEE) acceptable to consumers. To this end, two WEEE sorting centres (SC) will open for the first time in the country, in the Attika and the Central Macedonia regions. Their core activity is the collection,

storage and sorting of WEEE, depending on their condition, and then their preparation for reuse or treatment. Moreover, a measurement tool is being developed to facilitate the assessment of WEEE reuse in Greece. In addition, a web platform has been set up to facilitate the donation and exchange of EEE. It targets households, companies and public services. Among the project partners are RReuse, the Hellenic Recycling Agency, the Ecological Recycling Society, the Green Fund and the Harokopio University of Athens.
- Sporos Platform (Sporos Platform), launched in 2019, aims to become the first platform for impact investment and collaboration that adds value to SMEs in Greece and Southeastern Europe, on the basis of the Circular Economy. Sporos utilises sustainable financing tools, combined with consultancy services and environmental, social and governance (ESG) and sustainable development goal (SDG) metrics.

5.1.8 Ireland Factsheet

5.1.8.1 Policy and Regulatory Framework

Governance

Ireland's ambition is to move beyond target compliance and to become a circular economy leader among its European peers. In order to achieve that, Ireland found it necessary to adopt a whole-of-Government approach, with the development of the circular economy led by the Minister for the Environment, Climate and Communications, but involving all Ministers and Government Departments, as well as local government and relevant state agencies. Delivering a circular economy is expected to have positive environmental, economic and social impacts, and a well-designed circular policy framework can maximise these impacts and identify cobenefits so that environmental "wins" also provide economic and social opportunities and vice versa. This is why the Programme for Government, Our Shared Future, committed to a range of actions to support the transition to a circular economy, including the establishment of a cross-Government Circular Economy Unit in the Department of the Environment, Climate and Communications (DECC) and the adoption of a Circular Economy Strategy [50]. Following the publication of the Waste Action Plan for a Circular Economy (WAPCE), a Circular Economy Unit was established within the DECC to lead circular economy policy in Ireland by developing a circular economy strategy published in December 2021. Following the Strategy's publication, the Circular Economy Unit is setting up an inter-departmental Circular Economy Working Group with relevant ministers, government departments, state agencies and local governments [251].

The Environmental Protection Agency (EPA) is in charge of leading and implementing the Circular Economy Programme (CEP) which incorporates and builds upon the previous National Waste Prevention Programme to support national-level, strategic programmes to prevent waste and drive the Circular Economy in Ireland [62]. Construction and buildings are one of seven priority areas in the CEP aligned with the EU Green Deal. The EPA's new Circular Economy Programme will co-ordinate circular economy initiatives across levels of government and stakeholders [251].

Legislation

Ireland passed the Circular Economy and Miscellaneous Provisions Act 2022 (CE Act) [71] and defined the circular economy for the first time in Irish domestic law (Part 2, Interpretation). The legislation provides that the statutory version of the CES may feature both economy-wide and sectoral targets and specify sectors which its future versions must adopt as targets in relation to construction, agriculture, packaging, and electronic and electrical equipment [87]. According to the Sect. 7 of the CE Act, the Minister for the Environment, Climate and Communications shall, prepare and submit to the Government for their approval, with such modifications (if any) as they consider appropriate, a circular economy (CE) strategy setting out the policy, objectives and priorities for the time being of the Government in relation to the circular economy. In making a circular economy the Minister shall consult relevant documents related to national (action) plans and government programs, especially those covering waste, climate action, and sustainability. The circular economy strategy shall promote the use of criteria relating to the circular economy in public procurement and shell set out targets, in respect of the following sectors of the economy: construction, agriculture; retail; packaging; textiles; and electronic equipment. The circular economy strategy shall set out necessary actions to support Government policy on the circular economy, including measures to inform, and promote dialogue with the public regarding the challenges and opportunities in the transition to a circular economy, and actions necessary to meet the targets set out in the strategy. The CE strategy shall be publicly available together with the annual report on the implementation of the strategy and the progress made in reaching the targets and taking the necessary actions. According to Sect. 10 of the CE Act, the Environmental Protection Agency (EPA) shall establish a circular economy programme setting out measures to be taken by the EPA, including measures in respect of reporting and the provision of funding and education, to give effect to the objectives set out in the circular economy strategy. A circular economy programme shall include the waste prevention programmes established by the Agency, and shall be integrated into waste management plans or other environmental policy programmes, or operate as a separate programme.

Circular Economy Strategy, Roadmap or Action Plan

The Circular Economy Strategy (CES) and the EPA's Circular Economy Programme were published in 2021. The Whole of Government Circular Economy Strategy 2022–2023: Living More, Using Less is the first national CE strategy with the following key objectives [50]:

1. To provide a national policy framework for Ireland's transition to a circular economy and to promote public sector leadership in adopting circular policies and practices.
2. To support and implement measures that significantly reduce Ireland's circularity gap, in both absolute terms and in comparison with other EU Member States, so that Ireland's rate is above the EU average by 2030; such measures to address facets of sustainable production and consumption most impactful in an Irish context.
3. To raise awareness amongst households, businesses and individuals about the circular economy and how it can improve their lives.
4. To support and promote increased investment in the circular economy in Ireland, with a view to delivering sustainable, regionally balanced economic growth and employment; and
5. To identify and address the economic, regulatory and social barriers to Ireland's transition to a more circular economy.

Subsequent iterations of the CES will include targets and detailed actions, including specific targets for the construction and demolition sector. Potential actions in the construction sector include:

- Increased use of offsite design and manufacture
- Modular building design
- Refurbishment and retrofitting of existing stock
- Tackling dereliction and bringing stock back into occupancy
- Increase use of Construction & Demolition Waste as a secondary construction material.

The Circular Built Environment Playbook or 'CircularBuild' is a 3-year project funded by the EPA under the 2022 EPA Research Call. The project is led by the Irish Green Building Council (IGBC) in collaboration with Atlantic Technological University (ATU), Technological University Dublin (TU Dublin) and the University of Galway (UG). The project aims to:

- Co-design and develop a National Circular Built Environment Roadmap to 2040 in collaboration with industry stakeholders representing the full built environment value chain.
- Co-design, develop, and pilot an evidence-based and research-informed Circular Built Environment Toolkit in collaboration with industry across a range of selected case studies.

- Establish a dedicated Circular Built Environment Resource and Training Hub for industry, policy makers and educational audiences.

Construction and Demolition Waste Management Regulatory Framework

The primary waste legislation is the Waste Management Act 1996 [72], as amended, and Sect. 32 of the Act places a general obligation on the holder of waste to comply with legislation and ensure all wastes are managed within the requirements of the Act. In short, the obligation to manage waste legally lies with the holder of waste, which means the waste producer or the person who is in possession of the waste. At a construction site, the mandatory obligation to appropriately manage waste generated at the construction site lies with the Client and the Contractor. Under Sect. 3(1) of the Act, the requirements do not apply to the following materials, which hence are not considered 'waste' [63: 3–4]:

- Land (in-situ) including unexcavated contaminated soil and buildings permanently connected with land—relates to land and buildings prior to any construction or demolition where material remains untouched. Once it has been excavated or otherwise removed, the material may enter into the control regime set down by the Waste Management Acts.
- Uncontaminated soil and other naturally occurring material excavated in the course of construction activities where it is certain that the material will be used for the purposes of construction in its natural state on the site from which it was excavated.

There are two more important provisions within the European Union (Waste Directive) Regulations 2011–2020 that are of relevance to the construction sector and the prevention of waste, and these allow for the classification of resources out of the waste regime as follows [63: 4]:

- Article 27 allows for the notification of a material as a by-product rather than a waste where certain criteria can be demonstrated by the legal person (i.e. further use is certain, no need for further processing, produced as part of a process and further use is lawful).
- Article 28 sets out the grounds by which a material, which is recovered or recycled from waste, can be deemed to be no longer a waste and complies with a set of end-of-waste criteria (substance/object to be used for specific purposes, a market or demand exists, fulfils technical requirements and no overall adverse impact to human health or the environment).

The Waste Action Plan for a Circular Economy (WAPCE) [49], published in September 2020 by DECC, paved the way for actions to [251]:

- Ensure that materials and products remain in use for longer by rewarding circularity and discouraging waste;
- Increase producer responsibility for products and packaging;
- Support sustainable business models;

- Promote a multi-sectoral approach with the voluntary sector, R&D, producers, manufacturers, regulatory bodies and civil society; and
- Clarify and strengthen institutional arrangements for the waste sector, including through a heightened role for local authorities.
- Over the coming years, the construction and demolition sector needs to:
- Promote waste prevention in the first instance;
- Follow best available techniques;
- Expand the range and use of recycled products;
- Create a market demand for recycled products and segregating more material on-site to allow for recycling; and
- Meet the target of preparing for reuse, recycling and other material recovery (incl. beneficial backfilling operations using waste as a substitute) of 70% by weight of C&D non-hazardous waste (excluding natural soils and stone).

There is a need to plan for CDW management at the earliest possible stage in a construction project, ideally at concept stage. The 2006 Best Practice Guidelines on the Preparation of Waste Management Plans for Construction and Demolition Waste Projects needed revision, so the EPA [63] published new Guidelines in 2021. The purpose of the Best practice guidelines for the preparation of resource & waste management plans for construction & demolition projects is to provide a practical approach which is informed by best practice in the prevention and management of construction and demolition wastes and resources from design through to construction and deconstruction.

5.1.8.2 Platforms and Networks Facilitating the Transition to a Circular Economy

Circular Economy Platforms

CES announced the establishment of a national online platform that will provide an authoritative source of information about what the circular economy is and how the circular transition is taking place in Ireland, identifying key actors, ongoing initiatives at the local, regional, and national levels and areas of likely future progress.

The Rediscovery Centre [289] is the National Centre for the Circular Economy in Ireland that supports re-use of materials, shares circular skills and provides information on circular economy developments here in Ireland and abroad. The Centre receives funding from a range of public bodies. The Rediscovery Centre boiler house is used as a state-of-the-art ecological building to demonstrate best practice building design, construction and operation, and as a tool to inspire action with respect to resource management and efficiency [287].

CIRCULÉIRE [26] is the National Platform for Circular Manufacturing with a mission to demystify, derisk, and deliver circular business model innovation by unlocking the value that resides in an Irish circular economy. It is a €4.5 m public–private partnership co-created by Irish Manufacturing Research and the Department of the Environment, Climate and Communications (DECC), the Environmental Protection Agency (EPA), and EIT Climate-KIC with 25 Founding Industry Members. CIRCULÉIRE's Circular Economy Knowledge Library seeks to support industry, entrepreneurs and policy-makers by providing an open-access knowledge repository of circular economy resources ranging from business case studies to key design guides, to sectoral reports, to circular economy policy documents for example, national and EU circular economy strategies and action plans.

Circular Buildings Platform

The Build360 Hub [19] is a unique collaboration between academia, industry, and the public sector, which aims to provide leadership in transitioning the Irish construction sector towards a decarbonized and circular built environment. The Atlantic Technological University (ATU) Build360 research group, the Regional Waste Management Planning Office and the Environmental Protection Agency (EPA) have developed this resource for a range of stakeholders who wish to learn more about how to embed circular economy principles into their everyday working practice. Considering that education is key to supporting industry professionals, academics, and students on their circular economy journey, the Build360 Hub presents educational opportunities ranging from short Continuing Professional Development programmes to dedicated postgraduate programmes.

Building Materials Platform

Irish Green Building Council created the Construction Materials Exchange (CMEx) platform [157] to unlock the potential of excess construction material. CMEx is a user friendly, digital platform that connects organisations with each other so that they can exchange or trade excess construction materials between them. The platform showcases excess materials from projects where they can be seen, traded, exchanged and reused, ensuring that materials continue to have a value beyond their initial lifecycle. Organisations can choose to have an internal account to exchange materials within the organisation or an external account where reclaimed materials can be exchanged between different organisations. The CMEx platform supports three types of users: an organisation that would like to upload materials for exchange from their projects, an organisation that would like to find materials for their projects, and an individual that would like to find material for their project.

Public Procurement Platform

The EPA has published new Green Public Procurement: Guidance for the Public Sector in September 2021 [60]. This guidance and accompanying ten criteria sets are aimed primarily at public sector procurers in central and local government, state agencies and other public bodies such as universities, hospitals and schools. They are also relevant for utility sector procurers and may be of interest to private companies whether they are responding to tenders or applying green criteria in their own procurement. They have been fully updated to reflect legislation and policy in place as of March 2021. One of the criteria sets in the EPA's Green Public Procurement: Guidance for the Public Sector is the Design, Construction and Management of Office Buildings. This sets out criteria developed based on detailed lifecycle analysis studies carried out at EU level, identification of best practice and state-of-the-art technologies, and consideration of costs over the lifetime of a building. Circular economy considerations, such as the use of recycled concrete and recovery of demolition waste, are also addressed. The role of project managers, design teams, construction and specialist contractors is addressed, as is the role of facility managers.

To facilitate use of these criteria, the Office of Government Procurement (OGP) has launched GPP Criteria Search [114]. This online search tool allows the user to rapidly find, select, and download the GPP criteria relevant to a specific procurement. This tool was developed in collaboration with the EPA and the DECC, with support from the Public Service Innovation Fund. Within the sector "Office building design, construction and management" there are seven subsectors:

1. Environmental management of the project manager, design team, and contractors;
2. Detailed design and performance requirements;
3. Strip-out, demolition, and site preparation works;
4. Construction or major renovation works;
5. Heating, lighting, and energy-related products;
6. Completion and handover;
7. Facilities management.

It is expected that future GPP criteria related to construction will explicitly incorporate the Level(s) framework to ensure all environmental impacts across the life cycle of built assets are addressed in a consistent and measurable way. The Office of Public Works is developing a roadmap to promote the greater use of lower-carbon building materials in construction, which will further support GPP in this sector and cover a broader set of building types [63].

5.1.8.3 Funding Opportunities

Circular Economy Innovation Grant Scheme (CEIGS)

The Circular Economy Innovation Grant Scheme (CEIGS) is a Government of Ireland initiative led by the Department of the Environment, Climate and Communications

(DECC). The CEIGS directly supports the growth of the circular economy in Ireland by funding circular economy projects and enterprises and provides high-profile examples of best practice. The Scheme provides annual funding for innovative medium-to-large scale projects which can support development of the circular economy at the national or regional level. The purpose of this grants scheme is to provide support to projects which work in the Circular Economy space, with the aim of advancing the Circular Economy in Ireland and raising awareness of the need to transition to a Circular Economy. Government will continue to provide direct support to circular economy initiatives, and although the support is not limited to social enterprises and voluntary and community based organisations, DECC is particularly interested in submissions from organisations of these types [115].

Environmental Protection Agency (EPA) Green Enterprise Scheme

The EPA's Green Enterprise: Innovation for a Circular Economy is an annual funding call to support innovators in Ireland to develop, demonstrate and implement circular economy approaches in their business models. It is managed through the EPA's National Circular Economy Programme and is co-funded by EPA Research. The fund is aimed at innovative projects and the maximum amount of funding available is €100,000 per grant award. All organisations applying for funding must provide written declarations that they have not been in receipt of more than €200,000 in state aid in the last three years, including funding applied for as part of this programme [61].

Green Start/Plus Programmes

Green Start and Green Plus are national Enterprise Ireland funded programmes that aim to enable companies to develop sustainable business models [50].

Climate Action Fund

The Climate Action Fund was set up under Ireland's National Development Plan (2018–2027) and is administered by Department of the Environment, Climate and Communications. In order to support innovation and capacity building for climate change mitigation, it makes €500 m available for projects over 10 years (up to €10 m per project) [40].

Disruptive Technologies Innovation Fund

The Disruptive Technologies Innovation Fund was set up under Ireland's National Development Plan (2018–2027) and is administered by Department of Business,

Enterprise and Innovation. In order to inspire enterprise participation in the development of disruptive technologies it makes €500 m available for projects over 10 years (between €5 m–€10 m per project).

EU LIFE Programme

The LIFE Programme provides funding for the support of Environment, Nature Conservation and Climate Action projects. LIFE projects under the sub-programme "circular economy and quality of life" will develop technologies and solutions to enhance the circular economy.

Horizon Europe

The EU's Horizon Europe programme funds large scale international research projects including projects related to the circular economy and sustainability.

Circular Bio-Based Europe Industry Joint Undertaking Europe

The Bio-based Industries Joint Undertaking (BBIJU) is a €3.7 billion partnership between the EU and the Bio-based Industries Consortium with the objective of developing new technologies to transform renewable natural resources into bio-based products.

C-Voucher

C-VoUCHER is a European funding programme that uses design thinking to develop new cross-sectoral and cross-border circular value chains demonstrating circularity.

The Whole of Government Circular Economy Strategy 2022–2023: Living More, Using Less sets out in the Annex 8 the range of supports to business, enterprise and third level knowledge institutions available through the Ireland's Foreign Direct Investment Agency IDA, Enterprise Ireland, Local Enterprise Offices, DECC, EPA and the EU [50].

5.1.8.4 International Cooperation and Support from International Organizations

Irish Green Building Council collaborated with Excess Materials Exchange (EME) of Netherlands to create the CMEx platform with funding from the Circular Economy Innovation Grant Scheme of the Department of the Environment, Climate & Communications [148].

Holland Circular Hotspot (HCH) [148] and The Rediscovery Centre have entered a partnership by signing the Memorandum of Understanding on Monday the 29th of May, 2023 to accelerate the transition towards a circular economy in Ireland, the Netherlands and beyond. Under the agreement, both parties provide their experience, knowledge and expertise to intensify bilateral cooperation in the circular economy space. Areas for collaboration are, for instance, sharing best practices in circularity, promoting European and international collaboration on circular economy initiatives and circular hotspots, and exploring opportunities for global initiatives and proposals [288].

5.1.8.5 Challenges, Barriers and Potential Improvements

As identified in the CES, one of major social barriers to the circular transition is lack of public awareness about what the circular economy is, how it relates to everyday purchasing and consumption decisions and the potential environmental, health and economic benefits associated with increased circularity. This lack of awareness can translate into lack of demand and then act as a drag on investment in the circular economy. A related barrier is found where consumers and/or businesses are aware of the circular economy as a concept but lack confidence to invest in circular products or services. Moreover, the CES states that challenges to improving Ireland's circularity relate to increased investment requirements, the need for more skills and training and removing regulatory barriers [50].

OECD [251: 13–14] highlights the main governance challenges for Ireland's transition towards the circular economy. Ireland has a sectoral view of the circular economy, mainly based on waste, rather than a broader view of resource management and a holistic approach to leverage the circular economy as a cross-sectoral driver of economic growth, job creation, social well-being and environmental protection. Ireland's current approach tends to focus on recycling and recovery rather than preventing, repairing and reusing. Ireland is also facing regulatory gaps related to lengthy and unreliable licensing processes, a lack of circular criteria in green public procurement, and limited government oversight of the waste sector. Funding gaps relate to the absence of a clear and coherent funding framework for the circular economy that also considers private investment, and limited use of price-based incentives. Relevant government departments, agencies and local authorities are still building the technical expertise and human resources required to lead the circular transition, while capacity gaps among Irish SMEs are hampering the adoption of circular practices and applications for funding. Finally, there is limited awareness and understanding of the costs and benefits of a circular economy among Irish businesses and civil society. Across sectors, businesses are facing operational, informational and legislative barriers, while knowledge institutions are confronted with challenges in applying for research and innovation grants and limited collaboration with policy maker. OECD [251: 14–16] highlights the significant role of the government of Ireland in promoting, facilitating, and enabling the shift toward a circular economy.

To promote the circular economy, the national government can [251: 14]:

- Clarify roles and responsibilities for circular economy policy making and implementation across levels of government by: identifying the entities responsible for the achievement of the different objectives of the Whole of Government Circular Economy Strategy and enforcement mechanisms; considering placing the Strategy under the oversight of the Department of the Taoiseach, the Prime Minister of Ireland, as is the case for the Climate Action Plan; making sure that other government departments are involved in circular economy policy making, notably via the Circular Economy Working Group; consulting the Department of Finance on decisions related to the implementation of the Strategy; clarifying the role of Regional Assemblies and broadening the role of local authorities in circular economy policy implementation.
- Complement the Whole of Government Circular Economy Strategy with (i) an analysis of stocks and flows of materials; (ii) a mapping of existing circular economy-related initiatives; (iii) clear objectives, timelines and actions; (iv) a dedicated budget; (v) a shared vision with stakeholders; (vi) a monitoring framework to measure progress and allow adjustments in the second phase based on evidence on what worked, what did not work and what can be improved.
- Raise awareness of and build trust in the circular economy through a national circular economy online platform, targeted communication campaigns, and incentives for behavioural change, such as certificates, labels and nudging.

To facilitate collaboration among a wide range of actors to realise the full potential of the circular economy, the national government can [251: 14–15]:

- Establish horizontal co-ordination to align objectives, share information, identify synergies and avoid duplications, through the activation of the inter-departmental Circular Economy Working Group to identify how circular economy principles can be applied to different sectoral policies to reduce waste, improve resource efficiency, create jobs and improve access to services; as well as better co-ordination with government departments and agencies setting standards.
- Implement formal co-ordination mechanisms between local authorities and the DECC and EPA respectively, building on current initiatives such as the Local Authority Prevention Network (LAPN) but going beyond the focus on waste; and adapt and extend current regional waste management planning co-ordination mechanisms between the DECC, EPA and local authorities, such as Regional Steering Committees and Regional Operations and Task Groups to the circular economy for effective multi-level co-ordination.
- Foster policy coherence and transcend silos by enhancing synergies and complementarity between the Circular Economy Strategy and the Climate Action Plan, as well as other relevant policy documents (e.g. the National Planning Framework, Enterprise strategy and Bioeconomy policy).
- Facilitate collaboration between public, private and non-profit actors on the circular economy by introducing stakeholder coordination mechanisms that

are broader than public consultations (e.g. representation, partnerships or co-decision); and broaden the current sectoral focus on waste by formally including new stakeholder groups such as designers, standard-setters and the research community in the Circular Economy Advisory Group.
- Leverage territorial specificities for the circular economy by facilitating local demonstration projects and urban–rural linkages, for example through establishing reuse facilities (e.g. reuse centres) among other initiatives that make reuse easy.

To enable the necessary governance and economic conditions for the circular economy transition, the national government can [251: 15–16]:

- Make the legislative and regulatory framework fit for the circular economy by making regulation conducive to ecodesign, repair, reuse and remanufacturing; streamlining regulatory processes for reusing material considered as waste through end-of-waste and by-product processes; broadening extended producer responsibility (EPR) to new waste streams and improving existing EPR schemes to facilitate reuse; mandating Resource and Waste Management Plans, which are currently recommended on a voluntary basis, for all construction and demolition projects; and ensuring that circular criteria are included alongside green criteria in GPP.
- Use economic tools to incentivise the transition to a circular economy, and identify and remove environmentally harmful subsidies preventing the application of the waste hierarchy across sectors.
- Set up a clear funding framework for the circular economy and assist local authorities in supporting and scaling up small circular initiatives by setting up dedicated local funding. The DECC could support private investment in the circular economy with alternative funding mechanisms, such as crowdfunding, leasing, equity participation, grants, loan guarantees, green bonds and loans for circular economy projects and businesses, and leverage EU funds.
- Adapt human resources to the challenges at hand by extending current capacity building programmes to the DECC and its Circular Economy Unit, as well as strengthening local authorities' capacity for local circular economy policy implementation, to enable them to take an active role in the transition. For businesses, workers and private investors, the government can design and implement sectoral circular economy training and toolkits for workers and businesses (especially SMEs) on the circular economy, in partnership with CIRCULÉIRE, the Rediscovery Centre, Ibec and Chambers Ireland, for instance. For civil society, particularly youth, the government can integrate circularity into first, second and third-level education curricula to build awareness of and foster skills for a circular economy.
- Support market innovation and business development with different initiatives, including the creation of innovation labs as open innovation ecosystems, the establishment of a one-stop-shop for information on the circular economy for businesses, especially SMEs, and the introduction of tailored advisory services to

support project promoters in making circular economy initiatives commercially viable.

- Develop a national circular economy information system to monitor and adjust policy, by harmonising data collection among data providers, expanding data collection from waste-related data to environmental, economic and social data, and collecting locally disaggregated and sectoral data, particularly in key sectors for the circular transition in Ireland (e.g. food and the built environment) to inform circular economy policy.
- Track progress on the achievement of the targets defined by the Strategy, as well as on the implementation of governance frameworks required for the circular economy transition, by inviting stakeholders to take part in regular (e.g. annual) assessments.

5.1.8.6 Examples of Successful Implementation

Public Policy Initiatives

The Atlantic Technological University (ATU) Build360 research group, the Regional Waste Management Planning Offices and the Environmental Protection Agency have developed the Build360 Hub for a range of stakeholders (industry professional, academics and students) who wish to learn more about how to embed circular economy principles into their everyday working practice.

Private Policy Initiatives

The new European Headquarters for Kerry Group's Global Technology and Innovation Centre is located in Millennium Park, Naas, Co. Kildare, Ireland and serves as a point of reference across the fields of building engineering and architecture. The facility is made up of 25,000 m^2 over four storeys, and presents a mixture of office accommodation, customer care areas, meeting rooms, research and development areas, as well as ancillary services areas to accommodate up to 895 staff. Kerry Group and Gilbane, along with the design team of RKD Architects, Arup and Bruce Shaw cost consultants visited several existing international facilities in order to form design principles and Arup provided a collaborative BIM design during the fast track design period. The project targeted a LEED system Silver accreditation from an early stage. The mechanical systems were designed to be highly efficient through state-of-the-art equipment efficiencies and also intelligent control systems to minimise energy consumption. Intelligent lighting control systems combine with an energy efficient façade to reduce overall energy consumption throughout the facility. The building façade includes twin-skin elevations with motorised interstitial blinds and high specification g and u values. This assists greatly in reducing the building's energy consumption by minimising unwanted solar gains. The system works in tandem with an intelligent lighting control system which actively dims any light fittings in an area of plentiful daylight. Overall water demand to the site is decreased greatly by the specification of low flow type water fittings throughout. LEAN construction concepts were utilised throughout the procurement and construction phases. Weekly work plans were formulated to track the number of specific tasks to be undertaken by each sub-contractor and the task completion rate was monitored on a weekly basis. The project was delivered ahead of schedule in early 2015 and at 5% below the initial project budget. Finally, it won the Engineering Project of the Year Award at the Engineers Ireland Excellence Awards 2015 [14].

5.1.9 Italy Factsheet

5.1.9.1 Policy and Regulatory Framework

Governance

In June 2022, the Italian Ministry of Ecological Transition [206] proposed a "National strategy for the circular economy" that was accepted and promoted by the following Ministry of Environment and Energy Security. The "National strategy for the circular economy" is, therefore, a programmatic document, within which the actions, objectives and measures to be pursued in defining policies are identified as well as the institutions aimed at ensuring an effective transition towards a circular economy.

With the National strategy for the circular economy, they intend, in particular, to define the new tools, administrative and fiscal aspects to enhance the market for secondary raw materials, so that they are competitive in terms availability, performance and cost compared to virgin raw materials. To this end, the Strategy acts on the chain of purchase of materials (Minimum Environmental Criteria for green purchases in the Public Administration), on the criteria for the cessation of the status of waste (End of Waste), on the extended responsibility of the producer and on the role of the consumer, on the diffusion of sharing practices and of "product as a service". The strategy also constitutes a fundamental tool for achieving the objectives of climate neutrality and defines a roadmap of measurable actions and targets between now and 2035.

The Ministry of Enterprises and Made in Italy created a section targeted to elaborate proposals for the development of products, technologies and production processes with a lower environmental impact, the sustainable development, the transition towards a circular economy, and for the promotion of bioeconomy sectors. Moreover, the Ministry works on connection actions with state and regional administrations and with other public entities that implement programs and interventions in favor of companies for eco-sustainable development. It wants to define proposals and evaluations relating to the use of public demand for industrial policy purposes with particular reference to the circular economy and research and innovation.

Legislation

The decree n. 116/2020 (Official Gazette Series n. 226, 11 September 2020) [263]—in force since 26 September 2020—modifies Legislative Decree 152/2006 by transposing the European directives on waste EU 2018/851 and on packaging and packaging waste 2018/852. The decree is an implementation of directive (EU) 2018/851 amending directive 2008/98/EC on waste and implementation of directive (EU) 2018/852 amending directive 1994/62/EC on packaging and packaging waste. The main obligations of the manufacturer change.

This decree substantially modifies the fourth part of Legislative Decree 152/2006, the so-called Consolidated Environmental Text. All public and private entities that produce waste and that operate in the field of waste, packaging and packaging waste management will have to adapt to this new text.

Circular Economy Strategy, Roadmap or Action Plan

The main national programmatic document is the National Strategy for Circular Economy (2022) which has its basis in the National Recovery and Resilience Plan.

Moreover, the Minimum Environmental Criteria are the environmental requirements defined for the various phases of the purchasing process, aimed at identifying the best design solution, product or service from an environmental point of view along the life cycle, taking into account market availability. The Minimum Environmental Criteria have an effect on various product sectors and are widely used in the construction sector.

At regional level, there are some Regions (Lombardia, Puglia, etc.) which have a specific strategy for sustainable development. The implementation of strategies of circular economy is still framed in this context.

Construction and Demolition Waste Management Regulatory Framework

The Legislative Decree n. 76/2020 [264] (converted into Law no. 220/2020, Official Gazette Series 228, 14 September 2020) regulated the matter in terms of demolition and reconstruction of buildings, with the introduction of amendments to the rules contained in the consolidated construction text, n. 380/2001. The decree introduces definitions of demolition and reconstruction of existing buildings in order to comply with the legislation concerning the distances between buildings for demolition and reconstruction between existing buildings.

Regarding the Minimum Environmental Criteria applied to the construction sector, the main objective is to put into circulation materials and services with a high sustainable value, calculated through the use of the life cycle analysis (LCA) of the materials involved.

Italy adopted the Agenda 2030 targets set at the European levels, sharing the plan developed by the Buildings Performance Institute Europe [59].

5.1.9.2 Platforms and Networks Facilitating the Transition to a Circular Economy

Circular Economy Platforms and Networks

The Italian Circular Economy Stakeholders Platform was created in 2018 to bring together initiatives, share experiences, highlight critical issues, and indicate perspectives to represent in Europe the Italian specificities in terms of circular economy and to promote the circular economy in Italy through dedicated actions. ICESP is promoted by the National Agency for New Technologies, Energy and Sustainable Economic Development (ENEA) as an initiative integrated with the European Platform for the Circular Economy (ECESP) with the aim of spreading knowledge, mapping good practices, and fostering multi-stakeholder dialogue. The 275 members of the ICESP are representatives of local and national public institutions in education, research and innovation, but also of companies, trade associations and representatives of civil society.

Moreover, in 2017, ENEA promoted the the first Italian industrial symbiosis network, SUN—Symbiosis Users Network, which involves 42 partners including universities, political institutions, research bodies, private companies, technological networks and local bodies. The SUN network is proposed as an Italian reference for operators who want to apply industrial symbiosis, at an industrial, research and territorial level.

The Sustainable Development Foundation was established in 2008 as a point of reference for the main sectors and actors of the green economy in Italy. The foundation collects experiences and knowledge to support businesses and organizations in sharing a common vision towards the ecological transition. The Foundation has established a structure, the Circular Economy Network, for the development of studies and proposals on efficient use of materials within production, distribution, consumption and in the area of waste management. The activities target both private and public sector and include reports and assessments, prevention programs and waste management plans, evaluation and analysis of governance models, studies and legal/regulatory assistance.

Circular Buildings Platform

No circular buildings platform has been established to date.

Building Materials Passport Platform

No building materials platform has been established to date.

Public Procurement Platform

In Italy, a first signal towards the use of Green Public Procurement comes with the approval by the Inter-ministerial Committee for Economic Planning (CIPE) of resolution n. 57 of 2 August 2002 "Environmental action strategy for sustainable development in Italy", which establishes that "at least 30% of the goods purchased must also meet ecological requirements; 30–40% of the durable goods stock must be reduced energy consumption, taking into account replacement and using the scrapping mechanism".

With the decree 8 May 2003 n. 203 [262] (Official Gazette Series n. 180, 5 August 2003), the Ministry of the Environment and Land Protection has identified "rules and definitions for the regions to adopt provisions, intended for public bodies and companies with predominantly public capital, also for service management, which guarantee that manufactured goods and goods made with recycled material cover at least 30% of the annual requirement".

Funding Opportunities

The National Recovery and Resilience Plan (NRRP) is the package of investments and reforms prepared by the Italian government as part of "Next Generation EU", the programme established by the European Union. One of the aim of this program is also to make Building more modern and sustainable roads, railways, ports and airports throughout the country within five years. €25.40 billion are dedicated to this mission.

With the opportunities offered by the NRRP, tax incentives are foreseen, rewards in terms of scores for virtuous companies and certainly calls for tenders for the financing of research and development projects with the aim of increasing the scientific and technological-industry level in this area.

Moreover, funding opportunities are offered by public/private organizations such as:

- Cariplo that is a philanthropic association providing grants to social enterprises or cooperatives active in sustainable projects;
- The Italian bank Intesa Sanpaolo devoted EUR 5 billion in 2018–2021 to support projects in the circular economy in partnership with Ellen MacArthur Foundation. The bank is a major partner of the European Investment Bank (EIB) with two credit lines for the circular economy worth EUR 1 billion.

5.1.9.3 International Cooperation and Support from International Organizations

The Italian Circular Economy Stakeholders Platform (ICESP) has been selected by a jury chaired by the Holland Circular Hotspot to be the protagonist of an exclusive video at the World Circular Economy Forum (WCEF2023), one of the main global events on the circular economy (Helsinki, Finland, from 30 May to June 2, 2023).

5.1.9.4 Challenges, Barriers and Potential Improvements

The ETC CE Report [88] highlights that Italian policy still uses voluntary instruments. As a consequence, regulations, standards, and consumer behaviour are not in line with effective circular production and consumption patterns. Moreover, primary raw materials are often cheaper than secondary raw ones, market demand for recovered products is still limited as circular products do not present clear guarantee schemes and can be more expensive than other products.

Further barriers can be individuated in:

- poor reliance on recycled building materials/products compared to traditional materials and the limited ability of companies to seize opportunities have a considerable weight;
- absence of a well-structured, monitored and participated value chain;
- limited communication between institutions and citizens;
- lack of awareness in companies, public authorities and among citizens of circular economy concepts;
- policy and regulatory;
- lack of financial support and fiscal incentives.

The report of the European project SocKETs [329] highlighted that the building and construction sector in Italy is still characterized by a linear approach. Almost all the involved subjects declared that steps ahead have been made in the field of energy efficiency, but they underlined that a major part of the way towards a circular economy in the sector is still to be covered, in almost all the other aspects, such as implementation of ecodesign principles, reduction of resources consumption, extension of product life cycle, and proper use of waste and secondary raw materials.

Potential improvements can be obtained through:

- introducing materials passport and appropriate national databases that facilitate the analysis of the life cycle on a national scale;
- publicizing good practices in terms of secondary raw materials and introducing into procurement, not only reward scores, but advantages in terms of tax deductions and incentives;
- the Italian construction sector is mainly fragmented into small and medium-sized enterprises, especially family-run ones. Thus, it is important to find an effective method to transfer what is quite easily achieved in large enterprises to the myriad of SMEs;
- adopting a secondary raw materials policy;
- training of end users on the concepts and benefits of the circular economy.

5.1.9.5 Examples of Successful Implementation

Public Policy Initiatives

Two Italian Region, Puglia and Piemonte, implemented the Regional Strategy for Sustainable Development and are partners in the pilot project "Region2030: Monitoring the SDGs in EU regions—bridging the data gap" of the Joint Research Center (JRC) of the European Commission [298].

Moreover, with Resolution n. 98-9007 of 16 May 2019, the Regional Council approved the "Technical document setting up and first guidelines of the Regional Strategy for Sustainable Development of Piedmont" identifying the Circular Economy as a model to be promoted for the sustainable development of the "regional system" (Regione [297].

The Lombardy Region, in collaboration with Unioncamere Lombardia, has allocated over 4 million euros to support SMEs and micro-enterprises in the transition towards a circular economy model to be applied, in particular, at the supply chain level (Lombardy Region Council Resolution no. 6402 of 23 May 2022) [295].

The Lombardy Region created in 2018 the Observatory for the circular economy and the energy transition for sharing the strategic objectives of regional climate policies and for the sustainability of the use of resources with all the players in the territory [294].

The Marche Region has approved the regional law (17 March 2022, n. 4) "Promotion of investments, innovation and digital transformation of the Marche entrepreneurial system" with particular focus on the implementation of a circular economy [296].

The Reginal project "Imprese green" promoted by the Lazio Region supports the green conversion of small and medium-sized enterprises (SMEs) and agricultural businesses in Lazio and develop the circular economy model. In particular, 23.3 million euros have been allocated for sustainable and intelligent mobility, green building and smart building, circular economy and energy [292].

The Lazio Region is the first in Italy that approved the Regional Ecological Transition Plan with a resolution of the Department of Ecological Transition and Digital Transformation (January 2023). The document plans public spending to achieve the global sustainability objectives between now and the next few years and up to 2050 for a total of 5.9 billion in total [293].

Private Policy Initiatives

An example of application of the circularity model in the construction sector is offered by Webuild Group, which reuses 100% of the excavated materials and developed models to achieve the recycling of 90% of the steel used. For this Ialian company, the development of the circular economy is one of the pillars on which its sustainability policy is based. For this reason, Webuild adopts practices aimed at minimizing the use of natural resources and intensifying the recovery of waste materials within the same work or in neighboring areas. The circularity of the practices adopted by Webuild allowed the company to reduce the waste produced by 46% in 2020 compared to 2019, while the waste involved in recovery processes was 69% of the total generated. Furthermore, 100% of the excavated materials were reused and the volume of water consumed on construction sites fell by 10% compared to 2019.

Several near-zero energy buildings (NZEB) constructed in Italy gain a wider resonance and acknowledgement as made of recycled materials or from natural origin, with low environmental impact. Good examples are: Fiorita Passive House in Cesena, Biocasa_82 in Treviso, Kindergarten in Guastalla, LAGO Campus in Padova [249].

5.1.10 Latvia Factsheet

In Latvia in 2019, 70% of the collected construction debris was a mixture of construction demolition waste and inert materials, while 20% consisted of only inert materials, total construction and demolition waste accounts for around 380,000 tons per year. Latvia has developed the Waste Management Plan 2021–2028 to address the waste management system and EU objectives in the long-term, but already now it is obvious that there are several obstacles to their implementation, especially in the household sector.

In this regard, Latvian policymakers hesitate to issue strict regulations on the construction sector and react with rather soft actions. Nevertheless, the Latvian Ministry of Economy and several state administrative institutions, non-governmental organizations, and state capital companies agreed on joint cooperation to promote the production and use of wooden building materials and construction products with high added value in construction, promoting sustainable construction and the growth of Latvia's economy, by signing the Memorandum of cooperation on promoting the use of wood in construction on April 23, 2021. In addition, in order to promote the availability of low-cost, including operating costs, ergonomic, easy-to-use, and energy-efficient housing, a demo project for a multi-apartment residential building based on modular wooden construction has been developed by order of the Ministry of Economics in 2022. At the time of writing, it has not been exercised by any private or public developer.

According to Association of Latvian Construction Contractors, the construction industry in Latvia (as in most EU countries) plays a significant role in the growth of the national economy, as it is one of the largest industries in terms of material consumption. Therefore, the construction industry, by increasing the demand for circular as well as sustainable building materials and construction products produced from local renewable natural resources or recycled, downcycled materials, could endorse the pace at which Latvia achieves its sustainable development goals. As a result of the research of the current situation in the Latvia, it can be generally concluded that the principles of circular economy in construction are not currently regulated in Latvia. Article 4 of the Construction Law defines the principles to be observed in planning and construction, including the observance of economic aspects and public interests in architectural solutions, technological and economic efficiency in engineering solutions, and the principle of sustainable construction, including circularity. However, the general regulation does not have direct technical application and implementation norms in the standards and/or regulations of the Cabinet of Ministers.

Some statistical data:

- Construction and demolition waste (CDW) volume in Latvia accounts for more than 350,000 tonnes per annum (including around 30,000 tonnes of hazardous CDW).
- CDW forms 16% of total waste generated in Latvia vs EU average 36% of total waste stream.
- Up to 95% of CDW can be recycled or reused, although much of CDW is still landfilled, mostly being mixed with municipal waste.

Although construction and sustainability specialists are familiar with the criteria proposed by BREEAM, LEED, DGNB, WELL, as well as some other sustainable certification systems and the requirements for proving their compliance, industry experts admit that the regulatory environment in Latvia actually does not allow for the requirements for obtaining this certificate in certain circulation criteria, because the process of proving and certifying the properties is not clear, as well as there are no technical standards or building regulations for working with construction waste recycling products.

Within the framework of the EU interregional cooperation program URBACT III project "Transition to a circular economy in urban construction (URGE)", the municipality of Riga City Council has developed Guidelines for sustainable construction based on EC sustainable construction principles, which define the principles of sustainable design, construction, management, and governance. For now, however, this compilation of good practices and the manual of principled solutions do not fully answer the questions about the methodology for proving, testing, checking technical compliance and certifying compliance, which would also be harmonized with the regulatory enactments regulating construction in Latvia. There is no clear regulation in Latvia on obtaining "end-of-waste status" for construction recycling products or recovered materials, as well as comprehensive technical standards for work with construction waste recycling products, all participants in the process have limited opportunities to increase the circulation of construction materials.

5.1.10.1 Policy and Regulatory Framework

Governance

Ministry of Climate and Energy of the Republic of Latvia (former Ministry of Environmental Protection and Regional Development of the Republic of Latvia) is responsible for implementing policy in three areas—environment protection, regional development as well as information and communication technologies.

With the order of the Minister of March 21, 2017 no. 1–2/47 approved the operational strategy of the Ministry of Environmental Protection and Regional Development 2017–2019 for the year, determining the goals to be achieved in the next 3 years, directions of action and tasks for achieving the goals. The second line of action envisaged the development of a policy for the transition to a circular economy.

The Latvian Cabinet of Ministers has approved Latvia's vision for achieving climate neutrality by 2050 (https://ec.europa.eu/clima/sites/lts/lts_lv_lv.pdf) and Latvia's climate change adaptation plans for the period until 2030, in order to help Latvian citizens and the national economy better adapt to the already ongoing climate change and thus reduce the damages caused by climate change. The main goal of "Latvia's climate change adaptation plan until 2030" is to reduce the vulnerability of Latvia's people, economy, infrastructure, buildings and nature in relation to the effects of climate change and to promote the use of opportunities provided by climate change. The construction sector has been assigned the third of five strategic objectives: infrastructure and construction are climate-resistant and planned according to possible climate risks.

The Ministry of Environmental Protection and Regional Development outlined 5 priorities in the Operational strategy for 2020—2022. For the second priority—Economically efficient and thoughtful use of resources as a task for the implementation of the direction of action was to introduce the principle of circular economy in Latvia. The achievable result with the deadline of 30.12.2022 was formulated as follows: "Promoting the development of the circular economy in Latvia, determining the main directions of action and measures to contribute to the development of the national economy, competitiveness, while ensuring the fulfillment of the country's international obligations". In addition, 3 sub-results were defined:

- Determining the main circular economy action directions and measures (Circular Economy Strategy)
- Implementation of the measures specified in the circular economy strategy for Latvia
- Development of the regulatory framework for the implementation of circular economy principles.

Legislation

"Latvian National Energy and Climate Plan for the 2021–2030" the seventh course of action determines measures in the development of waste management, with the main focus on reducing the volume of waste and recycling waste, especially based on the principles of the circular economy. As one of the benefits as a result of the successful implementation of this document, the introduction of the circular economy and the promotion of the emergence of industrial symbioses are mentioned. The energy consumed by the building sector accounts for up to 40% of the entire energy balance, so in this planning document, in relation to the construction sector, the greatest attention is paid directly to energy efficiency issues and the promotion of the use of emission-free technologies. The need to increase the use of timber in construction and the integration of green infrastructure into the urban environment has also been emphasized.

In the "Guidelines of the national industrial policy for the 2021–2027" it is emphasized that in ensuring the sustainability of Latvia's economy and creating innovations, it is essential to take into account the established Latvian environmental protection and climate goals, paying special attention to the integration and development of circular economy and climate issues both in business and science.

Latvian Waste Management Law was adopted on 28.10.2010 by the Saima. The purpose of this law is to determine the procedure for waste management in order to protect the environment, human life and health, preventing the generation of waste, ensuring separate collection and regeneration of waste generated on the territory of Latvia, as well as promoting the effective use of natural resources and reducing the amount of waste to be buried. The law also mentions construction waste and its management requirements.

Circular Economy Strategy, Roadmap or Action Plan

The Ministry of Environmental Protection and Regional Development developed an Action plan for the transition to the circular economy for 2020–2027, which was adopted on 04.09.2020 by the order of the Cabinet of Ministers no. 489. The purpose of the development of the plan is to provide a policy framework for the provision of such an operating environment that would promote the country's transition to a circular economy, while contributing to the implementation of the European green course and the achievement of global sustainable development goals. The plan was developed so that the circular economy would be thoughtfully introduced in the national economy and society of Latvia, with the aim of promoting society's progress towards more thoughtful and responsible resource planning, use, sustainable production and consumption, integrating these basic considerations into all sectoral policies, regarding the priority stages of the circular cycle and resource flows.

The implementation of circular economy requires effective cooperation at all levels of state administration and cooperation between different sectors of the national economy. The Action Plan contains seven initiatives:

1. The transition from waste management to resource management. Some of the key measures include developing secondary material markets and expanding and improving extended producer responsibility (EPR) schemes.
2. Improving resource productivity in all sectors of the economy by encouraging the development of research and innovation. This initiative includes the fostering of industrial symbiosis and support for redesigning existing products and designing new ones in line with the eco-design principle. These measures are necessary for the resource productivity target value of EUR 1.55 per kilogram to be reached by 2027, up from EUR 0.90 per kilogram in 2020.
3. The establishment of pre-conditions for the reuse of goods. Measures include support for social innovation and entrepreneurship.

4. Promotion of the transition from the purchase of goods to services, i.e., the promotion of rental services and green public procurement (GPP).
5. Improving the management of materials, processes, and waste in priority sectors. This includes the preparation and distribution of informative materials on waste minimisation, especially food, textile, and furniture waste minimisation.
6. Strengthening the role of municipalities in the implementation of the principles of a CE.
7. Engagement, information, and education of the public.

At the local level, the "Riga Sustainable Development Strategy 2030" is in effect, the main goal of which is the need to use municipal resources wisely and sustainably. For example, with regard to construction, the document states: in order to promote the attractiveness of neighborhoods for living, it is necessary to prioritize the revitalization of the living environment in micro-districts designed during the Soviet era. The preference is not for the construction of new residential buildings. Requirements for sustainable use and construction of the territory should be determined, promoting the emergence of high-quality architecture and the use of modern construction technologies.

Based on the document of the European Commission "Circular economy—principles of building design", guidelines for circular construction have been developed, which, by providing examples and specific influences, are suitable for the context of Latvia and Riga. The aim of the guidelines is to provide knowledge and practical support in the construction, maintenance, renovation and demolition of buildings based on circular economy principles. They will also be useful for building and construction management and procurement specialists, as well as residents.

The developed guidelines offer eight general principles that inform and support the members of the construction chain, as well as provide principles for orderly planning of buildings:

1. Circular economy and sustainable building design principles apply to all parties involved in the value chain;
2. When making sustainable choices, total life cycle costs, financial and non-financial returns on investment should be considered;
3. A viable business model must be present at each stage of the circular construction value chain;
4. The principles should be applied with proportionality in mind—the benefits should outweigh the costs;
5. Better knowledge of construction methods is needed to facilitate deconstruction and increase building resilience and adaptability;
6. The durability of buildings depends on better design, improved performance of building products and information sharing;
7. It is necessary to develop a new design culture, not allowing premature demolition of buildings;
8. Products and systems should be chosen so that they can be easily reused, repaired or recycled.

At the regional level, the Integrated Action Plan for the transition to a circular economy in urban construction in the municipality of the city of Riga was developed with the aim to provide a framework for the next five years for the implementation of circular economy principles in urban construction in the municipality of Riga State City and to determine those responsible for the implementation and monitoring of the plan. Although the IAP is the first local planning document that directly addresses the transition to a circular economy, there are other sustainability-promoting initiatives in housing and construction. For example, a municipal support program has been established for the sustainable and energy-efficient renovation of a private multi-apartment housing stock. Also, work is underway to establish the Riga Energy Efficiency Fund in the city to finance the renovation of buildings and the production of renewable energy, as well as the creation of waste sorting areas, which will become a significant support for the development of circular economy infrastructure.

Various structural units of the council are involved in promoting the sustainability of construction, such as REA, RD PAD, RD ĪD, RD MVD, RD ĪK and others. In 2025, Riga municipality adopted the Circular Economy Strategy 2040. The strategy identifies priority directions until 2040, including the development of sustainable urban mobility, the promotion of circular construction and building renovation, waste prevention and valorisation, support for green entrepreneurship, and the creation of innovation platforms for industrial symbiosis. The main goal of the Strategy is to transition from a linear to a circular and systemic approach, thus fostering resource conservation and creating environmental and economic benefits for the municipality and its citizens. The Strategy also points out the importance of citizen engagement, education, and cooperation between municipalities, businesses, and academia in driving systemic change. This Strategy is seen as an efficient aligner of local priorities with the European Green Deal and Latvia's national circular economy policy framework. Thus, the Strategy provides both a strategic roadmap and actionable measures for accelerating the transition towards a climate-neutral and resilient urban economy. Based on this document, Riga municipality has developed Riga Circular Economy Action Plan for 2026–2030, that sets out specific measures that will promote the circularity of resources (sharing, reusing, repairing, renewing and recycling of materials and products), citizen engagement, circular entrepreneurship, as well as the adaptation of municipal governance and procurement processes to the goals of the circular economy.

Construction and Demolition Waste Management Regulatory Framework

It is not a common regulation for construction and demolition waste and its management. However, in the municipality level, different municipalieties have their own regulations. For example, the key regulations regarding construction and demolition waste management in Saldus is in the Waste management regulations, which was adopted on 28.04.2022 by the order of the Saldus Dome no. 24.

"State waste management plan for the 2021–2028" includes the principles of the circular economy and in one of the analyzed development scenarios, the implementation of the principles of the circular economy is also envisaged. The 2nd overarching goal of the document stipulates ensuring the rational use of waste as a resource based on the basic principles of the circular economy and, as far as possible, promoting the return of resources into the economic cycle in a way that is useful for the national economy. Among other things, the document envisages improving technological processes in order to prepare construction waste flows for recycling, requires municipalities to ensure separate collection of construction waste, reuse of support materials and the use of technologies that reduce the generation of waste in construction.

Construction waste management guidelines have been developed in 2023. Their purpose is to create a common understanding and provide information about the procedures for managing waste generated during construction and demolition of buildings, the activities to be performed, the division of duties and responsibilities. The guidelines are intended for those working in the construction sector, providers of waste management services, institutions supervising and controlling construction processes and the waste management sector. A separate section is devoted to the management of construction waste generated in households. The guidelines consider construction waste circulation from the moment of development of the construction project, which reflects the planned amounts of waste generated during construction and demolition of buildings and their management, until the final processing, regeneration or disposal of this waste.

5.1.10.2 Platforms and Networks Facilitating the Transition to a Circular Economy

Circular Economy Platforms

Since 2021, Latvia has been implementing the European Union LIFE programme integrated project "Waste To Resources Latvia – boosting regional sustainability and circularity" (LIFE Waste To Resources IP, No. LIFE20 IPE/LV/000014), which, as one of its pillars, has industrial symbiosis and the development of an industrial symbiosis platform. The platform is still under development, but it already provides an interactive map with companies and their production leftovers. It is planned that this platform will be used by industrial symbiosis facilitators and also by businesses to seek symbiotic connections in order to optimise resource flow, contribute to waste prevention and foster the circular economy. The web page of the platform is https://sinergia.lv/.

At the same time, few platforms exist which are working based on circular economy principles. For example, Andelemandele, is an advertising platform. The Portal is not a financial intermediary for sale/purchase transactions. The actual sale/purchase transaction takes place between the direct buyer and the seller, with both parties mutually agreeing on the most convenient way to deliver the item and pay for it.

Or, for example, OtraĒlpa, their activity is based on the interaction of three basic values—charity, conservation of the environment and social responsibility. In each of the charity shops, customers can purchase various 'second life' items donated by private persons and organizations, as well as give away things that have unnecessary to yourself.

Circular Buildings Platform

Within the project Life project "Waste to resources IP", the creation of a national digital platform for the circular economy in construction was planned, which would also contain a user portal for monitoring progress in achieving the goals of the circular economy set by European and national legislation. The platform would allow to have an accessible data system for current monitoring and trade of construction debris and its processing products.

First pilot projects have been launched on establishing construction material exchange platforms. Up to now a short-term pilot project has been launched in Riga and an ongoin project is observed in Kuldiga municipality. The idea of the project is to offer a possibility to private individuals to bring or pick up various repair materials and construction materials free of charge. Materials such as tile, paint, varnish, plumbing, cement, plasterboard, veneer, chipboard, stone wool, insulation materials and more are accepted. The main condition for handing over materials—they must be usable.

Building Materials Passport Platform

No building materials platform has been established to date.

Public Procurement Platform

Latvia has official public procurement platform. The public procurement notices are published by the Procurement Monitoring Bureau free of charge and, if necessary, are sent for publication in the Official Journal of the European Union. The platform also contains information on construction procurement.

In Latvia, the Ministry of Environmental Protection and Regional Development (VARAM) is responsible for promoting green procurement and policy development. Explanations and regulatory acts, frequently asked questions, as well as the Green Procurement Calculators are available on the VARAM website:

- GPP life cycle cost calculators
- GPP cycle calculator mobile application for energy consuming devices
- Construction Life Cycle Cost (LCC) calculator.

Requirements for green public procurement and their application procedure was developed in 20.06.2017 and were accepted by the Cabinet of Ministers.

For the construction, a couple of guidelines for green public procurement were developed:

- GPP guidelines for design, construction and management of office buildings
- GPP guidelines for recreation and sports infrastructure
- GPP guidelines for road design, construction and maintenance
- GPP guidelines for transport.

Mandatory criteria for "green public procurement" are planned to be applied in the procurement of construction works. Currently, the application of these criteria to construction works is voluntary.

No national green public procurement platform has been established to date.

5.1.10.3 Funding Opportunities

Latvia uses different financial or technical support from an international institution or EU funding programme for the process of transition towards the circular economy, mostly through the implementation of different scientific projects: both local and international.

International Cooperation and Support from International Organizations

No data to this point.

5.1.10.4 Challenges, Barriers and Potential Improvements

There are several challenges and barriers related to different areas of implementation of circular economy in the built environment. The most important in Latvia are:

Policy and regulatory barriers; market barriers for recycled or reused resources, i.e. not economic efficient, lack of demand; lack of quality standards for the use of secondary raw materials during the construction of new buildings; lack of infrastructure etc.

The concept of a circular economy is continuously gaining popularity among academics, business community and decision-makers. In recent years, the concept of "circular ecosystems" has been developed where multiple actors and stakeholders are involved to implement circularity at intra-industry and cross-industry levels. Even the role of actors within circular ecosystem differs, key players within the ecosystem are public authorities both at legislative level and municipal levels responsible for the setting up and implementation of circular economy policies including in construction waste management.

In 2021 Latvia received a significant grant under LIFE IP project—LIFE Waste to Resources IP, LIFE20 IPE/LV/000014 "Waste To Resources Latvia—boosting regional sustainability and circularity", that is being implemented within the time-frame of 2022 to 2028. Under this project, which gathers a consortium of over 20 partners from waste management, municipalities, the construction industry, etc., it was possible to undertake a study to explore the main challenges for municipalities in facilitating the circularity of construction and demolition waste streams from households. The study is based on 24 municipalities in Latvia survey and follows in-depth interviews to tackle the main barriers and possible solutions to minimize illegal dumping and to promote the circularity of construction and demolition waste generated in households. Previous academic studies on barriers to circular ecosystem development mainly focused on challenges for companies, however, the research on the local governments' view is limited. The research demonstrated that more than 60% of municipalities see illegal dumping of construction and demolition waste in Latvia as problematic and 70% of municipalities see households as the main polluters. Insufficient resources (financial and human), underdeveloped infrastructure (e.g., no specific waste acceptance areas), and inadequate legislative framework including a lack of motivation people to sort construction and demolition waste are a few of the identified main barriers towards circularity. The conclusions of the study and recommendations for the action will contribute to further academic research and policy development in the area of construction and demolition waste circularity and the enablement of a circular ecosystem (Fig. 5.1).

24 out of 43 Latvia's municipalities took part in the survey. Valuable information was obtained regarding CDW management challenges across different regions. The use of open-ended questions helped to define short-term and mid-term priorities from the public authorities' perspective. More than 60% of municipalities consider environmental pollution with CDW as an issue within the municipality. 70% of municipalities recognize households as the main source of illegal dumping. 60%

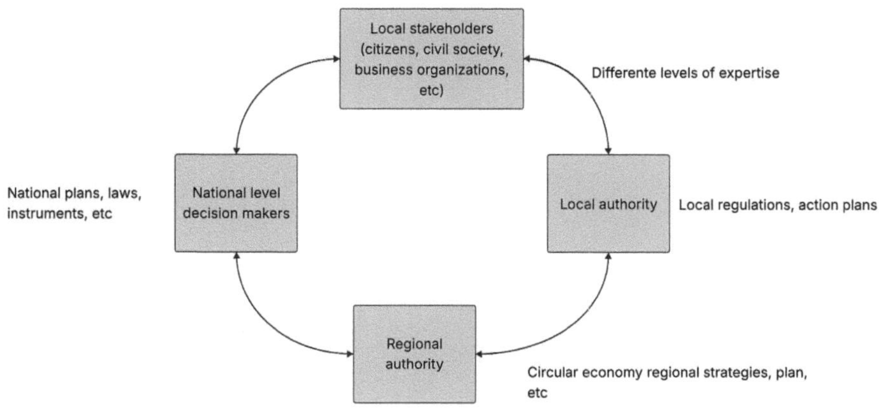

Fig. 5.1 Circular ecosystem governance

of municipalities see the management of hazardous construction waste as an issue within the municipality.

An external information on circular economy principles in construction (ec.europa.eu), according to the EC guidelines "Circular economy principles for building design", the contractor can contribute by choosing solutions with greater adaptability (e.g. modular solutions), increasing recycling and regeneration potential (for example by tracing the origin) as well as following Ecodesign principles and following life cycle costs. However, if you look at the construction process in terms of the responsibilities of its participants, then in Latvia these choices are made by the designer. It is the responsibility of the contractor to supply appropriate materials and services and to certify their compliance, as well as, if determined by the compliance certification system for the specific material or solution, to perform technical tests and inspections. By the end of 2028, the following must be achieved:

- reduction of illegal dumping of construction debris and increase of separate collection and sorting;
- reduction of unprocessed or unrecovered construction debris is planned by 30% compared to the total amount of construction waste produced in 2019.

A more detailed analysis of the data on where the sorted, recycled and reused construction waste ends up shows that most of it does not return to the industry, but is "embedded" in temporary solutions. Such actions prevent an increase in the volume of landfilled waste, but as a result, they do not reduce the initial use of resources. Taking into account the complexity of the construction industry, researchers and experts evaluated decision-making, material flows, regulation, including the requirements and opportunities of technical regulations in five stages of the construction process—design (architectural and engineering sections), construction works, demolition, construction debris management and production of construction materials.

Key Challenges as Defined by Municipalities:

- Lack of resources [human and financial].
- Underdeveloped infrastructure (e.g. no designated place to collect CDW from households).
- Lack of motivation for people to sort CDW.
- Legal framework and implementation shall be improved.

Below priority actions to address CDW management have been summarized:

1. Households: Public awareness activities (incl. on CDW sorting); Changes in the laws; Strengthening administrative and control capacity on the municipal level.
2. Businesses: Special financial support programmes for the development of infrastructure and CDW recycling capabilities; Strengthening administrative and control capacity on the state level; Development of the integrated system for the tracking of recycled CDW; Identification of challenges and barriers at national levels.

The successful development of CDW sector requires:

- Short term activities—public information/awareness raising; special events (e.g. free of charge asbestos roofing collection); control enforcment etc.).
- Mid term activities—programme for the CDW management infrastructure development; legislative changes; extended PRO discussion.

According to the OECD and the European Topic Centre on Circular economy and resource use [89], the main challenges identified were:

1. a lack of cross-sectoral coordination and of a powerful driving force to move economic ideas forward in Latvia;
2. developing and implementing sectoral strategies without considering smart natural resource management objectspainsoain ives and changing production and consumption patterns, or insufficiently integrating them;
3. insufficient use of economic and other instruments to influence consumption;
4. lack of awareness and support for social innovation and dissemination of good practice;
5. public attitudes towards consumption: it is difficult for citizens to change their habits and to start sorting waste, in addition to insufficient infrastructure availability;
6. use of environmental and natural capital in an unbalanced way;
7. lack of control—non-compliance of actual activity with the capacity specified in the documents.

5.1.10.5 Examples of Successful Implementation

Public Policy Initiatives

Circularity strategy in construction is related to the choice of materials, construction waste management and functional solutions. More use of recycled, renewable or reusable construction products promotes the implementation of circularity principles already at the design stage. Also the material important conditions in the choice are their origin, production of raw materials from renewable resources, processing techniques. On the other hand, long-term focused and thoughtful functional solutions facilitates the adaptation of the building during its operation with less resources use. In Latvia, the concept of regularity is not provided for in the regulatory acts that would lay it as a basis designers to use these principles in the development of the construction project. The experience so far is mainly based on stimulating factors of external conditions, not on direct and deliberate actions of the customer and the designer in promoting traffic.

Ziepju Street 11 renovation according to circular economy principles. The first construction project based on circular economy principles was implemented in Riga and Latvia. It is a five-story building at Ziepju street 11, which since 1970 served as a service hotel for employees of the trolleybus park. Later, the building was used as a shared house for different social groups. Since 2018, the building has not been uninhabited, its communications have been disconnected and the windows have been nailed up. After reconstruction, municipal rental apartments have been installed in it. Since the internal layout of the building was significantly changed, a large amount of construction debris was created. The resulting construction debris was maximally used on site for building renovation and area improvement works. The remaining resources (which would be called waste in the circular economy) were used in other objects and sent for recycling (https://www.riga.lv/lv/jaunums/ziepju-iela-11-parbuves-eku-pec-aprites-ekonomikas-principiem). Transition to circular economy in urban construction "URGE" is the interregional cooperation program of the European Union "URBACT III 2014–2020". for the year" project. More information: https://urbact.eu/urge.

Reconstruction of the Riga Circus. Reuse of material—the maximum of the dome structure of the historical arena interior use. Exposing the historical and modern connection, preserving the historical scope.

Temporary auditoriums of Riga Stradins University. Possibilities of adjusting and moving structures The customer's need to provide a temporary solution to the learning process for provision. After evaluating various alternatives, the most sustainable solution was found chosen adaptation of modular containers to provide educational functions. After upon completion of construction works, the temporary auditoriums will be moved to another building site, providing a learning function. At the end of the cycle, the containers will be placed in Riga On the property of Stradiņi University to provide a recreational function.

New building of Ogre Central Library. Use of construction products made from recycled materials. Participation in the open tender of projects financed by the emission allowance auction instrument “Reducing greenhouse gas emissions by developing an energy self-sufficient building construction”, where one of the quality evaluation criteria was “Recycled materials assessment of use” and by including in the project that the construction of the building provides to use recycled (recycled) materials, evaluation points were received. In addition it is stated that the building will be certified according to the BREEAM standard.

Reconstruction of the New Riga Theater. Reuse of cultural and historical building elements in the rebuilt building, construction products use produced from recycling. The place of operation of the theater is a cultural monument of national importance and is located in the historical center of Riga in the center area, the goal is to create an architecturally high-quality and functional building that meets for modern theater needs. Reconstruction of the teaching building of the Faculty of Civil Engineering.

Use of recycled materials, preservation of existing building elements. Economic aspects, namely to use existing resources in case of limited funding, which is of good quality and meets the technical requirements.

Mežaparka’s Great Stage. Preservation of historical building elements, development of temporary solutions for modifications. Exposition of dismantled building elements as historical evidence; construction of a temporary solution, to provide a stage function during the construction phase.

At the design stage, the most common initiative is to apply the principles of circularity related to the attraction of external funding and the reconstruction of cultural and historical buildings. External in the case of fundraising, additional requirements are set in the tender regulations, which are binding on both the customer and the designer. On the other hand, cultural and historical buildings in the case of reconstruction, the normative acts are binding, which are the basis for the construction of the building and for preservation and/or restoration of building elements. There are significant requirements for construction debris management and thoughtful selection of building materials determined in the international building sustainability certification standards. Separately Projects in the BREEAM certification process include requirements for waste management and efficient use of resources, but it is more related to reducing the future waste stream.

Private Policy Initiatives

No data to this point.

5.1.11 Malta Factsheet

5.1.11.1 Policy and Regulatory Framework

Governance

The main government institutions in charge of implementing the circular economy in Malta is the Ministry for the Environment, Energy and Enterprise [234]. The Circular Economy Malta (CE Malta) is the designated competent entity, set up by Government, for the Circular Economy under the Environment Protection Act [74]. The legal persona of the Agency is provided through the Public Administration. CE Malta was established through Subsidiary Legislation SL.595.28 [36]. The entity, originally set up as Resource Recovery and Recycling Agency (RRRA) works to create a business environment whereby producers would be responsible for the products placed on the market, from placement on market until the end of product life. The agency is active on a number of projects and initiatives, including the beverage container refund scheme, end-of-life waste tyres, construction waste, food waste, and textile waste. Overall, Malta is working towards Circular Economy and the transition can be facilitated through government action to create a favourable climate for responsible, positive impact investment by providing technical assistance and mobilising financial resources for sound investment in sustainable innovations. This has been happening in Malta, through visions, policies and regulations. The transition towards a CE is dependent on the stakeholders' willingness and capacity to collaborate and forge long-term relationships with one another.

Legislation

The Waste Regulations, S.L. 549.63, are the overarching national legislation on waste and aim to transpose the provisions of the Waste Framework Directive 2008/98/EC. The Waste Regulations, S.L. 549.63 [361] provide a framework to protect the environment and human health by preventing or reducing the generation of waste, the adverse impacts of the generation and management of waste and by reducing overall impacts of resource use and improving the efficiency of such use, which are crucial for the transition to a circular economy. The Waste Regulations set the basic concepts and definitions related to waste management as well as portray the main waste management principles such as:

- The waste hierarchy;
- The polluter pays principle;
- Extended Producer Responsibility (EPR);
- When waste ceases to be waste to become a secondary raw material (end-of-waste criteria);
- How to distinguish between waste and by-products;
- The list of waste;

- Hazardous waste;
- The difference between municipal waste, household waste and biowaste;
- Waste Management Plan.

The waste hierarchy is the cornerstone of European and national waste policies and legislation, ranking waste management options in a specific order according to their environmental impact. Its primary purpose is to minimise adverse environmental effects from waste and to increase and optimise resource efficiency in waste management and policy.

Reference is also made to the Environment Protection Act, Malta CAP 549: Circular Economy Malta (CE Malta), as the designated competent entity in Malta was set up by the Government under the Environment Protection Act.

The Action 3 of the Circular Economy Strategic Vision: Towards a Circular Economy 2020–2030 [183] refers to the deposit-refund schemes and other measures to encourage the efficient collection of used products and materials.

SL. 623.08 Construction Site Management Regulations [44] refer to the standard SM810 for Recycling oriented Deconstruction and Classification of Waste, therefore incorporating the standard within the regulatory framework. Compliance with the standards is considered as an essential requirement prior to the issuance of an executable Development Permit. Any deconstruction or excavation works on a site shall be carried out in line with sections five (5), seven (7) and eight (8) of the national standard SM 810:2022 'Recycling-oriented Deconstruction, Controlled Excavation Works and Classification of Waste' [317]. Furthermore, any apertures utilised in a residential building shall by design follow the standardisation of dimensions of internal and external apertures as specified in the Technical Document titled 'standardisation of Apertures for Residential Buildings in Malta'.

Circular Economy Strategy, Roadmap or Action Plan

Circular Economy Strategic Vision: Towards a Circular Economy 2020–2030. Malta has a Circular Economy Strategic Vision: Towards a Circular Economy 2020–2030 [183]. The Circular Economy Strategy Vision 2020–2030 was developed to go along with the government's plans to build the country's first waste-to-energy plant and continue its efforts to reduce landfilling. The vision aims to establish an environment that will lead to the development of a sustainable, low-carbon, resource-efficient, and competitive economy, in line with the EU Commission's Circular Economy Strategy. The following actions have been identified:

1. fiscal incentives for the donation of products;
2. extended producer responsibility (EPR) schemes for various types of waste and measures to increase their effectiveness, cost efficiency and governance;
3. deposit-refund schemes and other measures to encourage the efficient collection of used products and materials;
4. sound planning of investment in waste management infrastructure, including through European funds;

5. promote sustainable public procurement to encourage better waste management and the use of recycled products and materials;
6. use of fiscal or other means to promote the uptake of products and materials that are prepared for reuse or recycled; and
7. support for research and innovation in advanced recycling technologies and remanufacturing.

Construction and Demolition Waste Strategy for Malta 2021–2030: Managing Construction & Demolition Resources

There are targets and strategic goals related to the implementation of circular economy in construction sector in Malta through the Construction and Demolition Waste Strategy for Malta 2021–2030 Managing Construction & Demolition Resources [45]. This Strategy is a framework acting as a driver intended to bring about a cultural and behavioural shift within the sector in terms of its attitude towards excavation, demolition and construction methods. The Construction and Demolition Waste Strategy for Malta will further Malta's transition to a CE and hence close the loop of products' lifecycles. The CE policy element refers to the setting up of standards for the construction industry. These standards include: best practice for (de)construction; classification of construction and demolition waste; appropriate excavation activities; dimensions of internal and external apertures of residential dwellings. The Construction and Demolition Waste Strategy for Malta 2021–2030 recognises the need to set specific targets related to the reuse and recycling of construction and demolition waste. The main aims of such targets are:

- to decrease the dependency on natural resources;
- to reduce the need for virgin aggregates; and
- to move towards a CE.

The Strategy identifies the following targets to facilitate the achievement of the above aims:

- a minimum of 40% of excavated material shall be reused or recycled;
- a minimum of 15% of construction material shall be made up of reused material or materials recycled locally, with a possibility of further reuse or recycling at a building's end of life;
- at least 25% of the granular material used for construction shall be made up of aggregates recycled locally in order to decrease the dependency on virgin materials. The percentage share of recycled aggregates used will contribute to the 15% target for reuse and recycled materials highlighted above; and
- a minimum of 40% by weight of non-hazardous waste generated during demolition activities shall be prepared for reuse, recycling and/or other forms of material recovery (excluding backfilling operations), with a view to increase the minimum target to 55%. Information on the intended reuse and materials destined for recycling will be included in pre-demolitions audits to be submitted prior work starting.

The above mentioned measure is to be implemented between 2026 and 2030.

Long Term Waste Management Plan 2021–2030. The Long-Term Waste Management Plan 2021–2030 (LTWMP) [173] aims to maximise the resource value of waste through holistic waste management solutions by adopting a collaborative approach whilst fostering the necessary behavioural change. This Plan is intended to be the cornerstone of a process that will strengthen the transition to a CE. The Circular Economy Policy Element refers to support for the establishment of a reuse and repair centre to promote such practices. The Long Term Waste Management Plan, through the waste prevention programme, aims to facilitate the creation of a centre to which people can take items that are suitable for repair, upgrade or reuse. Furthermore, reuse, upgrading and repair activities will be promoted through positive economic incentives, quality assurances and warranties to improve confidence in second-hand goods.

Single-Use Plastic Products Strategy for Malta 2021–2030. The objective of the Single-Use Plastics Strategy for Malta is to reduce environmental harm from certain plastic products, and to promote the transition to a CE with innovative and multi-use alternatives. The Circular Economy Policy Element refers to the setting up of "green corners", whereby the strategy aims to encourage schemes for grocery shops to set up "green corners" for packaging-free foods: to promote responsible consumer behaviour; to reduce the generation of plastic waste; to inform consumers about the availability of reusable alternatives.

Standards for the construction industry. The objectives of the standards proposed was to:

- Establish best practice for (de)construction, aimed at reducing the generation of construction and demolition waste and purifying the resulting waste streams;
- Establish a classification for construction and demolition waste by type, material, composition and weight to encourage on-site separation as well as improve the quality of waste streams for subsequent reuse or recycling;
- Establish standards for appropriate excavation work, with the aim of reusing excavated rock for the purposes of construction.
- Establish standards for the dimensions of internal and external apertures of residential dwellings aimed at encouraging the re-use of fittings as well as reducing diversity, bringing about economies of scale.

These standards are incorporated within the regulatory framework and compliance with these standards is an essential requirement prior to the issuance of an executable Development Permit.

The above standards aim to classify construction and demolition waste according to the material type and composition in order to improve the quality of waste streams for subsequent treatment. The standards were organized into: two standards for C&D Waste (SM810: Recycling-oriented Deconstruction, Controlled Excavation Works and Classification of Waste; SM820 Classification of Recycled Aggregate (In consultation phase); and a technical guideline for apertures.

Recycling Oriented Deconstruction of Buildings Standard SM810:2022. The Recycling-oriented Deconstruction, Controlled Excavation Works and Classification of Waste was published in 2022 by MCCAA as an important and fundamental National Standard, following a public Consultation. The standard is incorporated within the regulatory framework through SL. 623.08 Construction Site Management Regulations. Compliance with the standards is considered as an essential requirement prior to the issuance of an executable Development Permit.

5.1.11.2 Platforms and Networks Facilitating the Transition to a Circular Economy

Circular Economy Platforms

The main Circular Economy Platform in Malta is CE Malta, the Government agency promoting CE through the National Strategy.

Following the European and National action plans, various initiatives have started to promote the Circular Economy in Malta. Initiatives are taken by different stakeholders, including Industry, Academia, NGOs, and other partners, including Government Agencies. The Building Industry Consultative Council's Research and Innivation work group promotes Circular Economy in the Construction Industry as a key initiative and worked with the University of Malta and MCCAA towards the definition of new Standards in the Industry (SM810 and SM820).

The University of Malta has been leading initiatives and research in Circular Economy primarily in the Construction sector for several years, strengthening through funded international and national research projects. Other initiatives are taken by Non-Government Organisations, including iiSBE (International Initiative for a Sustainable Built Environment) [313], SBE Malta, Sustainable Built Environment Malta, iiSBE National Chapter, and ZIBEL environment NGOs.

The Ecohive Reuse centres support CE. These Centres offer used items that still hold value and can be reused. Unwanted items that are still in good condition can also be donated in special containers that have been placed at every Civic Amenity Site. By donating or picking items for reuse, stop them from going to the landfill, save the energy needed to recycle them, reduce the natural resources and energy used to create new products and contribute to a circular economy.

Circular Buildings Platform

EcoBuild was a new tool developed by the Building Industry Consultative Council, intended to promote Green Building Technology in Malta [17]. It is a first database in Malta for available Green Building Products and is based on an online system. The platform increased product visibility and competition, while providing building industry professionals with a wider range of information and tools to implement better performing buildings. The platform supported also circularity in the construction industry.

Building Materials Passport Platform

A building materials passport platform at a national level was assessed and proposed by the Research and Innovation Work Group of the Building Industry Consultative Council towards implementation in the Maltese Islands.

Public Procurement Platform

The Ministry for the Environment, Climate Change and Planning (MECP) published the Green Public Procurement Second National Action Plan (2022–2027) [323], which came into force as of 01 January 2022. The scope of the Green Public Procurement Second National Action Plan (GPP 2ndNAP) is primarily to strengthen the achievements as well as address any identified challenges encountered during the implementation of the First National Action Plan. It also aims to promote, motivate and incentivise the implementation of Green Procurement practices, production and consumption, thus, reinforcing and integrating the concept of Circular Public Procurement. The mission of the GPP 2nd NAP is to 'enhance the Greener Public Procurement Function, recognising the opportunity to limit further the environmental footprint whilst driving markets towards greener products and services'. In fact, its vision is to progressively increase the share of Government's Green Public Procurement to 90% by 2027.

5.1.11.3 Funding Opportunities

EU Cohesion Funds

EU funds through the EU Cohesion Policy are key to implementing circular economy in Malta. In 2014–2020, Malta benefited from Cohesion Funds worth more than €1 billion, through which the country has improved waste management systems and sustainability practices, among others [65]. Waste management capacity has improved in Malta through the EU Cohesion Policy thanks to EU Cohesion funds [66].

Malta's Recovery and Resilience Plan

Malta's Recovery and Resilience Plan includes various measures that aim to shift Malta towards a more CE [184]. In 2021, Malta implemented measures from the Recovery and Resilience Plan [180], in particular:

- the prohibition on the imports, production, sale and distribution of certain single-use plastic items, including lightweight plastic carrier bags; and
- the adoption of the Construction and Demolition Waste Strategy, which is in line with the main aims of EU waste legislation and the EU Construction and Demolition Waste Protocol. The Strategy provides for measures to encourage options that deliver the best overall environmental outcome by promoting a shift towards a more CE and hence closing the loop of products' lifecycles.

During 2022, Malta implemented one of the measures in the Construction and Demolition Waste Strategy for Malta 2021–2030, which has also been included as a national Resilience and Recovery Plan: that is to establish standards for the construction industry in the form of mandatory guidelines for economic operators within the sector (SM 810:2022).

Malta Council for Science and Technology

The XJENZA Malta (Previously the Malta Council for Science and Technology MCST) [181] manages national funds for R&I through the development and operation of programmes, which in turn, create vehicles for research across the public, academic and private sectors. A portfolio of funding programmes provides researchers with the opportunity to translate their ideas into tangible projects and eventual products or services. Projects funded by MCST include a number of research programmes focusing on Circular Economy and Waste Recycling in particular addressing C&D Waste: RECP (Circular Economy through Recycling Poultry Waste for value-added products in Construction and Agriculture) [286] and ReCON (Large Volume Waste Recycling for Low-Impact High-Performance Concrete) [166] and 3DCONCRETE (3D Printing of Concete for Sustainable Construction).

Malta Smart Specialisation Strategy

The Malta Smart Specialisation Strategy presents Circular Economy as an important area. Niche Areas of Focus are presented and the area of sustainable use of resources for climate change mitigation and adaptation is selected as a smart specialisation area for 2021–2027 [185]. In particular, there is a strong need for R&I investment to develop tailor-made solutions for Malta. This may in turn create additional opportunities for tapping into international collaboration and markets. Any type of tailor-made solution proposed for Malta should be developed with current and future climate change scenarios in mind. Based on this, investments in niche areas have the most potential in providing such needed solutions in the coming years. A niche area presented refers to: "Turning Waste into a Resource": While the shift from a linear to a circular economy is indeed a global one, the challenges that different countries face and the solutions required may well be very diverse. Malta's insularity and lack of economies of scale are perhaps the greatest obstacles to reach the defined targets.

International Cooperation and Support from International Organizations

The University of Malta is a member of *fib* (International Federation for Structural Concrete) [93] and academics are active in the *fib* Commission 7: Sustainability with active leadership of TG7.8—Recycled Materials and industrial by-products for high performance reinforced concrete structures [332].

Malta is also active in Standardisation through the MCCAA (Malta Communication and Consumer Affairs Authority) [182] and participates in Technical Committees for standards on Circular Economy, including CEN TC350/SC1 Circular Economy in the Construction sector.

5.1.11.4 Challenges, Barriers and Potential Improvements

Challenge 1: Institutional Challenge for a Cross-Sectoral Development

Being a small island, Malta has limited economies of scale. Nonetheless, small countries like Malta do not find it economically feasible to develop treatment facilities for all waste and all types of treatment, meaning that some waste has to be shipped abroad for treatment. Community and international policies that allow small island countries like Malta to be circular even beyond national borders can certainly be of help. Actions that could help include:

- co-ordination across different levels of government to align priorities, goals, regulation and funding sources;
- system thinking to ensure policy coherence across different sectors, such as waste, water, energy and transport, to maximise synergies and ensure a coherent set of incentives;
- collaboration and dialogue between the public sector, not-for-profit actors and businesses to stimulate innovation for more sustainable production and consumption patterns.
- development within local administrations and across businesses.

Challenge 2: Companies' Ability to Grasp Opportunities

The main concerns and challenges that companies experience are:

- maintaining consumers' expectations;
- implementing government regulation;
- lack of waste treatment facilities;
- lack of recycling technologies;
- level of costs for implementing CE concepts.

Challenge 3: Market Barriers

A CE approach can support the prevention of waste at all stages of a product's lifecycle, from design, production and distribution right through to use and disposal. Designing a product to be easily repaired or upgraded can maximise its use before it reaches its end of life. As indicated in the Circular Economy Action Plan, although 80% of products' environmental impacts are determined at the design phase, the linear pattern of take-make-use-dispose does not provide producers with sufficient incentives to make their products more circular. As many goods available on the market in Malta are imported. Malta is limited in terms of influencing the products' entire lifecycles as it does not have much influence on the design and production stage.

Challenge 4: Consumer Behaviour

Consumption patterns and choices shape the amount and type of waste generated. Current consumption levels are unsustainable. A reason for this low engagement in CE practices could be that consumers lack information about product durability and reparability, as well as the lack of sufficiently developed markets, for, amongst others, second-hand products, renting, leasing or sharing services.

Challenge 5: Indicators and Targets

- Limited data on material stocks and waste created
- Necessity of investigating not only existing CE targets but also possible or advisable targets
- Identifying and establishing CE indicators and targets
- The need to examine targets beyond specific solutions, economic sectors, or geographical characteristics.

A number of barriers and issues are being addressed with the proposed regulations and their implementation will accelerate the implementation of circular economy. A key challenge refers to the economies of scale of a small Islands State with limited natural resources—where the management of resources and waste is critical.

Several challenges and barriers refer to specific areas of implementation of Circular economy in the Construction Industry and the Built Environment, including: Need of new regulations; poor implementation of existing regulations and lack of enforcement; Collaboration between partners including academic/professional and industry partners; development if new products based on recycled materials and their uptake on the market; adoption of design for deconstruction principles in new building stock to address lifetime engineering principles including building and construction techniques; regulations supporting uptake of percentage of recycled CDW materials; attitudes in the construction sector and production industry; cost of dumping and illegal dumping; skills in the industry in the circular economy sector; upscaling of research innovation to industry scale; perceived value of C&D Waste and market dynamics where new products are preferred instead of recycling; lack of demand for products based on waste; knowledge of professionals and partners in the industry; gap in standards for the classification of recycled aggregate as national standards in Malta; lack of confidence in the quality of recycled materials; lack of space for Waste separation depots; space for building disassembly; the nature building typologies in the case of existing building stock; lack of incentives for recycling/disincentives for the use of natural products.

Particularly important areas for improving the implementation of the circular economy in the built environment in Malta refer to improved regulations for C&D Waste and stricter control, application of the classification of aggregate, the building of infrastructure (recycling plants and recycling depots) and raising awareness with the key stakeholders.

5.1.11:5 Examples of Successful Implementation

Public Policy Initiatives

The Long Term Waste Management Plan 2021–2030 aims to maximise the resource value from waste through holistic waste management solutions by adopting a collaborative approach whilst fostering the necessary behavioural change. This Plan is intended to be the cornerstone of a process that will strengthen the transition towards a CE. The Plan identifies various measures to promote resource efficiency and reduce waste generation across sectors. These measures seek to incentivise greener business processes, and prompt societal change towards smarter consumption patterns. Furthermore, such measures seek to maximise the intrinsic resource value of waste as well as reducing pressure on Malta's waste logistics and infrastructure and the islands' dependence on either exporting waste or landfilling.

- **Product-related policies, including on the R-strategies (repair, reuse, remanufacturing, etc.)**. The establishment of a reuse and repair centre: The Waste Management Plan identifies the need to the establish reuse centres, creating places to which people can take items that are suitable for repair, upgrade or reuse. In view of this, during 2022, Wasteserv Malta opened its first reuse centre, at which the public can donate items that are still in good condition for reuse/resale. It is envisaged that in the coming years more reuse centres will be opened with the money raised through them used for other environmental initiatives.
- **Change in consumption patterns and consumer behaviour. Green corners**. This initiative involves the setting up of incentives for supermarkets (at national level) to provide packaging-free areas where customers can only buy the selected food products without plastic packaging. The aim of such initiative is both to encourage consumers to move towards reusable containers as well as reducing the amount of plastic waste generation.

- **Green Public Procurement 2nd National Action Plan**. Malta's Second Green Public Procurement (GPP) National Action Plan came into force on 1 January 2022. It provides a series of targets and measures that sees a level of ambition that has never before existed in Malta. Implementing the Plan and achieving its targets will require the support and commitment of economic operators as well as those responsible for public procurement. Building on the success achieved in the first National Action Plan, the Second GPP Action Plan aims to progressively increase the share of greener products and services in government procurement to 90% by 2025—a target which although not set by the European Commission, will drive action on the ground, promoting sustainable consumption and production. An ambitious but realistic approach is considered the best way forward in order to coax the market into a smooth transition. Consequently, the Plan sets out targets for 17 product and service groups, 14 of which will be mandatory. Green public procurement will also further work towards greener consumption and contributing to increased circularity by implementing nine new initiatives to increase GPP uptake.

Private Policy Initiatives

- **Construction and Demolition, Excavation Waste Materials Recycling**. The Xrobb l-Ghagin Wildlife rehabilitation Centre (ex-Deutsche Welle Radio Relay Station) [366] Adaptive reuse of an ex Deutsche Welle Radio Relay Station, into a Wildlife Rehabilitation Centre. The project included the recycling of stone and exploitation of reconstituted stone that is reuse of stone from dismantled buildings in the construction of 1 km long boundary wall with planters surrounding the Nature Park; the recycling of dismantled materials and elements; together with the adaptive reuse of an existing building following repair and strengthening instead of reconstruction, to promote circular economy. (ERDF Funded project).
- **ReCON**: The large volumes of Construction and Demolition Waste, Excavated Waste limestone, Quarry Waste generated in the construction industry, present significant challenges in disposal with environmental impacts. Recycling presents opportunities in reducing the disposal of waste material and also resulting in lower demands on the extraction of new resources. However low-quality inert waste which is generated in construction activities, presents limited opportunity in recycling, resulting in large volume of waste disposal. The increase in construction activity is leading to higher demands for construction products, with increased pressures on natural resource extraction including the extraction of aggregate and cement for the production of concrete and concrete products. In particular there is a large demand for concrete blocks in the construction industry. Cement as a binder in concrete, has a high embodied energy with negative impacts on the environment. The project refers to the recycling of large volumes of excavation waste consisting primarily of lower quality limestone and other materials, normally considered inadequate if used as aggregate in civil engineering applications, primarily due to

low mechanical characteristics and impurities. The new product consists of a low-impact high-performance concrete, an eco-construction product based on waste excavation material. The innovation lies in the transformation of the waste into a high-performance construction product. The new product is produced through a new technological process based on a production methodology consisting of key steps and resulting in premium quality products for the construction industry. The final product is strong, stable and durable for construction applications. The new product, can effectively provide for the increasing demand for construction products. It can be presented in the form of building blocks, cladding panels or in other geometries and forms of high-performance construction products. The new product is based largely on waste, resulting in a reduction in consumption of raw materials and less waste disposed, and lower environmental impact. The new technology effectively transforms large quantities of waste into a resource, a low-impact high-performance material.

- **RECP Circular Economy through Recycling Poultry Waste for value-added products in Construction and Agriculture** The project addresses recycling options of waste generated from the poultry industry and processing techniques, with reference to bone waste and feathers. Waste from the poultry industry primarily offal, bone, skin and meat are transformed into a fertilizer for the improvement of soil fertility. The physicochemical changes during the processing of the waste render nutrients more available for plant uptake. In addition the organic matter content of the soil shall improve hence leading to the long term release of nutrients and an improvement in the structure of the soil leading to better soil aeration and water holding capacity. Feathers are processed and exploited as fibres in cement-based materials, for crack control in concrete.
- **SORĠI** is a research project on the potential of circularity in Malta, the first result of which is an outdoor furniture collection for public areas that is very critical of the increasing building sector [318]. Six seats, each inspired by a different structure damaged by the local construction boom, will serve as a reminder of today's decisions as well as a recommendation for future acts. All the bench pieces are created from recycled construction waste, inviting visitors to consider the island's growth and its influence on the environment, as well as its social and historical significance. After mapping out the building sector and waste-producing locations, numerous materials were investigated and tested, including glass fragments, recycled concrete and asphalt, and marble offcuts: all significant resources that are now being discarded in landfill sites that are already at capacity. For example, the use of pigments was tested as a way of adding colour and creating remarkable textures, increasing the uniqueness of each bench, during the development and production phases. All leftover components and possible rubbish were gathered and recycled during the production process to make small products such as coasters and pots, which are also available for purchase.

5.1.12 Moldova Factsheet

5.1.12.1 Policy and Regulatory Framework

Governance

Sustainable development and the promotion of the green economy are development priorities for the Republic of Moldova. These are reflected in the main policy documents of the Government, especially in the National Development Strategy of Moldova 2030, the Energy Strategy-2030, the Environmental Strategy 2014–2023, the Program for the promotion of the Green Economy in the Republic of Moldova for 2018–2020. These are also part of the commitments of the Republic of Moldova towards the country's development partners from outside.

The Association Agreement with the European Union expressly aims to ensure sustainable development and promote the green economy in our country. By signing the Agreement, the Republic of Moldova undertakes to harmonize national legislation with European legislation and to ensure the integration of environmental protection provisions, rational use of resources and energy efficiency, eco-labelling, eco-innovations, in all sectors of the national economy and social life.

In the process of promoting the green economy, the Ministry of Economy and Infrastructure joined international programs and platforms in the field of green economy, such as the EU Program "Greening the economy in the Eastern Neighborhood", the Eastern Partnership (Greening economies in the Eastern Neighborhood Countries, EaP GREEN Program 2014–2017), Green Industry Platform, UNIDO Country Program (launched in 2019), participates in the GREEN Action Program, OECD and, in partnership with the Ministry of Agriculture, Regional Development and Environment, is the coordinator and beneficiary of the EU Action—EU4Environment—European Union for the Environment (2019–2022).

Currently, the Ministry of the Environment announces, starting from 25.04.2023, the initiation of the development of project of the Program for the promotion of the green and circular economy for the years 2023–2027. The first policy document, which supported the implementation of green economy principles in the Republic of Moldova, was the "Green" Economy Promotion Program in the Republic of Moldova for the years 2018–2020, approved in 2018.

The institution responsible for developing the project of the Program for the promotion of the green and circular economy for the years 2023–2027 is the Ministry of the Environment with its internal structural subdivisions and subordinate institutions, as well as a number of other institutions from different sectors, among which the Ministry of Infrastructure and Regional Development with the structures its subordinates (Energy Efficiency Agency, National Office for Regional and Local Development), but also representatives of the business environment, representatives of civil society in the environmental field (EcoContact, Ecovisio, National Environment Center, etc.), representatives of development partners.

The process of developing the Program for the promotion of green and circular economy for the years 2023–2027 will be coordinated by the interministerial Working Group for the promotion of sustainable development and the green economy, created by the joint Order between the Ministry of Economy and the Ministry of the Environment "Regarding the creation of the Interministerial Working Group for the promotion of sustainable development and the green economy" (format in 2015 and updated in 2019).

The program, like the Environmental Strategy project, comes to directly support the implementation process of the key documents of the national strategic development framework reflected in National Development Strategy of Moldova 2030. The program will contribute to the development of the general objective "Ensuring a healthy and safe environment" and the specific objective 10.4.: Active transition to the green and circular economy. In addition, priority policies and interventions will be implemented in accordance with point 5 of the NDS, in particular with regard to policies and management in the field of environmental protection: Promotion of climate change adaptation and mitigation policies of the footprint of illegal human and economic activities on the environment, including through activities based on research and development.

At the same time, the priority directions and interventions included in the NDS will contribute to the achievement of the set objectives and priorities corresponding to the commitments from the Association Agreement Republic of Moldova—European Union and from other agreements concluded with the European Union, including the measures aimed at preparing the country for accession to European union.

In a broader perspective, the provisions of the Program ensure synergy with other public policy documents in force or in the process of elaboration, as follows:

- Waste management strategy in the Republic of Moldova for the years 2013–2027;
- Project of the national strategy for agricultural and rural development 2023–2030 [279],
- The Draft Strategy for an inclusive, sustainable and digital economy until 2030;—Greening program for small and medium enterprises;
- The re-technological and energy efficiency program of small and medium-sized enterprises;
- The 2030 Low Emissions Development Program;
- Water supply and sewerage strategy for the years 2014–2028;
- Project of The national forest expansion and rehabilitation program for the period 2023–2032;

At the same time, the Program will respect and take into account the commitments of the 2030 Agenda on sustainable development and will contribute to achieving the following national targets of Sustainable Development:

In the field of constructions, the institution responsible for the regulation and management of the sector is the Ministry of Infrastructure and Regional Development, which aims to approve the unitary regulatory framework for territorial development, authorization and execution of construction works, ensuring the quality of constructions.

Legislation

In the Law on quality in constructions of 02.02.1996, in article 6, it is specified that in order to obtain constructions of appropriate quality, it is mandatory to achieve and maintain for the entire duration of the constructions some essential requirements, among which *by letter G—sustainable use of natural resources.*

At the same time, the Law on the energy performance of buildings specifies that the Government exercises the following powers in the field of energy efficiency of buildings (art. 5):

> promoting the construction of new buildings whose energy consumption is almost equal to zero and the transformation of existing buildings into buildings whose energy consumption is almost equal to zero.

At the same time, by art. 7 of the Law, the public authority in the field of energy efficiency has the following basic attributions:

1. provides assistance to local public authorities in the integration, in local energy efficiency programs and plans, of actions to improve the energy performance of buildings;
2. ensures the record of coordinated national projects and projects implemented by local public authorities regarding the improvement of the energy performance of buildings;
3. promotes the implementation of international experience and practices regarding the improvement of the energy performance of buildings;
4. provides the necessary informational support for promoting the improvement of the energy performance of buildings.

Local public authorities, within the limits of financial possibilities and in the order of their priority hierarchy, have the following basic attributions in the field of energy efficiency of buildings (art. 8):

1. ensures the integration, in local energy efficiency programs and plans, of actions to improve the energy performance of buildings;
2. contribute to the co-financing of national programs regarding the improvement of the energy performance of buildings;
3. contribute to informing the local community to promote the improvement of the energy performance of buildings;
4. Contributes, at the local level, to the implementation of the state policy in the field of energy performance of buildings.

At the same time, by art. 14. It is specified that (1) New buildings and existing buildings undergoing major renovations *must use a minimum amount of energy from renewable sources*, established by the central specialized body of the public administration in the field of constructions. (2) The requirements regarding the use of energy from renewable sources are established differently, depending on the categories of buildings, and are applied if it is technically, economically and functionally feasible.

In the DECISION No. HG248/2013 of 10.04.2013 regarding the approval of the Waste Management Strategy in the Republic of Moldova for the years 2013–2027 it is mentioned that it is expected that in the period 2010–2027 the amount of construction and demolition waste per capita in the Republic of Moldova will increase in relation to the growth of the real GDP in the country. As a result, the total amount of construction and demolition waste will increase from 1.4 million tons in 2010 to 2.6 million tons in 2027. Taking into account the regulations aimed at the reuse, recycling or recovery of construction and demolition waste in the period 2013–2015, legal and organizational measures need to be taken. Consequently, the share of reuse, recycling or recovery of the respective waste will increase from 165 thousand tons in 2015 to 1520 thousand tons in 2020 and to 1850 thousand tons in 2027. Also in the strategy it is mentioned that one of the problems is *chaotic abandonment of construction and demolition waste, animal droppings, street waste,* etc.

In the ACTION PLAN regarding the implementation of the Waste Management Strategy in the Republic of Moldova for the years 2013–2027 several general and specific objectives with multiple actions aimed at the management of construction waste are given.

Specific Objective 2: Establishing the national normative framework in the field of waste management from General Objective 1 provides for the development of construction norms for the design, construction and operation of waste deposits. Those responsible are the Ministry of Economy and Infrastructure; Ministry of Agriculture, Regional Development and Environment. For this action, 155,079 MDL is allocated from the National Ecological Fund.

General objective 3: Development of systems for the collection and treatment of specific waste flows (packaging, electrical and electronic equipment waste, tires, batteries, etc.) by promoting and implementing the "producer's responsibility" principle, including hazardous ones (medical waste, oils used etc.); each collection point at the regional level includes several specific objectives, of which Specific Objective 1: Creation of the infrastructure for the management of the following waste flows: vegetable waste, manure, sawdust, waste from the industrialization of wood, packaging waste and uncontaminated waste from constructions and demolitions stipulates the following actions, out of which:

1. Creation of networks for the separate collection of plant waste, animal waste, wood processing waste. Those responsible are—the Ministry of Economy and Infrastructure; Ministry of Agriculture, Regional Development and Environment; local public authorities; non-governmental organizations. For this action, 77,539,500 MDL will be allocated from the National Ecological Fund.
2. At the same time, within Specific Objective 2: Creating the infrastructure for the management of the following hazardous waste flows: waste resulting from medical activity and from research institutions, used oils, used tires, used batteries and accumulators, electrical and electronic equipment waste and contaminated waste with hazardous substances from constructions and demolitions, point 9 stipulates—Creation of capacities to treat waste contaminated with hazardous

substances from constructions and demolitions, with a view to recovery or elimination. Responsible—Ministry of Agriculture, Regional Development and Environment; local public authorities. For this action, 38,769,750 MDL will be allocated from the National Ecological Fund.

From the point of view of legislation in the field of sustainability in construction, the Republic of Moldova has some normative acts such as: Code of Practice in Construction, Code of Practice in Construction CP A.09.04:2014—Construction and demolition waste management, Code of Practice in Construction CP B.01.01:2021—Systematization of the territory and localities. Guide to urban design in accordance with the principles of environmental protection. Access to these documents cannot be achieved online.

The provisions of the Practical Code have been adjusted with the provisions of the Community legislation such as: Directive no. 2008/98/EC on waste; Directive no. 91/689/EEC on hazardous waste; Directive no. 99/31/EC on waste storage; Decision no. 2000/532/EC on the list of waste, amended by Commission Decision no. 2001/119 (which replaces Decision No. 94/3/EC on the list of waste and Decision No. 94/904/EC on the list of hazardous waste); Regulation (EU) no. 305/2011 of the European Parliament and of the Council of March 9, 2011 establishing harmonized conditions for the marketing of construction products and repealing Council Directive 89/106/EEC. At the same time, I want to mention the draft of the Urban Planning and Construction Code, developed by the Ministry of Infrastructure and Regional Development and which is waiting its examination in the reading of the Parliament, which also provides requirements regarding the management of these types of waste.

5.1.12.2 Platforms and Networks Facilitating the Transition to a Circular Economy

Circular Economy Platforms

No circular economy platforms has been established to date.

Circular Buildings Platform

No circular buildings platform has been established to date.

Building Materials Passport Platform

No building materials platform has been established to date.

Public Procurement Platform

No national green public procurement platform.

5.1.12.3 Funding Opportunities

- In the domain of green economy promotion: European Union for the Environment Action (EU4Environment)—Assisting six partner countries to preserve their natural capital and enhance people's ecological well-being by supporting environmental action, demonstrating and creating opportunities for greener growth and establishing mechanisms for better risk and environmental impact management. The budget of the program—About 20 million euros (EU contribution: 19.5 million euros). Duration of the program—2019–2024.
- In construction: No fund available for promotion of sustainability of buildings.

The National Ecological Fund was created by the Decision of the Government of the Republic of Moldova no. 988 of 26.09.98 in accordance with the Law on the Protection of the Environment (No. 1515 of 16.06.1993), the Law on Amending and Supplementing the Law on the Protection of the Environment (No. 1539-XIII of 25.02.1998), the Law on Payment for Pollution of the environment (no. 1540-XIII of 25.02.1998) with the aim of accumulating additional means for financing activities in the field of the environment. But within the financing domains there are no circular economy or sustainability practices in construction.

5.1.12.4 Challenges, Barriers and Potential Improvements

Landfilling remains the basic method of waste management. Unfortunately, the state does not have sufficient technological capacities for the collection and recycling of waste, as well as for their use as secondary raw material for other technological processes. Under these conditions, most of the generated waste ends up in landfills, containing useful materials such as glass, metal, paper, plastic, construction materials.

There are several companies in the country that offer construction waste disposal services, but their representatives are not willing to communicate about where the construction waste is deposited and what happens to the collected waste. In the Republic of Moldova, data on the quantities of construction waste are missing, and as long as there are no legislative regulations on construction waste, most of it will continue to end up in landfills. Significant changes are needed in the field of regulation of these activities and the organization of an integrated waste management system. Following the examination, a gap was also detected in the monitoring of the situation regarding waste management related to waste from construction and demolition activities, as well as the lack of a regulation regarding their management, or some other acts that regulate this sector.

In the Republic of Moldova, some data on the quantities of construction and demolition waste are missing, this is due to the lack of strict centralized records and an inadequate method of disposal at existing warehouses, without weighing, or illegal burial or storage on surfaces undeveloped such as roadsides, riverbanks and ravines, on agricultural and forested lands, etc., which leads to the appearance of natural garbage.

At the same time, there is an internal reuse in the own household or a commercialization on an undeclared market. The current system for the regulation and management of solid waste collection and disposal in the Republic of Moldova is not effective and does not minimize the negative effects on the environment following waste disposal.

5.1.12.5 Examples of Successful Implementation

Public Policy Initiatives

No data to this point.

Private Policy Initiatives

Moldova Green Building Council is a non-profit association that encourages the emergence of the market, educational and legislative conditions necessary for the transformation of the construction industry and those adjacent to the creation and renovation of high-performing, sustainable and profitable buildings. Some achievements:

- The Council for Green Buildings from the Republic of Moldova became the partner of BC "ProCredit Bank" S.A. Through this strategic partnership, BC "ProCredit Bank" S.A. has developed a special financing program for customers who want to purchase or build certified "Green Homes" real estate. We are talking about a new type of mortgage "ECO Housing loan" intended for the purchase of green homes, through which the bank's customers are supported to make smart choices, with minimal impact on the environment.
- Through the GREEN HOMES&GREEN MORTGAGE program, the Council for Green Buildings has launched a new trend in residential housing in Moldova. Being a voluntary certification and verified by independent auditors, the Green Homes certification fully covers all the criteria necessary for a construction to be considered sustainable.

5.1.13 *Netherlands Factsheet*

5.1.13.1 Policy and Regulatory Framework

Governance

The government is responsible for policy on the transition to a circular economy and is accountable to the House of Representatives in this regard. Within the government the State Secretary for Infrastructure and Water Management is responsible for coordinating the transition, while the members of the government at the line ministries most closely connected with this issue are each responsible for circularity within their own remit [222]. The Ministry of Infrastructure and Water Management [116] consists of three sections: policy, implementation and inspection. In the policy section, four Directorates-General are concerned with developing policy in the areas of mobility, water management, aviation and maritime affairs and the environment. The Directorate-General for Public Works and Water Management (Rijkswaterstaat—RWS) [306] ensures that policy is implemented. Human Environment and Transport Inspectorate oversees compliance with statutory regulations by private individuals and companies. The Directorate-General for Public Works and Water Management, Human Environment and Transport Inspectorate, the Netherlands Environmental Assessment Agency [245] and the Royal Netherlands Meteorological Institute are agencies of the ministry. The Netherlands Environmental Assessment Agency is the national institute for strategic policy analysis on environment, nature, and spatial planning issues of national and international significance. The agency contributes to political and administrative decision-making by conducting outlook studies, analyses and evaluations that take an integrated approach. The Ministry of Infrastructure and Water Management [116] is also committed to organising an annual National Circular Economy Conference at which stakeholders discuss progress with the circular economy. That might lead to adjustments to the Circular Economy Implementation Programme together with the findings and recommendations of the progress report published by the Netherlands Environmental Assessment Agency.

Legislation

In the Netherlands, there is an obligation to provide information about the environmental performance of building materials that are covered by the Buildings Decree 2012 [22]. The objective of this obligation is to regulate the minimal environmental performance of materials. In the environmental performance of building materials, the emphasis is on the climatic effects and there is no requirement in the Buildings Decree concerning the degree of circularity. The environmental performance of a building is already measured as standard over a single cycle, but the stakeholders want to have a clear method for its assessment over multiple life cycles [307]. The

Dutch Soil Quality Decree (*Besluit bodemkwaliteit*) imposes conditions on the use of soil, dredging spoil and building materials on or in the ground or in surface water [21] in order to achieve a balance between a healthy living environment and the use of the soil. Anyone planning to refurbish, build, demolish or occupy a building must comply with the Building Decree 2012. All new buildings must also meet the Almost Energy Neutral Building requirements (*Bijna Energieneutrale Gebouwen*—BENG) that result from the European Energy Performance of Buildings Directive [79]. There are different requirements for different building types and there is a separate requirement for the exterior of buildings. The municipal building rules, which can differ from one municipality to another, relate to urban planning, building on contaminated land and requirements regarding the external appearance of buildings [22].

Circular Economy Strategy, Roadmap or Action Plan

National. The Netherlands' strategy A Circular Economy in the Netherlands by 2050 [117], adopted in autumn 2016, outlines a way to transform the Dutch economy into a sustainable, fully circular economy by 2050. The programme describes what needs to be done to use raw materials, products and services in a smarter and more efficient way and sets the goal of a 50% reduction in the use of primary raw materials by 2030. Priority sectors identified in the Strategy include biomass and food; plastics; the manufacturing industry; construction sector; and consumer goods.

The construction sector is distinguished in Commercial and Non-residential Building (Burgelijke en Utiliteitsbouw—C&NRB) on the one hand, and Soil and Civil Engineering (Grond-, Weg- en Waterbouw—S&CE) on the other. C&NRB largely involves private funding and a large, diverse playing field with many (major and minor) stakeholders, while S&CE often involves public commissioning and features a smaller number of more homogenous stakeholders. The implementation of a circular economy in the construction sector is related to the following vision:

> By 2050, the construction industry will be organised in such a way, with respect to the design, development, operation, management, and disassembly of buildings, as to ensure the sustainable construction, use, reuse, maintenance, and dismantling of these objects. Sustainable materials will be used in the construction process, and designs will be geared to the dynamic wishes of the users. The aim is for the built-up environment to be energy-neutral by 2050, in keeping with the European agreements. Buildings will utilise eco system services wherever possible (natural capital, such as the water storage capacity of the sub-soil) [117].

The Cabinet has set four strategic goals for construction sector [117]:

1. C&NRB and S&CE to use (mainly) renewable raw materials;
2. Optimising the use of material throughout the building's life cycle (value preservation, cost savings, more reuse, lower environmental impact);
3. Minimising CO_2 emissions by the construction sector, in both the construction and the operational phases;
4. As an innovative sector, the construction sector must proactively respond to changes in society as well as private sector and consumer demands.

As the general support for a circular economy increased, in January 2017, the National Raw Materials Agreement was signed by 180 parties from government, industry, trade unions and environmental organisations. The agreement sets out what is to be done to accelerate the transition towards a Circular Economy and to ensure the support and commitment through the whole Dutch society. The agreement started the multistakeholder dialogue on how to organise the transition and resulted in 5 transition agendas drawn up by the government and the signatories [126, 127]. Circular Construction Economy Transition Agenda describes the strategy for achieving a circular construction economy by 2050 and contains the Agenda for the 2018–2021 period [335]. This Agenda has been drawn up by a transition team made up of experts from the fields of science, the government, and market participants. The Transition Team formulated four focal points for the 2018–2021 Agenda [335]:

1. market development
2. measuring method
3. policy, legislation and regulations
4. knowledge and awareness.

Selected proposed actions and interventions based on these focal points are the following [335]:

- all public tenders will be circular in 2030
- action plan for CO_2 emission reduction in the construction industry
- decision regarding mandatory materials passport by 2020
- subsidy for circular business and revenue models
- development of a uniform circularity measuring method
- processing circularity in construction government standards
- international positioning and cooperation
- circular construction a comprehensive part of education in 2021
- creating a circular construction knowledge institute
- creating a circular construction awareness campaign.

The Circular Economy Implementation Programme translated the five transition agendas developed for priority sectors into concrete actions and projects to be put into effect between 2019 and 2023. The Ministry of Infrastructure and Water Management updated the Circular Economy Implementation Programme in 2020 and 2021. The National Circular Economy Programme 2023–2030 (NPCE) was released in February 2023 and it elaborates the ambitious circular economy goal, which is simultaneously a climate goal. Instead of focusing on voluntary measures the government decided to guide and compel measures to achieve these goals. The National Circular Economy Program 2023–2030 contains general measures, specific measures for main product chains and supporting measures. The program outlines various actions, including developing innovative business models, waste reduction strategies, and collaboration with businesses and knowledge institutions to create a sustainable economy [150].

Regional. Two relevant documents have been launched at the regional level—Northern Netherlands Circular—roadmap to a circular northern Netherlands (Noord-Nederland Circulair: Routekaart naar een circulair Noord-Nederland) (2018) and the Brabant's region Bouwstenennotitie circulaire economie 2019–2028, Brabant beweegt in kringlopen (2019, in Dutch).

Local. With the Amsterdam Circular 2020–2025 Strategy, the City of Amsterdam is setting the course towards a circular city by 2050 [33]. They aim to halve the use of new raw materials by 2030 and their ambitious targets can only be achieved if everyone plays their part. As the municipal authority, they are taking the lead and want to set a good example for Amsterdam's residents and businesses. The Strategy focuses on the three value chains to shape the circular economy—Food & Organic Waste Streams, Consumer Goods, and Built Environment. For the Built Environment chain, the City can propose further-reaching policy than for other chains thanks to its influence on the design of areas, its role as a commissioning authority for the public space and in the realisation of its own accommodation and the granting of permits for construction and demolition. Because circular chains go beyond municipal boundaries, the City of Amterdam as a local authority cooperates with the region and the national government so that together to achieve a 100% circular city in a 100% circular country by 2050.

Courses of action to foster transition to circular built environment include [33]:

- Use recycled and biobased materials in construction to reduce the need for primary raw materials in the short term
- Determine the value of the current built environment
- Defining and safeguarding the circular ambitions at the city and district levels
- Joint knowledge as a starting point (a central municipal expertise centre)
- New forms of value assessment (the total cost of ownership and total cost of use of the built environment)
- Extending the useful life of existing buildings
- Tighten internal municipal processes to encourage circularity
- Stimulate innovative projects
- Formulating and implementing agreements on circular ambitions with extra-municipal parties, etc.

At the local level the City of Rotterdam and the City of Groningen have launched circular economy roadmaps (Routekaart Circulaire Economie–Gemeente Groningen, in Dutch) [106]. The Rotterdam Circularity Programme 2019–2023 [39] is focused on four sectors with construction being the most important one. Circular construction will be promoted by creating a materials passport, a circular concrete covenant, building hubs and a digital market place for building materials.

Construction and Demolition Waste Management Regulatory Framework

In the Netherlands 77% of their waste is recycled and the residual waste is mainly used for energy production. The main five elements that helped to achieve these results are: order of preference for waste management (waste hierarchy), stringent waste treatment standards, planning on national level, producer responsibility, and use of various (economic) instruments to stimulate prevention and recycling. In order to reduce the environmental pressure arising from waste management, stringent standards were introduced such as standards for soil protection from landfilling, standards for the quality of secondary materials derived from waste (building materials), a ban on landfill for 35 waste-streams (waste streams suitable for recovery or incineration are not allowed on landfills), etc. [308]. The majority of the building and demolition waste (more than 95% in 2014) is already recycled. However, the construction sector in the Netherlands is not circular yet as a large portion of most materials used in buildings (the residential and non-residential building sector) after demolition and recycling is used in civil engineering, often as a road base material or as filler material to raise the level of industrial estates. Although in civil engineering more than 50% of the materials used consists of recycled materials, which are functionally used and replace primary raw materials, in the building/construction sector a very small proportion of secondary material is used [307]. According to the Circularity Gap Report: Built Environment the Netherlands published in 2022 [31], only 8% of the Dutch built environment's material use comes from secondary sources.

The National Waste Management Plan (NWMP) is intended for waste which is subject to the Environmental Management Act and this Act stipulates that all authorities must must take into account the National Waste Management Plan when dealing with aspects of waste management. The NWMP sets out the policy for waste management in the Netherlands. The Third Waste Management Plan [310] covers the period 2017–2023, looking ahead to the period up to 2029. The NWMP consists of a policy framework that sets out the main points of waste policy and sector plans containing the policy framework for specific categories of waste. The NWMP contains 85 sector plans and each sector plan describes the policy for the relevant waste as well as particular aspects relating to licensing and imports and exports. Construction and demolition waste is released during the construction, renovation and demolition of buildings, structures and roads. This includes not only the construction and demolition waste itself, but also the waste materials that are also released during these activities but do not directly fall under the heading 'construction and demolition waste', such as cable remnants, paint packaging, etc. Out of 85 sector plans, 19 deal with construction and demolition waste [309].

5.1.13.2 Platforms and Networks Facilitating the Transition to a Circular Economy

Circular Economy Platforms

The Circular Economy Accelerator portal (Versnellingshuis) helps businesses with its plans or ideas for the circular economy, or respond to questions about knowledge, network partners, legislation or financing. The portal is a partnership of the Confederation of Netherlands Industry and Employers (VNO/NCW), the Royal Association MKB-Nederland for SMEs, sustainability think-tank the Green Brain and the Ministry of Infrastructure and Water Management [127].

Circular Buildings Platform

Platform CB'23 (Circular Construction 2023) [270] is committed to drafting agreements to anchor circular thinking and actions in daily construction practice for the entire Dutch construction sector: both residential and non-residential construction and civil engineering. Teams consisting of professionals from different parts of the Dutch construction sector, have laid a solid basis for circular agreements on important circular topics. At this stage, they essentially are working agreements, laid down in guides, rather than formal standards. The guides 'Measuring circularity in the construction sector' [271] and 'Passports for the construction sector' [272] have been drawn up for the topics Measuring Circularity and Information and Data. Platform CB'23 also created guidelines about Circular design, Circular tendering and Future re-use (in Dutch).

Building Materials Passport Platform

Madaster serves as an online library of information on materials and products [177]. For registered buildings and infrastructure objects, this platform provides insight into the materials and products used and their location, as well as their impact on circularity and the environment. In Madaster, a digital twin, or digital copy, of a building or other construction object is created, providing a clear overview of the materials used in it, the amount of CO_2 these materials contain and the extent to which they could be reused. Madaster enables companies to create material passports by uploading a BIM or Excel document [177]. Madaster received funding through the EU Horizon 2020 framework programme in 2018, and in the following year it began its international expansion to Germany, Norway, Switzerland and Belgium [178].

Public Procurement Platform

PIANOo is the Dutch Public Procurement Expertise Centre that was set up to professionalise procurement and tendering in all government departments, with a view to improving efficiency and compliance with the rules. As of 1 January 2017 PIANOo is part of the Netherlands Enterprise Agency (RVO.nl). Netherlands Enterprise Agency (RVO.nl) is part of the Ministry of Economic Affairs and Climate Policy. PIANOo brings experts in specific areas together, pools knowledge and experience and provides advice. It also fosters dialogue between government contracting authorities and private sector companies [267].

5.1.13.3 Funding Opportunities

MIA and Vamil for Entrepreneurs

All entrepreneurs in the Netherlands who pay income or company tax can use the Environmental investment deduction (MIA) and Arbitrary depreciation of environmental investments (Vamil) schemes. For 2023, the budget for MIA was €192 million and for Vamil €25 million. The tax scheme is interesting for entrepreneurs in different sectors, such as agriculture, industry, catering companies, business services and waste processing, but also for those who invest in circular economy, sustainable transport, sustainable recreation and sustainable buildings [246].

Different types of financial support such as subsidies, loans and tax benefits for circular entrepreneurship are available at https://english.rvo.nl/subsidy-guide

International Cooperation and Support from International Organizations

Holland Circular Hotspot (HCH) is a private–public platform in which the Holland Circular Hotspot foundation, (local) government authorities, knowledge institutes and companies intensively and internationally collaborate and exchange knowledge with the aim of stimulating entrepreneurship in the field of circular economy.

The activities of Holland Circular Hotspot are [149]:

- Offer insights in and access to the network of Dutch circular pioneers;
- Develop and exchange knowledge on international market opportunities for circular economy;
- Create circular opportunities internationally by matching offer and demand;
- Support companies and organizations that want to contribute to internationalization of circular economy;
- Stimulate cooperation between the private sector, knowledge institutions, governments and other relevant parties;
- Provide international visibility of Dutch CE innovations/best practices;

- Facilitate access to Dutch and international (financing) instruments and programmes.

HCH and the Rediscovery Centre have entered a partnership to accelerate the transition towards a circular economy in Ireland, the Netherlands and beyond. The signing of the Memorandum of Understanding (MoU) took place on Monday the 29th of May, 2023, coinciding with the Dublin Circular Economy Hotspot 2023 event. Under the agreement, both parties provide their experience, knowledge and expertise to intensify bilateral cooperation in the circular economy space. Areas for collaboration are sharing best practices in circularity, promoting European and international collaboration on circular economy initiatives and circular hotspots, and exploring opportunities for global initiatives and proposals [288].

The Dutch government is also active in multistakeholder platforms, where government, business and international organisations drive shared goals towards a circular economy. One example is the Platform for Accelerating the Circular Economy (PACE), with more than 40 members. PACE is a public–private collaboration platform for global changemakers and their organizations to accelerate the transition to a circular economy. PACE was launched by the World Economic Forum and is a delivery platform of the World Resources Institute. PACE is facilitated by a full-time team in The Hague, Netherlands [266].

Challenges, Barriers and Potential Improvements

Circle Economy and Metabolic [31] applied the Circularity Gap methodology to the Dutch building sector for the first time to identify challenges, barriers and opportunities as well as provide recommendations for the Dutch built environment. The Dutch building sector is faced with numerous challenges: labour shortages, an ageing workforce, and a lack of skills needed to go circular. A core barrier to realising material circularity is that virgin materials are far cheaper than their secondary counterparts, largely due to the lack of price incentives to promote or foster the uptake of secondary materials over virgin ones. The development of circular value chains requires the large-scale consumption of secondary and renewable materials and streamlined processes that allow for efficient disassembly and high-value material recovery, reuse and recycling. Such system calls for significant investments in physical infrastructure, advanced recycling technology, the knowledge and capacity to apply secondary materials, and new, circular business models. The investments needed for training and organisational change, as well as the infrastructure needed to facilitate recycling and reuse of materials at scale, currently represent significant barriers to actors in the sector. Considering the future cash flows (e.g., through the recuperation of materials after 50 or more years) contribute very little to current valuation, the long-term investments in circular solutions are hard to be broadly adopted within the industry. Another key barrier is the investment in the necessary physical and technical infrastructure to efficiently sort, process, and store large volumes of materials. A number of key barriers to innovation in the construction sector persist, such as a lack of information, knowledge and skills, and legal and regulatory challenges, as well as market complexities.

Based on the Circularity Gap Report: Built Environment, the Netherlands, there are several recommendations for actions needed by stakeholders in the built environment to overcome the identified barriers [31]:

- set a level playing field for circularity
- set explicit targets for secondary and biobased material use
- create long-term economic incentives
- write mandatory circular procurement into public tenders
- stimulate desired behaviour by rewarding frontrunners
- better monitor circularity in the built environment
- update regulations to remove barriers to circularity
- invest in physical infrastructure (material hubs and marketplaces)
- invest in digital infrastructure
- train employers and staff around circularity
- upgrade education and skills policy
- prioritise better working conditions throughout the transition
- address problematic labour market trends
- collaborate with private and public actors to promote career opportunities.

The Report highlights two aspects of these recommendations as key solutions—investing in human capital and creating a fair playing field. The construction sector employs 685,000 people, but it has difficulty finding new workers and current workers are getting closer to retiring age. In addition, the current workers do not have the right skills to shape the transition to be more circular. As many jobs will change towards the circular transition, upskilling and retraining will be necessary, as well as a culture shift towards lifelong learning. It is also important to invest in better working conditions, both financially and in terms of health and safety. At the moment, circular strategies have not yet been competitive with their linear counterparts, as negative environmental effects are not included in the pricing of linear materials and processes. Solutions with a lower impact on the environment should be rewarded, for example, through taxes or other economic incentives. The development of new circular business models requires more investments in innovation and the labour that is needed to build with secondary materials [30].

5.1.13.4 Examples of Successful Implementation

Public Policy Initiatives

The Central Government Real Estate Agency realised the temporary district court of Amsterdam with the DPCP consortium (an combination of Du PrieBouw & Ontwikkeling and developer Cepezedprojects) through a Design, Build, Maintain and Remove (DBMR) call for tender. The newly built construction on the grounds of the current Parnas complex at Parnassusweg is used as a temporary district court for the period of five years. The temporary nature of the building in no way harms the representativeness and quality in terms of equipment, complex logistics, acoustics, comfort, and safety. The focus was on preventing waste and maximising the residual value after this initial use period. This makes the building highly adaptable and allows for completely different uses by different occupants. At the end of its lifespan this building can be fully disassembled and reused elsewhere. The temporary district court of Amsterdam serves as a good practice example in the Circular Construction Economy Transition Agenda.

Private Policy Initiatives

Alliander company's head office in Duvian was renovated and opened in 2015. Everything about the building exudes style and newness, even though 83% of it is made up of recycled material [25]. The existing buildings were integrated into the new design, with a curved roof creating a large atrium which connects buildings and creates space where employees and visitors can meet. Circularity has been an integral part of the design in many ways: respecting the majority of the existing constructions, using waste wood for the facades, reusing the concrete from the parts which were demolished, reusing the steel construction for the extensions of the buildings, recycling the asphalt from the existing roofs, reusing the existing toilets and ceiling plates and converting the existing doors into new furniture, amongst other things [34].

5.1.14 North Macedonia Factsheet

5.1.14.1 Policy and Regulatory Framework

Governance

The Ministry of Environment and Physical Planning (MOEPP) in its National Plan for Waste Management for the period 2021–2031 (2021) and Waste Prevention Plan for the period 2022–2028 (2022) identifies as a goal, the move in the direction of developing and implementing collection and treatment systems in accordance with the waste hierarchy and to achieve environmentally safe processing and disposal, as well as a change from a linear to a circular economy, in a way that is best adapted to the conditions in North Macedonia. In the context of the aligning of North Macedonia with the EU's objectives for the introduction of a circular economy, the Plan provides a framework of measures and foresees a period in which the country's activities could move in accordance with the new action plan for circular economies for a cleaner and more competitive Europe. The Ministry of Environment and Physical Planning has also prepared the Long-term Strategy on Climate Action for the period 2021–2050 (2021) which focuses on boosting efficient use of resources and restored biodiversity, i.e.: clean sources of energy, circular economy and cut of pollution.

The Ministry of Economy in the Industrial Strategy with a Focus on Manufacturing (2018) presents an industrial policy with a focus on: "promoting industrialization, by stimulating the growth and development of the manufacturing industry, with the aim of increasing productivity, creating good jobs, increasing incomes and strengthening human capital, while solving the challenges of the circular economy."

The Ministry of Finance according to the Growth Acceleration Plan for the period 2022–2026 (2021) proposes investments in areas such as: industrial innovations and circular economy, green buildings, clean energy resources and sustainable mobility, sustainable land use and nature, and willingness to apply the EU Internal Market Standards.

Legislation

No current legislation address the implementation of circular economy principles directly. The Law on Public Procurement [259] includes relevant provisions on green public procurement, including on lifecycle costs. These costs cover parts or all of the following costs: (a) costs borne by the contracting authority or other users (e.g. costs relating to acquisition or purchase),costs of use, such as consumption of energy and other resources; maintenance costs; end of life costs, such as collection and recycling costs; (b) costs of eliminating the impact of the products, services or activities on the environment during their lifecycle, provided that their monetary value can be determined and verified, and which may include the cost of GHG emissions and of other pollutants, as well as other climate change mitigation costs. Application of lifecycle costs as a criterion for tender is optional and the data to be provided by the tenderers and the method for determining lifecycle costs should be indicated in the tender documentation.

The Law on Industrial and Green Zones [258] is in the process of amendment to include new provisions for regulating the benefits and opportunities for zone users, unifying applicable provisions for industrial and green zones, and better defining the conditions for performing activity in these areas. The amendment law has a specific focus on carrying out activities in line with circular economy principles in the Green and Industrial Zones.

The Law on Building Construction [257], in Arcticle 9, states that the building and its devices for heating, cooling and ventilation should be designed and performed in such a way that, depending on the climatic conditions of the location, they will ensure that the energy consumption during its use is equal to or lower than the prescribed level, as well as to meet the requirements for energy efficiency prescribed in the Regulations for energy characteristics on the buildings [258].

Circular Economy Strategy, Roadmap or Action Plan

The OECD's "A Roadmap towards Circular Economy of North Macedonia" was published on March 15, 2024, providing a strategic framework for the country's circular economy transition. The document outlines over 40 policy recommendations and a dedicated monitoring framework. It targets five priority areas: Circular business models for SMEs; Construction sector; Biomass and food; Textile industry; and Mining and metallurgy.

On a regional level North Macedonia has adopted the Green Agenda for the Western Balkans [91] and its Action Plan for the Implementation of the Sofia Declaration on the Green Agenda for the Western Balkans 2021–2030 [67]. The ten-year perspective of the Action Plan allows its alignment with the most important international and European policy objectives such as the United Nations (UN) Sustainable Development Goals (SDGs), the European Energy and Climate Policy Framework for 2030, the EU Biodiversity Strategy by 2030, the Farm to Fork Strategy, the Circular Economy and the Zero Pollution Action Plans. The Action Plan is structured so as to reflect the seven components of the Sofia Declaration (Climate Action, Energy, Transport, Circular Economy, Pollution, Sustainable Agriculture and Nature and Biodiversity Protection).

Construction and Demolition Waste Management Regulatory Framework

North Macedonia has adopted its new Law on Waste Management [260]. The new law is expected to enable the establishment of a functional system for regional waste management, such as regional collection, transport, sorting and recycling of waste, the construction of new regional landfills and closing of all non-standard landfills—in accordance with circular economy principles.

In addition a set of laws targeting specific waste streams and management schemes were adopted in 2021. These include the Law on Management of Batteries and Accumulators and Waste from Batteries and Accumulators, the Law on Management of Electrical and Electronic Equipment and Waste from Electrical and Electronic Equipment (WEEE), the Law on Packaging and Packaging Waste Management, the Law on additional Waste Flow Management and the Law on Extended Producer Responsibility (EPR). These new laws fully transpose the EU Waste Framework Directive and the Landfill Directive, and regulate EPR schemes for packaging, WEEE, batteries, waste oils and tiers, waste textile and End-of-Life Vehicles.

5.1.14.2 Platforms and Networks Facilitating the Transition to a Circular Economy

Circular Economy Platforms

No circular economy platforms have been established to date.

Circular Buildings Platform

No circular buildings platform has been established to date.

Building Materials Passport Platform

No building materials platform has been established to date.

Public Procurement Platform

No public procurement platform has been established to date. However, the Public Procurement Bureau and the Energy Agency have developed guidelines on energy efficiency measures to be used in public procurement procedures. The objective of these measures is to boost demand for resource/energy efficient, durable, recyclable, repairable products, and promote new business models based on offering functionalities and services instead of selling products. Green public procurement also allows local, regional and national authorities to set examples and standards for businesses and industries to follow. However, the introduction of green public procurement measures in tenders has remained limited since the adoption of these laws.

5.1.14.3 Funding Opportunities

Strategic Green Investment Fund (SGIF)

The establishment of the Strategic Green Investment Fund (SGIF) is also envisaged and aims at facilitating private sector financing for green transition to attain zero GHG emission target by 2050 and accelerating economic growth by attracting foreign and domestic investment in the green industries in Technological Industrial Development Zones (TIDZs). Moreover, the First Green Zone (TIDZ Gevgelija) was established in 2022, located near the border crossing to Greece [54].

Fund for Innovation and Technology Development

The Fund for Innovation and Technology Development (2021), which encourages innovation by providing additional resources for financing innovative activities, plans to include a green business facility, funded through the Instrument for Pre-accession Assistance (IPA) (the financial agreement has been signed at the end of 2021). It will provide direct subsidies and grants to SMEs, as well as awareness raising for green projects with a total budget of EUR 27 million.

Growth Acceleration Plan for the Period 2022–2026

The proposed financing instruments have yet to be established and include, among other elements, green bonds, a Hybrid National Green and Digital Fund for SMEs, Start-ups and Innovative Enterprises to invest in green and digital SMEs (with a total portfolio of EUR 27 million), an Energy Efficiency Fund (initial capital of EUR 5 million) and a Strategic Green Investment Fund (expected to be worth EUR 36.3 million) to accelerate investments in renewable energy sources and financing energy efficiency solutions [220].

International Cooperation and Support from International Organizations

The European Bank for Reconstruction and Development approved a country-wide sovereign loan of EUR 55 million for the establishment of regional landfills in five administrative regions, in line with the new set of laws on waste management. The loan covers reconstruction/construction of selected regional, sanitary landfills and procurement of bins and containers for recyclables and residual waste, transport vehicles and equipment.

The Swiss Agency for Development and Cooperation supports North Macedonia by engaging with the non-profit private sector packaging recovery organization Pakomak to enhance the effectiveness, efficiency and transparency of the national packaging waste recovery system through the digitalisation of the primary selection and collection process.

The twinning project, Supporting the implementation of the legal framework for waste management and the extended producer responsibility system (EPR) [215], which started in 2023, will enable the efficient and effective functioning of a modern and responsible industry for recycling and waste processing in North Macedonia, in accordance with the principles of circular economy. The project, for a period of 24 months, will be implemented by the Ministry of Environment and Physical Planning of North Macedonia, in partnership with the Foundation for Climate Protection "Blue Planet" as the main partner and the Republic of Austria as a junior partner, and aims to made it possible to strengthen the extended producer responsibility system.

The project for Development of the Environmental Monitoring Information System [211], funded by the EU through IPA 2014–2020, aims to improve the data collection process and introducing electronic systems for reporting,which will be capable of supporting the monitoring, processing, reporting and dissemination of environmental data.

A project for Promoting economic growth in North Macedonia, commissioned by the Federal German Ministry for Economic Affairs and Energy, aims at improving business support services for micro, small and medium-sized enterprises (MSMEs) and covers the circular and green economy [103]. Another project Developing human capital towards improved climate-change adaptation and mitigation was commissioned by German Federal Ministry for Economic Affairs and Climate Action (BMWK), in order to support the development of human capital towards improved climate-change adaptation and mitigation, and aims to mainstream climate change in the education sector and at boosting green jobs [102].

A knowledge Alliance in Eco-Innovation Entrepreneurship to Boost SMEs Competitiveness (SMecoMP) concerns the development of an educational framework and appropriate training tools to improve the skills of business executives in Eco-Innovation Entrepreneurship [11]. The consortium of the SMecoMP project is made up of organizations from four Balkan countries and seeks to achieve cross-border co-operation between businesses and universities in order to promote and raise awareness of environmental innovation and entrepreneurship.

The World Bank has funded the North Macedonia Public Sector Energy Efficiency Project aims to reduce energy consumption in the public sector, and support the development and implementation of a sustainable financing mechanism for energy efficiency in the public sector [334].

Challenges, Barriers and Potential Improvements

Implementing a circular economy in North Macedonia comes with its own set of challenges and barriers, that can be categorized into economic, regulatory, infrastructure, and awareness-related issues. Addressing these challenges and barriers in North Macedonia will require a coordinated effort involving government, private sector, non-governmental organizations, and international organizations. Developing a comprehensive circular economy strategy, updating regulations, and raising awareness are important steps toward fostering a more sustainable and circular economy in the country (Table 5.3).

5.1.14.4 Examples of Successful Implementation

Public Policy Initiatives

No data to this point.

Private Policy Initiatives

No data to this point.

Table 5.3 Overview of challenges and potential improvements for the implementation of the circular economy in North Macedonia

Challenges	Potential improvements
High dependency on oil and coal for energy generation. Oil and coal also represent two-thirds and renewable energy only accounts for 14% of North Macedonia's energy supply	Phasing-out coal from its energy mix taking into account underlying economic and social challenges will be a priority, by diversifying of the energy mix through other renewable sources and curbing demand growth through energy efficiency measures
Domestic material consumption levels have been constant in recent years and are mainly made up of fossil energy materials and biomass, exerting growing pressures on the environment	Efforts to improve resource efficiency in mining, smelting, and refining are needed to reduce environmental impact
Inadequate recycling and waste management infrastructure limits the ability to recover and recycle materials	Investments in recycling facilities and collection systems are necessary
Limited public awareness and understanding of circular economy principles and benefits hinder adoption	Educational programs and campaigns needed to raise awareness among businesses, policymakers, and the general public
A lack of expertise and knowledge about circular economy practices hinder businesses from transitioning to more sustainable models	Training and capacity building are required to develop the necessary skills and expertise
Outdated or inadequate regulations and policies do not provide sufficient incentives for businesses to adopt circular practices	Legal frameworks need to be updated to support circular economy initiatives and incentivize recycling, reuse, and sustainable production
Access to funding and investment for circular economy projects is limited, especially for small and medium-sized enterprises (SMEs)	Initiatives to attract private and public funding for circular projects are needed
Transitioning to a circular economy requires significant changes in supply chains, which can be costly and complex	Businesses need support in reconfiguring their supply chains to incorporate circular principles
Consumer behavior, preferences for disposable and cheap products can discourage businesses from adopting circular practices	Public awareness campaigns and incentives can change consumer behavior
Insufficient demand for recycled or refurbished products makes it challenging for businesses to invest in circular practices	Encouraging consumer and industrial demand for sustainable products is essential
Circular economy practices may not always be economically competitive in the short term	Policy measures, subsidies, or incentives are needed to level the playing field
Lack of data on resource consumption, waste generation, and circular economy performance hinder decision-making and policy formulation	Investment in data collection and monitoring systems is necessary to track progress
Collaboration among government, private sector, non-governmental organizations, and academia is crucial for successful circular economy implementation	Building partnerships and engaging stakeholders can be challenging but is essential for driving change

5.1.15 Portugal Factsheet

5.1.15.1 Policy and Regulatory Framework

Governance at a National Level

- Ministry of Environment and Climatic Action/Ministério do Ambiente e da Ação Climática. In Portugal, the Circular Economy is under the responsibility of the Ministry of Environment and Climatic Action. Within this Ministry, there are four secretaries of state in different key fields: Energy and Climate, Environment, Nature Conservation and Forests, and Urban Mobility. The circular economy is a transversal subject to three of the four secretaries of state.
- **Environmental Fund/Fundo Ambiental**. The Environmental Fund concentrates the resources of previously existing funds (Permanent Forestry Fund, the Innovation Support Fund, the Energy Efficiency Fund and the Energy Sector Systemic Sustainability Fund) to obtain an instrument with greater financial capacity and with greater adaptability to the challenges posed. The Environmental Fund has an annual plan for allocating support to specific environmental and climate action projects using environmental-related revenue. The supported projects are grouped into seven different clusters:

  1. waste management and transition for a circular economy,
  2. climate change mitigation,
  3. environmental protection, radiologic protection and environmental risk and damage management;
  4. protection and conservation of nature and biodiversity;
  5. adaptation to climate change;
  6. international cooperation
  7. energy efficiency.

  Moreover, recently this Fund also manages six components of the National Recovery and Resilience Plan (PRR):

- C8. Forestes
- C9. Hidric management
- C12. Energy efficiency in buildings
- C14. Hydrogen and renewables
- C15. Sustainable mobility,

Portuguese Environment Agency/Agencia Portuguesa do Ambiente (APA). The Portuguese Environment Agency is a public institute integrated into the indirect administration of the State, under the supervision of the Ministry of the Environment and Climate Action and endowed with administrative and financial autonomy and its own assets.

The Portuguese Environment Agency is the entity responsible for implementing environmental policies in Portugal aiming to contribute to a high level of protection and enhancement of the environment through the provision of quality services to citizens. The Portuguese Environment Agency has six main fields of action:

- **Climate**, with the role of proposing and developing climate policy and designing and implementing mitigation and adaptation strategies to climate change, particularly to achieve the goals defined in national, community and international commitments.
- **Water**, being the National Water Authority, defines policies and management instruments that ensure the application of these principles.
- **Air and noise**, being is the competent authority for air, responsible for implementing the air quality assessment and management policy, particularly with regard to the quality control of information and its dissemination to the European Commission and the general public; and competencies within the scope of the noise pollution prevention and control regime, the General Noise Regulation (RGR).
- **Waste**, being the National Waste Authority, ensuring its planning and management, in order to prevent or reduce its production, its harmful nature and possible adverse impacts. On the other hand, it seeks to promote efficiency in the use of resources, based on the principles of the waste hierarchy and Circular Economy.
- **Environmental evaluation and management**, being the entity responsible for the development and implementation of environmental evaluation and management instruments, also ensuring dialogue with the European Commission.
- **Risk prevention and management**, with the role defined within the scope of each risk, which may include monitoring, the development of risk reduction strategies, the definition and implementation of rules in relevant sectors of activity, the implementation of risk reduction actions or support in emergency response.

In the Portuguese Environment Agency the framing of the circular economy within one of the fields of action is somehow unclear: currently being framed under a trans-sectional subject, has also been under the waste management field.

- **ADENE**. ADENE is an associative entity with public utility status with the purpose of promoting and carrying out activities of public interest in the energy area and its interfaces with other sectoral policies, in conjunction with other entities with responsibilities in these areas, and also promoting and carrying out activities of public interest in the areas of efficient use of water and energy efficiency in mobility. Currently has added Circular Economy to the four other interest fields: industry, buildings, mobility and hydric.
- **LNEG**. The National Laboratory for Energy and Geology and Geological Resources (LNEG) carries out research with a view to its application in advanced solutions that allow us to boost our Economy. One of the main research areas is circular economy and life cycle management towards sustainability. Among other circular economy projects, LNEG runs the Circo Hub Portugal. Creating Business through Circular Design) that emerged in the Netherlands in 2015 and is based on the application of design and design thinking to the development of circular products, services and business models. It is a program that activates and

facilitates the implementation of circular businesses by industrial companies and designers. With an exclusive and proven method, the CIRCO program offers each company a perspective to initiate circular activities and solutions, in cooperation with its value chain.

Governance at a Regional Level

- **Regional Coordination and Development Commissions/Comissões de Coordenação e Desenvolvimento Regional (CCDR)**

Portugal has 5 Regional Coordination and Development Commissions (CCDR) that correspond to the 5 regions of mainland Portugal:

- North region: Comissão de Coordenação e Desenvolvimento Regional do Norte (CCDR-N), with a Regional Agenda For A Circular Economy (2019) that has identified by the regional metabolism study four main areas: agro-food, construction, construction and demolition waste, textile, transports and cities (as an anchor for the circular economy)
- Central region: Comissão de Coordenação e Desenvolvimento Regional do Centro (CCDR-C), with a voluntary institutional pact for the valorization of circular economy in the central region (2019) inviting players to join assuming some variable commitments towards a more circular economy
- Lisbon and Tagus Valley: Comissão de Coordenação e Desenvolvimento Regional de Lisboa e Vale Tejo (CCDR-LVT), with a Regional Agenda for a Circular Economy in the LVT Region (2019), including challenges (consumption, stock and recycling); goals (leveragers and transversal); approaches (productive cycle for agro-food, productive sector for construction, industrial symbioses for industrial agglomerations) and measures focussing on institutions, companies and consumers.
- Alentejo region: Comissão de Coordenação e Desenvolvimento Regional do Alentejo (CCDR-A), with a Regional Agenda for Circular Economy in Alentejo (2019) with several projects in the CE field,
- Algarve region: Comissão de Coordenação e Desenvolvimento Regional do Algarve (CCDR-ALG) with an analysis and an Action Plan for the CDW in Algarve (2019).

- Regional Directorate for the Environment and Climate Change of the autonomous region of Madeira/Direção Regional do Ambiente e Alterações Climáticas da região autónoma da Madeira (DRAAC Madeira). In the Madeira autonomous region, the Direção Regional do Ambiente e Alterações Climáticas (madeira.gov.pt) has an interesting field named waste and circular economy with waste being the main focus of action.
- Regional Directorate for the Environment and Climate Change of the autonomous region of the Azores/Direção Regional do Ambiente e Alterações Climáticas da região autónoma dos Açores (DRAAC Açores). In the Azores autonomous

region, the Regional Directorate for the Environment and Climate Change of the autonomous region of the Azores has published a Regional Plan For Climate Change and a Roadmap for the Circular Economy in the Agricultural and Forestry Sector of the Autonomous Region of the Azores. It has participated in some Circular Economy projects with the construction sector and together with other municipalities, Ponta Delgada is part of the National Initiative Circular Cities.

Governance at a Municipal Level

At the municipal level, some actions towards a circular economy are being carried out by several municipalities. Just to cite some examples:

- **Porto Municipality** has presented the Porto Pact for Climate, which aims to awaken the action of citizens and organizations and create a large community of learning, sharing and mutual support, with a voluntary, non-binding and free-of-charge subscription.,
- **Gaia Municipality** is the leading partner of the Sustainable and Circular Cities Network and also includes Mangualde, Mértola, Oliveira de Frades, Ponta Delgada, Ponte de Sor, Ribeira Brava and Valongo municipalities.
- **Lisbon Municipality** presents the Lisbon Sustainable Companies Platform as a commitment for the 2020–2030 decade in which companies and organizations commit, within the scope of their activity, to provide measures that contribute to achieving ESG sustainability goals in their city

5.1.15.2 Legislation

Although there are several strategic documents closely related to the circular economy and a waste management regime has been established, there is currently no specific law in Portugal related to the circular economy. A legal framework focusing on buildings within the scope of energy efficiency and Building Energy Certification System (list can be found here) transposes regulations at the European level.

Energy Performance of Buildings Directive. Transposing the European Energy Performance of Buildings Directive 2002/91/CE (EPBD) the following Portuguese laws were published:

- 78/2006 (SCE), approves the National Energy Certification and Indoor Air Quality System in Buildings;
- 79/2006 (RSECE) approves the Regulation on Energy Air Conditioning Systems in Buildings;
- 80/2006 (SCE) approves the Regulation on the Thermal Behavior Characteristics of Buildings.

Following the EPBD revision 2010/31/UE national legislation was published.

- 118/2013 (SCE, REH and RECS), approves the Building Energy Certification System, the Energy Performance Regulation for Residential Buildings and the Energy Performance Regulation for Commercial and Service Buildings,

  More recently, and as a consequence of EPBD last revision 2018/844/UE.

- 101-D/2020, establishes the requirements applicable to buildings to improve their energy performance and regulates the Building Energy Certification System.

In the 101-D/2020 current law, the concept of nZEB is defined as "almost zero energy needs building" means a building with a very high energy performance (…) and in which almost zero or very small energy needs are covered, to a large extent, by energy from from renewable sources, preferably local or originating close to the building, when that is not sufficient". Moreover it is stated that new buildings must be "buildings with almost zero energy needs" (art.6) and that buildings energy certification will serve "as a support element in accessing financing instruments"; and "serve as a support element for the attribution of tax benefits, encouraging, in particular, the implementation of improvement measures, as well as the design or renovation of buildings with a view to achieving high energy performance" (art. 17).

5.1.15.3 Circular Economy Strategy, Roadmap or Action Plan

Action Plan for Circular Economy/Plano de Acção Para a Economia Circular (PAEC)

Portugal is one of the EU member states that have ongoing action plans for a Circular Economy, in line with the ambitions of the European Commission, being named Plano de Acção para a Economia Circular (PAEC) (Presidência de Conselho de Ministros n.o 190-A/2017, 2017). The content of the plan acts on the culture, market and political barriers, targeting three specific fields:

- **Political field** with political instruments that promote the efficient use of resources, from the design of the product/service to the recovery of by-products and waste;
- **Knowledge field** disseminating information about best practices, case studies, and funding opportunities, among others, and promoting the development of collaborative-based R&D initiatives in this area;
- **Economic field** through specific interventions in existing financial instruments to value initiatives that contribute effectively to the CE, namely through sectoral and intersectoral projects in this matter.

And presents actions aligned with the European pillars of Action for the Circular Economy and 3 levels of operationalization:

- **Macro level**: actions of a structural scope, which produce transversal and systemic effects that enhance the appropriation of circular economy principles by society;

- **Meso (or sectoral) level**: actions or initiatives defined and undertaken by a group of stakeholders in the value chain of relevant sectors to increase productivity and efficient use of the country's resources, capturing economic, social and environmental benefits;
- **Micro (or regional/local) level**: actions or initiatives defined and undertaken by a set of governmental, economic and social, regional and/or local agents, which incorporate the local economic profile and value it in addressing social challenges.

The summary of the plan (in English) and the report about its implementation (only in Portuguese) were publicly presented. Currently, the PAEC is under revision and in the public discussion, and comments can be submitted until the 10th of November 2023.

Action Plan for Circularity in Construction in Portugal/Plano de Ação Para Circularidade na Construção em Portugal (PACCO)

The Action Plan for Circularity in Construction in Portugal (PACCO) [4] has the objective of identifying and discussing measures that can accelerate the transition to a circular economy in the construction sector in Portugal, responding to the need identified by PAEC (Presidência de Conselho de Ministros n.o 190-A/2017, 2017) to develop sectorial plans. The executive summary of the PACCO presents the analyses done of the sector, main barriers and opportunities, the strategy of the plan including the four pillars (political, technological, market, and cultural) and sub-pillars, nine vectors (political context, certification schemes, fiscal instruments, databases, methodologies and digital tools, training and accreditation, communication plan, digital platform and e-commerce, and circular economy entity) and the 30 measures towards a circular construction. Each one of the 30 measures was presented with specific actions, implementation timeline, sustainable development goals (SDG), targets, indicators, vector of implementation, actors and financing mechanism.

Roadmap for Carbon Neutrality 2050/Roteiro Para a Neutralidade Carbonica 2050 (RNC2050)

Besides specific legislation on circular economy and waste management Portugal adopted in 2019 a Roadmap for carbon neutrality, named "Roteiro para a neutralidade carbonica 2050 (RNC2050)" (Ambiental and Ambiente 2019). This roadmap supported the transition towards a competitive, circular, resilient and carbon–neutral economy. It defined the trajectories towards a carbon–neutral economy in 2050, assessing the energy system (including eletroprodotor, mobility and transports, industry and buildings); the agro-florest role, waste and water, and circular economy.

National Plan for Energy and Climate 2030/Plano Nacional Energia e Clima 2030 (PNEC 2030)

This Plan (PNEC 2030) [276] aims to establish objectives and respective policies and measures in terms of reducing greenhouse gas emissions, incorporating energy from renewable sources, energy efficiency, energy security, internal market and research, innovation and competitiveness, as well as a clear approach to achieving the aforementioned objectives and goals. In conjunction with the objectives of RNC2050, the National Energy and Climate Plan 2021–2030 (PNEC 2030) was developed, which constitutes the main instrument of national energy and climate policy for the next decade towards a carbon–neutral future, which is now approved. The PNEC 2030 establishes ambitious but achievable goals for the 2030 horizon and implements the policies and measures for the effective application of the guidelines contained in the RNC2050 and for achieving the defined goals. The main objectives of the National Plan for Energy and Climate 2030 are:

1. Decarbonize the national economy
2. Give priority to energy efficiency
3. Strengthen the focus on renewable energy and reduce the country's energy dependence
4. Ensure security of supply
5. Promote sustainable mobility
6. Promote sustainable agriculture and forestry and enhance carbon sequestration
7. Develop an innovative and competitive industry
8. Ensure a fair, democratic and cohesive transition.

PNEC 2030 is currently under review.

5.1.15.4 Construction and Demolition Waste Management Regulatory Framework

General Waste Management Regime/Regime Geral Da Gestão De Resíduos (RGGR)

The General Waste Management Regime (RGGR) [277] was approved in 20020 in line with EU Waste directives, and as policies relating to waste management have evolved towards the sustainable management of materials, to protect, preserve and improve the quality of the environment, protect human health, ensure prudent, efficient and rational use of natural resources, reduce pressure on the regenerative capacity of ecosystems, promote the principles of the circular economy, reinforce the use of renewable energy, increase energy efficiency, reduce dependence on imported resources, provide new economic opportunities and contribute to long-term competitiveness. This document regulates the legal regime for landfilling waste and amends the management regime for specific waste streams. It includes five sections: I General provisions and principles; II Regulation of waste management; III Registration of information and monitoring of waste management; IV Economic and financial regime for waste management; and V Administrative offence regime.

In the Article 21. Prevention objectives and targets, to decouple economic growth from the impacts on human health and the environment associated with waste production, the following calendar of targets relating to the prevention and reduction of waste production and its dangerousness is established:

- reduce the amount of urban waste produced per inhabitant by 5% (2025) and 15% (2030) compared to 2019 values
- reduce the amount of food waste in restaurant establishments collective and commercial feed and production and supply chains, including industries agri-food companies, catering companies, supermarkets and hypermarkets, by 25% (2025) and 50% (2030) compared to 2020 values;
- reduce the amount of non-urban waste per unit of gross domestic product (GDP) by 5% (2025) and 10% (2030) particularly in the civil construction and public works sector, compared to 2018 values;

Targets relating to preparation for reuse, recycling and recovery in article 27.º and aiming at promoting the transition to a circular economy with a high level of resource efficiency, entities responsible for waste management must adopt the necessary measures, through waste management plans and programs, to ensure compliance with following goals:

- From the date of entry into force of this regime, a minimum global increase of 50%, by weight, about the preparation for reuse and recycling of urban waste;
- From the date of entry into force of this regime, a minimum increase to 70% by weight about preparation for reuse, recycling and other forms of material recovery, including filling operations using waste as a substitute for other materials, of non-hazardous CDW, excluding natural materials defined in category 17 05 04 of the LER;
- A minimum increase to 55% (2025), 60% (2030) and 65% (2035) by weight of preparation for reuse and recycling of municipal waste, where at least 5% (2025), 10% (2030) and 15% (2035) is resulting from preparation for reuse of textiles, electrical and electronic equipment, furniture and other waste suitable for preparation for reuse.

Finally in article 28 about the design, production and distribution of products that generate waste, it is stated that it is mandatory to use at least 10% of recycled materials or those that incorporate recycled materials in relation to the total amount of raw materials used in construction, within the scope of contracting construction and infrastructure maintenance contracts under the Contract Code Public (Decree-Law no. 18/2008, CCP).

5.1.15.5 Platforms and Networks Facilitating the Transition to a Circular Economy

Circular Economy Platforms

- **ECO.NOMIA portal**. The ECO.NOMIA portal is one of the components of the Ministry of the Environment's action plan, serving as a space for sharing knowledge about the Circular Ecomomy. On the one hand, publicizing the advantages and financing opportunities to consumers and companies and, on the other, launching an interaction forum for collaborative investment projects in the Circular Economy.
- **eCIRCULAR**. ADENE has recently launched eCIRCULAR, the Classification System that allows the evaluation of performance and improvement of management practices in the circular economy of organizations. Focusing on the whole economy, and specifically in the first stage in the production companies, this evaluation system currently in a Beta stage, proposes the evaluation of the performance relative to resources management—energy, water and materials—and strategies and practices—internal and external focus.
- **MyWaste**. MyWaste is an online platform for hosting a national exchange that consists of a business-to-business (B2B) network for sharing waste/by-products/RES that can be recovered. This aims to inspire the circular economy, waste reduction and reuse, with each entity being able to manage and optimize its resources, offering features for managing the waste generated.

Circular Buildings Platform

Resulting from a Circular Construction research project (PRCD—Prevention of Construction and Demolition Waste), promoted by ASWP and financed by the Environmental Fund, this platform hosted at https://www.construcaocircular.pt/ aims to promote educational actions aimed at the various agents along the chain associated with RCD, supporting their interaction, to promote a more sustainable value chain organization, in line with the principles of the circular economy.

Building Materials Passport Platform

No building materials platform has been established to date, but some research projects are focused on this theme: Circular EcoBIM (3Drivers), and Edifícios Circulares (Associação Smart Waste Portugal).

Public Procurement Platform

- **BaseGOV managed by IMPIC**. The Institute of Public Markets, Real Estate and Construction, I.P. (IMPIC, I.P.) is a public institute with legal personality, administrative and financial autonomy and its assets, which carries out tasks under the supervision of the Ministry of Housing in the areas of construction, real estate, including the regulation of public contracts. The Portal BASE centralizes the information on public contracts celebrated in Portugal's mainland and autonomous regions.
- **National strategy for green procurement/Estratégia Nacional para as compras públicas ecológicas (ENCPE)**. National strategy for green procurement (Resolução do Conselho de Ministros n.o 65/2007, 2007) applies to the State, namely to bodies under its direct and indirect administration and to the State's business sector, and also, on an optional basis, to autonomous administration and other legal entities governed by public law, whenever the acquisition of goods, services or the preparation of projects for the execution of public works that are part of the list of priority goods and services and is presented in this website: https://encpe.apambiente.pt/. It had two main goals:

  1. Increase the percentage of public pre-contractual procedures for the acquisition of goods and services covered by ENCPE 2020 that include environmental criteria, in the direct and indirect administration of the State by 55% (by 2017–2018) and 60% (by 2019–2020) and in the state business sector 35%(by 2017–2018) and 40% (by 2019–2020)
  2. Increase the percentage of the financial amount associated with public pre-contractual procedures for the acquisition of goods and services covered by ENCPE 2020 that include environmental criteria in the direct and indirect administration of the State by 55% (by 2017–2018) and 60% (by 2019–2020) and in the state business sector 35% (by 2017–2018) and 40% (by 2019–2020).

  A revised national strategy for green procurement is currently under revision.

5.1.15.6 Funding Opportunities

Recovery and Resilience Plan

The Recovery and Resilience Plan (Plano de Recuperação e Resiliência—PRR) is a national program, with an implementation period of 2026, which will implement a set of reforms and investments aimed at propelling the country on the path to recovery, sustained economic growth and convergence with Europe over the next decade, guided by a concept of sustainability inspired by the United Nations Sustainable Development Goals (SDGs). The PRR is aligned with the six relevant pillars of the European 2030 strategy:

- Green transition;

- Digital transformation;
- Smart, sustainable and inclusive growth, including economic cohesion, employment, productivity, competitiveness, research, development and innovation, as well as a well-functioning Single Market with strong small and medium-sized enterprises (SMEs);
- Social and territorial cohesion;
- Health and economic, social and institutional resilience, including to increase the capacity to react and prepare for crises;
- Policies for the next generation, children and young people, including education and skills.

This programme has 20 components, grouped into three main clusters:

- **Resilience** with nine components: C01—National health service; C02—Housing C03—Social responses C04—Culture C05—Capitalisation and business innovation C06—Qualifications and competencies C07—Infrastructure, C08—Forests; and C09—Water management
- **Climate transition** with six components: C10—Sea C11—Decarbonisation of industry C12—Sustainable bioeconomy C13—Energy efficiency in buildings C14—Hydrogen and renewables C15—Sustainable mobility
- **Digital Transition** with five components: C16—Business 4.0 C17—Quality and sustainability of public finances C18—Economic justice and business environment C19—Public administration—digitalisation, interoperability and cybersecurity C20—Digital schools.

Component 11_Decarbonizing industry focuses on the promotion of the circular economy supporting the industry to make the transition. It aims to develop innovative and competitive industries, through the promotion of circular and low-carbon economies, industrial symbioses and new circular, low-carbon products and services.

Portugal 2030

Portugal 2030 puts into practice the Partnership Agreement between Portugal and the European Commission to invest 23 billion euros of European funds in projects that stimulate and develop the Portuguese economy, between 2021 and 2027. The total amount available to finance projects will be distributed across programs, organized by themes and regions. This program will support Private companies; Local authorities and other public bodies; Cooperatives, associations, private social solidarity institutions, and other private entities.

Portugal 2030 will support projects in six main areas:

1. **Portugal + Smart**: Investing in research and innovation, digitalization, the competitiveness and internationalization of companies, skills for smart specialization, industrial transition and entrepreneurship.
2. **Portugal + Connected**: With strategic transport networks, based on a strong commitment to railways, enhancing the mobility of people and goods, as well as

the qualification of territories, guaranteeing their attractiveness, competitiveness and insertion in national and international markets.

3. **Portugal + Close**: Close to citizens, supporting development strategies at the local level, promoting social and territorial cohesion, and supporting sustainable urban development, based on the concept of network interconnection, centred on people's needs.
4. **Portugal + Green:** Oriented towards the green transition, following the climate emergency and incorporating the goals of decarbonization, energy efficiency and reinforcement of renewable energy, and supporting innovation, the circular economy and sustainable mobility.
5. **Portugal + Social**: Supporting the improvement of the population's qualifications, equal access to healthcare, promoting quality employment and social inclusion, following the priorities established in the European Pillar of Social Rights.
6. **Portugal + Just transition**: To ensure that the transition to a sustainable and carbon–neutral economy is carried out fairly.

The circular economy agenda is framed under the areas Portugal + Green and Portugal + Just Transition. Some Circular Economy Business Dynamics, Training and Incentives are summarized here.

International Cooperation and Support from International Organizations

No data to this point.

Challenges, Barriers and Potential Improvements

There are several challenges and barriers related to the implementation of a circular economy in the construction sector. The Plan for Circularity in Construction in Portugal has identified the main challenges and proposed several measures and actions for implementing the Circular Economy in this sector (Table 5.4).

5.1.15.7 Examples of Successful Implementation

Public Policy Initiatives

Research projects promoted by FCT and other entities, to respond to the emergency associated with this transition. Financing through EEA grants has proven to be relevant in the area of circularity and sustainability in the AEC sector. Some of the research projects that have been supported to push the Circular Economy are:

- C + D (CERIS), Close the loop by Disclosing the benefits of buildings' deconstruction and materials re-use, aiming to develop a web-based platform to calculate

Table 5.4 Overview of challenges, recommended measures and actions for the implementation of the circular economy in the Construction sector in Portugal

Challenges	Recommended measures	Recommended actions
Political and regulatory barriers: – Complex legislation – Project without specifications circularity – Cost of environmental degradation and non-internalized social – Lack of (in)formation when hiring ecological – Lack of incentives for projects circular – Lack of actions, goals, metrics and schedule for transition	1. Proposal to reform the legislative and regulatory framework	Reform the legal framework
		Defining goals and incentives
		Define a Sectorial Agreement
	2. Ex-ante evaluation and monitoring of measures	Building the base scenario
		Assessment of consequential impact
		Monitoring of measures and data collection
	3. Develop and implement digital guides	The digital guide to support circular design
		Digital user guide
		The digital guide to support deconstruction
		Digital waste management guide
	4. Proposal for a political and sectoral support structure	(Infra)structures to support reverse logistics
		Waste sorting, storage and disposal sites
		Increase inspection capacity
	5. Proposal to create certification schemes	Certification of products, solutions and constructions
		Implement guarantee schemes
		Standards for assessing circularity and sustainability
	6. Proposal for material/building passports	Policies that promote material passports
		Guidelines applicable to construction materials and elements
		Standardized measures for the eco-design of materials and products and the (de)construction
	7. Proposal to support green contracting	Proposal for policies that establish increasing needs for the use of materials circulars in public and later private procurement
		Promote voluntary agreements for ecological and innovative public procurement

(continued)

Table 5.4 (continued)

Challenges	Recommended measures	Recommended actions
		Progressive implementation of sustainability and circularity criteria
	8. Transition financing proposal	IDI financing proposal
		Tax incentives to support the retraining and training of HR
		Tax incentives to support new business models
		Tax incentives to support the transition of existing businesses
	9. Tax penalty	Tax penalties on the use of natural resources
		The tax penalty for waste disposal in landfills
		Finance the transition through penalized revenue
	10. Tax cuts	Reduce labor tax
		Reduce tax on secondary, recycled, and other materials or with the incorporation of recycled materials
	11. Internalization of environmental and social impacts	Policies that internalize environmental and social impacts
Technological barreiers: – Technology for deconstruction does not develop – Lack of knowledge of materials alternative and innovative – Lack of information about the performance Lack of a national database – Unique and unrepeatable works – Computer skills insufficient or inappropriate – Lack of marking including criteria circularity	12. Proposal for platform(s) and digital tools	Knowledge sharing platform
		National database platform (costs and impacts)
		Platform to support public and private procurement (e-commerce) in conjunction with the existing Portal Base
		Project support platform (incl. quality criteria, circularity and sustainability)
		Building management platform (based on Digital Twin)
		Waste/secondary materials marketplace

(continued)

Table 5.4 (continued)

Challenges	Recommended measures	Recommended actions
		Construction and updating of digital databases
		Development of digital tools to support design, construction, management and demolition of buildings
	13. Develop and disseminate collaborative methodologies	Develop collaboration methodologies based on BIM, LCA, LCC and circularity
		Develop methodologies for exchanging, verifying and sharing information
	14. Proposal to support case studies and databases	Support demonstrator pilot cases
		Support database development
	15. Proposal to support R&D new processes and technologies	Support R&D for new manufacturing processes
		Support R&D of new solutions and systems
		Support R&D for new forms of transport
		Support R&D of new technologies to support the project
		Support R&D for new building management systems
		Evaluating new processes and technologies (holistic multi-criteria approach
	16. Support R&D of new materials and products	Support R&D of materials and products with useful life extension
		Support R&D of new materials and products that incorporate recycled materials
		Support R&D of new materials and products that enhance recycling or reuse
		Evaluating materials and products (holistic multi-criteria approach)

(continued)

Table 5.4 (continued)

Challenges	Recommended measures	Recommended actions
Market barriers: – Lack of demonstrator projects – No market for CDWs – High price of recycled materials – Lack of awareness among manufacturers – Lifespan and difficult maintenance – Lack of training of sector actors – Financing that excludes environmental criteria	17. Support circular business models	Support new, more circular and/or energy-based business service models
		Support the transition of existing business models (preferential support for micro and SMEs)
	18. Foster an environment that supports R&D	Support innovation centres for the study of topics related to sustainability and circularity and serve as a knowledge transfer interface between entities' research and industry
	19. Strengthen partner networks	Support for mobilizing partners along the value chain and logistics for sustainability and circularity
	20. Regulate and support training, capacity building and accreditation	(Re)qualify the human resources of public and private entities (focus on micro and SMEs)
		(Re)qualify human resources in public administration as a regulator and construction owner
		(Re)qualify human resources in public and private vocational education and training entities
	21. Regulate and support new professions	Regulate and support new professions in a Circular Economy
		Regulate and support new courses (technical, secondary, polytechnic and university students)
		Regulate and support professional training
	22. Minimize resources, costs, impacts and waste	Minimize the use of resources, impacts and waste generation encouraging durability, and repair instead of replacement, in contractual provisions

(continued)

Table 5.4 (continued)

Challenges	Recommended measures	Recommended actions
		Minimize lifecycle cost due to performance-optimized and adequate maintenance and opting for supply chains local supplies
	23 Maximize durability, adaptability, flexibility and reuse	Maximize durability, adaptability, flexibility and products and buildings
		Maximize the reuse of materials/products (the "buildings as materials bank"
Cultural barriers: – Lack of market acceptance of recycled materials – Non-existent or unknown practical examples – Lack of awareness/training of the construction owner – Limited knowledge of society – Lack of understanding of benefits and difference	24. Promote a business culture of innovation and sustainability	Promote innovation and sustainability through the development of internal innovation, sustainability and circularity strategies
		Encourage systems thinking by challenging "business-as-usual"
		Promote business culture contrary to a fragmented approach
	25. Encourage HR and market transition	Proposal for incentives for the transition of senior staff and managers("lead-by-example" concept)
		Proposal for incentives for employee transition
		Proposal for incentives for market/demand transition ("demand-pull" concept
	26. Involve the logistics chain	Involvement of partners throughout the entire chain
		Strengthen inter-company and institutional relationships
		Develop demonstrative success stories (in companies and the state)
	27. Promote knowledge sharing	Promote knowledge sharing between the various actors within the construction sector

(continued)

Table 5.4 (continued)

Challenges	Recommended measures	Recommended actions
		Centralization and management of knowledge of entities reference in the AEC sector on CE and sustainability
		Support transparency and traceability of information (between entities and throughout the life cycle)
	28. Proposal to create the entity for Circular Construction	Create an entity responsible for transition, management, monitoring and dissemination of knowledge for CE in the AEC sector
		Define support instruments for monitoring the implementation, validation and monitoring of measures
		Defining and promoting simplex practices
	29. Communicate the plan	Communication plan on the principles of economics circular in the construction sector
		Communication plan regarding environmental benefits, social and economic
		Communication plan for a positive social image
	30. Strengthen trust	Reinforce confidence in political decisions on circularity, sustainability, and quality of the built environment
		Reinforce social awareness of causes and effects associated with climate change
		Reinforce the construction sector's contribution to environmental objectives

Source Acordo Circular [4]

the economic and environmental benefits (2E) associated with the deconstruction and reuse process.

- Circular EcoBIM (3Drivers), aims to create a BIM standard-like Product Data template, i.e., a data structure for BIM objects, and a set of applications that allow the calculation of circularity passports for buildings and components, integration and development of DAP of products and the calculation of LEVEL(S) system indicators for buildings, based on building models, thus contributing to the implementation of circularity practices in construction.
- (Des)construir para an Economia Circular (Comunidade Intermunicipal do Baixo Alentejo) made it possible to provide the Baixo Alentejo region with a strategy and an institutional model to implement a system for collecting, storing, sorting, processing and marketing RCD, to maximize the reuse of materials and recycling.
- ReBuild17 (Secret. Regional dos Transportes e Obras Públicas-LREC) is a project that aims to contribute decisively to the development of circularity in value chains, through the creation of a platform for the recovery of waste originating in the construction sector, in which the various stakeholders associated with the sector participate.
- Circular2B (FEUP), aims to combine energy efficiency with waste recovery, focusing on the development of two new materials: the insulating core of SIP panels (structural insulated panels) and the exterior finishing plaster, replacing current materials with functionally equivalent materials, the result of the recovery of plastics, RCD and slag. The focus on prefabrication (modular construction) is supported by the fact that this type of construction considerably reduces RCD.
- CirMAt (Domingos da Silva Teixeira, S.A), CIRcular aggregates for sustainable road and building MATerials, aims at industrial development and the promotion of multiple products/materials with a high degree of incorporation of waste from the construction and steel industry sectors
- Edifícios Circulares (Associação Smart Waste Portugal) aims to develop tools to support the design and construction of circular buildings, namely a set of standards for defining material passports for buildings, for defining and calculating material, water and energy efficiency metrics and, finally, to define circularity indicators in the DAP.
- UAveiroGreen Buildings (UA), a project that aims to provide UA with a methodology that allows for increased application of Circular Economy principles
- Circular Build (LNEC), development and validation of the concept of circularity applied to modular prefabricated construction
- CLOSER (LNEC), Close to Resources Recovery seeks to develop a Portuguese guide for preliminary demolition and/or renovation audits applicable to the Portuguese reality.
- GrowingCircle (Instituto de Construção), Integrated Data for Efficient and Sustainable Construction, a project that uses the digitization trends in the construction sector point to processes and technologies that allow systematizing, aggregating, managing, tracking and maintaining information about built objects and the construction products that make them up.

- Projeto 3R-2CE (CLEAN), three roads to a circular economy, aims at impacting three key areas that contribute to moving towards a circular economy in Portugal—Reduce, Reuse, and Recycle—by strengthening material efficiency, exploring options for the improved management of specific waste streams, and assessing urban waste infrastructures capacities.

Private Policy Initiatives

Some private initiatives that are examples of success are:

- Costa Almeida Ambiente headquarters, the rehabilitation of an unfinished industrial unit with two floors, with a reinforced concrete structure and brick partitions that make up the administrative area, that maximized components reuse and materials recycling, using CE principles;
- CREE buildings, implemented by Casais, are capable of reducing the carbon footprint by more than 60%, by using engineered wood and just 1/3 of the concrete of a traditional building, and considering circular economy principles (design for disassembly)
- Examples of successful implementation available in:
  - http://agendacircular.ccdrc.pt/boas-praticas
  - https://eco.nomia.pt/pt/exemplos
  - https://smartwasteportugal.com/boas-praticas/

5.1.16 Romania Factsheet

5.1.16.1 Policy and Regulatory Framework

Governance

Under the current governance structure, the circular economy lacks a dedicated process and/or a specific ministry or unit responsible for it. Instead, the circular economy is managed alongside other related topics such as sustainable development or climate change, while the responsibilities for the circular economy in Romania are currently shared between several authorities. The Ministry of Environment, Water and Forests (MoEWF), within its domain of expertise, plays a pivotal role in formulating national policies, regulations, strategies, and plans related to waste prevention and management. In contrast, the Ministry of Economy (MoE) spearheads the development of mid-term and long-term strategies and reforms to facilitate Romania's economic transition from a linear model to a circular model. Several other government ministries are responsible for issues related to the circular economy, which are also intertwined with larger environmental mandates, ministries such as the Ministry of Development, Public Works and Administration, the Ministry of Energy and the

Ministry of Research, Innovation and Digitalization. Depending on the theme, other ministries have also responsibilities and mandates on environmental issues.

In addition to those mentioned above, there are other relevant governmental bodies that support circular economy:

- under the aegis of the Prime Minister's Office, the Department for Sustainable Development (DSD) is responsible for the implementation and monitoring of the Sustainable Development Goals (SDGs) since 2017. Additionally, DSD has played a crucial role in the development of sustainable development strategies. Joining forces with the MoEWF, the DSD collaborated with the "Ernest Lupan" Institute for Circular Economy and Environmental Research—IRCEM to create ROCES2030 (a project named "Strategy for Romania's transition to the circular economy 2020–2030", part of the Partnership for a Green and Zero Waste Economy in Romania until 2030, in the framework of which a series of events took place, including sectoral conferences and workshops, in the main cities of Romania).
- the Interdepartmental Committee for Sustainable Development (ICSD) plays a central role in overseeing the implementation, monitoring, evaluation, and revision of Romania's National Strategy for Sustainable Development 2030 (RNSSD 2030). Headed by the Prime Minister and comprising all ministers, the ICSD ensures that the RNSSD 2030 objectives are integrated into sectoral policies and strategies, and it upholds Romania's international and European reporting requirements in the field of sustainable development.
- the National Commission on Climate Change, an inter-ministerial body without legal personality, serves as an advisory council to the Romanian government on climate change matters. It is coordinated by the MoEWF and falls under the purview of the Interministerial Council for Agriculture, Rural Development and Environment.
- The Interministerial Committee on Climate Change, established as a relatively recent addition to the Romanian government's structure, is chaired by the Prime Minister and serves as a key coordinating body for climate change initiatives. It plays a crucial role in aligning policies across various sectors that influence climate change and monitoring the implementation of these policies. The committee's main responsibility lies in amending and complementing national climate change policies to ensure their effectiveness and align with international commitments.

The Consultative Council on Sustainable Development (CCSD) was established by Government Decision 114/February 4, 2020 and it was settled at the level of academia, the scientific community and civil society to monitor the impact of policies in the field of sustainable development. The Council consists of 34 members whose role is to propose the initiation and elaboration of documents and methodologies for the implementation of the RNSSD 2030, to provide advice, scientific and technical support to the DSD and to support and encourage the implementation of good practices relevant to Romania within the mechanisms of public administrations involved in the sustainable development process [64].

Legislation

Although Romania has a large body of legislation related to the circular economy, most of them focus on waste management. In 2017, Romania took a comprehensive approach to waste management by enacting the National Waste Prevention and Management Plan (NWPMP), alongside a suite of supporting legislation including 16 laws, 23 government emergency ordinances (GEOs), and over 40 government decisions and ministerial orders. Despite a comprehensive legislative framework, effectiveness remains limited due to insufficient enforcement and monitoring. This stems from factors like resource constraints, fragmented responsibilities across diverse government bodies, and lack of long-term strategic vision [341].

Two key laws constitute the foundation of Romania's legal framework for promoting a circular economy: EO no. 92/2021 regarding the waste regime (replacing General Regime of Waste Management Law no 211/2011) and Law no. 249/2015 regarding the framework of managing packaging and packaging waste. June 2021 saw a major update to Romania's waste laws by adopting Emergency Ordinance no. 74/2018. This ordinance amended and supplemented key laws, including Law no. 211/2011 on the waste regime, Law no. 249/2015 on packaging and packaging waste management, and Government Emergency Ordinance no. 196/2005 on the Environmental Fund. EO 74 advances the goals of the Circular Economy package by integrating within national legislation various economic instruments, including pay-as-you-throw (PAYT), Extended Producer Responsibility (EPR), and the introduction of a landfill tax. In September 2021, EO no. 92/2021 on the waste regime superseded Law no. 211/2011 and amended EO 74/2018. Despite these changes, the underlying objectives of promoting a circular economy remain unchanged.

Other policies that include circular economy policy elements are Industry *De minimis* Aid Scheme (adopted through Order no. 27/August 24, 2022) for the transition to the circular economy, Waste landfilling policy (adopted through Order no. 2/August 11, 2021), Control of waste transfers from other countries policy (adopted through Order no. 1647/June 10, 2022), the policy that provides Making waste management more efficient (adopted through Emergency Ordinance no. 38/April 6, 2022) and the proposed Code of Territorial Planning, Urbanism and Construction.

Circular Economy Roadmap, Strategy and Action Plan

Romania's roadmap towards circular economy started recently compared to other European countries:

- on September 25, 2015, Romania, alongside 192 other countries, embraced the 2030 Agenda for Sustainable Development;
- on May 25, 2017, the Romanian government has decided to establish the Department of Sustainable Development (DSD), under the authority of the Prime Minister, as the national coordinator for the implementation of the 17 goals of the

2030 Agenda. In its capacity, DSD plays a pivotal role in integrating and monitoring data related to sustainable development and circular economy, while also collaborating with public institutions to achieve these goals;

- Romania, through the Ministry for Environment and Climate Change (MfECC), submitted its first Voluntary National Review (VNR) at the High-Level Political Forum for Sustainable Development in New York in July, 2018, marking a significant step in Romania's journey towards achieving the 2030 Agenda;
- on November 9, 2018, the Romanian government has adopted the National Strategy for the Sustainable Development of Romania 2030 (RNSSD 2030), developed under the guidance of the DSD, providing at the national level, among others, elements of transition to a circular economy;
- the Romanian government established on May 8, 2019, the Interdepartmental Committee for Sustainable Development (ICSD), a high-level body led by the Prime Minister to oversee and facilitate inter-ministerial collaboration;
- one year after establishing the ICSD, the Romanian government established on May 2, 2020, the Consultative Council for Sustainable Development, a body comprising 34 experts with expertise in all 17 Sustainable Development Goals (SDGs);
- empowering diverse perspectives and expertise, the DSD, on April 13, 2021, initiated the Coalition Sustainable Romania, a collective effort involving civil society organizations and private sector representatives. In an effort to foster cross-sector collaboration and expertise, the DSD, on May 25, 2021, created a network of 22 sustainable development hubs within all ministries, encompassing approximately 90 specialists.
- the Romanian government has formally adopted on June 8, 2022, the National Action Plan (NAP- RNSSD), developed under the direct guidance of the SDS, for implementing RNSSD 2030;
- on September 21, 2022, the Romanian government adopted the National Strategy for the Circular Economy (NSCE) through Government Decision no. 1.172. The Strategy was developed by the DSD, in conjunction with the MoEWF and the MoE, under the aegis of the Technical Support Instrument project.
- on October 5, 2023, the Romanian government adopted the National Action Plan for the Circular Economy Strategy (NAP-CES) through Government Decision no. 927.

In a step towards sustainable resource management, Romania's government adopted the National Strategy for the Circular Economy (NSCE) on September 21, 2022, through Government Decision no. 1.172. The overarching aim of the Strategy aligns seamlessly with the United Nations Sustainable Development Goals (SDGs) and global climate objectives embracing the principles and initiatives fostered under the EU Green Deal; its core objective is to lay a robust framework for implementing the Action Plan, fostering a clear direction for all stakeholders involved. The Strategy dives deep into 14 economic sectors, analysing their circularity potential and it identifies seven sectors with the highest potential for circular transformation: agriculture and forestry, construction, automotive, consumer goods such as food and beverages, textiles, packaging and electrical and electronic equipment (EEE).

Through Government Decision no. 927/October 5, 2023, Romania's government adopted the National Action Plan for the Circular Economy Startegy (NAP-CES). This Action Plan targets five key areas to jumpstart Romania's circular economy journey: (1) reducing the consumption of virgin raw materials through sustainable extraction, recycling and recovery of raw materials; (2) reducing the consumption of consumer goods by extending the life of products; (3) reducing the impact of production activities on ecosystems and the environment; (4) reducing the impact of waste and wastewater management activities on ecosystems and the environment; (5) improving policy coherence and governance, communication and collaboration between local, regional, and national authorities.

At the national level, the transition to a circular economy is also addressed within Romania's National Strategy for Sustainable Development 2030 (RNSSD 2030) adopted through Government Decision no. 877/November 9, 2018, the National Action Plan for the Implementation of Sustainable Development Strategy (NAP-RNSSD) adopted through Government Decision no. 754/June 8, 2022 and the National Recovery and Resilience Plan (NRRP). This policies include circular economy policy elements such as Resource efficiency (within NRRP), Reuse of materials and reduction of waste generation (within NRRP), Improving waste management governance for the acceleration of the transition to the circular economy (within NRRP), Increasing the energy efficiency of historic buildings (within NRRP).

The National Strategy for the Circular Economy (NSCE) establishes the following overall objectives:

- Prioritizing local production: encourage the use of locally sourced products and materials over imported alternatives, supporting domestic economic activity.
- Enhancing economic competitiveness and workforce development: strengthen the competitiveness of Romanian businesses within the circular economy through investments in workforce training and upskilling initiatives.
- Responsible and sustainable raw material supply: promote the responsible and sustainable sourcing of raw materials, prioritizing recycled and recovered materials whenever possible.
- Fostering innovation and research: elevate innovation and research in the field of circular economy as a national priority, driving advancements in sustainable technologies and practices.
- Preservation, conservation and sustainable use of natural resources: ensure the preservation, conservation, and sustainable use of natural resources for present and future generations.
- Minimizing waste generation and promoting sustainable waste management: implement strategies to prevent waste generation at the source and promote sustainable waste management practices.

- Encouraging responsible consumption and environmental education: foster responsible consumption patterns among citizens and implement comprehensive environmental education programs to raise awareness and understanding of circular economy principles.
- Protecting ecosystems and public health: ensure the protection of ecosystems and safeguard public health by mitigating the environmental impact of economic activities within the circular economy framework.

In the construction sector, National Action Plan for the Circular Economy Strategy (NAP-CES), which accompanies NSCE, highlights several key areas requiring efforts to leverage the circular model and achieve improvements. These areas include:

- Stimulating demand for circular products and solutions: promoting the market uptake of circular products and solutions within the construction sector to incentivize their adoption and integration into building practices.
- Facilitating enhanced material recovery: implementing effective strategies to increase the recovery of construction materials for reuse, minimizing waste generation and maximizing resource utilization.
- Preventing illegal waste disposal: strengthening enforcement measures and fostering responsible waste management practices to eradicate the illegal dumping of construction materials, ensuring environmental protection and resource conservation.
- Developing digital infrastructure: establishing a robust digital infrastructure to support the introduction of innovative intervention practices and non-invasive methodologies. This may encompass the development of standards, technical recommendations and guidelines addressing the use of sustainable materials and solutions to improve building energy efficiency and promote circularity within the construction industry.
- Building capacity and knowledge: implementing capacity-building and knowledge-sharing initiatives to create an enabling environment for the effective dissemination and adoption of circular economy principles throughout the construction sector.

Construction and Demolition Waste Management Regulatory Framework

The 2018 proposal for a government decision on the management of construction and demolition waste (CDW) from and/or dismantling was materialised three years later in Government Emergency Ordinance No. 92/August 19, 2021 on the Waste Regime, which sets out new obligations for holders of construction and demolition permits, new recycling quotas and special waste management plans. The Ordinance on Waste Regime provides in Art. 17(4) that "the holder of the construction/dismantling permit issued by the local or central public administration authority or by the institutions authorized to authorize construction works of a special nature is obliged to have a waste management plan for construction and/or demolition activities, where applicable, establishing sorting systems for waste from construction and demolition activities, at least for wood, mineral materials—concrete, brick, sandstone and ceramics, stone, metal, glass, plastic and gypsum for on-site recycling/reuse, as far as economically feasible, environmentally sound and safe construction, and to take measures to promote selective demolition to enable the safe disposal and handling of hazardous substances to facilitate high quality reuse and recycling through the disposal of non-valuable materials". The same regulatory framework provides in Art. 17(7) that "holders to whom construction and/or demolition permits have been issued in accordance with the provisions of Law No 50/1991 on the authorisation of construction works are required to manage construction and demolition waste in such a way as to achieve a level of preparedness for reuse, recycling and other material recovery operations, including backfilling operations using waste to replace other materials, of at least 70% of the mass of non-hazardous waste from construction and demolition activities".

Component 5—The Renovation Wave of the National Recovery and Resilience Plan (NRRP) seeks to integrate key elements of the European Green Deal and other relevant European and national strategies within the Code of Territorial Planning, Urbanism, and Construction (CTPUC). These key elements will encompass strategies such as climate change, resource efficiency, energy efficiency, circular economy and nature-based solutions. Regarding the management of CDW, the CTPUC establishes specific regulations similar to the regulations provided by the Ordinance on Waste Regime:

- "The management of construction and demolition waste shall be carried out under conditions of protection of the health of the population and the environment, in accordance with the applicable special law"—Art. 373 (1);
- "In order to prevent the accumulation and management of waste generated on the site during construction/demolition works and during the operation of constructions, in accordance with the specific legislation, the following measures shall be implemented:

    1. producers and holders of waste legal entities are obliged to classify each type of waste generated from their own activity in the list of waste, according to the special law;
    2. The development and implementation of a programme for the reduction of waste generated;
    3. the development and implementation of a waste management plan;
    4. producers and holders of waste are obliged to collect separately at least the following categories of waste: paper, metal, plastic and glass;
    5. waste producers and waste holders are obliged to recover the waste in accordance with the legal provisions;
    6. re-use, recycling and other material recovery operations, including backfilling operations using waste to replace other materials;
    7. the obligation to subject waste that has not been recovered to a safe disposal operation.
    8. the costs of waste management operations shall be borne by the waste producer"—Art. 373 (2);

- "Responsibility for waste management shall rest with the developer, producer and contractor"—Art. 373 (3).

The implementation of the waste management within the National Recovery and Resilience Plan (NRRP), settles targets regarding setting up Integrated Separate Collection Centres for urban agglomerations and Voluntary Collection Centres to ensure separate collection of household waste that cannot be collected in a "door-to-door" system, including CDW. According to Component C3—Waste management of the National Recovery and Resilience Plan (NRRP), seven integrated separate collection centres and 250 voluntary collection centres will be established in 2024, while 15 integrated separate collection centres and 565 voluntary collection centres will be established in 2025.

5.1.16.2 Platforms and Networks Facilitating the Transition to a Circular Economy

Circular Economy Platforms

Romanian Circular Economy Stakeholder Platform—ROCESP is a platform founded by "Ernest Lupan" Circular Economy and Environment Research Institute—IRCEM and it is financed by the European Social Fund. As a network of networks, the platform aspires to serve as a national hub for convergence on circular economy initiatives, experiences, challenges, perspectives, and expectations. The objectives of ROCESP are to drive knowledge dissemination across Romanian actors, creating a shared understanding of circular economy practices, cultivate collaborative networks to unlock the power of joint action within the circular economy landscape, advocate for the cross-sectoral integration of circular economy initiatives across Romania, develop a permanent and operational tool to facilitate dialogue and interaction between diverse stakeholders and to disseminate Romania's unique approach to the circular economy, drawing on its traditions, cultural context, and entrepreneurial spirit. The ROCESP Platform fosters collaboration between a wide range of stakeholders including representatives from local and central public administration, academic institutions, research and innovation centers, businesses, professional associations, and civil society organizations.

Led by industry, the Circular Economy Coalition (CEC) offers Romania a platform for policy monitoring, research development, and program implementation. Its members actively promote circularity at national and international levels through various activities, including advocating for effective legislation, contributing to technical standards, and participating in international networks. CEC fosters knowledge sharing, expertise exchange, and regulatory support, enabling a sustainable and resource-efficient economy.

- Ecoteca is a platform providing specialised information on waste management, circular economy and climate impact by informing, educating and supporting individuals and organisations on the prevention, reduction, separate collection and proper management of waste and ensuring the transition to the circular economy and the reduction of climate impact.

- Harta Reciclării (Recycling Map) is an interactive platform that serves as the first national resource dedicated to identifying and pinpointing the locations of various recycling collection points throughout Romania. It unites individuals in a shared mission to promote waste reduction, reuse, and recycling across all categories, from paper and plastic to appliances, tires, and even hazardous materials.
- Romanian Sustainability Code is a platform launched by DSD to support companies in reporting and achieving the goals outlined in Romania's Sustainable Development Strategy. Developed in 2022, the platform and its accompanying Guide provide clear standards and guidance for companies to produce sustainability reports and offers a central hub for publishing company reports, best practices, and open data resources empowering investors and stakeholders to make informed decisions based on sustainability performance.

Circular Buildings Platform

CircularBIM is a free, open-source educational platform dedicated to empowering professionals and students with advanced strategies for reusing building materials within the construction industry. By providing access to BIM software and learning resources, it empowers learners to minimize construction waste (by discovering techniques and technologies for reducing waste generation at the source), reintegrate discarded materials (by learning how to reintroduce used materials back into the construction value chain) unlock material potential (by gaining insights into the potential revaluation of each building component), embrace circular construction (by raising awareness about circular economy principles within the construction sector) and master BIM for reusability (by leveraging BIM formats and tools to facilitate material reuse throughout the building lifecycle).

Building Materials Passport Platform

No building materials platform has been established to date.

Public Procurement Platform

No green public procurement platform has been established to date. The Romanian Electronic Public Procurement System does not yet include any functionality related to the promotion of green procurement.

5.1.16.3 Funding Opportunities

De minimis Aid Scheme for the Transition to the Circular Economy

In a move to accelerate its transition to a circular economy, Romania has launched a 'de minimis' aid scheme through Ordinance no. 27/August 24, 2022. This aid scheme offers financial support to businesses and organizations working to: develop new, innovative industries (this includes fostering the creation of new sectors focused on circularity, such as repair and refurbishment, recycling, and resource efficiency), revitalize existing resources (the scheme encourages the revaluation and reuse of existing materials and products, minimizing waste generation and pollution), drive research and development (grants will be available for projects focused on ecological design, innovative technologies, and services that promote circular principles). The scheme offers grants ranging from the equivalent of EUR 15.000 to EUR 200.000, with 85% funded by the state budget and 15% by the beneficiary. Key areas of focus include

- tech transfer for circular economy: supporting companies that bring new circular technologies to market;
- research and innovation activities: funding projects that develop eco-friendly products and services within the circular economy principles;
- repair and refurbishment centers: establishing and equipping facilities for extending the lifespan of goods;
- Waste recycling activities: encouraging the development of efficient and sustainable waste management practices [64].

National Recovery and Resilience Plan (NRRP)

Romania's recovery and resilience plan provides investment in green technologies and capacities, including energy efficiency, building renovation, circular economy, and biodiversity, while contributing to the Union's climate objectives, promoting sustainable growth, creating jobs, and maintaining energy security. Under Pillar I—Green Transition, within PNRR funding is granted (among others) for activities related to waste management, development of centres for the recovery of historic building materials and their reuse, moderate/deep energy renovation of approx. 4.3 million sqm of multi-family residential buildings and 2.3 million sqm of public buildings respectively, installation and connection to the grid of at least an additional 3000 MW of renewable energy capacity (wind and solar).

Energy-Efficient House

Starting in 2020, the Energy Efficient House Programme offers financial support to individuals to increase the energy efficiency of their buildings. The objective of the programme is to increase the energy performance and/or the use of energy from renewable sources in existing single-family households located in a building with a maximum of three floors. In 2024, the programme was funded from the NRRP, and through it 60,000 houses in Romania will benefit from a voucher of LEI 72,500 including VAT—the equivalent of EUR 14,500 (for houses with a useful surface area between 50 and 100 m^2) or LEI 100,000 including VAT—the equivalent of EUR 20,000 (for houses with a useful surface area of more than 100 m^2). Beneficiaries of the programme will be able to choose from several types of energy rehabilitation works, including thermal insulation of facades, roofs and floors, installation of an energy efficient heating and domestic hot water system, replacement of windows and doors with double glazed windows and doors, or installation of photovoltaic panels.

Programme for Increased Energy Efficiency and Smart Energy Management in Public Buildings

The aim of the programme is to increase the energy efficiency of public buildings and improve the quality of the environment by reducing greenhouse gas emissions, reducing annual primary energy consumption and promoting the use of renewable energy sources. Within a funding session, the maximum amount that can be granted to an applicant, through one or more funding applications, starts from 6,000,000 lei (the equivalent of EUR 1,200,000) for municipalities with a population up to and including 5000 inhabitants and goes up to 24,000,000 lei (the equivalent of EUR 4,800,000) for administrative-territorial subdivisions of Bucharest municipality. In 2024, the programme was funded from the NRRP.

Funding for the Circular Economy Through the Bucuresti Environment Platform

The Environmental Platform for Bucharest, initiated by the Bucharest Community Foundation and ING Romania, regularly launches funding rounds for waste management, sustainable development and circular economy projects, with the support of Nhood Romania. The allocated funds aim to support projects that address infrastructure, education and advocacy challenges and opportunities in the field of consumption, biodiversity, air quality or waste. Funding is awarded to projects that, in addition to the proposed solutions, have a solid plan for scalability and involvement of public authorities in the adoption of the proposed measures, so that the impact goes beyond a one-off intervention and provides long-term benefits.

Photovoltaic Green House

Starting in 2019, the Photovoltaic Green House Programme offers financial support to individuals for the purchase and installation of photovoltaic panel systems using renewable, non-polluting energy sources for the production and use of electricity by consumers connected to the national electricity grid. The aim of the programme is to increase energy efficiency, improve air quality and reduce greenhouse gas emissions through the use of photovoltaic panel systems with a minimum output of 3 kW for the production of electricity for own consumption and the delivery of the surplus to the national energy system. Funding is granted up to a maximum of 20,000 lei (approximately 4000 euro), subject to the beneficiary's own contribution of 2000 lei (approximately 400 euro).

Horizon Europe

The EU's Horizon Europe programme funds large scale international research projects including projects related to the circular economy.

International Cooperation and Support from International Organizations

The International Finance Corporation (IFC), a member of the World Bank Group, significantly increased its support through strategic investments in key areas in Romania. These included landmark investments aimed at strengthening the capital markets and promoting climate-friendly financing. In 2023, a €100 million investment in Banca Comercială Română S.A.'s MREL green bonds issuance allocated proceeds towards renewable energy, green building projects, and green mortgages. In addition to the aforementioned investment, IFC committed another €40 million loan to Garanti BBVA Romania specifically to expand access to green housing financing through green mortgages and home improvement loans. Furthermore, a $48 million investment in UniCredit Bank S.A.'s senior bonds issuances aimed to reinforce the green housing loan market in Romania.

5.1.16.4 Challenges, Barriers and Potential Improvements

Although Romania has established a comprehensive framework of green transition strategies, national plans, and policies over the past decade, its effectiveness hinges on robust implementation and enforcement. The transition to circular economy in Romania is slowed by multiple factors, as presented in the World Bank Diagnostic Analysis for Circular Economy Interventions in Romania:

- The lack of standards in the field of circular economy development: firstly, the absence of a national dashboard hinders progress by obscuring key data and trends; secondly, the private sector lacks fiscal incentives to embrace circular practices, limiting its participation. Finally, Extended Producer Responsibility (EPR) schemes are inadequately implemented for certain materials like CDW and textile, creating loopholes in waste management. Addressing these issues with high priority is crucial, especially for supporting SMEs in implementing circular standards through close collaboration and effective monitoring.

- Low level of enforcement and proper monitoring of Romanian policies and legislation: while Romania has developed numerous environmental strategies and legislation in recent years, concerns remain regarding their effectiveness. Enforcement and monitoring often fall short, hindering progress towards sustainability goals. Additionally, resource allocation may not always match ambitions, and action plans may lack long-term vision and coordinated efforts across government institutions.
- The proper implementation of legislation is hampered by the weak institutional capacity of central and local government: for example, despite being mandatory for over a decade, separate waste collection remains elusive, existing only in pilot areas. This exemplifies a broader challenge: Romania's public administration lacks dedicated human resources for circular economy, sustainable development, and climate change. Instead of proactive leadership, the government often adopts a reactive stance towards waste management. Legislation struggles to reach the local level, hindering effective implementation. Furthermore, waste prevention languishes as a non-priority, evidenced by the unimplemented laws on anti-food waste, composting, and Green Public Procurement (GPP). The absence of supporting regulations for their application and enforcement further exacerbates the issue.
- Low absorption capacity of EU funds: While Romania boasts a 70% absorption rate for EU structural funds in the 2014–2020 period, significant challenges remain. Delays in implementation and a lack of qualified staff to manage projects are major hurdles. This is particularly true for local public institutions, which often struggle with weak budgets, complex administrative processes, and unclear regulations. Adding to the burden are opaque co-financing requirements, limited ability to secure matching funds, and few opportunities for project collaboration and regional coordination. These issues are further magnified by vast regional disparities: counties like Bihor, Bucharest, Constanta, and Cluj demonstrate high absorption rates, while others like Giurgiu, Ialomita, and Teleorman lag behind, widening economic gaps within the country.
- Lack of technical capacity and low level of institutional efficiency: the 2021–2027 EU funding for Romania risks suffering from the same administrative and technical limitations as the previous period. Local authorities' capacity to design and implement new projects remains restricted, potentially hindering the efficient utilization of available resources.
- The development of circular businesses is limited by a lack of government vision in recent years, current legislation and fiscal policy on circular economy: Industrial symbiosis and sustainable partnerships with private actors receive little attention from most Romanian municipalities. Non-mandatory GPP further hinders progress towards more eco-friendly local procurement.
- Poor or missing data: Concerns about the accuracy of waste management data in Romania have been voiced by the EU, urging improvements in both quality and

transparency. This data deficiency presents a significant obstacle to measuring progress and developing effective strategies.
- The demand for products that have a low impact on the environment is still low, especially as these products tend to be more expensive than their conventional counterparts: current Romanian consumer behavior doesn't prioritize circular options. Initial steps should focus on shifting attitudes away from a "throw-away" mentality through strategic interventions such as national and local policies, targeted communication campaigns, comprehensive media coverage, and consistent, clear messaging from political leaders [364].

To accelerate Romania's transition towards a circular economy, a multifaceted approach is needed. This includes enacting comprehensive circular economy legislation, fostering technical and institutional capacity development, and driving behavioral change. World Bank outlined specific actions within each of these key areas in the Diagnostic Analysis for Circular Economy Interventions in Romania:

1. Consolidate and simplify the regulatory landscape:
    - Harmonize and streamline existing waste legislation to eliminate inconsistencies and ensure clarity for all stakeholders.
    - Orient legislation towards long-term circular economy goals, aligning with EU directives and focusing on prevention, reuse, and resource recovery.
    - Enforce existing legislation vigorously and consistently to deter non-compliance and promote responsible waste management.
2. Prioritize waste prevention and circular practices:
    - Establish a GPP Action Plan with mandatory targets for public institutions, focusing on circular procurement and resource efficiency.
    - Develop a comprehensive food waste prevention strategy as part of the broader circular economy package.
3. Empower Local Action and Collaboration:
    - Allocate resources to local authorities based on their commitment to circular practices and industrial symbiosis initiatives.
    - Establish financial incentives for circular businesses to promote innovation and investment in sustainable solutions.
    - Create a national compost policy that empowers local authorities while ensuring strong central oversight and monitoring.
4. Empowering Institutions for Effective Implementation:
    - Conduct a comprehensive assessment of the MoE and its regional/local branches, evaluating resource needs, training gaps, and implementation challenges related to circular economy.
    - Address lags in EU law transposition, insufficient legislation, and weak enforcement through targeted capacity building and reorganization.
5. Upskilling the Workforce:

- Deliver targeted training programs on circular economy, sustainable development, and climate change for civil servants at all levels.
- Offer specialized training for key personnel in Parliament, the Prime Minister's office, and ministries to guide national leadership.
- Foster stronger collaboration between relevant ministries and agencies to ensure coordinated action.

6. Strengthening Data Management:
   - Implement an open-source, reliable centralized waste reporting system with robust verification mechanisms involving authorities, research bodies, and producers.
   - Improve data availability and transparency to inform decision-making and track progress towards circular goals.
7. Fostering Innovation and Knowledge Sharing:
   - Establish partnerships with universities to promote research on circular solutions, product design, and closed-loop practices.
   - Develop advanced circular economy studies to create a pool of national experts and foster innovation clusters.
   - Create a national database highlighting best practices and successful circular economy initiatives to inspire broader adoption.
8. Setting Standards and Incentives:
   - Establish national circular economy standards for both public institutions and businesses, addressing resource use reduction and promoting circular practices.
   - Link these standards to financial incentives to motivate investment and adoption by public and private actors.
9. Empowering Consumers:
   - Promote informed choices: educate consumers about sustainable products, cost savings, and their role in a circular economy.
   - Facilitate cost-effective sustainability: encourage eco-friendly choices through incentives, subsidies, and accessible green alternatives.
10. Educating and Engaging:
    - Raise awareness: implement broad-based education campaigns to foster environmental consciousness and responsible consumption.
    - Empower communities: support civil society initiatives and encourage citizen participation in sustainability efforts.
    - Plan for the future: integrate long-term obsolescence management strategies into product design and waste management plans.
11. Boosting SME Resource Efficiency:

- Invest in knowledge: expand training and education programs for SMEs on resource efficiency, leveraging EU best practices.
- Unlock export potential: enhance resource efficiency to improve product competitiveness and access new markets.

12. Fostering Innovation and Collaboration:

    - Strengthen research infrastructure: develop regional research hubs within universities and research centers focused on circular economy solutions.
    - Engage with the EU: Actively participate in the EU Stakeholders Platform for circular economy to share and learn from best practices.

13. Designing for Sustainability:

    - Invest in future-proof products: upskill designers and manufacturers to create products for reuse and extended lifespans.
    - Minimize waste: implement obsolescence management strategies to reduce product waste and promote repair and refurbishment.

14. Sustainable Business Models:

    - Incentivize sustainable practices: promote and reward businesses that adopt circular economy models.
    - Drive innovation: encourage the development and adoption of innovative, sustainable business models.

15. Enforcing Extended Producer Responsibility (EPR) Schemes:

    - Strengthen enforcement: ensure effective implementation and enforcement of EPR schemes [364].

Following specific strategies and actions should be implemented in the Romanian construction sector in order to encourage the transition towards a circular economy in Romania:

- Procurement: accelerate market uptake of circular construction solutions through innovative procurement approaches that prioritize resource efficiency and closed-loop material management.
- Competence: enhance the human capital of the construction sector by fostering the development of knowledge, skills, and competencies relevant to circular economy practices.
- Recovery: develop a comprehensive legal framework and supporting infrastructure to enable efficient recovery of construction materials.
- Accountability: establish a comprehensive CDW register to track and manage waste materials, coupled with a digital marketplace to facilitate efficient trading and reuse.
- Reinforcement: strengthen enforcement mechanisms and implement stricter penalties to deter illegal dumping of construction waste, promoting compliance with environmental regulations (IFC 2023).

5.1.16.5 Examples of Successful Implementation

Romania's construction sector witnesses a growing embrace of sustainability and circular economy principles, evidenced by the rising number of BREEAM and LEED certified projects. More than 730 buildings in Romania have obtained BREEAM or LEED certifications by the end of 2023, including buildings from educational sector, healthcare, industrial, offices, residential and retail. From these, 32 buildings have an "Outstanding" rating score in the BREEAM certification and 65 buildings have a "Platinum" award in LEED certification.

Public Policy Initiatives

EFdeN Signature House. The EFdeN Signature House, a prototype solar house designed by Romanian students, proudly represented their country at the prestigious Solar Decathlon Middle East 2018 competition in Dubai. This remarkable journey involved constructing the house in Romania, meticulously dismantling it for transport, then reassembling and testing it in the scorching Dubai climate.

The house's success was undeniable, achieving an impressive 4th place finish globally and securing 4 podium positions: 1st in communication, 2nd in engineering and construction, 2nd in comfort conditions, and 3rd in sustainability. This remarkable performance solidified its reputation as a leader in sustainable design and circular economy. Upon returning from Dubai, securing LEED for Homes v4.1 certification became a top priority. The certification process ran alongside the final reconstruction of the house on the Faculty of Building Services Engineering campus in Bucharest, taking two years to complete. The EFdeN Signature House rose to the challenge, scoring a remarkable 105 points, the highest ever achieved by a LEED project in Europe [105].

Buzau City. Buzau city has emerged as a pioneer in Romania's transition towards a circular economy, developing a strategy toward circular economy with seven key directions for circularity for 2030:

1. Materials: Buzau is already leading the charge with zero waste and a 60% circular materials flow, demonstrating its commitment to resource efficiency and responsible consumption. The city is the proud host of Green Group, the largest private recycling conglomerate in Romania. This European leader operates the continent's largest integrated recycling park, specializing in recyclable waste management, collection, and diverse waste recovery. Green Group's mission is to transform waste into valuable resources, closing the loop and conserving resources and energy. Green Group actively collaborates with local authorities. This partnership also yielded a revised County Plan for Waste Management, solidifying their joint efforts towards a greener future.
2. Energy: Transitioning towards a renewable energy future, Buzau prioritizes local production, ensuring energy security and reducing carbon footprint.

3. Biodiversity: Recognizing the value of its natural resources, Buzau prioritizes regenerating ecosystems and protecting its natural capital through strategic actions.
4. Health: the city of Buzau promotes healthy living through flexible and efficient infrastructure, prioritizing sustainable mobility with near-zero emissions.
5. Society and Culture: Building a vibrant and inclusive community, Buzau fosters a healthy, safe, and attractive environment with recreational spaces for all residents.
6. Generation of Value: Buzau cultivates a strong local economy that empowers entrepreneurs and encourages the development of circular businesses, driving sustainable economic growth.
7. Resilience: Buzau emerges as a leader in CE implementation, showcasing its commitment to adaptation and long-term sustainability.
8. The strategy offers valuable insights into the current economic landscape and explores potential pathways towards circular development, though detailed action plans are pending.

Private Policy Initiatives

MEX DAY hall. The building, originally designed in 2008 for flexibility in location and use, was initially erected 60 km from Timisoara, in 2009. The structure remained unclad until 2012, when plans for a "Pasta fabrication" plant emerged. However, these plans were ultimately shelved. In 2017, under new ownership, the structure was planned to be repurposed as a Cereal Storage Unit, utilising the entire hall and incorporating a partial mezzanine for office and laboratory space. The new plan involved relocating the entire structure to a new site, demonstrating its adaptability and potential for reuse. Despite its initial design in 2008, this steel-structure required to be evaluated and strengthened. This was due to the stricter building codes in place in 2017, which mandated higher climatic (snow and wind) and seismic load requirements compared to the standards in place when the building was first designed. The structure underwent a resourceful transformation. While relocating the original structure, the layout was cleverly adapted to incorporate new office spaces. This required the addition of a complementary steel structure, ensuring seamless integration with the existing framework. Notably, both the main and secondary structure were entirely reused, minimizing waste and maximizing resource efficiency [56, 57].

HIDROTIM offices building. Constructed in the 1960s in Timisoara, Romania, this existing steel structure was originally designed to house a hydraulic testing laboratory for hydraulic power steel equipment. Comprising a single-story industrial hall, the building was characterized by its steel structural elements, integrated crane, and thermally insulated roofing and wall cladding. In 2004, the investor, Siemens Ltd, through its subsidiary HIDROTIM Ltd Timisoara, requested the transformation of the existing single-story industrial hall into a five-story office building (ground floor + four additional stories). The plan also included a partial basement for additional space. Due to the restrictions imposed by the 2004 construction permit for the central Timisoara site, which primarily consists of multi-story concrete dwellings, complete demolition of the former hydraulic laboratory building was not authorized. Therefore, the deconstruction/demolition process was divided into several phases, adhering to the permit stipulations. This necessitated the reuse of the existing steel structure, requiring additional structural measures to ensure its suitability for the new five-story office building function. More than 50% from the steel needed for the new building was provided by reusing the existing elements of the industrial hall. The project leveraged existing structural elements, carefully integrating them with newly designed components to achieve the desired functionality. The successfully revitalized building is now fully operational, standing as a testament to the project's innovative approach and commitment to sustainable urban development. The successfully revitalized building is in function at present time, standing as a testament to the project's innovative approach and commitment to sustainable urban development [56, 57].

ALUM. ALUM, a Romanian alumina producer belonging to the Alro Group, joined forces with LafargeHolcim and other partners in the innovative ReActiv project. Funded by the European Commission's Horizon 2020 program with €8.8 million, the project aimed to create low-carbon circular industries by transforming waste into valuable resources [94].

TenarisSilcotub. TenarisSilcotub, a Romanian a pipes manufacturer belonging to the Tenaris Group, achieved a remarkable 100% recovery rate for iron waste generated in its production processes. This commitment to sustainability is further underscored by their reliance on scrap metal, which constitutes 96% of their raw materials at the Calarasi facility. Calarasi facility processes 500,000 tonnes of scrap per year, demonstrating their dedication to closed-loop manufacturing and circular economy principles.

CONCORDIA Employers' Confederation. CONCORDIA is a Romanian organization that brings together 15 industries essential for the Romanian economy, representing almost 2200 companies and a total contribution of 26% to Romania's GDP. Concordia holds a unique position as the sole Romanian representative within influential organizations such as Business Europe, Business at the OECD (BIAC), and the International Organization of Employers (IOE). This distinction grants CONCORDIA a legitimate voice in public discussions at both government and parliamentary levels, as well as the authority to participate in decision-making processes within Romania. Unveiling innovative practices in Romania, [Organization name] published a guide showcasing over 20 member companies' circular economy initiatives across numerous sectors, from hospitality to energy. This insightful resource, "Circular Economy in Romanian Business", identifies common practices for sustainable progress.

5.1.17 Serbia Factsheet

5.1.17.1 Policy and Regulatory Framework

Governance

As stipulated by the first strategic planning document related to circular economy [130] and adopted by the government in December 01, 2022, the Ministry of Environmental Protection (MEP) of the Republic of Serbia (Serbia, RS) is the government agency in charge of overseeing and controlling the implementation of the circular economy in Serbia. The MEP and its administration body, the Serbian Agency for Environmental Protection (SEPA), also have main regulatory authority regarding waste management, with the exception of radioactive waste [244]. The national policies and rules established by the MEP are implemented by local authorities which are in charge of planning and carrying out services for waste collection, transport, and disposal within their respective territories.

Along with MEP, the Ministry for Construction, Transport, and Infrastructure (MCTI) is in charge of overall policy, laws, regulations, and strategic research pertaining to construction, construction land, urban and spatial planning, housing and architectural policy, communal activities, and energy efficiency, as well as road, rail, water, and air transport, and infrastructure. As such, it plays a significant role in the application of circular economy principles in the built environment in Serbia.

Legislation

Currently, aside from the aforementioned strategic document and the closely related waste management programme [131] and Waste Management Act [132] and Planning and Building Act [129], there is no specific act related to circular economy in Serbia. Instead there are two acts with a certain degree of circular economy principles incorporated in them. When it comes to waste management, one of the main objectives of the Waste Management Act (Article 2) is to provide and secure conditions for the prevention of waste generation through rational use of natural resources and conditions for recovery and recycling of waste and the use of secondary raw resources. One of the guiding principles for providing such conditions is the hierarchy of waste treatment, which means that the least desirable option is disposal and the most ideal options are prevention, reuse, recycling, and other forms of recovery.

Similarly to the Waste Management Act the circular economy related provisions in the Planing and Building Act in its Article 3 stipulated that the activities of planning, development and the use of space, should be guided by sustainable development and rational and sustainable use of non-renewable resources and optimal use of renewable resources. In addition the Planning and Building Act's by-law, the Rulebook on the content, method and procedure of technical documentation preparation and control according to the class and purpose of the objects [202], stipulates that the rational use of energy and sustainable use of natural resources shoul be one of the basic requirements of each object.

Circular Economy Strategy, Roadmap or Action Plan

The implementation of the "Circular Economy Platform for Sustainable Development in Serbia" project, initiated by United Nations Development Programme (UNDP) in partnership with the Government of Republic of Serbia (RS), marked the start of the circular economy's ascent in Serbia in 2018. One of the main result of the project was the creation of a "Roadmap for Circular Economy in Serbia" [218]. The roadmap aimed to gather, promote and connect stakeholders which can contribute to a faster transition to circular economy. Additionaly it identified several priority sectors, construction has been selected as one of the priority sectors in Serbia's transition to a circular economy by the working group for the circular economy, which includes representatives of various stakeholders.

Analysis of the ex-ante effects of the circular economy initiated by the Ministry of European Integration and supported by the European Union was the first step in Serbia's strategic planning in this area [355]. The analysis highlighted the need for harmonization with the EU regulatory framework, a special circular economy agency, institutional coordination, fiscal incentives, national media campaign for circular economy; payment system for waste disposal; easier institutional and infrastructural use of RES; a mandatory share of recycled materials in the final product; a mandatory minimum of green public procurement in public procurements.

According to these findings, the MEP developed and adopted the Circular Economy Development Programme in the Republic of Serbia for the period from 2022 to 2024 (CE Programme) with the overarching goal of fostering the growth of the circular economy and the green transition in Serbia economy [130]. The most crucial areas for a circular economy are covered in this document, including waste management, water management, chemical management, renewable energy sources and energy efficiency, green public procurement, innovation, and awareness-building. The document also contains a three-year action plan with general and more specific objectives as well as key performance indicators.

For the built environment and the construction sector, the CE Programme envisioned improvements to the waste management system to enable efficient use of non-hazardous construction and demolition waste (CDW). The program's initiatives, however, are raising awareness oriented, and focused only on the analysis of the current situation and the creation of a construction and demolition waste management guideline. In contrast to the CE Programme, Waste Management Programme for the period 2022–2031 includes more specific objectives for the construction and demolition waste. These include a specific share of 40% of construction and demolition waste prepared for re-use by the end of 2029 and 70% of CDW pretreated CDW by the end of 2034 (soil excluded). The programme even highlights the need for the determination of legal frameworks which would enable the mandatory use of recycled aggregates (with the share of 10%, for example) in green public procurement.

Construction and Demolition Waste Management Regulatory Framework

Aside from several designated section in the waste management programe and the waste management act, currently there is no CDW specific legislation in Serbia. Only the definition of CDW is included in the act when it comes to CDW; while all other waste management provisions such as mandatory separation at source, reporting on generation and treatment, may be interpreted to apply to CDW as well. However, both the circular economy and waste management program highlight the need for the CDW specific by-law. Additionally, the draft of the Planning and Building Act Amendments, which is awaiting National Assembly adoption, contains various provisions focused on improved monitoring and reporting on the amounts of CDW generted and treated.

5.1.17.2 Platforms and Networks Facilitating the Transition to a Circular Economy

Circular Economy Platforms

There is no official platform, run by government or any of its administration bodies, for circular economy in Serbia. However there are two informal platforms created to promote circular economy prinicples in the national and local context. The first is Digital Platform for Circular Economy (CE HUB) created by the Chamber of Commerce and Industry of Serbia and the second "Circular Communities" is created in partnership of MEP and UNDP with the support of Global Environmental Fund (GEF). By displaying circular economyrelated information on best practices, potential grant opportunities, financial assistance, and business models, CE HUB strives to advance knowledge, enhance practice, and connect stakeholders from many sectors at the national level. On the other hand "Circular Communities" platform was created within the five year project "Reducing the carbon footprint of local communities by applying the principles of the circular economy in the Republic of Serbia" which has been implementing since March 2022. So far it has provided financial assistance for 19 local projects in the total amount of about USD 1.5 million.

In addition to these, two other circular economy related initiatives are identified: "Circular Economy Balkans Beacons" project and Centre for the promotion of Circular Economy. They are formed up of interested individuals and experts who primarily gather other individual experts in the field of circular economy and utilise their knowledge to advance new business models and raise awareness of the concept.

Circular Buildings Platform

No circular buildings platform has been established to date in Serbia. However, Serbia Green Budiling Council member of World Green Building Council and Serbian Demolition Association (SDA) member of European Demolition Association (EDA), which promote selective deconstruction, primary separation, re-use and recycling of CDW and use of secondary raw materials in new buildings.

Building Materials Passport Platform

No building materials platform has been established to date in Serbia.

Public Procurement Platform

The public procurement in Serbia is governed by the Public Procurement Office and regulated by the Public Procurement Act [128]. In this act there are no specific provisions regarding the green public procurement criteria and procedures. However, Article 132 provides the posibility that the contract awarding criteria include cost-effectiveness approach such as life cycle costing or the price-quality ratio which shall be assessed on the basis of criteria, including qualitative, environmental and/or social aspects.

5.1.17.3 Funding Opportunities

Some potential funding sources that may support circular economy projects in Serbia include EU funds, international organisations and programmes such as UNDP, WB and GEF, as well as China Export and Credit Insurance Coorporation, etc.

European Union (EU). Serbia has been a candidate country for EU membership, and as such, there are various EU funding programs that may support circular economy projects. EU with its grants and favourable loans for 17 economy sectors is the biggest donor in Serbia. In cooperation with the RS Government, the EU donated and invested EUR 892 million in the environment sector in since 2007. Over EUR 192 million was invested to improve the waste management sector which directly supports the transition to circular economy. The current investment portfolio is based on the European Green Deal and the Economic and Investment Plan for the Western Balkans. The "EU for Green Agenda in Serbia" initiative was launched in 2021 by the EU Delegation and MEP and is implemented by UNDP, in cooperation with the Embassy of Sweden and the European Investment Bank (EIB) with additional funding from the Governments of Sweden, Switzerland and Serbia. Its objective is to contribute to the efficient, inclusive and sustainable implementation of the Green Agenda for the Western Balkans and its Action Plan.

European Bank for Reconstruction and Development (EBRD). In 2021, EBRD with the support of GEF and Austrian Fedreal Ministry of Finance have launched the program "Circular Economy Regional Initiative" in Turkey and Western Balkans. A USD 154.76 million will be used to support investments in the private sector particulary SMEs to implement resource-efficient technologies and adopt circular business model. In addition to this programme EBRD provides credit lines to partner banks in Serbia under the Western Balkans Sustainable Energy Financing Facility (WeBSEFF) is a financing facility under which the EBRD provides credit lines to partner banks to lend to businesses and municipalities wanting to invest in energy efficiency and small-scale renewable energy projects.

The Deutsche Gesellschaft für Internationale Zusammenarbeit (GIZ). Since 2021, Germany and Serbia cooperates on sustainable climate policy in line with the climate objectives and the Green Agenda. In this partnership, GIZ helps to improve the legal, institutional and technical conditions for the use of renewable energies, improve energy efficiency, and supports the transition towards a green economy.

Global Environmental Fund (GEF). GEF endorse environmental related projects in Serbia either alone or in partnership with the United Nations (UN) or the World Bank (WB) since 2007. So far it has supported 26 projects which focused on climate change, land degradation, biodiversity, chemicals and waste.

World Bank. First Serbia Green Transition Programmatic Development Policy Loan providing EUR 149.9 million was developed in cooperation of the World Bank with French Development Agency (Agence Française de Développement, AFD) and the German Development Bank (Kreditanstalt für Wiederaufbau, KfW). The loan was approved in March 2023 for the acceleration of the clean energy transition and align harmonisation of domestic legislation with European Union standards on climate and environment action, with emphasis on waste and air quality management.

International Cooperation and Support from International Organizations

Aside from the aforementioned initiatives by the international financing institutions and organisations there are no other types of wider cooperation on this matter in Serbia.

5.1.17.4 Challenges, Barriers and Potential Improvements

So far challenges and barriers for the implementation of circular economy prinicples in Serbia were only identified in the Roadmap for Circular Economy in Serbia. The document included challenges from the regulatory, institutional and financial aspects one one side and lack of infrastructure and poor knowledge on possible implementation on another. More particular, the current legislation calls for harmonisation of national documents like strategic documents, laws and bylaws, technical regulations, etc. as well as documents in connection to the EU framework. While certain aspects of the regulations, such as the strategic documents for waste management, have been updated, the implementation of the existing regulations has remained a significant problem. This is particularly relted to the required primary sorting and high rates of illegal dumping, is also one of Serbia's main regulatory obstacles. Even if these issues had not existed, the absence of a waste management infrastructure, however, would still hinder the circularity of CDW in Serbia as there are no other recycling facilities and organised recycling of CDW in Serbia save one central plant in Belgrade. This is something that the aforementioned waste management programme, which calls for the establishment of 26 mobile recycling facilities (one in each region), is acutely aware of. However, the finance for these facilities has been allocated to the private sector, which still lacks both financial incentives and knowledge of the circular economy implementation advantages.

5.1.17.5 Examples of Successful Implementation

Public Policy Initiatives

As part of a new development project funded by the Republic of Serbia, the abandoned industrial historic site "Ložionica" (Furnace Room) will be revitalised and readapted into a multifunctional centre for creativity and innovation. The 3600 m^2 Ložonica building complex has been standing for almost a century. It includes a water tower, a railroad roundabout and the furnace chamber, which is made of reinforced concrete. The project's fundamental principle is the preservation of the cultural and historical heritage in the present with reconstructed furnance room and water tower, with a clear emphasis on new elements and structures such as public open space around the roundabout and new office building.

Private Initiatives

FEPLO is a Serbian company manufacturing waterproof eco-panels that are used as construction material for rooftops and floor construction. To manufacture the waterproof panels, they use the carton packaging from communal and packaging waste, while instead of glue they add 10% of mouldable polymers. These panels have a wide range of use, as they're employed for rooftops, floors, walls and flooring. They are used as an alternative to wooden products.

The cement factories in the Republic of Serbia: Lafarge Serbia, Beocin;, CRH Serbia, Popovac, Paracin; Titan Cement Factory use dangerous and hazardous waste as an alternative fuel. In addition, this industry in the Republic of Serbia, uses more than 300,000 tons of non-hazardous and hazardous waste per year (mostly fly ash and granulated blast furnace slag) as an alternative raw material.

5.1.18 Slovakia Factsheet

The construction industry is one of the key sectors of the Slovak economy and the largest producer of industrial waste in the country. All industrial construction activities combined account for 16% of total industrial waste generation. Meanwhile, the recovery rate of construction and demolition waste was only 51% in 2018. The demand for living space is increasing that projects a growing trend of demolishing and reconstruction of old building stock. The renovation of apartment buildings in Slovakia reaches 3% per year, which is an above-average rate in Europe-wide comparison. The renovation of buildings is a top priority of Slovakia's National Recovery Plan. Planned investments are around EUR 500 million for family houses and EUR 200 million for historical buildings. The environmental policy goals of the Slovakia and the current trends of the construction market create a need for new waste disposal sites with advanced recycling solutions [151].

5.1.18.1 Policy and Regulatory Framework

Governance

Slovakia does not have specific government institutions solely dedicated to implementing the circular economy. However, various government bodies and ministries are involved in promoting and implementing policies related to sustainability and resource efficiency. Key institutions and ministries in Slovakia that play a role in promoting the principles of the circular economy include:

- Ministry of Environment of the Slovak Republic (Ministerstvo životného prostredia Slovenskej republiky). The Circular Economy Section of the Ministry of the Environment—Department of Waste Management is looking more closely at this topic. The Slovak Environmental Agency (SAŽP), under the competence of the Ministry of the Environment, evaluates and monitors the indicators of the circular economy at national level (revised framework for monitoring the circular economy).
- Ministry of Economy of the Slovak Republic (Ministerstvo hospodárstva Slovenskej republiky). It is partly possible to identify efforts to implement elements of circularity in selected documents and activities of the Ministry of Economy of the Slovak Republic (e.g. in the energy policy and renewable resources areas).
- Ministry of the Transport of the Slovak Republic (Ministerstvo dopravy Slovenskej republiky). In Slovakia, the Ministry of Transport of the Slovak Republic is the body responsible for building administration, housing policy and construction. Its strategic documents do not explicitly include an elaborated strategy for the transition to circularity in the construction sector.

Legislation

In the Slovak Republic, the Construction Act (Act No. 201/2022 Coll.) primarily addresses construction-related matters. While the Act does not explicitly focus on circularity in the construction industry, it does regulate specific methods for handling construction waste and outlines the responsibilities of various stakeholders involved in the construction process. For instance, according to §23, one of the obligations of the construction contractor is the "placement of construction products on the construction site and removal of waste from the construction site." Similarly, §26 mandates that the construction manager coordinate, receive, control, and document the delivery of construction products to the site, as well as the removal of waste, while maintaining records of such deliveries and removals.

The Act No. 133/2013 Coll. on Construction Products and on Amendments and Additions to Certain Acts, as amended by Act No. 91/2016 Coll., establishes the requirements governing the placement and availability of construction products in the market of the Slovak Republic. The legislation covers various aspects such as the authorization process for legal entities to perform technical assessments and assess and verify the consistency of the manufacturer's declared parameters regarding the essential characteristics of the products. Additionally, it outlines the rights and obligations of authorized individuals, manufacturers, and the responsibilities of state administrative bodies in this regard.

Many of the existing policy interventions and plans at the other life cycle stages introduce approaches with indirect circularity impacts. These include, for instance, policies related to the objectives of energy efficiency and renovation that are relevant for the use phase (i.e. Energy Efficiency Act, the Energy Performance of Buildings Act, and the Long-Term Building Renovation Strategy).

Circular Economy Strategy, Roadmap or Action Plan

The Slovak Republic has developed several cross-cutting strategies and regulations, which aim to set out a vision towards climate neutrality and sustainable development. The six policies directly or indirectly related to the circular economy in the construction sector include:

- the Sustainable Development Strategy to 2030 [223],
- the 2030 Environmental Strategy [216],
- the Recovery and Resilience Plan (RRP) [219],
- the Economic Policy Strategy until 2030 [208],
- the Low-Carbon Development Strategy [217],
- the National Energy and Climate Plan [328].

Besides the Recovery and Resilience Plan, which demonstrates the country's national commitment to steer the construction sector towards a circular economy around waste management, the policy targets and goals of the other cross-cutting strategies (related to the circular economy) remain rather limited and defined at a relatively high level. Moreover, many of the circular economy objectives are being addressed by targets and policies indirectly related to the circular economy, such as those concerning energy efficiency and renewable energy.

The Slovak Ministry of the Environment, in cooperation with the European Commission and the OECD, has worked on the technical basis for the preparation of a roadmap for a circular economy in Slovakia since 2020. "A Roadmap Towards Circularity For Competitiveness, Eco-Innovation And Sustainability. Closing The Loop In The Slovak Republic. OECD Environment Policy Paper No. 30" (2022) provides a thorough overview of the business opportunities of the circular economy in Slovakia. The main objectives of the Roadmap are to improve waste management, decrease the environmental footprint, use natural resources effectively, support innovation progress, and increase the country's competitiveness [151]. A Roadmap Towards Circularity For Competitiveness, Eco-Innovation And Sustainability (2022) also includes the role of the construction sector in transitioning to a circular economy, giving an overview of the current situation and policy framework. The gap analysis informs policy recommendations to shift away from the focus on energy efficiency or construction and demolition waste (CDW) towards the integration of circular economy principles along the entire construction value chain. Key recommendations on how to transition to a circular construction sector in the Slovak Republic by 2040 are complemented with some concluding reflections.

The Waste Management Programme of the Slovak Republic (WMP SR) is a significant strategic document that guides waste management efforts in the Slovak Republic (SR) from 2021 to 2025. It has been developed in alignment with the principles of sustainable growth, as presented in the EU Action Plan for Circular Economy, the European Green Deal, and the New EU Action Plan for Circular Economy. The Programme also specifically addresses the management of construction and demolition waste, with the aim of increasing the preparation for re-use and recycling of such waste, including backfilling, to a target of 70%.

To achieve this target, the Programme establishes measurable indicators and outlines the following measures:

- Conduct analysis of construction waste, considering both organic carbon content and recycling and recovery possibilities.
- Incorporate obligations into decision-making authorizations of competent authorities (via legislation, guidelines, or methodological guides), including:
  a. Sorting requirements for expanded polystyrene used for building insulation.
  b. Implementation of selective demolition practices during building removal, along with the submission of photo documentation of the demolition process.
  c. Submission of evidence demonstrating the recovery of waste to the authorizing authority.
- Encourage research and development activities focused on new technologies for construction waste management.
- Promote the financing of technologies and projects aimed at recycling construction waste into higher value-added output products, utilizing recyclates as feedstock. However, funding for technologies primarily focused on shredding construction and demolition waste should not be supported.
- Consider the implementation of standards for wood waste management activities in line with the waste hierarchy principles.

- These measures reflect the commitment of the WMP SR to enhancing the circularity of the construction and demolition waste sector while promoting sustainable waste management practices.

The New Waste prevention program of the Slovak Republic for the years 2019–2025 was established in 2018. The new waste prevention measures include environmental education; application of the principles of green public procurement; implementation of voluntary instruments in the field of environmental policy (e.g., Eco-design, EMAS); adequate organizational, personnel, and financial coverage via increase landfill fees (2022); linking the activity with the 2030 Agenda for Sustainable Development and the preparation of the National Investment Plan of the Slovak Republic for the years 2018–2030 [151].

Construction and Demolition Waste Management Regulatory Framework

Waste management, including construction waste, is explicitly addressed in the Waste Act No. 79/2015, as amended, with subsequent acts such as Act 230/2022. Article 77 of this legislation governs the "Management of construction and demolition waste." The law provides a definition for construction waste and outlines the obligations of waste producers. It also introduces the concept of "selective demolition," which involves planning demolition activities to enable the separation and sorting of removed construction materials and waste. This provision can be considered a legislative framework aimed at enhancing the efficient sorting and management of construction waste, as well as incorporating elements of the circular economy into the construction industry.

The Waste Act No. 79/2015 sets the basis for a circular economy in the end-use-phase by introducing the principles of waste management hierarchy, applicable to all sectors. It also includes a construction-specific target that demands to recover 70% of non-hazardous Construction and Demolition Waste (CDW) by 2020 (derived from the corresponding EU obligation for 2020). The Waste Act also imposes a general obligation for responsible entities to reuse the produced CDW in constructions, renovations or infrastructure maintenance. At the same time, a large part of the recovered CDW still ends up as backfilling material rather than being recycled. The Waste Prevention Programme (WPP) [216] (https://www.minzp.sk/files/sekcia-enviromentalneho-hodnotenia-riadenia/odpady-a-obaly/registre-a-zoznamy/ppvo-sr-19-25.pdf) sets high-level targets and goals for nine waste streams, including for CDW, among others. The most recent Waste Management Plan for 2021–2025 singles out "circular economy" as one of its key priorities and extends the 70% CDW material recovery target to 2025. Further policy reform ambitions have recently been announced for the end-of-life stage as a part of the Slovak RRP, with a primary focus on CDW management reform. However, other priority objectives, related to the use of buildings in the RRP, remain at the level of improving building energy efficiency and decarbonisation.

Construction waste in Slovakia has traditionally been an untapped resource of valuable raw materials that can be reused and fully utilized within the construction industry. Regrettably, in 2018, almost half of the treated mineral waste from construction and demolition activities ended up in landfills in Slovakia. Only 37% of mineral waste underwent recycling, with approximately 13% utilized for backfill purposes. These statistics indicate one of the lowest performances within the European Union. Recognizing the urgency to address this issue, the Ministry of the Environment (MoE) has taken proactive measures. They have proposed an amendment to the Waste Act (Law 230/2022) for legislative consideration. The amendment specifically aims to promote the recovery and recycling of construction waste, aligning with the objectives outlined in the Recovery and Resilience Plan of the Slovak Republic, specifically COMPONENT 2: Building Renewal 1.1 Policy Area—Green Economy—Reform 3: Reform of construction waste management. The primary goal of this amendment to the Act is to enhance the potential of the circular economy within the construction waste and construction sectors. By doing so, it is expected to raise recycling rates significantly and prevent the generation of excessive construction waste. As an example, the amendment introduces modifications to waste management requirements for construction sites larger than 300 m^2. It mandates that a minimum of 70% of non-hazardous construction and demolition waste must be recycled, emphasizing the importance of sustainable waste management practices.

5.1.18.2 Platforms and Networks Facilitating the Transition to a Circular Economy

Circular Economy Platforms

The country's environmental policies consider waste management and recycling key tools to stimulate Slovakia's transition towards a circular economy. The current state of the country's waste management sector is desperate, more than half of the waste ends up in landfills, making Slovakia one of the most backward countries in the field of waste management. The utilization of secondary raw materials shows an emerging trend in the country. The platforms, associations of processors, recyclers, and manufacturers providing virtual space for trading are becoming significant players in the market. However, there is no uniform database in Slovakia comparing processing capacities across all classes of materials. Companies are seeking such databases to find a suitable technology for processing or suppliers for their material [151].

In Slovakia, there are currently several active platforms supporting and promoting the circular economy and its principles:

The Slovak Green Building Council (SKGBC) is the main ambassador of sustainable constructions in Slovakia. SKGBC brings together companies from different sectors of the economy and promotes quality construction and building renovation by organising meetings focused on sustainable solutions in the construction or renovation of buildings.

Circular Slovakia is a public–private platform for the promotion of a circular economy that brings together business companies, government and knowledge institutions, associations and NGOs with the ambition to accelerate the transition to a greener and circular Slovakia. Actually the platform brings together 66 entities.

The Institute of Circular Economy (INCIEN) is a non-profit organization that focuses on innovative environmental solutions in the field of waste prevention, environmental management. INCIEN collaborates with its partners on projects that represent the transition from a linear to a circular economic system. Within the framework of its projects, it introduces and implements circular economy in practice in the conditions of the Slovak Republic. INCIEN is composed of a team of experts who focus on waste management and circular economy issues. The basic effort is to create cooperation with research, but also governmental institutions, non-profit sector and experts in various fields. INCIEN is looking for a sustainable path from an environmental but especially economic perspective.

In Slovakia, the Green Economy Information Platform has been established as a hub for sharing and publishing solutions related to the circular economy, sustainable resource usage, energy efficiency, green buildings, and housing, among other topics. This platform serves as a valuable resource for information dissemination.

The Ministry of Economy of the Slovak Republic actively supports initiatives geared towards transitioning to a circular economy. One notable initiative is their participation in the MOVECO project (Mobilising Institutional Learning for Better Exploitation of Research and Innovation for the Circular Economy). This project aims to facilitate the transition to a circular economy within the countries of the Danube Region, while also promoting awareness of this important issue. In Slovakia, the Slovak Business Agency oversees the implementation of the MOVECO project, working in collaboration with various stakeholders to drive the circular economy agenda forward. The Ministry of Economy's involvement in this project demonstrates their commitment to fostering a sustainable and circular economy within the country.

Public Procurement Platform

Based on legislative changes, technological advancements, and the development of green public procurement practices in both the European Union and the Slovak Republic, the National Action Plan for Green Public Procurement in the Slovak Republic for 2016–2020 (NAP GPP III) was formulated. This action plan received approval from the Government of the Slovak Republic on 14th December 2016 through Resolution No. 590. At the national level, the **NAP GPP III** represents the third consecutive action plan focusing on concrete commitments by the State to promote the integration of environmentally friendly practices in public procurement. It serves as a guiding document for the implementation of green public procurement in the Slovak Republic.

The primary objective of the NAP GPP III is to establish a clear path for green public procurement within the context of the Slovak Republic. This is achieved by outlining specific measures and activities that, when implemented through a proactive approach by relevant institutions, aim to achieve more ambitious targets for green public procurement from 2016 to 2020. The NAP GPP III demonstrates the commitment of the Slovak Republic to enhance sustainability and environmental considerations in public procurement processes, aligning with the broader goals of promoting a greener and more sustainable economy.

5.1.18.3 Funding Opportunities

National Recovery and Resilience Plan

The greening of the Slovak economy, including the construction sector, is primarily facilitated through funding allocated under the National Recovery and Resilience Plan. This funding is then directed towards national initiatives, such as the Household Greening and Climate Change Adaptation Programme (Green for Households, Renew the House, Renewal of Public Historic and Listed Buildings). Approximately €2.17 million has been designated through the Plan to support the development of the green economy.

Environmental Fund

The Environmental Fund was established to implement state-supported initiatives for environmental care and the creation of a sustainable environment. Its core mission is to provide financial resources, in the form of grants or loans, to applicants aiming to achieve the objectives of national environmental policies at various levels: national, regional, and local. Currently, subsidies can also be obtained from the Environmental Fund (administered by the Ministry of the Environment of the Slovak Republic) for circular activities and more efficient waste management. The fund supports various activities, including the construction of collection yards that facilitate preparation for re-use and re-use of products or product components that are not considered waste for the same purpose for which they were originally intended. It also supports the establishment and operation of re-use centers for collection, sorting, and re-use of materials.

Operational Programme Quality of the Environment (OP EQE)

Another funding opportunity for initiatives promoting the circular economy is the Operational Programme on Environmental Quality (OP EQE). It offers support in the area of Waste, specifically Investment Priority 1, which focuses on investing in the waste management sector to meet the requirements of the Union's environmental regulations and to address additional needs identified by Member States that go beyond these requirements. Within the context of waste management, the OP EQE prioritizes activities related to the preparation for re-use and recycling of waste materials.

International Cooperation and Support from International Organizations

The Slovak Ministry of the Environment has established collaborative partnerships with the European Commission and the OECD in the pursuit of a circular economy. Since 2020, they have been actively engaged in developing the technical framework for a circular economy roadmap in Slovakia. You can find more information on this topic in the following document:

Furthermore, Slovakia has engaged in cooperation with the **World Bank** to advance circularity efforts.

One noteworthy example of international cooperation is the collaboration between the Dutch organization Circle Economy and the capital city of Bratislava. They are currently working together to conduct a scan of the city's material flow, with the aim of providing recommendations to enhance circularity within the city.

5.1.18.4 Challenges, Barriers and Potential Improvements

The Slovak construction sector faces notable obstacles and unexplored possibilities when it comes to reducing the reliance on virgin raw materials and encouraging the utilization of recycled building materials. Not only is construction a vital economic sector in the Slovak Republic, but it also accounts for over half of the country's domestic raw material consumption and generates a substantial amount of waste. Until recently, efforts to promote circularity within this sector have been limited, with most of the emphasis placed on enhancing the energy efficiency of buildings.

However, both new construction projects and renovations present significant opportunities for applying principles of the circular economy. Currently, there is a lack of incentives in Slovakia that actively promote circular design and the adoption of secondary building materials, which possess immense potential in reducing material consumption and mitigating associated environmental impacts.

The transition to a circular economy in the Slovak Republic holds significant potential for reducing greenhouse gas (GHG) emissions. The construction sector, in particular, presents the greatest opportunity for GHG emissions reduction through the application of circular economy principles in steel and cement production. Slovakia could consider several key activities to contribute to its climate targets, such as increasing the use of alternative materials to steel, promoting space-sharing practices in non-residential buildings, and improving the recycling, recovery, reuse, and upcycling of construction and demolition waste.

By reducing the use of concrete, cement, and steel throughout the life cycle of buildings, emissions from buildings in the EU could potentially be reduced by 61% in 2050 compared to 2015 levels (Ramboll, Fraunhofer ISI, and Ecologic Institute

2020). These activities align with the goal of closing the loop and achieving a more circular construction sector in Slovakia [252].

To further support the implementation of circular economy principles, the establishment of material banks becomes crucial across different sectors of the national economy, including construction. In environmentally friendly demolition practices within the construction sector, materials can be separated for reuse or used as inputs for producing new products. Building materials should be viewed as valuable resources that can be stored and utilized throughout their life cycle. An excellent example of such practices in Slovakia is Cyrkl, an organization that operates a free digital marketplace. Cyrkl enables the selling and buying of industrial waste, by-products, secondary raw materials, and used materials. The marketplace also provides circular solutions and optimization of material flows and management, utilizing tools such as the Circular Waste Scan. Cyrkl also conducts projects focused on material recovery from the demolition process and optimizing material flows for the production of construction materials.

Key policy recommendations to help the country transition to a circular construction sector are:

- Strengthening evidence-based policy making and multi-stakeholder collaboration
  - Improve the measurement and monitoring of CDW flows from producers to final waste processors.
  - Develop a comprehensive overview of the national building stock and its renovation needs.
  - Strengthen the collaboration and partnerships between stakeholders from across the value chain.
- Policy recommendations related to extraction, design and construction phases
  - Consider introducing an aggregates tax to discourage the extraction of construction minerals.
  - Introduce a quality standard for recycled construction materials.
  - Introduce minimum recycled content requirements for specific construction products.
  - Increase the use of GPP criteria in the construction sector to stimulate the market for the construction and renovation of sustainable buildings, possibly including minimum recycled content requirements as a GPP criterion.
  - Encourage business model innovation for accessing secondary materials and for developing new recovery processes and technologies.
  - Use future construction projects as pilots to test and apply circular economy principles and innovations.
- Overview of buildings use and renovation-related policy recommendations
  - Investigate the potential of using fiscal instruments (tax benefits or subsidies) to stimulate the use of secondary and renewable materials in renovations.
  - Examine the potential of revising some of the zoning codes to include more flexibility in space distribution and utilisation.

- Overview of policy recommendations related to end-of-life, reuse and recycling
- Gradually increase the landfill taxes (in place from July 2022), and reform the redistribution of proceeds from the landfill tax in combination with greater waste management enforcement.
- Remove legal obstacles to the use of recycled materials by implementing EoW criteria for some additional construction and building materials (in place from June 2022).
- Introduce mandatory selective demolition, including a system of inspection/audit before and after demolitions take place (in place from June 2022).
- Consider expanding EPR to construction products or specific renovation waste.

5.1.19 Spain Factsheet

5.1.19.1 Policy and Regulatory Framework

Governance

The Spanish Circular Economy Strategy (*España Circular 2030*) [192] was approved in June 2020 by the Ministry for Ecological Transition and Demographic Challenge (known as MITERD, previously MITECO). The Strategy, aligned with the European Green Deal (among other international regulations, strategies and plans) has the main purpose of promoting a new production and consumption model in a country, Spain, with a GDP of EUR 1 122.0 billion (which represents the 8.4% of EU27 total in 2020, according the aforementioned Strategy), in order to contribute to achieve a sustainable and decarbonised economy. The Strategy identifies **six priority sectors**, including the construction sector, which is a very significant sector within the Spanish economy (the sector contributed 6.5% of Spanish GPD in 2018, according the Strategy), with a substantial impact in the environment. Therefore, the fulfilment of the circular target set by the Plan for the construction sector "which involves both environmental and financial benefits".

The Spanish Government, again through the MITERD, is also in charge of developing and approving triennial action plans to implement this Strategy, whose progress will be assessed through a group of indicators include on the basis of those introduced in the EU CE Monitoring Framework. Moreover, sectoral action plans are pointed out as another tool to implement the Spanish Strategy, being also aligned with the climate goals set by the Paris Agreement and the objectives of the European Commission's circular economy strategy.

The first of these plans, already adopted, is the Circular Economy Action Plan I (2021–2023) [195]. This plan was approved on 25 May 2021 and is divided into five axes and three lines of action. By the successive action plans a series of quantifiable goals are expected to be met by 2030. Among them, it could be highlighted the goal of "reducing greenhouse gas emissions to less than 10 million tons of carbon dioxide equivalent", with the construction sector playing a key role in achieving that goal.

The Spanish governance model outlined to assess the progress of the Strategy implementation includes three difffent bodies that introduce the perspective of the different administrations involved (at national, regional, and municipal levels). In fact, the regional governments (known in Spain as autonomous communities) are empowered to develop their own Circular Economy strategies, plans and regulations, aligned with the national ones; indeed, some of these autonomous communities have already done it (for example, the Basque Country Region).

Legislation

Apart from the above, no other specific national Circular Economy legislation has been established to date. However, several Circular Economy policy elements are included in different national policies already in place. In fact, the Law 7/2022, of April 8th, on waste and contaminated soils for a circular economy (in Spanish), Ley 7/2022, de 8 de abril, de residuos y suelos contaminados para una economía circular [159] can be considered as an important instrument to support the Spanish Circular Economy Strategy since it includes some fiscal measures to promote the circular economy (among other measures aligned with the Strategy).

Another policy element to highlight is the Spanish Urban Agenda [188], aligned with the Urban Agenda for the EU, which included measures to promote circularity within materials cycle, for example. It should be highlithed another legal document, the Orden PCI/86/2019 [190], by which was approved the Green Public Procurement Plan (2018–2025), including legal criteria for sustainable production and waste management (for example, the use of sustainable construction materials, taking into account their lifespan). Moreover, the long-term decarbonisation strategy [193], published in November 2020, also includes measures linked to circularity, such as the adaptation of infrastructure to efficient and circular consumption systems in order to support decarbonisation goals.

Regarding the activities in the field of building work, within the document known as Just Transition Strategy [203], the need of a Circular Economuy thinking in building refurbishment plans was noted. The Law 7/2021, of May 20th, on climate change and energy transition (in Spanish, Ley 7/2021, de 20 de mayo, de cambio climático y transición energética, [158]) includes an article (Article 8) focused on the energy efficiency and building rehabilitation that introduce Circular Economy measures to be applied in construction materials (for example, the usage of "materials with the smallest carbon footprint") or building services (for example, "the management of demand and the use of energy from renewable sources in the field of building").

Construction and Demolition Waste Management Regulatory Framework

The conditions for the construction waste management were specifically addressed by the Royal Decree 105/2008, of February 1st, regulating the production and management of construction and demolition waste—CDW—(in Spanish, Real Decreto 105/2008, de 1 de febrero, por el que se regula la producción y gestión de los residuos de construcción y demolición [189]). Coming into force in 2008, it was developed to establish the legal framework for CDW, to promote "their prevention, reuse, recycling, and other forms of valorization, ensuring that those destined for disposal operations receive appropriate treatment, contributing to a sustainable development of the construction activity". This regulation has been complemented by the aforementioned Law 7/2022. The Law 7/2022 set as an objetive that reaching a minimum of 70% by weight of the total waste generated of non-hazardous CDW being prepared for reuse, recycling, and other material recovery. Moreover, this Law has an article (Article 30) focused on the CDW which, apart from regulating its management, includes some circular measures as is the case of the obligation (on 2024) for the demolition to be preferably carried out selectively.

5.1.19.2 Platforms and Networks Facilitating the Transition to a Circular Economy

Circular Economy Platforms

The organizational model of the Spanish Circular Economy Strategy establishes the creation of Council for Circular Economy. The main purpose of this Council is "to promote collaboration between public and private sectors so as to speed up the transition towards a circular economy both in the public and private spheres" [194]. To achieve that goal, the Council must develop, implement, monitor, and review annual proposals in the framework of the Spanish Strategy.

Circular Buildings Platform

No circular buildings platform has been established to date. Nevertheless, the Green Building Council of Spain (GBCe) is taking a lead role in the promotion of Circular Economy in the building environment. It can be mentioned the report "Circular Economy in Construction" [356], developed with the objective of diffusion, mobilization of the sector, and raising awareness among professionals to promote a transition of the sector toward a Circular Economy.

Building Materials Passport Platform

No building materials platform has been established to date at national level.

Public Procurement Platform

The aforementioned Green Public Procurement Plan [190], approved in 2019, includes some legal criteria for green public procurement of products and services considered as priorities by the EU. Under this Plan, the official public procurement platform will have a section dedicated to helping develop specifications with the criteria approved to promote a sustainable production and waste management in the construction sector.

To the best of our knowledge, this section is not yet fully operational.

5.1.19.3 Funding Opportunities

Although European funds have supported various projects and initiatives in Spain related to energy-efficient buildings and improving waste management in the construction environment during the last few years (for example, through the European Regional Development Fund), nowadays, the most relevant ones are supported via the Recovery and Resilience Mechanism funds, the European funds mobilized to help rebuild a post-COVID-19 Europe.

National Recovery and Resilience Plan

The Spanish **Recovery, Transformation and Resilience Plan** (PRTR) [107] is the instrument that allocates the European funds linked with the Next Generation EU Instrument. This plan set up some measures aimed at facilitating the economic recovery needed after the coronavirus pandemic by speeding up the transition to a more digital and greener economic model. Among the 30 components included in the Plan, the number 12 adresses the objective of promoting the Circular Economy in various industrial sectors, for instance, by supporting the implementation of the Spanish Circular Economy Strategy and waste regulations (addressed in Investment 3 of the Component 12) [90].

In addition, the Circular Economy has been introduced as a cross-section in several components of the plan; the Component 2 directly addresses this topic in the construction sector by including CE requirements in the building refurbishment programmes. The Royal Decree 853/2021, of October 5th, regulating assistance programs for residential rehabilitation and social housing as part for the Component 2 of the PRTR (in Spanish, Real Decreto 853/2021, de 5 de octubre, por el que se regulan los programas de ayuda en materia de rehabilitación residencial y vivienda social del Plan de Recuperación, Transformación y Resiliencia, [191]) sets some circular measures as a requirement for receipt of the aid.

Finally, it is worth noting that the PRTR contemplates the implementation of various Strategic Projects for Recovery and Economic Transformation (PERTEs) to bring together all levels of government and the private sector. Among these PERTEs there is one on Circular Economy with a budget of EUR 492 million seeking to increase the competitiveness of industrial sectors and companies in general in this area [108].

International Cooperation and Support from International Organizations

As already noted, the Green Building Council of Spain (GBCe) is providing technical support, taking a lead role in the promotion of Circular Economy in the building environment. We lack awareness of any official organization offering direct financial assistance in this particular domain.

5.1.19.4 Challenges, Barriers and Potential Improvements

As already mentioned, there are several challenges and barriers related to different areas of implementation of circular economy in the built environment. In the case of Spain, two Catalogues of Best Practice in Circular Economy has been developed by the MITERD [204, 205], where the companies participating in them laid down the main difficulties encountered to carry out circular economy activities. These companies assessed a list of non-country specific circular economy barriers as they perceived, obtaining the ranking of those barriers shown below (Table 5.5). As a result of the analysis carried out, various policy initiatives have been developed to address the main obstacles and challenges have been encountered.

5.1.19.5 Examples of Successful Implementation

Although the Spanish data are still below the most advanced European countries in this particular area, they show a positive trend in the last few years [90]. This is mainly because of the policy initiatives already implemented at different levels.

The already mentioned Catalogues of Best Practice in Circular Economy [204] also collect relevant initiatives coming from the private sector. The General Sub-Directorate for the Circular Economy of the Ministry in charge of developing these cataloghes is working on the implementation of a database with the examples of CE activities already mapped.

Public Policy Initiatives

Apart from the legislation already in place before mentioned, work is now being carried out on new legislation in the Circular Economy area. These new legislation initiatives aim to introduce the new European regulations, including the 2035 targets [90].

Table 5.5 Ranking of circular economy barriers after the companies assessment

Ranking (Rk.n)	Circular economy barriers
Rk.1	Lack of circular regulation
Rk.2	Behavioral change/lack of awareness or cooperation
Rk.3	Recognition of by-products/secondary raw materials
Rk.4	Lack of circular infrastructure/technical or logistical...
Rk.5	Absence of legal standards and definitions
Rk.6	Administrative burden
Rk.7	Complex process to make it circular
Rk.8	High initial investments
Rk.9	Quality problems
Rk.10	Cooperation with authorities
Rk.11	Useful applications of recycled materials
Rk.12	Insufficient demand
Rk.13	Access to relevant information and applicable assessments
Rk.14	Low return on investment
Rk.15	Lack of incentives
Rk.16	Time-consuming processes
Rk.17	Organizational structures
Rk.18	Other
Rk.19	Quantity problems
Rk.20	Access to financing
Rk.21	Transboundary waste shipments
Rk.22	Price volatility
Rk.23	Lack of implementation

Source MITERD [204, 205]

There are also regional and local public initiatives to be highlighted, such as the Circular Thinking Initiative [154] developed by the Basque Department of Environment of Basque Country region or the initiatives in this area implemented by the cities of Valladolid or Gijón.

Private Policy Initiatives

The Spanish Association for Standardisation (UNE) lead a study on standards techniques capable of supporting different aspects of the Circular Economy [340], in order to highlight the availability of reliable tools for reducing the use of raw materials and promoting a better waste management [90].

5.1.20 Türkiye Factsheet

5.1.20.1 Policy and Regulatory Framework

Governance

With the spread of the circular economy concept in recent years, important studies have been carried out by the Ministry of Environment, Urbanization and Climate Change (MoEUCC), especially on effective resource and waste management. In this context, the concept of "circular economy" serves as a guide for many of the MoEUCC's studies in the field of legislation and implementation. Within the scope of the "Green Deal Action Plan" published by the Ministry of Commerce, an action has been defined for "Determining Priority Sectors in the Field of Circular Economy" and "Conducting Detailed Impact and Needs Analyzes" and MoEUCC has been assigned as the responsible institution for this action. The main output of the action; "Preparation of Needs and Impact Assessment Reports" for sectors. While studies on prioritization of sectors are carried out; Areas such as textile, plastic, construction and manufacturing will be examined in detail. Another action that is under the responsibility of MoEUCC is "Second Phase of 2023". "Preparation of the Circular Economy National Action Plan by the 2nd Quarter" [175]. The main output of the action, there will be a "Circular Economy National Action Plan" and "determination of technical criteria for the use of recycled secondary materials". With the amendment made to the Environmental Law in December 2020; General principles regarding the dissemination of "zero waste" and "circular economy" practices and the protection, improvement and prevention of pollution of the environment are also included in the Law. Another important step taken with this change is; It is a regulation regarding the "compulsory use of waste or recycled materials obtained from waste", which will contribute greatly to our country's circular economy vision in the coming period. To ensure that all these studies are carried out systematically, to determine our country's road map for the circular economy and to strengthen Türkiye's institutional and technical capacity in the transition to the circular economy in line with the EU Circular Economy Model, the "Technical Assistance Project for the Assessment of Türkiye's Transition to the Circular Economy Potential" 07.02.2022. It was officially implemented on the project is carried out by the Department of Circular Economy and Waste Management, where the Department of European Union Investments of the General Directorate of European Union and Foreign Relations of the Ministry of Environment, Urbanization and Climate Change is the Contracting Authority, and is carried out by the European Union Pre-Accession Assistance Instrument (IPA) II. It is supported under the Environment and Climate Action Sector Operational Program within its period.

Legislation

"Law No. 7261 on the establishment of National Environmental Agency and related amendments:" This Law lays down rules for the establishment of Turkish Environmental Agency for the prevention of environmental pollution; the contribution to the protection, improvement and development of green areas; the increase in resource efficiency in line with the circular economy and zero waste approach; and the establishment, operation, monitoring and control of deposit management system on a national scale. This Law sets forth provisions on the activities, structure and budget of the Agency. This Law makes amendments in the Environment Law No. 2872 in the fields of strengthened waste management, recycling, and clean technologies. Furthermore, Article 20 is amended regarding the activities without environmental permit or environmental permits and licenses and related penalties. Provisional Clause 14 is added regarding the establishment of eco-labelling system.

Circular Economy Strategy, Roadmap or Action Plan

National. According to the "Republic of Türkiye Updated First Nationally Determined Contribution" document published in April 2023, the national circular economy action plan and circular economy strategy and action plan was stated as being drafted.

In order to reach the net zero target by 2053, protect the natural environment and improve competitiveness, a resource efficient and sustainable circular economy is important. In this sense, The "Green Deal Action Plan" (GDAP) of Türkiye was published on 16th of July 2021 with the Presidential Circular numbered 2021/15. The Action Plan, including a total of 32 objectives and 81 actions under nine main headings, dwells on the green transformation of Türkiye's industries and the adoption of measures especially in areas related to trade and industry. The GDAP includes actions to be taken in a wide range of areas, such as Combating Climate Change, a Green and Circular Economy, Green Finance, Carbon Border Adjustments, Clean, Affordable and Secure Energy Supply, Sustainable Agriculture, Sustainable Smart Transportation and Diplomacy in order to facilitate the green transformation towards a lowcarbon, resource-efficient and circular economic structure. As a country that understands the value of adaptation policies, Türkiye has acted decisively in this area and is dedicated to the pursuit of its goals across a wide range of initiatives, including impact, vulnerability, and risk assessments, information systems, legal and political instruments at the national and local levels, capacity building, financing, monitoring, and implementation. Some important adaptation efforts are National Climate Change Adaptation Strategy and Action Plan (NASAP), National Drought Management Strategy Paper and Action Plan (2017–2023), National Report on Land Degradation Neutrality for Türkiye (2016–2019), National Basin Management Strategy and Action Plans (2014–2023) In accordance with international standards, Türkiye is enhancing its green financial ecosystem to foster investment opportunities for more ambitious climate action. Financial institutions are key for greening the financial system. There have been number of developments to improve the green financial ecosystem in Türkiye. One of the most important of these developments is "Sustainable Banking Strategic Plan 2022–2025".

According to the The National Action Plan for Waste Management (NWMAP) 2016–2023 in Türkiye, the key targets are included [174]:

- Recycling rate to be increased by 35% by 2023. The recycling rate for glass, plastic, metal and paper are to increase to 60% by 2023;
- Collect 4.6 million tonnes of packaging waste at the point of origin by 2023; and
- Construction of three new incinerators for hazardous waste.

Regional. Türkiye, one of the parties to the Barcelona Agreement, has submitted its commitments to the "Sustainable Consumption and Production Regional Action Plan for the Mediterranean," which is built around: "By 2027 a prosperous Mediterranean region is established, with nonpollutant, circular, socially inclusive economies based on sustainable consumption and production patterns, preserving natural resources and energy, ensuring the well-being of societies and contributing to clean environment and healthy ecosystems that provide goods and services for present and future generations." [339]. The SCP Regional Action Plan has been implemented under regional programs like the EU-funded SwitchMed Programme, however, Türkiye could not participate in such regional programs and only be integrated as an observer. In addition, although Türkiye has a variety of policy instruments, including strategies and measures to reduce the adverse influences of high-impact sectors resulting from carbon-, energy-, waste-intensive current consumption and production techniques, indicator 12.1.1 of SDGs, "Countries with sustainable consumption and production (SCP) national action plans or SCP mainstreamed as a priority or a target into national policies" still has not reached a satisfactory level.

Construction and Demolition Waste Management Regulatory Framework

"Regulation on Demolition of Buildings" was published in the Official Gazette on 13 October 2021 by the Ministry of Environment and Urbanization (Official Gazette Directive 31627). This regulation mandates selective demolition in buildings to facilitate the implementation of the circular economy. It involves the separation of materials that can be reused during selective demolition, thus preventing waste generation. Throughout the demolition process, waste is segregated and stored separately, ensuring their recycling. Waste that cannot be recycled is disposed of in compliance with applicable regulations.

Regulation of the "Control of Excavation Soil and Construction and Demolition Waste" number 25406 enacted by the Ministry of the Environment and Forestry which came into force on March 18th, 2004 (Official Gazette Directive 25406). The purpose of this Regulation is; to regulate the technical and administrative issues and the general rules to be followed regarding the reduction, collection, temporary accumulation, transportation, recovery, evaluation and disposal of excavation soil and construction and demolition wastes at the source in a way that does not harm the environment.

5.1.20.2 Platforms and Networks Facilitating the Transition to a Circular Economy

Circular Economy Platforms

Since 2016, BCSD Türkiye and EBRD have joined forces to create awareness and accelerate the transition to circular economy in Türkiye by providing tools and technical support that enable businesses to move away from the traditional ineffective way/concept to a more powerful way of doing business. The journey started with Türkiye Materials Marketplace that was instrumental in creating an ecosystem around circular economy. Through the course of the past 4 years BCSD Türkiye felt the need to create a space where anything and everything on circular economy is explained in detail. With the establishment of Türkiye Circular Economy Platform, BCSD Türkiye aim to provide practical solutions, incentives, news and opportunities in the field of circular economy. The platform at hand includes a knowledge hub, an e-commerce platform (Türkiye Materials Marketplace), measurement tools, and offers training, financial opportunities, and consultancy services for companies that are truly looking to accelerate their transition to circular. BCSD Türkiye believes this joint effort will support businesses to identify opportunities and provide solutions to turn them into reality.

Circular Buildings Platform

No circular buildings platform has been established to date.

Building Materials Passport Platform

No building materials platform has been established to date.

Public Procurement Platform

No public procurement platform has been established to date.

5.1.20.3 Funding Opportunities

TÜBİTAK (The Scientific and Technological Research Council of Türkiye): is a prominent government agency in Türkiye that supports research and innovation projects across various sectors, including construction and sustainability. TÜBİTAK often offers research grants and funding opportunities for projects that promote environmental sustainability, resource efficiency, and circular economy practices. Researchers, universities, and businesses in Türkiye can apply for TÜBİTAK grants and programs related to sustainable construction and circular economy research and development.

International Cooperation and Support from International Organizations

Horizon 2020 (European Union Program): While Horizon 2020 is an EU program, Turkish researchers and organizations can participate in Horizon 2020 projects and access funding opportunities through collaboration with European partners. Horizon 2020 has funded research projects focused on sustainable construction, circular economy, and resource efficiency. Turkish entities can join these consortia to benefit from funding and expertise. COSME Program (EU Program for Small and Medium-sized Enterprises): COSME is another EU program that supports small and medium-sized enterprises (SMEs). Turkish SMEs involved in construction or related industries can explore funding and support opportunities under COSME, particularly in projects related to sustainability and circular economy. European Structural and Investment Funds (ESIF): Türkiye, as a candidate country for EU membership, can access European Structural and Investment Funds (ESIF) to promote regional development and sustainable growth. ESIF funds may be allocated to projects aimed at improving environmental sustainability, including circular economy practices in construction, within eligible regions in Türkiye. National Development Agencies: In addition to TÜBİTAK, Türkiye has several regional development agencies that offer funding and support programs for projects contributing to regional economic development and sustainability. These agencies may support circular economy initiatives in construction on a regional level.

5.1.20.4 Challenges, Barriers and Potential Improvements

The main challenges and barriers to the implementation of circular economy in the built environment in Türkiye can be defined as follows:

- Lack of financial and economic aid;
- Deficiencies in the legislation and standards on waste;
- Lack of awareness regarding the environment;
- Lack of cooperation between stakeholders;
- Insufficient effective areas for waste collection and separation;
- Lack of innovation in recycling and low efficiency in recycling systems;
- Uncertainty about circular economy targets and strategies;
- Lack of temporary holding sites;
- Insufficient methods for sorting, transporting and recycling waste;

- Weaknesses in environmental regulations and insufficient monetary sanctions;
- Insufficiency in awareness regarding environmental protection;
- Insufficient advertising and media coverage for waste management.

The particularly important areas for improving the implementation of the circular economy in the built environment in Türkiye can be defined as follows:

- Governmental measures, landfill bans, taxes, green public procurement supporting recycling;
- Development of new standards;
- Develop further life cycle analysis indicators for the saving of natural resources;
- Adopting a secondary raw materials policy;
- New technological development/new business models.

5.1.20.5 Examples of Successful Implementation

Public Policy Initiatives

The UK-Türkiye GreenLegOuse project has created a flexible, Lego-like construction system that can be quickly built to provide affordable accommodation for low income communities, including the homeless, slum dwellers and refugees, not only in Newton Fund partner countries but countries all around the world [247]. The system can also be disassembled and reassembled, increasing its reusability and recyclability with easy-to-implement, low-cost and energy-efficient techniques, and removing the need for heavy demolition processes and the pollution this creates. Within the scope of project, the first-ever real-time demonstration of a fully demountable 1-story building constructed in 2021 with construction and demolition waste (CDW)-based geopolymer concrete has been constructed and completed. After the successful construction of the green building, it is believed that real-time demonstration of the study's outcomes in the field will help the impact of this study to last for years to come and will be used as a viable tool and advertisement element for anybody who is interested in making collaborations/learning more about CO_2 capture/storage, innovative CDW recycling and design-for-demountability for future reuse.

Private Policy Initiatives

No data to this point.

5.1.21 United Kingdom Factsheet

5.1.21.1 Policy and Regulatory Framework

Governance

The United Kingdom (UK) Government has committed to the circular economy. It has been horizon scanning to identify best practice internationally, for example, the EU's Circular Economy Package (CEP). Under its transition out of the EU, the UK is required to transpose the CEP into UK law. It also recognises the trade benefits of alignment with the EU bloc. Governance is spread across various departments and through the devolved institutions in Scotland, Wales and Northern Ireland, see below [135]. Details can be found on the UK Government portal.

At national and English level, the Department for Environment, Food and Rural Affairs (Defra) is taking the lead. In 2018 A Green Future: Our 25 Year Plan to Improve Environment [139] for England was published, with a sub-section of which is the Resources and Waste Strategy (RWS). Within the devolved administratons, the Department of Agriculture, Environmental and Rural Affairs (DAERA) is overseeing this and the Climate Change Act in Northern Ireland. The Scottish Government was the first of the UK nations to publish its circular economy strategy [12]. Zero Waste Scotland (ZWS) is the lead body, funded by the Scottish Government, responsible for promoting and advising on all circular economy issues. ZWS is a not-for-profit environmental organisation that works alongside government, local councils and other bodies to jointly deliver a more circular economy [351]. The Welsh Government has expressed its commitment to accelerating the transition to a circular and low carbon economy in the circular economy strategy. The Department for the Economy launched a public consultation on the draft Circular Economy Strategy for Northern Ireland in January 2023 [51].

Legislation

The Environmental Act [140], introduces legislation around waste and resource efficiency within England and Wales. The CEP Directive was transposed into UK law through *The Waste (Circular Economy) (Amendment) Regulations 2020* [136]. In 2023 the Scottish Government published the Circular Economy Bill [10].

Measures have been implemented by each UK nation:

- Technical guidance WM3 has applied in the UK from 1st July 2015. This was updated in 2021. This document provides guidance on waste classification. It is a comprehensive reference manual for anyone involved in producing, managing and regulating waste [137].
- The Welsh Government has created a portal outlining planning policy and guidance on waste [357].

- Scotland waste legislation may be accessed from their Waste Legislation portal [321].
- In North Ireland waste is overseen by the Department of Agriculture, Environment and Rural Affairs, who along with the Northern Ireland Environment Agency oversee the provinces waste legislation, which may be accessed through their portal [48].

In addition, the British Standards Institution launched a guide for organisations in adopting a circular economy approach to their business model—BS 8001:2017 [16]. It outlines what the circular economy is and how to implement its principles in the delivery of both products and services.

Circular Economy Strategy, Roadmap or Action Plan

The Scottish Government published *Making Things Last*, their CE strategy, in 2016 [12], which sets out a vision to move Scotland from a linear to a circular economy. In this strategy four areas are prioritised:

1. Food and drink, and the broader bio-economy
2. Remanufacture
3. Construction and the built environment
4. Energy infrastructure.

Construction and demolition waste accounts for about 50% of all waste in Scotland and influencing the built environment has an impact on wider resource efficiency. The construction sector is the biggest user of materials responsible for over half of total carbon emissions when the operation of buildings is included. The strategy sets the following priorities related to the construction sector:

- ensure building designs consider waste reduction in both new build and refurbishment, while also enabling more reuse and recycling at end of life
- supporting SMEs to deliver building projects with the potential to deliver significant impact that can be scaled up and repeated
- build capacity to deliver change in the construction sector in collaboration with the Construction Scotland Innovation Centre and other partners
- avoid depletion of primary aggregates and timber resources through enhanced recycling of demolition materials.

In Wales, *Beyond Recycling* has been launched in 2021 [358] to make Wales a circular, low carbon economy with zero waste by 2050. They commit to prioritising the use of sustainable and low carbon materials in construction in Wales to support progress towards whole life net zero carbon for construction projects. These metarials are intended to be used first for public sector construction, the refurbishment of the social housing stock, and the construction of new schools.

In 2022, the UK's Competition and Markets Authority (CMA) gave advice to the UK Government [27]. This advice included changes to consumer law so that consumers can make more sustainable choices, effective enforcement where breaches of consumer law lead to environmental harm and identifying other ways of promoting more sustainable consumption.

The Northern Ireland Executive has been developing a Circular Economy Strategy. This started with a construction demolition/waste policy in 2017, that was updated in 2024 [47]. Their vision is to have an innovative, inclusive and competitive economy by 2050 where business, people and planet flourish, with responsible production and consumption at its core. They acknowledge the need to increase the provision, quality and accessibility of secondary and regenerative comnstruction materials. They highlight the role of stakeholder engagement to make the secondary building materials market competitive and the need for collaboration across the entire value chain from architects, manufacturers, designers and regulators to contractors. It had a consultation which concluded in March 2023, led by the Department of the Economy [51]. This was based upon *The Circularity Gap Report* [9] which provided a snapshot by applying circularity metrics to the Northern Ireland economy. One significant finding was that the Northern Ireland economy was less than 8% circular, meaning that over 92% of the economy consumes virgin material to maintain itself. The construction and agriculture sectors were highlighted for focus for change.

5.1.21.2 Platforms and Networks Facilitating the Transition to a Circular Economy

5.1.21.3 Circular Economy Platforms

The Climate Change Committee and the UK's Green Building Council (UKGBC) have been pro-active in promoting the Circular Econony. The London Mayor's office have launched a Good Design Guide for the Circular Economy [104]. The UKRI have established a hub for five national centres for research on the circular economy [338].

Circular Buildings Platform

The UKGBC has established a platform to promote CE [336]. This contains significant guidance on the transition required.

Building Materials Passport Platform

A materials passport for buildings is being trialled in London [161]. The New Civil Engineer reports that this could form the basis of a UK wide materials database. They are working in conjunction with Circuland to produce the Circuland Platform.

Public Procurement Platform

The Scottish Government has launched a platform to inform and support circular procurement [322].

5.1.21.4 Funding Opportunities

UK Government has issued a roadmap on Sustainable Investing [138].

International Cooperation and Support from International Organizations

No data to this point.

5.1.21.5 Challenges, Barriers and Potential Improvements

Stakeholders in the construction industry cite a number of challenges and barriers to the implmentation of a circular economy in the UK. The UK Government and the devolved insitutions have set out clear strategies that point towards the adoption of a circular economy, but there is little legislative leverage to help. The UK Government is primarily focused on a market driven adoption using levers of government to promote CE, with one stakeholder saying that there was a '... reliance on manufacturers to do the *right thing* with no enforcement, and a lack of meaningful choices'.

Challenges raised include cost of recycled material, in-built supply chains, a lack of knowledge of the circular economy, or the need to become sustainable. Some cite a lack of government funding to develop products and a lack of meaningful regulation. A lack of clear quantifiable performance metrics was raised. This shows a lack of dissemination of material developed by Royal Institution of Chartered Surveyors (RICS) [284] and the UKGBC—however, it does point out a lack of knowledge of how recycled or reused materials will perform when used a second time. In terms of reusing buildings, UK tax legislation, primarily VAT on building products, makes retrofit expensive to deliver compared to new build. There are also concerns as to who undertakes quality assessment, verification and validation of secondary use products.

A full report on overcoming the barriers to the implementation of CE in the built environment has been published by the UK GBC [337]. These have also beein addressed by the Ellen MacArthur Foundation and Arup report—*First steps towards a circular built environment* [3].

In Northern Ireland, the recent consultation [51] into barriers to circularity highlighted the following:

- Lack of data collection and measurement to understand material flows and management of waste.
- Material possession of new products is perceived as a sign of wealth and wellbeing.

- The current perception of growth and approach to capital investment is to build new.
- Lack of CE awareness and understanding of the opportunities CE can present for businesses
- Inadequate research and innovation to identify circular solutions to current linear practices.
- Public procurement currently supports linear business models.
- Lack of leadership, vision and funding from government
- Lack of incentives to source regenerative materials
- Connections are lost between the demand for raw materials and the impact it has on climate change, biodiversity loss and inequality.
- Inadequate provision of networks to foster collaboration across sectors.
- Lack of the knowledge and appropriate skills to enable circularity across industries
- The existing waste regulatory framework and waste classifications impede the circularity of materials and valorisation opportunities.

From this report eight scenarios for action were identified, those which are relevant in this report include:

- Create a circular built environment.
- Champion circular manufacturing
- Power clean mobility
- Leverage public sector procurement
- Welcome a circular lifestyle.

5.1.21.6 Examples of Successful Implementation

Public Policy Initiatives

Local Government in London has tended to lead in establishing initiatives, demonstrated by the Merton Rule, which required developers to go beyond published building regulations in embedding renewable energy technologies in buildings. As a result of initiatives by the London Mayor, ReLondon was established [285] with a wide-ranging brief to make London circular. It is a partner in a number of research projects. It has deployed several pilot projects to promote circularity. In the built environment it has established a Materials Reuse Portal (MRP) for surplus materials, with the aim of promoting what materials are in circulation [55]. Initiatives by Re-London have included matchmaking events, which introduced clients and contractors to each other who would not have otherwise met. They also undertook a retrofit feasibility study in Hounslow, interviewing stakeholders, and undertaking market assessments as to how circular principles could be adopted in retrofit projects to minimise embodied carbon, extend material/product lifespans and reduce waste.

Private Policy Initiatives

Construction materials supplier Amey worked with Staffordshire County Council to recycle 7250 tonnes of tar-bound planings and combined it with fly ash material. The tar was classified as a hazardous waste, costing around £140 per tonne to dispose of. The recycled material also saved £100,000 on materials.

Arup erected temporary 'circular' buildings for the London Design Festival. All the elements used had to be able of being reused after the festival closed. Conversations were required across the wider supply chains with consideration of new asset ownership models. It required a significant change in the design thinking to achieve the resultant building especially regarding how the components were assembled.

References

1. ADEME – The French Agency for Ecological Transition (n.d.)(20 23) Funding. Available from: https://www.ademe.fr/en/our-missions/funding/. Accessed 22 August 2023
2. ARSM (2023) Asociace pro rozvoj recyklace stavebních materiálů v České republice, ARSM. Available from: http://www.arsm.cz/recyklaty.php. Accessed 9 April 2018
3. Acharya D, Boyd R, Finch O (2018) From principles to practices: first steps towards a circular built environment, cowes. Ellen MacArthur Foundation. Available from: https://www.ellenmacarthurfoundation.org/articles/first-steps-towards-a-circular-built-environment. Accessed 21 Aug 2025
4. Acordo Circular (2023) Plano de Ação para a Circularidade na Construção (PACCO). Available from: https://circularidade.builtcolab.pt/?jet_download=2310. Accessed 21 October 2023
5. Act No. 133/2013 Coll. on construction products and on amendments and additions to certain Acts, as amended by Act No. 91/2016 Coll. (2016) Available from: https://www.slov-lex.sk/vyhladavanie-pravnych-predpisov?text=133%2F2013. Accessed 10 July 2023
6. Alpha Bank economic research, sectors in focus: circular economy and applications in various sectors—Greece, December 2021, Circular Economy_Sectors in Focus_December 2021 (alpha.gr). Accessed 23 Nov 2023
7. Andelemandele.lv (2023) Available from: Andelemandele. Accessed 01 Sept 2023
8. Anon (2016) Making things last: a circular economy strategy for Scotland, Edinburgh, Scottish Government. Available from: https://www.gov.scot/publications/making-things-last-circular-economy-strategy-scotland/. Accessed 19 Sept 2023
9. Anon (2022) The circularity gap report Northern Ireland, Department of the Economy, Northern Ireland Executive, Belfast. Available from: https://www.economy-ni.gov.uk/publications/circularity-gap-report-northern-ireland. Accessed 19 September 2023
10. Anon (2023) The circular economy (Scotland) Bill Edinburgh, Scottish Government. Available from: https://www.parliament.scot/bills-and-laws/bills/circular-economy-scotland-bill/overview. Accessed 19 Sept 2023
11. Anon (2019) The SMecoMP project. Available from: https://www.smecomp.eu/actions. Accessed 2022 Sept 2023
12. Anon (2016) Making Things Last: a circular economy strategy for Scotland, Edinbugh, Scottish Government. Available from: https://www.gov.scot/publications/making-things-last-circular-economy-strategy-scotland/. Accessed 19 Sept 2023
13. Arup (2023) New passenger terminal, Zagreb International Airport. Available from: https://www.arup.com/projects/new-passenger-terminal-zagreb-international-airport. Accessed 14 Sept 2023

14. Arup (2023) Kerry Group's global technology and innovation centre. Available from: https://www.arup.com/projects/kerry-foods-advance-facility?utm_source=LinkedIn&utm_medium=Social&utm_campaign=Project&utm_content=E0000118. Accessed 7 July 2023
15. Association of Latvian Construction Contractors (2023) Construction industry indicators. Available from: https://www.latvijasbuvnieki.lv/statistika/. Accessed 07 Sept 2023
16. BSI (2021) Framework for implementing the principles of a circular economy in organisations. Guide. London, British Standards Institution, Available from: https://doi.org/10.3403/30334443. Accessed 21 Aug 2025
17. Borg RP, Bartolo I (2016) Green building products and technology: EcoBuild. SBE 16 Malta: Europe and the Mediterranean towards a sustainable built environment, Malta, pp 179-182
18. Bouygues Bâtiment International (2017) A leading airport for the Croatian capital. Available from: https://www.bouyguesbatimentinternational.com/reference/zagreb-airport/. Accessed 14 Sept 2023
19. Build360 (n.d.) Leading the transition towards a Circular Built Environment. Available from: https://build360.ie/. Accessed 5 July 2023
20. Bulgarian Association of Recycling (2021) Action Plan for Circular Economy: Bulgaria, under Interreg Europe programme project Smart Circular Procurement (CircPro). The Smart Circular Procurement (CircPro) project
21. Business.gov.nl (2023a) Building materials, soil, and dredging spoil. Available from: https://business.gov.nl/regulation/building-materials-soil-dredging-spoil/. Accessed 12 July 2023
22. Business.gov.nl (2023b) Building regulations. Available from: https://business.gov.nl/regulation/building-regulations/. Accessed 12 July 2023
23. CEIC Data, 'Albania Real GDP Growth', 2020. https://www.ceicdata.com/en/indicator/albania/real-gdp-growth
24. CEN TC350/SC1 'Circular Economy in the Construction sector'. © 2022 CEN-CENELEC – European Standardisation. https://standards.cencenelec.eu/dyn/www/f?p=205:7:0::::FSP_ORG_ID:2868394&cs=1E483B462A3001B254B70E3969CA6730E
25. CFP (2022) Our top 5 circular buildings in the Netherlands. Available from: https://cfp.nl/en/news-and-cases/our-top-5-circular-buildings-in-the-netherlands/. Accessed 12 July 2023
26. CIRCULÉIRE (n.d.) The National Platform for Circular Innovation. Available from: https://circuleire.ie/. Accessed 5 July 2023
27. CMA (2022) Correspondence, Environmental Sustainability and the UK competition and consumer regimes: CMA advice to the Government, London, Her Majesty's Government, Available from: https://www.gov.uk/government/publications/environmental-sustainability-and-the-uk-competition-and-consumer-regimes-cma-advice-to-the-government, Accessed 21 Aug 2025
28. CZGBC (2023) Cirkulární stavebnictví a šetrné materiály | CZGBC. Available from: www.czgbc.org/cs/pracovni-skupiny/cirkularni-stavebnictvi-a-setrne-materialy. Accessed 16 Nov 2023
29. Cabinet of Ministers (2017) Requirements for green public procurement and their application procedure available from: Prasības zaļajam publiskajam iepirkumam un to piemērošanas kārtība (likumi.lv). Accessed 15 Sept 2023
30. Circle Economy (2022) Only 8% of used building materials are circular in the Netherlands. Available from: https://www.circle-economy.com/news/only-8-of-used-building-materials-are-circular-in-the-netherlands. Accessed 15 July 2023
31. Circle Economy and Metabolic (2022) Circularity gap report: built environment, the Netherlands. Available from: https://assets-global.website-files.com/5d26d80e8836af2d12ed1269/637375bea39b387c57eca11f_20221115%20-%20CGR%20BEN%20-%20Report%20-%20210x297mm.pdf. Accessed 12 Sept 2023
32. Circul'R (n.d.) Circul'R–Enter the loop. Available from: https://www.circul-r.com/. Accessed 22 August 2023
33. Circular Economy and City of Amsterdam (eds) (2020) Amsterdam circular 2020–2025 strategy. City of Amsterdam, Amsterdam

34. Circular PP (2018) Study tour to Alliander company in Duiven. Available from: http://circularpp.eu/study-tour-to-alliander-company-in-duiven/. Accessed 12 July 2023
35. Circular Economy Finland (KiSu) (n.d.) Circular Economy Finland (KiSu). Available from: https://kiertotaloussuomi.fi/en/. Accessed 22 August 2023
36. Circular Economy Malta (Establishment) Order S.L. 595.28. https://legislation.mt/eli/sl/595.28/eng/pdf
37. City of Athens, Resilience strategy for 2030, https://resilientathens.wordpress.com/, 2019. Accessed 23 Oct 2023
38. City of Helsinki (2020) The City of Helsinki's Roadmap for Circular and Sharing Economy. Urban Environment Publications 2/2020. Available from: https://circulareconomy.europa.eu/platform/sites/default/files/the-city-of-helsinkis-roadmap-for-circular-and-sharing-economy.pdf. Accessed 22 August 2023
39. City of Rotterdam (2019) From trash to treasure–Rotterdam circularity programme 2019-2023. Available from: https://rotterdamcirculair.nl/wp-content/uploads/2019/05/Rotterdam_Circularity_Programme_2019-2023.pdf. Accessed 5 July 2023
40. Climate Action Fund, Disruptive Technologies Innovation Fund, EU LIFE Programme, Horizon Europe, Circular Bio-Based Europe Industry Joint Undertaking Europe and C-Voucher - CIRCULÉIRE (n.d.) Funding. Available from: https://circuleire.ie/funding/. Accessed 5 July 2023
41. Co-PLAN, Institute for Habitat Development (2020) Regional circular economy country specific report. Available at: https://www.co-plan.org/wp-content/uploads/2021/01/Co-PLAN-Annual-Report_-2019-.pdf
42. Co-PLAN, Institute for Habitat Development, Regional Circular Economy Country Specific Report (2020). http://env-net.org/wp-content/uploads/2020/04/Circular-Economy-RegionalReport-2019.pdf
43. Construction Act (Act No. 201/2022 Coll.) (2022) Available from: https://www.slov-lex.sk/pravne-predpisy/SK/ZZ/2022/201/. Accessed 10 July 2023
44. Construction Site Management Regulations S.L. 623.08. https://legislation.mt/eli/sl/623.8/eng
45. Construction and Demolition Waste Strategy for Malta 2021–2030: Managing Construction & Demolition Resources – 2021 Environment & Resources Authority, Malta, 34pp. https://era.org.mt/construction-and-demolition-waste-strategy-for-malta-2021-2030-managing-construction-demolition-resources/ (PDF: https://era.org.mt/wp-content/uploads/2021/10/Construction-and-Demolition-Waste-Strategy-for-Malta-2021-2030-Managing-Construction-Demolition.pdf)
46. Council of Ministers (2021) National Development Programme BULGARIA 2030. https://www.minfin.bg/upload/46720/National+Development+Programme+BULGARIA+2030.pdf. Accessed 28 Aug 2025
47. DAERA (2024) Duty of Care – Construction/demolition waste and recycled aggregate, Belfast, Northern Ireland Executive, 2nd edn. Available from: https://www.daera-ni.gov.uk/publications/duty-care-construction-demolition-waste-and-recycled-aggregate. Accessed 21 Aug 2025
48. DEARA (n.d.) Resource and waste policy, Belfast, Northern Ireland executive. Available from: https://www.daera-ni.gov.uk/topics/resource-waste-policy. Accessed 21 Aug 2025
49. Department of Communications, Climate Action and Environment of the Government of Ireland (2022). A Waste Action Plan for a Circular Economy–Ireland's National Waste Policy 2020-2025. Available from: https://www.gov.ie/pdf/?file=https://assets.gov.ie/86647/dcf554a4-0fb7-4d9c-9714-0b1fbe7dbc1a.pdf#page=null. Accessed 5 July 2023
50. Department of the Environment, Climate and Communications of the Government of Ireland (2021) The whole of government circular economy strategy 2022–2023: living more, using less. Available from: https://assets.gov.ie/207622/bd90130d-494e-4d32-8757-46d36c77b912.pdf. Accessed 7 July 2023

51. Department for the Economy (2023) Draft circular economy strategy for Northern Ireland. Available from: https://www.economy-ni.gov.uk/sites/default/files/consultations/economy/draft-circular-economy-strategy-for-northern-ireland-main-report.pdf. Accessed 19 Sept 2023
52. Dianeosis, Circular economy: opportunities, challenges and impacts on Greek economy (2022) Home—Dianeosis. Accessed 23 Oct 2023
53. Diemer A et al (2022) Waste management and circular economy in the French building and construction sector. Front Sustain 3. https://doi.org/10.3389/frsus.2022.840091
54. Directorate for Technological Industrial Development Zones (2021) About us. Available from: https://fez.gov.mk/en/about-us-2/. Accessed 22 Sept 2023
55. Dsposal (2023) Materials Reuse Portal Case Study, Dsposal, London, Available from: https://dsposal.uk/articles/materials-reuse-portal-case-study. Accessed 28 Aug 2025
56. ECCS—European Convention for Constructional Steelwork (2020) European recommendations for reuse of steel products in single-storey buildings. Factsheet no. 6: MEXX DAY hall, Timisoara, Romania. Available from: https://www.steelconstruct.com/wp-content/uploads/FS06-Mexx-Day-hall-Timisoara.pdf
57. ECCS—European Convention for Constructional Steelwork (2020) European recommendations for reuse of steel products in single-storey buildings. Factsheet no. 3: HIDROTIM office, Timisoara, Romania. Available from: https://www.steelconstruct.com/wp-content/uploads/CS03-Hidrotim-office-Timisoara.pdf
58. EIT Climate-KIC (2019) Circular economy transition (CE) in smart specialization strategy (S3), source at https://ae4ria.org/eit-climate-kic-circular-economy-transition-ce-in-smart-specialization-strategy-s3/
59. ENEA (2019) Report on circular economy in Italy, Available from: https://circulareconomynetwork.it/wp-content/uploads/2019/04/Proposals-and-Research-Summary-Report-on-circular-economy-in-Italy-2019.pdf. Accessed 21 Aug 2023
60. EPA (2021). Green Public Procurement: Guidance for the Public Sector. Available from: https://www.epa.ie/publications/circular-economy/resources/GPP-Guidance-for-the-Irish-Public-Sector.pdf. Accessed 5 July 2023
61. EPA (n.d.) Green Enterprise. Available from: https://www.epa.ie/our-services/monitoring-assessment/circular-economy/green-enterprise/. Accessed 5 July 2023
62. EPA (2021a) The Circular Economy Programme 2021–2027. Environmental Protection Agency. Available from: https://www.epa.ie/publications/circular-economy/resources/EPA_Circular_Economy_2021_Programme_Apr22_Web.pdf. Accessed 2 July 2023
63. EPA (2021) Best practice guidelines for the preparation of resource & waste management plans for construction & demolition projects. Environmental Protection Agency, Wexford
64. ETC CE—European Topic Centre on Circular economy and resource use (2022/5) Circular economy country profile—Romania. Available from: https://www.eionet.europa.eu/etcs/etc-ce/products/etc-ce-products/etc-ce-report-5-2022-country-profiles-on-circular-economy/romania-ce-country-profile-2022_for-publication.pdf
65. EU Cohesion Policy: €817 million to Malta to support a more competitive, inclusive, green, and digital economy. Directorate-General for Communication, European Commission (2022) https://ec.europa.eu/commission/presscorner/detail/en/ip_22_5500
66. EU Cohesion Policy: Waste management capacity has improved in Malta thanks to EU funds. Directorate-General for EU regional and urban policy, European Commission (2023). https://ec.europa.eu/regional_policy/whats-new/newsroom/15-06-2023-eu-cohesion-policy-waste-management-capacity-has-improved-in-malta-thanks-to-eu-funds_en
67. EU Regional Cooperation Council (2021) Action plan for the implementation of the Sofia declaration on the green agenda for the Western Balkans 2021–2030. s.l.:s.n
68. Eclaira (n.d.) Eclaira – Réseau de l'économie circulaire en Auvergne-Rhône-Alpes. Available from: https://www.eclaira.org/. Accessed 22 August 2023
69. EcoUp Oyj (n.d.) Construction materials & solutions for circular economy. Available from: https://ecoup.fi/en/. Accessed 22 August 2023

70. Economiecirculaire.org (n.d.) Economiecirculaire.org – Plateforme internationale de l'économie circulaire. Available from: https://www.economiecirculaire.org/. Accessed 22 August 2023
71. electronic Irish Statute Book (eISB) (n.d.) Circular Economy and Miscellaneous Provisions Act 2022. No. 26 of 2022. Available from: https://www.irishstatutebook.ie/eli/2022/act/26/enacted/en/html. Accessed 10 September 2023
72. electronic Irish Statute Book (eISB) (n.d.) Waste Management Act, 1996. No. 10 of 1996. Available from: https://www.irishstatutebook.ie/eli/1996/act/10/enacted/en/pdf. Accessed 10 September 2023
73. Ellen MacArthur Foundation (2021) France's antiwaste and circular economy law: eliminating waste and promoting social inclusion. Available from: https://emf.thirdlight.com/file/24/kLSzgopkL.2CJxQkLb3XkLQlS7_/Case%20Studies%20-%20French%20Anti%20Waste%20Law.pdf. Accessed 7 July 2023
74. Environment Protection Act, Malta CAP. 549. https://legislation.mt/eli/cap/549/eng/pdf
75. Environmental Technology Verification ETC4Innovation, GREEN SYNERGY cluster. Available at: https://greensynergycluster.eu/pf/etv4innovation/?lang=en
76. European Cluster Collaboration Platform (n.d.) WeBSEFF — Western Balkans Sustainable Energy Financing Facility. Available from: https://www.clustercollaboration.eu/content/webseff-western-balkans-sustainable-energy-financing-facility. Accessed 12 June 2023
77. European Circular Economy Stakeholder Platform (n.d.) Circul'R. Available from: https://circulareconomy.europa.eu/platform/en/dialogue/existing-eu-platforms/circulr. Accessed 22 August 2023
78. European Circular Economy Stakeholder Platform (n.d.) Green Deal on Circular Procurement in France. Available from: https://circulareconomy.europa.eu/platform/en/about/cg-activities-documents/green-deal-circular-procurement-france. Accessed 22 August 2023
79. European Commission (n.d.) Energy Performance of Buildings Directive. Available from: https://energy.ec.europa.eu/topics/energy-efficiency/energy-efficient-buildings/energy-performance-buildings-directive_en. Accessed 5 July 2023
80. European Circular Economy Stakeholder Platform—Bulgaria. https://circulareconomy.europa.eu/platform/en/main-language/bulgarian
81. European Commission (2022) European semester country report for Bulgaria. https://commission.europa.eu/document/download/fdf2c3a2-9dc8-4dd2-a1df-41f88e4ccf41_en?filename=2022-european-semester-country-report-bulgaria_en.pdf. Accessed 28 August 2025
82. European Commission (n.d.) Cyprus' recovery and resilience plan. Available from: https://commission.europa.eu/business-economy-euro/economic-recovery/recovery-and-resilience-facility/cyprus-recovery-and-resilience-plan_en#:~:text=The%20plan%20will%20foster%20economic,States%2C%20for%20instance%20through%20exports. Accessed 18 Sept 2023
83. European Commission (2020) A new circular economy action plan for a cleaner and more competitive Europe. Communication from the Commission to the European Parliament, the Council, the European Economic and Social Committee and the Committee of the Regions. Available from: https://eur-lex.europa.eu/legal-content/EN/TXT/PDF/?uri=CELEX:52020DC0098. Accessed 18 Sept 2023
84. European Environment Agency (2021) Municipal waste management—country fact sheet North Macedonia. s.l.: EEA, Copenhagen
85. European Topic Centre on Circular economy and resource use (2022a) Circular economy country profile—Croatia. European Environment Agency. Available from: https://www.eionet.europa.eu/etcs/etc-ce/products/etc-ce-products/etc-ce-report-5-2022-country-profiles-on-circular-economy/croatia-ce-country-profile-2022_for-publication.pdf. Accessed 12 June 2023
86. European Topic Centre on Circular economy and resource use (2022f) Circular economy country profile—Bulgaria. Available from: https://www.eionet.europa.eu/etcs/etc-ce. Accessed 12 June 2023

87. European Topic Centre on Circular economy and resource use (2022b) Circular economy country profile—Ireland. European Environment Agency. Available from: https://www.eionet.europa.eu/etcs/etc-ce/products/etc-ce-products/etc-ce-report-5-2022-country-profiles-on-circular-economy/ireland-ce-country-profile-2022_for-publication.pdf. Accessed 12 June 2023
88. European Topic Centre on Circular economy and resource use (2022c) Circular economy country profile—Italy. Available from: https://www.eionet.europa.eu/etcs/etc-ce. Accessed 28 Aug 2023
89. European Topic Centre on Circular economy and resource use (2022d) Circular economy country profile—Latvia. Available from: https://www.eionet.europa.eu/etcs/etc-ce/products/etc-ce-products/etc-ce-report-5-2022-country-profiles-on-circular-economy/latvia-ce-country-profile-2022_for-publication.pdf. Accessed 13 Sept 2023
90. European Topic Centre on Circular economy and resource use (2022e) Circular economy country profile—Spain. European Environment Agency. Available from: https://www.eionet.europa.eu/etcs/etc-ce/products/etc-ce-products/etc-ce-report-5-2022-country-profiles-on-circular-economy/spain-ce-country-profile-2022_for-publication.pdf. Accessed 8 Oct 2023
91. European Union: European Commission (2022) Guidelines for the Implementation of the Green Agenda for the Western Balkans. Brussels: s.n
92. European Union, Albanie 2019 Report, 2019. https://ec.europa.eu/neighbourhoodenlargement/sites/near/files/20190529-albania-report.pdf
93. fib (International Federation for Structural Concrete). https://www.fib-international.org/
94. FRD Center (2021) Circular economy in Romania. Available from: https://www.frdcenter.ro/frd-center-circular-economy-in-romania-2021/. Accessed 5 Feb 2024
95. Filipova M, Zheleva I, Hvarchilkova D (2020) Preliminary analysis of construction and demolition waste management in Bulgaria. In: Proceedings of 12th international online conference for promoting the application of mathematics in technical and natural sciences—AMiTaNS'20, 24–29 June 2020, Albena, Bulgaria
96. Finnish Environment Institute (SYKE) (2025) Circwaste – Kiertotalousratkaisuja. Available from: https://kiertotalousratkaisuja.syke.fi/en/circwaste/. Accessed 22 August 2023
97. Fisu Network (n.d.) Finnish sustainable communities. Available from: https://fisunetwork.fi/en/. Accessed 22 August 2023
98. France (2020) LOI n° 2020-105 du 10 février 2020 relative à la lutte contre le gaspillage et à l'économie circulaire. Available from: https://www.legifrance.gouv.fr/jorf/id/JORFTEXT000041553759/. Accessed 22 August 2023
99. from: download (varam.gov.lv). Accessed 01 Sept 2023
100. Fund for Innovation and Technological Development (2021) Mission and goals. Available from: https://fitr.mk/en/mission-and-goals/. Accessed 22 Sept 2023
101. Fundo Ambiental and Agência Portuguesa do Ambiente (2019) Long-term strategy for carbon neutrality of the Portuguese economy by 2050. Available from: https://unfccc.int/sites/default/files/resource/RNC2050_EN_PT%20Long%20Term%20Strategy.pdf. Accessed 21 Oct 2023
102. GIZ (2022) Promoting economic growth in North Macedonia. Available from: https://www.giz.de/en/worldwide/117373.html. Accessed 22 Sept 2023
103. GIZ (2023) Promoting economic growth in North Macedonia. Available from: https://www.giz.de/en/worldwide/132043.html. Accessed 22 Sept 2023
104. GLA (n.d.) Design for a Circular Economy. Primer, London, Greater London Authority, Available from: https://www.london.gov.uk/sites/default/files/design_for_a_circular_economy_web_2.pdf. Accessed 21 Aug 2025
105. Gaman A (2022) Casa EFdeN Signature—Cea mai sustenabilă casă din Europa, LEED for Homes platinum. Green Report [Online], 1 February. Available from: https://green-report.ro/casa-efden-signature-cea-mai-sustenabila-casa-din-europa-leed-for-homes-platinum/. Accessed 1 Feb 2024
106. Gemeente Groningen (2020) Routekaart Circulaire Economie—Gemeente Groningen. Available at: https://gemeenteraad.groningen.nl/Documenten/Bijlage-1-Circulaire-Economie-1.pdf. Accessed 5 July 2023

107. Gobierno de España (2021) Plan de Recuperación, Transformación y Resiliencia. Available from: https://www.lamoncloa.gob.es/temas/fondos-recuperacion/Documents/160621-Plan_Recuperacion_Transformacion_Resiliencia.pdf. Accessed 9 Oct 2023
108. Gobierno de España (2022) PERTE en economía circular. Available from: https://www.miteco.gob.es/content/dam/miteco/es/calidad-y-evaluacion-ambiental/temas/economia-circular/perteenec_tcm30-537854.pdf. Accessed 8 Oct 2023
109. Government of the Republic of Croatia (2024) Building act (In Croatian). Official Gazette of the Republic of Croatia, No. 153/13, 20/17, 39/19, 125/19, 145/24
110. Government of the Republic of Croatia (2024) Waste management act (In Croatian). Official Gazette of the Republic of Croatia No. 84/21, 142/23
111. Government of the Republic of Croatia (2023) Decision on the adoption of the Waste management plan of the Republic of Croatia for the period 2023–2028 (In Croatian). Official Gazette of the Republic of Croatia, No. 84/23. Available from: https://narodne-novine.nn.hr/clanci/sluzbeni/2023_07_84_1334.html. Accessed 12 May 2023
112. Government of the Republic of Croatia (2020) National Reform Programme 2020. Available from: https://vlada.gov.hr/UserDocsImages/Europski%20semestar%202020/National%20reform%20programme%202020.pdf. Accessed 12 May 2023
113. Government of the Republic of Croatia (2021) Decision on Green Public Procurement in the Central Public Procurement Procedures (In Croatian). Official Gazette of the Republic of Croatia, No. 49/21. Available from: https://zelenanabava.hr/dokumenti/Odluka-VRH-ZeJN-NN-49-2021.pdf. Accessed 12 May 2023
114. Government of Ireland (n.d.) GPP Criteria Search. Available from: https://gppcriteria.gov.ie/. Accessed 5 July 2023
115. Government of Ireland (n.d.) Circular Economy Innovation Grant Scheme (CEIGS). Available from: https://www.gov.ie/en/service/b3faa-apply-for-the-circular-economy-innovation-grants-scheme-ceigs/. Accessed 5 July 2023
116. Government of the Netherlands (n.d.) Ministry of Infrastructure and Water Management. Available from: https://www.government.nl/ministries/ministry-of-infrastructure-and-water-management. Accessed 5 July 2023
117. Government of the Netherlands (2016). A circular economy in the Netherlands by 2050. Available from: https://circulareconomy.europa.eu/platform/sites/default/files/17037circulaireeconomie_en.pdf. Accessed 12 July 2023
118. Government Decision 102 from 05.02.2013 on the Energy Strategy-2030 of the Republic of Moldova [2013] MO nr. 27–30 art.147 from February 8 2013
119. Government Decision 107 from 28.07.2014 on the Code of Practice in Construction CP A.09.04:2014—Construction and demolition waste management [2014] MO nr. 249–255 art.1211 from August 22 2014
120. Government Decision 160 from 21.02.2018 on the Program for the promotion of the Green Economy in the Republic of Moldova for 2018–2020 and of the Action Plan for its implementation [2018] MO nr. 68–76 art.76 from March 3 2018
121. Government Decision 199 from 20.03.2014 on the Water supply and sewerage strategy for the years 2014–2028 [2014] MO nr. 72–77 art.222 from 28 Feb 2014
122. Government Decision 3248 from 10.04.2013 on the Waste management strategy in the Republic of Moldova for the years 2013–2027 [2013] MO nr. 82 art.5306 from 12 Apr 2013
123. Government Decision 386 from 17.06.2020 on the planning, development, approval, implementation, monitoring and evaluation public policy documents. Project of the Program for the promotion of the green and circular economy for the years 2023–2027 [2020] MO nr. 153–158 art.509 from 26 June 2020
124. Government Decision 592 from 27.11.2019 on the Greening program of small and medium enterprises [2019] MO nr. 360–366 art.907 from 12 Dec 2019
125. Government Decision 721 from 02.02.1996 on the Law on quality in constructions of [1996] MO nr. 25 art.259 from 25 Apr 1996

126. Government of the Netherlands (2023a) Circular Dutch economy by 2050. Available from: https://www.government.nl/topics/circular-economy/circular-dutch-economy-by-2050#:~:text=The%20government%2Dwide%20programme%20for,smarter%20and%20more%20efficient%20way. Accessed 12 July 2023
127. Government of the Netherlands (2023b) Accelerating the transition to a circular economy. Available from: https://www.government.nl/topics/circular-economy/accelerating-the-transition-to-a-circular-economy. Accessed 12 July 2023
128. Government of the Republic of Serbia (GOV RS) (2019) Public procurement act (In Serbian). The Official Gazette of the Republic of Serbia No. 91/2019
129. Government of the Republic of Serbia (GOV RS) (2021) Planning and building act (In Serbian). The Official Gazette of the Republic of Serbia No. 72 and 81/2009, 64/2010, 24/2011, 121/12, 42, 50 and 98/2013, 132 and 145/2014, 83/2018, 31 and 37/2019, 9/2020, 52/2021, 2021
130. Government of the Republic of Serbia (GOV RS) (2022a) Circular economy development program for Republic of Serbia for period 2022–2024 (In Serbian). The Official Gazette of the Republic of Serbia No. 137/2022
131. Government of the Republic of Serbia (GOV RS) (2022b) Waste management program in the Republic of Serbia for the Period 2022–2031 (In Serbian). The Official Gazette of the Republic of Serbia No. 12/2022
132. Government of the Republic of Serbia (GOV RS) (2023) Waste management act (in Serbian). The Official Gazette of the Republic of Serbia No. 36/2009, 88/2010, 14/2016, 95/2018, and 35/2023
133. Greek Association for Local Development and Government Administration, Circular Economy and local government administration, September 2020, 13102020_odhgos_kyklikhs_oikonomias.pdf (eetaa.gr). Accessed 23 Nov 2023
134. Green Building Council Finland (FIGBC) (n.d.) Green Building Council Finland. Available from: https://figbc.fi/en/front-page. Accessed 22 August 2023
135. HMG (2020) Circular Economy Package policy statement: Policy Paper, Department for Environment, Food and Rural Affairs, London, Whitehall, Her Majesty's Government. Available from: https://www.gov.uk/government/publications/circular-economy-package-policy-statement/circular-economy-package-policy-statement. Accessed 21 August 2025
136. HMG (2020) The Waste (Circular Economy) (Amendment) Regulations, London, Whitehall, Her Majesty's Government. Available from: https://www.legislation.gov.uk/uksi/2020/904/made. Accessed 21 Aug 2025
137. HMG (2021) Technical Guidance WM3: Waste Classification, London, Environmental Agency, Available from: https://www.gov.uk/government/publications/waste-classification-technical-guidance. Accessed [21 August 2025]
138. HMG (2021) Green Financing: A Roadmap to Sustainable Investing, London, Whitehall, Her Majesty's Government, Available from: https://assets.publishing.service.gov.uk/government/uploads/system/uploads/attachment_data/file/1031805/CCS0821102722-006_Green_Finance_Paper_2021_v6_Web_Accessible.pdf. Accessed 21 Aug 2025
139. HMG (2018) A green future: our 25 year plan to improve the environment, London, Whitehall, His Majesty's Government. Available from: https://www.gov.uk/government/publications/25-year-environment-plan. Accessed 19 Sept 2023
140. HMG (2021) The Environmental Act 2021, London, Whitehall, Her Majesty's Printers. Available from: https://www.legislation.gov.uk/ukpga/2021/30/pdfs/ukpga_20210030_en.pdf. Accessed 19 Sept 2023
141. Hadjieva-Zaharieva R, Dimitrova E, Buyle-Bodin F (2003) Building waste management in Bulgaria: challenges and opportunities. Waste Manage 23(8):749–761
142. Hankintakeino.fi (n.d.) Hankintakeino.fi—Sustainable and Innovative Public Procurement. Available from: https://www.hankintakeino.fi/en. Accessed 22 August 2023
143. Hellenic Republic, Ministry of Development (n.d.) Green Public Procurement Action Plan, GREEN PUBLIC PROCUREMENT (GPP)—ΓΕΝΙΚΗ ΓΡΑΜΜΑΤΕΙΑ ΕΜΠΟΡΙΟΥ. Available from: mindev.gov.gr. Accessed 15 October 2023

144. Hellenic Republic, Ministry of Environment and Energy, National Strategy for Circular Economy, 2018, Εθνική-Στρατηγική-Για-την-Κυκλική-Οικονομία.pdf (ypen.gov.gr). Accessed 30 July 2023
145. Hellenic Republic, Ministry of Environment and Energy, Action Plan for Circular Economy, C:\Users\obourka\AppData\Local\Temp\mso2B15.tmp (ypen.gov.gr). Accessed 30 July 2023
146. Hellenic Republic, Ministry of Environment and Energy, "National Waste Prevention Program". http://www.opengov.gr/minenv/?p=11636. Accessed 30 July 2023
147. Hellenic Republic, Ministry of Environment and Energy, Modified National Waste Management Plan 2020–2030. http://www.opengov.gr/minenv/?p=12569. Accessed 30 July 2023
148. Holland Circular Hotspot (n.d.) Mission. Available from: https://hollandcircularhotspot.nl/. Accessed 5 July 2023
149. Holland Circular Hotspot (n.d.) About Holland Circular Hotspot. Available from: https://hollandcircularhotspot.nl/about-us. Accessed 5 July 2023
150. Holland Circular Hotspot (2023a) Netherlands introduces National Circular Economy Program 2023—2030. Available from: https://hollandcircularhotspot.nl/news/netherlands-introduces-national-circular-economy-program-2023-2030/. Accessed 8 July 2023
151. Holland Circular Hotspot (2023b) Circular economy opportunities in Slovakia. Available from: https://hollandcircularhotspot.nl/slovakia/. Accessed 12 July 2023
152. IFC Romania & Fepra Group & CentrIQ (2023) Romanian circular economy—sectorial consultations. Available from: https://circularity-in-action.ro/
153. INCIEN (2023) Český cirkulární hotspot—Institut cirkulární ekonomiky. Available from: https://incien.org/hotspot/. Accessed 16 Nov 2023
154. Ihobe (2018) Economía circular. Available from: https://www.ihobe.eus/economia-circular. Accessed 9 October 2023
155. Institut National de l'Économie Circulaire (n.d.) Institut National de l'Économie Circulaire. Available from: https://institut-economie-circulaire.fr/. Accessed 22 August 2023
156. Institut National de l'Économie Circulaire (2023) LIFE Waste2Build, Toulouse Métropole se circularise. Available from: https://institut-economie-circulaire.fr/life-waste2build-toulouse-metropole-se-circularise/. Accessed 7 July 2023
157. Irish Green Building Council (n.d.) Construction Materials Exchange – a digital platform to unlock the potential of excess construction materials. Available from: https://www.igbc.ie/construction-materials-exchange/. Accessed 5 July 2023
158. Jefatura del Estado (2021) Ley 7/2021, de 20 de mayo, de cambio climático y transición energética. Available from: https://www.boe.es/eli/es/l/2021/05/20/7. Accessed 30 Sept 2023
159. Jefatura del Estado (2022) Ley 7/2022, de 8 de abril, de residuos y suelos contaminados para una economía circular. Available from: https://www.boe.es/eli/es/l/2022/04/08/7/con. Accessed 30 Sept 2023
160. KONE (2022) Scorching summers and extreme winters: how France is preparing its buildings for a new climate reality. Available from: https://www.kone.com/en/news-and-insights/stories/scorching-summers-and-extreme-winters-how-france-is-preparing-its-buildings-for-a-new-climate-reality.aspx. Accessed 10 July 2023
161. Kanaris S (2023) Materials passports trial in London could trigger UK materials database creation New Civil Engineer, 26 May, Institution of Civil Engineers, London. Available from: https://www.newcivilengineer.com/innovative-thinking/material-passports-trial-in-london-could-trigger-uk-materials-database-creation-26-05-2023/. Accessed 19 Sept 2023.
162. Kincl B, Neidhardt V, Radić J, Vlašić A, Mujkanović N (2012) Passenger terminal construction at Zagreb airport. Građevinar 64(6):475–484. https://doi.org/10.14256/JCE.669.2012
163. Knuts, T., Östberg, G. & Hautala, R., (2020) Ostrobothnia in transition – Roadmap for sustainable development and circular economy. Jakobstad Region Development Company Concordia Ltd., Vaasa Region Development Company VASEK, Yritystalo Dynamo, Kristiinankaupungin Elinkeinokeskus Oy. Available from: https://circulareconomy.europa.eu/platform/sites/default/files/2022-12/Roadmap%20Ostrobothnia.pdf. Accessed 22 August 2023

164. Kosta B, Memaj F (2019). Business models for the circular economy—case of Albania. Available from: https://www.researchgate.net/profile/Eglantina-Hysa/publication/338149864_Relationship_between_Governance_and_Human_Development_in_European_Countries_Panel_Regression_Analysis/links/5e02800a92851c8364992117/Relationship-between-Governance-and-Human-Development-in-European-Countries-Panel-Regression-Analysis.pdf#page=13. Accessed 3 April 2024
165. LIFE Waste To Resources IP, LIFE20 IPE/LV/000014 "Waste To Resources Latvia—boosting regional sustainability and circularity", 2022–2028. Available from: Home I Waste to Resources (varam.gov.lv). Accessed 13 Sept 2023
166. Large Volume Waste Recycling for Low-Impact High-Performance Concrete – ReCon (University of Malta: https://www.um.edu.mt/ben/constructmanage/ourprojects/recon/; Malta Council for Science and Technology: https://mcst.gov.mt/funded-projects-2017-till-date/#1673968269291-7cc82eee-805b)
167. Latvian Association of Waste Management Companies (2023) Construction waste management guidelines. Available from: BBNA_vadlīnijas_final_30.01.2023.pdf (cdn-website.com). Accessed 12 Sept 2023
168. Latvian Cabinet of Ministers (2020). Latvian National Energy and Climate Plan for the 2021–2030. Available from: Par Latvijas Nacionālo enerģētikas un klimata plānu 2021.–2030. gadam (likumi.lv). Accessed 12 Sept 2023
169. Latvian Cabinet of Ministers (2020a) Latvia's vision for achieving climate neutrality by 2050. Available from: https://ec.europa.eu/clima/sites/lts/lts_lv_lv.pdf. Accessed 15 Sept 2023
170. Latvian Cabinet of Ministers (2020b) Action plan for the transition to the circular economy for 2020–2027. Available from: https://likumi.lv/ta/id/317168-par-ricibas-planu-parejai-uz-aprites-ekonomiku-20202027-gadam. Accessed 12 Sept 2023
171. Latvian Cabinet of Ministers (2021) Guidelines of the national industrial policy for the 2021–2027. Available from: Par Nacionālās industriālās politikas pamatnostādnēm 2021.–2027. gadam (likumi.lv). Accessed 05 Sept 2023
172. Law 741-IX of 19 December 2022 on the approval of The National Development strategy "European Moldova 2030" [2022] MO nr. 409–410 (8453–8454) from 21 December 2022
173. Long Term Waste Management Plan 2021-2030. ERA, Ministry for the Environment, Climate Change and Planning (2021) https://era.org.mt/long-term-waste-management-plan-2021-2030/ (PDF: https://era.org.mt/wp-content/uploads/2022/02/Long-Term-Waste-Management-Plan-v1.4.3-Spreads-Digital-Version.pdf)
174. MEUCC (2018) National action plan for waste management 2016–2023. Available from: http://faolex.fao.org/docs/pdf/tur199747.pdf. Accessed 30 Sept 2023
175. MEUCC (2023) Republic of Türkiye updated first nationally determined contribution. Ministry of environment, urbanization and climate change. Available from: https://unfccc.int/sites/default/files/NDC/2023-04/T%C3%9CRK%C4%B0YE_UPDATED%201st%20NDC_EN.pdf. Accessed 30 Sept 2023
176. MPO (2023) Národní plán obnovy, Národní plán obnovy. Available from: www.planobnovycr.cz/. Accessed 16 Nov 2023
177. Madaster (2023a) Our platform. Available from: https://madaster.nl/en/platform/. Accessed 12 July 2023
178. Madaster (2023b) Our purpose. Available from: https://madaster.nl/en/our-purpose/. Accessed 12 July 2023
179. Madaster (n.d.) Material Passports for Circular Construction. Available from: https://madaster.com/material-passport/. Accessed 22 August 2023
180. Malta: Recovery and Resilience Plan 2023, FONDI.eu (SEM) https://fondi.eu/wp-content/uploads/2023/11/Malta-Recovery-and-Resilience-Plan-2023-7.11.2023.pdf
181. Malta Council for Science and Technology (MCST). https://mcst.gov.mt/
182. Malta Competition and Consumer Affairs Authority (MCCAA). https://mccaa.org.mt/
183. Malta's Circular Economy Strategy 2020 – 2030 (CE Malta). https://www.cemalta.gov.mt/strategy/

184. Malta's Recovery and Resilience Plan. Directorate-General for Communication, European Commission. https://commission.europa.eu/business-economy-euro/economic-recovery/recovery-and-resilience-facility/country-pages/maltas-recovery-and-resilience-plan_en
185. Malta's Research and Innovation Smart Specialisation Strategy (RIS3) 2021 – 2027. The Malta Council for Science & Technology, Ministry for Equality, Research and Innovation (2021) https://mcst.gov.mt/psi/ri-strategy-post-2020/ (PDF: https://mcst.gov.mt/wp-content/uploads/2022/01/RIS3-Strategy-2020-2027.pdf)
186. Marangos O, Nisiforou O, Yiatros S (2019) Build up circular Cyprus country survey report. Available from: https://cypruscircular.org.cy/wp-content/uploads/2021/07/Survey-Country-Report_FINAL.pdf. Accessed 18 September 2023
187. Marguerite (2013) Zagreb Airport concession reaches financial close. Available from: https://www.marguerite.com/2013/12/zagreb-airport-concession-reaches-financial-close/. Accessed 14 September 2023
188. Ministerio de Fomento (2018) Spanish urban agenda. Available from: https://cdn.mitma.gob.es/portal-web-drupal/AUE/doc/AUE_EN_140220.pdf. Accessed 30 Sept 2023
189. Ministerio de la Presidencia (2008) Real Decreto 105/2008, de 1 de febrero, por el que se regula la producción y gestión de los residuos de construcción y demolición. Available from: https://www.boe.es/eli/es/rd/2008/02/01/105/con
190. Ministerio de la Presidencia, Relaciones con las Cortes e Igualdad (2019) Orden PCI/86/2019, de 31 de enero, por la que se publica el Acuerdo del Consejo de Ministros de 7 de diciembre de 2018, por el que se aprueba el Plan de Contratación Pública Ecológica de la Administración General del Estado, sus organismos autónomos y las entidades gestoras de la Seguridad Social (2018–2025). Available from: https://www.boe.es/eli/es/o/2019/01/31/pci86. Accessed 30 Sept 2023
191. Ministerio de Transportes, Movilidad y Agenda Urbana (MITMA) (2021) Real Decreto 853/2021, de 5 de octubre, por el que se regulan los programas de ayuda en materia de rehabilitación residencial y vivienda social del Plan de Recuperación, Transformación y Resiliencia. Available from: https://www.boe.es/eli/es/rd/2021/10/05/853. Accessed 30 Sept 2023
192. Ministerio para la Transición Ecológica y el Reto Demográfico, (MITECO) (2020a) España Circular 2030. Estrategia Española de Economía Circular. Available from: https://www.miteco.gob.es/content/dam/miteco/es/calidad-y-evaluacion-ambiental/temas/economia-circular/espanacircular2030_def1_tcm30-509532_mod_tcm30-509532.pdf. Accessed 30 Sept 2023
193. Ministerio para la Transición Ecológica y el Reto Demográfico, (MITECO) (2020b) Estrategia a Largo Plazo para una Economía Española Moderna, Competitiva y Climáticamente Neutra en 2050. Available from: https://www.miteco.gob.es/content/dam/miteco/es/prensa/documentoelp_tcm30-516109.pdf. Accessed 30 Sept 2023
194. Ministerio para la Transición Ecológica y el Reto Demográfico, (MITECO) (2020c) España circular 2030. Circular economy Spanish strategy. Executive summary. Available from: https://www.miteco.gob.es/content/dam/miteco/es/calidad-y-evaluacion-ambiental/temas/economia-circular/200714eeec_resumenejecutivo_en_tcm30-510578.pdf. Accessed 30 Sept 2023
195. Ministerio para la Transición Ecológica y el Reto Demográfico, (MITECO) (2021) I Plan de Acción de Economía Circular 2021–2023. Estrategia Española de Economía Circular. Available from: https://www.miteco.gob.es/content/dam/miteco/es/calidad-y-evaluacion-ambiental/temas/economia-circular/plan_accion_eco_circular_def_nipo_tcm30-529618.pdf. Accessed 30 Sept 2023
196. Ministerstvo místního rozvoje (2023) 'socially responsible public procurement | SOVZ', 16 November. Available from: https://www.sovz.cz/en/. Accessed 16 Nov 2023
197. Ministerstvo řivotního prostředí (2023) Vyhláška o podrobnostech nakládání s odpady č. 273/2021 Sb. Ministerstvo životního prostředí. Available from: https://www.mzp.cz/www/platnalegislativa.nsf/%24%24OpenDominoDocument.xsp?documentId=DE9DB31723575A15C1258758002B40E3&action=openDocument. Accessed 16 Nov 2023

198. Ministerstvo živoního prostředí (2023) Zákon č. 541/2020 Sb. o odpadech a související předpisy. Available from: https://www.mzp.cz/www/platnalegislativa.nsf/%24%24OpenDominoDocument.xsp?documentId=E4044163A66CAA76C1258655002DE3C9&action=openDcument. Accessed 16 Nov 2023
199. Ministerstvo životního prostředí (2022) Cirkulární Česko 2040, Cirkulární Česko 2040. Available from: https://www.mzp.czcz/cirkularni_cesko. Accessed 16 Nov 2023
200. Ministerstvo životního prostředí (2023) 130/2019 Sb. Vyhláška o kritériích, při jejichž splnění je asfaltová směs vedlejším produktem nebo přestává být o..., Zákony pro lidi. Available from: https://www.zakonyprolidi.cz/cs/2019-130. Accessed 16 Nov 2023
201. Ministry of Economy and Industry, Meeting of the expert group at the Minister, Sofia, May 11, 2023, from The Transition to a Circular Economy was Discussed at a Meeting at the Ministry of Economy and Industry ● Ministry of Economy and Industry (government.bg)
202. Ministry for Construction, Transport, and Infrastructure of the Republic of Serbia (MCTI RS) (2019) The rulebook on the content, method and procedure of technical documentation preparation and control according to the class and purpose of the objects (In Serbian), The Official Gazette of the Republic of Serbia No. 73/2019
203. Ministry for Ecological Transition and the Demographic Challenge (2020) Just transition strategy. Instituto para la Transición Justa. Available from: https://www.miteco.gob.es/content/dam/miteco/es/ministerio/planes-estrategias/transicion-justa/Just%20Transition%20Strategy_ENG.pdf. Accessed 8 Oct 2023
204. Ministry for Ecological Transition and the Demographic Challenge (2021a) I Catalogue of best practices in circular economy. General sub-directorate for the circular economy. Available from: https://www.miteco.gob.es/content/dam/miteco/es/calidad-y-evaluacion-ambiental/temas/economia-circular/fichas-bpec-i-en-web_def_tcm30-525011.pdf. Accessed 8 Oct 2023
205. Ministry for Ecological Transition and the Demographic Challenge (2021b) II Catalogue of best practices in circular economy. General sub-directorate for the circular economy. Available from: https://www.miteco.gob.es/content/dam/miteco/es/calidad-y-evaluacion-ambiental/temas/economia-circular/fichas-bpec-ii_en-web_def2_tcm30-538142.pdf. Accessed 8 Oct 2023
206. Ministry of Ecological Transition (2022) National strategy for the circular economy. Available from: https://www.mase.gov.it/sites/default/files/archivio/allegati/PNRR/SEC_21.06.22.pdf. Accessed 9 Aug 2023
207. Ministry of Economics (2023). Cooperation memorandum on promoting the use of wood in construction. Available from: Parakstīs sadarbības memorandu par koka izmantošanas būvniecībā veicināšanu | Ekonomikas ministrija. Accessed 07 Sept 2023
208. Ministry of Economy of the Slovak Republic (2018) Economic policy strategy until 2030. Available from: https://rokovania.gov.sk/RVL/Material/23019/1. Accessed 07 July 2023
209. Ministry of Economy of the Slovak Republic (2019) Ocenenie pre produkty a služby cirkulárnej ekonomiky. Available from: https://www.economy.gov.sk/press/ocenenie-pre-produkty-a-sluzby-cirkularnej-ekonomiky. Accessed 10 July 2023
210. Ministry of Economy (2018) Industrial strategy with a focus on manufacturing, Skopje: s.n
211. Ministry of Environment and Physical Planning (2019) Development of the environmental monitoring information system. Available from: https://www.moepp.gov.mk/en/%D0%BF%D1%80%D0%BE%D0%B5%D0%BA%D1%82%D0%B8-2/%D0%BF%D0%BE%D0%B4%D0%B4%D1%80%D1%88%D0%BA%D0%B0-%D0%BD%D0%B0-%D1%80%D0%B5%D0%BF%D1%83%D0%B1%D0%BB%D0%B8%D0%BA%D0%B0-%D0%BC%D0%B0%D0%BA%D0%B5%D0%B4%D0%BE%D0%BD%D0%B8%D1%98%D0%B0-%D0. Accessed 22 Sept 2023
212. Ministry of Environment and Physical Planning (2021) Long-term strategy on climate action for the period 2021–2050, Skopje: s.n
213. Ministry of Environment and Physical Planning (2021) National plan for waste management for the period 2021–2031, Skopje: s.n

214. Ministry of Environment and Physical Planning (2022) Support in the implementation of the waste management legislation and extended producer responsibility (EPR) scheme. Available from: https://www.moepp.gov.mk/en/%D0%BF%D1%80%D0%BE%D0%B5%D0%BA%D1%82%D0%B8-2/%D0%BF%D0%BE%D0%B4%D0%B4%D1%80%D1%88%D0%BA%D0%B0-%D0%BD%D0%B0-%D1%80%D0%B5%D0%BF%D1%83%D0%B1%D0%BB%D0%B8%D0%BA%D0%B0-%D0%BC%D0%B0%D0%BA%D0%B5%D0%B4%D0%BE%D0%BD%D0%B8%D1%98%D0%B0-%D0. Accessed 22 Sept 2023
215. Ministry of Environment and Physical Planning (2022) Waste prevention plan for the period 2022–2028, Skopje: s.n
216. Ministry of Environment of the Slovak Republic (2019) 2030 Environmental strategy. Available from: https://www.minzp.sk/iep/strategicke-materialy/envirostrategia-2030/. Accessed 07 July 2023
217. Ministry of Environmental of the Slovak Republic (2020) Low-carbon development strategy. Available from: https://www.minzp.sk/klima/nizkouhlikova-strategia. Accessed 07 July 2023
218. Ministry of Environmental Protection of the Republic of Serbia (MEP RS) and United Nations Development Programme (UNDP) (2019) Roadmap for circular economy in Serbia. Belgrade 2020. Available from https://www.undp.org/serbia/publications/roadmap-circular-economy-serbia. Accessed 15 July 2023
219. Ministry of Finance of the Slovak Republic (2021) Recovery and resilience plan (RRP) Available from: https://www.planobnovy.sk/. Accessed 07 July 2023
220. Ministry of Finance (2021) Growth acceleration plan for the period 2022–2026, Skopje: s.n
221. Ministry of Foreign Affairs of the Netherlands (2023) Factsheet circular economy and waste management in Croatia—potential for new developments. Ministry of Foreign Affairs, Den Haag. Available from: https://www.rvo.nl/sites/default/files/2023-02/Factsheet%20Circular%20Economy%20and%20Waste%20Management%20in%20Croatia%20%282023%29.pdf. Accessed 22 June 2023
222. Ministry of Infrastructure and Water Management (2023) National circular economy programme 2023–2030. Available from: https://hollandcircularhotspot.nl/wp-content/uploads/2023/02/NationalCircularEconomyProgramme2023-2030Summary.pdf. Accessed 12 July 2023
223. Ministry of Investments, Regional Development and Informatization (2020) Sustainable development strategy to 2030. Available from: https://www.mirri.gov.sk/sekcie/vizia-a-strategia-rozvoja-slovenska-do-roku-2030/index.html. Accessed 07 July 2023
224. Ministry of the Environment and Ministry of Employment and the Economy (2021) Government resolution on the strategic programme for circular economy. Available from: https://ym.fi/documents/1410903/42733297/Government+resolution+on+the+Strategic+Programme+for+Circular+Economy+8.4.2021.pdf/309aa929-a36f-d565-99f8-fa565050e22e/Government+resolution+on+the+Strategic+Programme+for+Circular+Economy+8.4.2021.pdf?t=1619432219261. Accessed 10 July 2023
225. Ministry of the Transport of the Slovak Republic (2023) Available from: https://www.mindop.sk/uvod. Accessed 05 July 2023
226. Ministry of Economy and Sustainable Development (2022) Implementation program of the Ministry of Economy and Sustainable Development for the period 2021–2024 (In Croatian). Available from: https://mingo.gov.hr/UserDocsImages/GLAVNO%20TAJNI%C5%A0TVO/Strategija,%20planovi%20i%20ostali%20dokumenti/Provedbeni%20program%20Ministarstva%20gospodarstva%20i%20odr%C5%BEi2024.%20-II.%20Izmjene%20i%20dopune.pdf. Accessed 12 June 2023
227. Ministry of Physical Planning, Construction and State Assets (2021) Programme for the development of circular spatial and building management for the period 2021–2030 (In Croatian). Available from: https://faolex.fao.org/docs/pdf/cro213055programme.pdf. Accessed 12 June 2023
228. Ministry of Economy and Sustainable Development (2022) Ordinance on waste management (In Croatian). Official Gazette of the Republic of Croatia, No. 106/22. Available from: https://narodne-novine.nn.hr/clanci/sluzbeni/2022_09_106_1552.html. Accessed 12 May 2023

229. Ministry of Environmental and Nature Protection (2016) Ordinance on Construction Waste and Waste Containing Asbestos (in Croatian). Official Gazette of the Republic of Croatia, No. 69/16. Available from: https://narodne-novine.nn.hr/clanci/sluzbeni/2016_07_69_1650.html. Accessed 12 May 2023
230. Ministry of the Environment & Ministry of Economic Affairs and Employment (2021) Government resolution on the strategic programme for circular economy. Finnish Government. Available from: https://ym.fi/documents/1410903/42733297/Government+resolution+on+the+Strategic+Programme+for+Circular+Economy+8.4.2021.pdf/309aa929-a36f-d565-99f8-fa565050e22e/Government+resolution+on+the+Strategic+Programme+for+Circular+Economy+8.4.2021.pdf?t=1619432219261. Accessed 22 August 2023
231. Ministry of the Environment (n.d.) Circular economy. Finnish Government. Available from: https://ym.fi/en/circular-economy. Accessed 22 August 2023
232. Ministry of Economic Affairs and Employment (2021) Innovation and Skills in Finland 2021-2027 promotes regional vitality, employment and wellbeing. Available from: https://valtioneuvosto.fi/en/-/1410877/innovation-and-skills-in-finland-2021-2027-promotes-regional-vitality-employment-and-wellbeing. Accessed 22 August 2023
233. Ministry for an Ecological and Solidary Transition (2018) Roadmap for the circular economy: 50 measures for a 100% circular economy. Available from: https://www.ecologie.gouv.fr/sites/default/files/FREC%20anglais.pdf. Accessed 22 August 2023
234. Ministry for the Environment, Energy and Enterprise (Ministry for the Environment, Energy and Regeneration of the Grand Harbour). © Government of Malta 2024. https://sustainability.gov.mt/
235. Ministère de la Transition écologique (n.d.) Ministère de la Transition écologique. Available from: https://www.ecologie.gouv.fr/. Accessed 22 August 2023
236. Ministère de la Transition écologique (2021). Conseil national de l'économie circulaire (CNEC). Available from: https://www.ecologie.gouv.fr/politiques-publiques/conseil-national-leconomie-circulaire-cnec. Accessed 22 August 2023
237. Ministère de la Transition écologique (2020) Réglementation environnementale RE2020. Available from: https://www.ecologie.gouv.fr/politiques-publiques/reglementation-environnementale-re2020. Accessed 22 August 2023
238. Ministère de la Transition écologique (2022) Produits et matériaux de construction du secteur du bâtiment (PMCB). Available from: https://www.ecologie.gouv.fr/politiques-publiques/produits-materiaux-construction-du-secteur-du-batiment-pmcb. Accessed 22 August 2023
239. Ministère de la Transition écologique (2022) Produits et matériaux de construction du secteur du bâtiment (PMCB). Available from: https://www.ecologie.gouv.fr/politiques-publiques/produits-materiaux-construction-du-secteur-secteurdu-batiment-pmcb. Accessed 22 August 2023
240. Ministère de la Transition écologique (2018) Financement des projets. Available from: https://www.ecologie.gouv.fr/politiques-publiques/financement-projets. Accessed 22 August 2023
241. Ministère de la Transition écologique (2018) Roadmap for the circular economy: 50 measures for a 100% circular economy. Available from: https://www.ecologie.gouv.fr/sites/default/files/FREC%20anglais.pdf. Accessed 22 August 2023
242. NOMAD Architects (2022). Guidelines for circular construction. Available from: https://drive.google.com/file/d/164JeQ1NxZMCsdbvhjUs6mwXBsCVeNKYQ/view. Accessed 25 Sept 2023
243. National Sectorial Plan for Solid Waste Management (Sectorial Strategy), Albania Ministry of Tourism and Environment: The document of strategic policies and national integrated waste management plan in Albania 2020–2035
244. National Assembly of the Republic of Serbia (2022) The Ministries Act (In Serbian). The Official Gazette of the Republic of Serbia No.128/2020 and 116/2022
245. Netherlands Environmental Assessment Agency (n.d.) About PBL. Available from: https://www.pbl.nl/en/about-pbl. Accessed 5 July 2023
246. Netherlands Enterprise Agency (n.d.) MIA and Vamil for Entrepreneurs. Available from: https://english.rvo.nl/subsidies-financing/mia-vamil/mia-and-vamil-tax-scheme. Accessed 8 July 2023

247. Newton Prize (2020) UK-TURKEY: reducing CO_2 emissions with new green concrete. Available from: https://www.newton-gcrf.org/newton-fund/newton-prize/2020-newton-prize/reducing-co2-missions-with-new-green-concrete/. Accessed 30 Sept 2023
248. Nordic Circular Hotspot (n.d.) Nordic Circular Hotspot. Available from: https://nordiccircularhotspot.org/. Accessed 22 August 2023
249. Nss Magazine (2021) 7 sustainable buildings in Italy. Available from: https://www.nssmag.com/en/art-design/26679/7-edifici-green-italia. Accessed 2 Sept 2023
250. OECD (2019) Environmental performance reviews: Latvia 2019. Available from: OECD Environmental Performance Reviews: Latvia 2019 | en | OECD. Accessed 01 Sept 2023
251. OECD (2022a) The circular economy in Ireland. OECD urban studies. OECD Publishing, Paris. Available from: https://doi.org/10.1787/7d25e0bb-en. Accessed 5 September 2023
252. OECD (2022b) Closing the loop in the Slovak Republic. A roadmap towards circularity for competitiveness, eco-innovation and sustainability. Available from: https://www.oecd-ilibrary.org/docserver/acadd43a-en.pdf?expires=1689276578&id=id&accname=guest&checksum=885D677D4BC8C9C44FD29FB781B816C6). Accessed 09 July 2023
253. OECD (2024) A roadmap towards circular economy of Albania. OECD Publishing, Paris. https://doi.org/10.1787/8c970fdc-en
254. ORÉE (n.d.) ORÉE – Réseau de l'économie environnementale. Available from: http://www.oree.org/index.html. Accessed 22 August 2023
255. Official Gazette Directive 25406 of 18 March 2004 on the Control of Excavation Soil and Construction and Demolition Waste [2004]
256. Official Gazette Directive 31627 of 13 October 2021 on the Regulation on Demolition of Buildings [2021]
257. Official Gazette of the Republic of North Macedonia (2009) Law on building construction. 130 ed. s.l.:s.n
258. Official Gazette of the Republic of North Macedonia (2013) Regulations for energy characteristics on the buildings. 94 ed. s.l.:s.n.
259. Official Gazette of the Republic of North Macedonia (2019) Law of public procurement. 24 ed. s.l.:s.n
260. Official Gazette of the Republic of North Macedonia (2021) Law on waste management. 116 ed. s.l.:s.n
261. Official Gazette of tirelandhe Republic of North Macedonia (2013) Law on industrial and green zones. 119 ed. s.l.:s.n
262. Official Gazette Series n. 180 of 5 August 2003, Decree n. 203/2003, Norme affinché gli uffici pubblici e le societa' a prevalente capitale pubblico coprano il fabbisogno annuale di manufatti e beni con una quota di prodotti ottenuti da materiale riciclato nella misura non inferiore al 30% del fabbisogno medesimo
263. Official Gazette Series n. 226 of 11 September 2020, Decree n. 116/2020, Attuazione della direttiva (UE) 2018/851 che modifica la direttiva 2008/98/CE relativa ai rifiuti e attuazione della direttiva (UE) 2018/852 che modifica la direttiva 1994/62/CE sugli imballaggi e i rifiuti di imballaggio
264. Official Gazette Series n. 228 of 14 September 2020, Decree n. 76/2020 converted into Law no. 220/2020, Misure urgenti per la semplificazione e l'innovazione digitale
265. Otra Elpa (2023) Available from: OtraĒlpa. Accessed 01 Sept 2023
266. PACE (2023) What we do. Available from: https://pacecircular.org/about. Accessed 12 July 2023
267. PIANNOo (n.d.) About PIANOo. Available from: https://www.pianoo.nl/en/about-pianoo-0. Accessed 5 July 2023
268. Petrova, V. (2020), Circular Economy Practices in the Bulgarian Raw Material Industry, Sustainable Extraction And Processing Of Raw Materials Journal, pp 79–84.
269. Platform CB'23 (2023) About platform CB'23. Available from: https://platformcb23.nl/english. Accessed 8 July 2023
270. Platform CB'23 (n.d.). Platform CB'23. Available from: https://platformcb23.nl/english/. Accessed 5 July 2023

271. Platform CB'23 (2022)Measuring circularity guide–A measurement method for circular construction (Version 3.0). Available from: https://platformcb23.nl/images/downloads/2022/Measuring_Circularity_Guide.pdf. Accessed 5 July 2023
272. Platform CB'23 (2022) Guide to passports for the construction sector–Working agreements and substantiation for passports in a circular construction sector. Available from: https://platformcb23.nl/images/downloads/2022/Guide-to-Passports-for-the-Construction-Sector_3.pdf. Accessed 5 July 2023
273. Pontikakis D, Fernández Sirera T, Janssen M, Guy K, Marques Santos A, Boden JM, Moncada Paterno Castello P (2020) Projecting Opportunities for INdustrial Transitions (POINT): concepts, rationales and methodological guidelines for territorial reviews of industrial transition. Publications Office of the European Union, Luxembourg
274. Presidência de Conselho de Ministros n.º 190-A/2017 (2017) Plano de Ação para a Economia Circular (PAEC). Available from: https://files.diariodarepublica.pt/1s/2017/12/23602/0005400073.pdf. Accessed 21 Oct 2023
275. Presidência de Conselho de Ministros n.o 65/2007 (2007) Estratégia Nacional para as Compras Públicas Ecológicas 2008–2010 (ENCPE). Available from: https://files.diariodarepublica.pt/1s/2007/05/08700/29742981.pdf. Accessed 21 Oct 2023
276. Presidência do Conselho de Ministros n.º 53/2020 (2020) Plano Nacional Energia e Clima (PNEC2030). Available from: https://files.diariodarepublica.pt/1s/2020/07/13300/0000200158.pdf. Accessed 21 Oct 2023
277. Presidência do Conselho de Ministros nº 239/2020 (n.d.) Regime Geral da Gestão de Resíduos (RGGR). Available from: https://files.diariodarepublica.pt/1s/2020/12/23901/0000200269.pdf. Accessed 21 Oct 2023
278. Project of the national forest expansion and rehabilitation program for the period 2023–2032. Available from: https://particip.gov.md/ro/document/stages/proiectul-hotararii-de-guvern-cu-privire-la-aprobarea-programului-national-de-extindere-si-reabilitare-a-padurilor-pentru-perioada-2023-2032-si-planului-de-actiuni-pentru-implementarea-acestuia-pentru-perioada-2023-2027/9945. Accessed 10 July 2023
279. Project of the national strategy for agricultural and rural development 2023–2030. Available from: https://cancelaria.gov.md/sites/default/files/document/attachments/864_maia.pdf. Accessed 1 July 2023
280. Project on the 2030 low emissions development program. Available from: https://mediu.gov.md/ro/file/13372. Accessed 14 July 2023
281. Public Procurements (2023) Available from: https://www.iepirkumi.lv/lv/. Accessed 01 Sept 2023
282. Public Procurement Agency (2022) The practical handbook for green procurement is published on the legislation and methodology public procurement portal, methodology, Sofia
283. Public Procurement Portal (2019) Public Procurement Agency publishes Practical Handbook for Green Public Procurement. Available from: https://www2.aop.bg/en/public-procurement-agency-publishes-practical-handbook-for-green-public-procurement/. Accessed 28 Aug 2025
284. RICS (n.d.) Sustainability, London, Royal Institution of Chartered Surveyors. Available from: https://www.rics.org/news-insights/current-topics-campaigns/sustainability. Accessed 21 Aug 2025
285. ReLondon (n.d.) ReLondon, ReLondon, London. Available from: https://relondon.gov.uk/. Accessed 28 Aug 2025
286. Recycled Bio-polymer feather Fibre Reinforcement for Cement-Based Materials – RECP (Times of Malta: https://timesofmalta.com/article/creating-circular-economy-construction-agriculture.1058067)
287. Rediscovery Centre (2022) Transforming Ireland through the circular economy—rediscovery centre strategic plan 2022–2026. Available from: https://www.rediscoverycentre.ie/assets/uploads/2023/08/Rediscovery-Centre-Strategic-Plan-2022-2024.pdf. Accessed 10 Sept 2023
288. Rediscovery Centre (2023) The Rediscovery Centre and Holland Circular Hotspot join forces for a circular economy. Available from: https://www.rediscoverycentre.ie/2023/03/23/the-rediscovery-centre-and-holland-circular-hotspot-join-forces-for-a-circular-economy/. Accessed 5 July 2023

289. Rediscovery Centre (n.d.) The National Centre for the Circular Economy. Available from: http://www.rediscoverycentre.ie/. Accessed 5 July 2023
290. Region of Central Macedonia, Strategic Plan for Central Macedonia—investment in employment and development https://www.pepkm.gr/uploads/2023/pepkm/programma-kentriki-makedonia.pdf. Accessed 23 Oct 2023
291. Regional Conference on Circular Economy Habitat Development (2020) Regional conference on circular economy, Tirana 2020. In: Conference report
292. Regione Lazio (2016) Imprese Green. Available from: https://progetti.regione.lazio.it/laziogreen/imprese-green/. Accessed 31 Aug 2023
293. Regione Lazio (2023) Regione Lazio: Ok Al Piano Di Transizione Ecologica. Prima In Italia. Available from: https://www.regione.lazio.it/notizie/Regione-Lazio-Ok-Al-Piano-Di-Transizione-Ecologica-Prima-In-Italia. Accessed 31 Aug 2023
294. Regione Lombardia (2018) Osservatorio economia circolare e transizione energetica. Available from: https://www.svilupposostenibile.regione.lombardia.it/it/azioni-per-la-sostenibilit%C3%A0/osservatorioec&te/osservatorio. Accessed 31 Aug 2023
295. Regione Lombardia (2022) Economia circolare, da Regione Lombardia 4 milioni per innovare le filiere. Available from: https://www.svilupposostenibile.regione.lombardia.it/it/b/978/economia-circolare-da-regione-lombardia-4-milioni-per-innovare-le-fili. Accessed 31 Aug 2023
296. Regione Marche (2022) Leggi e regolamenti regionali. Available from: https://www.consiglio.marche.it/banche_dati_e_documentazione/leggi/dettaglio.php?arc=vig&idl=2240. Accessed 31 Aug 2023
297. Regione Piemonte (2023) Green e Circular Economy. Available from: https://www.regione.piemonte.it/web/temi/ambiente-territorio/green-economy/green-circular-economy. Accessed 31 Aug 2023
298. Regione Puglia (2023) La Strategia Regionale di Sviluppo Sostenibile. Available from: https://pugliacon.regione.puglia.it/web/sit-puglia-dipartimento/strategia-per-lo-sviluppo-sostenibile-della-regione-puglia. Accessed 31 Aug 2023
299. Remeo Oy (n.d.) Remeo in English. Available from: https://remeo.fi/remeo-in-english/. Accessed 22 August 2023
300. Republic of Cyprus (2021) National action plan for strengthening circular economy 2021–2027 [Greek]. Nicosia. Ministry of Energy, Commerce and Industry. Available from: https://meci.gov.cy/gr/%CE%B5%CE%BD%CE%B7%CE%BC%CE%AD%CF%81%CF%89%CF%83%CE%B7/%CF%83%CF%84%CF%81%CE%B1%CF%84%CE%B7%CE%B3%CE%B9%CE%BA%CF%8C%CF%82-%CF%83%CF%87%CE%B5%CE%B4%CE%B9%CE%B1%CF%83%CE%BC%CF%8C%CF%82/%CE%B5%CE%B8%CE%BD%CE%B9%CE%BA%CF%8C-%CF%83%CF%87%CE%AD%CE%B4%CE%B9%CE%BF-%CE%B4%CF%81%CE%AC%CF%83%CE%B7%CF%82-%CE%B3%CE%B9%CE%B1-%CF%84%CE%B7%CE%BD-%CE%BA%CF%85%CE%BA%CE%BB%CE%B9%CE%BA%CE%AE-%CE%BF%CE%B9%CE%BA%CE%BF%CE%BD%CE%BF%CE%BC%CE%AF%CE%B1. Accessed 18 Sept 2023]
301. Republic of Cyprus (2021) Το Σχέδιο Δράσης της Κύπρου για μετάβαση στην κυκλική οικονομία. Press release of the Deputy Ministry for Research, Innovation and Digital Transition. Available from: https://www.dmrid.gov.cy/dmrid/research.nsf/All/4FE913AEE9F7CF71C225871F0038E3F4?OpenDocument. Accessed 18 Sept 2023]
302. Republic of Cyprus Law (2022) 185(I)/2011 and its latest amendment 48(I)/2022 concerning construction and demolition waste management [in Greek]
303. Republic of Cyprus Regulation (2023) Κ.Δ.Π. 112/2023 The regulations for CDW management in 2023. [in Greek]. Available from : https://www.etek.org.cy/uploads/News/11f1b3e768.pdf. Accessed 18 Sept 2023
304. Riga City Council (2020) Riga sustainable development strategy 2030 Available from: STRATEGIJA_WEB.pdf (rdpad.lv). Accessed 25 Sept 2023
305. Rigas Energetikas Agentura (2021) Integrated Action Plan for the transition to a circular economy in urban construction in the municipality of the city of Riga. Available from: 215c841969ca61ff6d8358594264817325d67e00.pdf (riga.lv). Accessed 12 Sept 2023

306. Rijkswaterstaat (n.d.). About Rijkswaterstaat. Available from: https://www.rijkswaterstaat.nl/en/about-us. Accessed: 4 July 2023
307. Rijskwaterstaat (2015) Circular economy in the Dutch construction sector—A perspective for the market and government. Available from: https://www.rivm.nl/bibliotheek/rapporten/2016-0024.pdf. Accessed 8 July 2023
308. Rijskwaterstaat (2023a) Elements of Dutch waste management. Available from: https://rwsenvironment.eu/subjects/from-waste-resources/elements-dutch-waste/. Accessed 10 Sept 2023
309. Rijskwaterstaat (2023b) National waste management plan. Available from: https://rwsenvironment.eu/subjects/from-waste-resources/national-activities/national-waste/. Accessed 10 July 2023
310. Rijskwaterstaat (n.d.) National Waste Management Plan. Available from: https://lap3.nl/service/english/. Accessed 5 July 2023
311. Robert PM, Giannopoulou IP, Savva P, Sakkas KM, Petrou MF, Nicolaides D (2023) New eco-friendly inorganic polymeric materials for the passive fire protection of structures. In: TMS annual meeting & exhibition. Springer Nature Switzerland, Cham, pp 758–767
312. Romanian Parliament (2023) pr.Codul Amenajării Teritoriului, Urbanismului si Construcțiilor. Available from: http://sgglegis.gov.ro/legislativ/docs/2022/11/_0n1p98wvfxmskrt375c.pdf
313. SBE Malta – Sustainable Built Environment Malta. http://www.sbemalta.com/
314. SITRA (2016) Leading the cycle—Finnish road map to a circular economy 2016–2025. Available from: https://www.SITRA.fi/app/uploads/2017/02/Selvityksia121.pdf. Accessed 10 July 2023
315. SITRA (2019) Finland 2016–2025 circular economy roadmap 2.0. Available from: https://www.SITRA.fi/app/uploads/2019/03/renkiertotalouden-tiekartta-tiivistelma-en2.pdf. Accessed 10 July 2023
316. SITRA (2021) Remeo increases the reuse of waste material generated by the construction industry and the retail sector: "High-quality raw material is a prerequisite for the circular economy". Available from: https://www.sitra.fi/en/cases/remeo-increases-the-reuse-of-waste-material-generated-by-the-construction-industry-and-the-retail-sector-high-quality-raw-material-is-a-prerequisite-for-the-circular-economy/. Accessed 10 July 2023
317. SM 810:2022 – Recycling oriented deconstruction, controlled excavation works and classification of waste – Standards and Services – Malta Competition and Consumer Affairs Authority (MCCAA) (2023). https://mccaa.org.mt/Section/Content?contentId=1243
318. SORĠI (https://www.sorgi.mt/)
319. Saeima (2010) Waste management law. Available from: https://likumi.lv/ta/id/221378-atkritumu-apsaimniekosanas-likums. Accessed 01 Sept 2023
320. Saldus district (2022). Waste management regulations. Available from: Atkritumu apsaimniekošanas noteikumi (likumi.lv). Accessed 12 Sept 2023
321. Scottish Government (n.d.) Waste: Legislation, Edinburgh, Scottish Government. Available from: https://www.gov.scot/publications/waste-legislation/. Accessed 21 Aug 2025
322. Scottish Government (2023) Circular Procurement, Edinburgh, Scottish Government, Available from: https://www.zerowastescotland.org.uk/resources/circular-procurement. Accessed 21 Aug 2025
323. Second NAP. Green Public Procurement Second National Action Plan (2022-2027). Ministry for the Environment, Climate Change and Planning (2021) https://environmentcms.gov.mt/en/decc/Pages/environment/gpp/secondNap.aspx (PDF: https://environmentcms.gov.mt/en/decc/Documents/environment/gpp/secondNap/greenPlanBook.pdf)
324. Sitra (2016) Leading the cycle—Finnish road map to a circular economy 2016–2025 (Sitra Studies 121). Available at: https://www.sitra.fi/app/uploads/2017/02/Selvityksia121.pdf. Accessed 22 August 2023
325. Sitra (n.d.) Sitra – Finnish Innovation Fund. Available from: https://www.sitra.fi/en/. Accessed 22 August 2023
326. Sitra (2019) Leading the cycle – Finnish road map to a circular economy 2.0 Summary (Sitra Studies 121). Available from: https://www.sitra.fi/app/uploads/2019/03/kiertotalouden-tiekartta-tiivistelma-en2.pdf. Accessed 22 August 2023

327. Sitra (n.d.) Project funding. Available from: https://www.sitra.fi/en/topics/project-funding/. Accessed 22 August 2023
328. Slovak Ministry of Economy (2019) Integrated national energy and climate plan. Available from: https://energy.ec.europa.eu/system/files/2020-03/sk_final_necp_main_en_0.pdf. Accessed 07 July 2023
329. SocKETs (2021) Circular economy in the building and construction sector in Italy: towards sustainable production and consumption, Societal Engagement with key enabling technologies, Available from: https://www.airi.it/airi2020/wp-content/uploads/2021/06/SocKETs_D1.1_Report_Airi_Final_Website.pdf. Accessed 28 Aug 2023]
330. Stora Enso (2023). French laws are decarbonising the building sector—could they provide a way forward internationally? Available from: https://www.storaenso.com/en/newsroom/news/2023/9/re2020. Accessed 25 Sept 2023
331. Suomen ympäristökeskus (2022) Valtakunnallinen jätesuunnitelma vuoteen 2027 / National Waste Plan until 2027. Suomen ympäristökeskus. Available from: https://julkaisut.valtioneuvosto.fi/bitstream/handle/10024/163978/YM_2022_13.pdf. Accessed 22 August 2025
332. TG7.8 – Recycled Materials and industrial by-products for high performance reinforced concrete structures. COM7: Sustainability, fib. https://www.fib-international.org/commissions/com7-sustainability.html
333. The Draft Strategy for an inclusive, sustainable and digital economy until 2030—Greening program for small and medium enterprises. Available from: https://cancelaria.gov.md/sites/default/files/document/attachments/135.pdf. Accessed 14 July 2023
334. The World Bank Group (2023) North Macedonia public sector energy efficiency project. Available from: https://projects.worldbank.org/en/projects-operations/project-detail/P149990. Accessed 22 September 2023
335. Transition team of the Circular construction economy transition agenda (2018) Circular construction economy transition agenda. Available from: https://hollandcircularhotspot.nl/wp-content/uploads/2019/09/Circular-Construction-Economy.pdf. Accessed 4 July 2023
336. UKGBC (n.d.) Circular Economy, London, UK Green Building Council, Available from: https://ukgbc.org/our-work/topics/circular-economy/. Accessed [21 August 2025]
337. UKGBC (2023) System enables for a circular economy, London UK Green Building Council. Available from: https://ukgbc.org/resources/system-enablers-for-a-circular-economy/. Accessed 21 Aug 2025
338. UKRI (n.d.) Circular economy hub, Swindon, UKRI. Available from: https://ce-hub.org. Accessed 21 Aug 2025
339. UN Environment/MAP (2017) Regional action plan on sustainable consumption and production in the Mediterranean. Athens, Greece. Available from: https://wedocs.unep.org/bitstream/handle/20.500.11822/20731/unepmap_SCPAP_eng_web.p-df?sequence=1&isAllowed=y. Accessed 30 Sept 2023
340. UNE (2022) Estudio de la contribución de las normas técnicas a la economía circular. Available from: https://www.une.org/normalizacion_documentos/Estudio%20de%20la%20contribuci%C3%B3n%20de%20las%20normas%20t%C3%A9cnicas%20a%20la%20econom%C3%ADa%20circular.pdf. Accessed 9 Oct 2023
341. UNECE—United Nations Economic Commission for Europe (2021) Environmental performance reviews—Romania. Third review. Available from: https://unece.org/sites/default/files/2021-12/ECE_CEP_189_0.pdf
342. Urban Environment Publications and City of Helsinki (2020) The city of Helsinki's roadmap for circular and sharing economy. Available from: https://circulareconomy.europa.eu/platform/sites/default/files/the-city-of-helsinkis-roadmap-for-circular-and-sharing-economy.pdf. Accessed 10 July 2023
343. VARAM (2017a) GPP guidelines for design, construction and management of office buildings. Available from: download (varam.gov.lv). Accessed 01 Sept 2023
344. VARAM (2017b) GPP guidelines for recreation and sports infrastructure. Available from: download (varam.gov.lv). Accessed 01 Sept 2023
345. VARAM (2017c) GPP guidelines for road design, construction and maintenance

346. VARAM (2017d) GPP guidelines for transport Available from: download (varam.gov.lv). Accessed 01 Sept 2023
347. VARAM (2017e) Review of the operating strategy of the Ministry of Environmental Protection and Regional Development for 2017–2019. Available from: https://www.varam.gov.lv/sites/varam/files/content/files/varam_darb_strat_2017_vienlapis_final.pdf. Accessed 05 Sept 2023
348. VARAM (2020) Operational strategy for 2020—2022. Available from: varamstrategija_2020_2022_gala28022020.pdf. Accessed 15 Sept 2023
349. VARAM (2021) State waste management plan for 2021–2028. Available from: Atkritumu apsaimniekošanas valsts plāns 2021.-2028.gadam. | Vides aizsardzības un reģionālās attīstības ministrija (varam.gov.lv). Accessed 12 Sept 2023
350. VARAM (2023a) Available from: https://www.varam.gov.lv/en. Accessed 12 Sept 2023
351. VARAM (2023b) Green procurement calculators. Available from: ZPI aprites cikla izmaksu kalkulatori | Vides aizsardzības un reģionālās attīstības ministrija (varam.gov.lv). Accessed 15 Sept 2023
352. Valamar (2023) Valamar Amicor green resort. Available from: https://www.valamar.com/en/hotels-hvar/valamar-amicor-green-resort. Accessed 14 July 2023
353. Valobat (n.d.). The Valobat eco-organization project: valorize each waste of the building. Available from: https://www.valobat.fr/en/le-projet-eco-organisme-valobat-valoriser-chaque-dechet-du-batiment/. Accessed 25 Sept 2023
354. Valobat (n.d.) Valobat – Éco-organisme pour le recyclage des produits du bâtiment. Available from: https://www.valobat.fr/. Accessed 22 August 2023
355. Vasiljević D, Petrović D (2020) The report on ex-ante analysis of the circular economy effects (in Serbian). Policy and Legal Advice Centre (PLAC III). The project funded by EU. Available from: https://www.ekologija.gov.rs/sites/default/files/2021-01/exante-analiza_efekata-za-oblast-cirkularne-ekonomije.pdf. Accessed 15 July 2023
356. Ventura A, Manyes A, de Diego B, Figuerola E, Marrot J, Bolea J, Batlle T (2021) Economía circular en la edificación. Green Building Council España. Available from: https://gbce.es/documentos/Informe_Economia-Circular.pdf. Accessed 30 Sept 2023
357. WAG (n.d.) Planning policy and guidance: waste, Cardiff, Welsh Government, Available from: https://www.gov.wales/planning-policy-and-guidance-waste. Accessed 21 Aug 2025
358. WAG (2021) Beyond Recycling, Cardiff, Welsh Government, Available from: https://www.gov.wales/beyond-recycling. Accessed 21 Aug 2025.
359. Waste Act No. 79/2015 (2015) Available from: https://www.slov-lex.sk/pravne-predpisy/SK/ZZ/2015/79/20230701. Accessed 10 July 2023
360. Waste Management Programme of the Slovak Republic (2021). Available from: https://www.enviroportal.sk/odpady/program-odpadoveho-hospodarstva-slovenskej-republiky-2021-2025. Accessed 08 July 2023
361. Waste Regulations S.L. 549.63. https://legislation.mt/eli/sl/549.63/eng/pdf
362. World Bank (2022a) Croatia: proposed circular economy action plan on construction and demolition waste—deliverable 1.b: CE action plan and its roadmap. Available from: https://thedocs.worldbank.org/en/doc/c22f85216d0d23a3287250b550d05935-0080012022/original/Deliverable1bCroatiaProposedCEactionPlan.pdf. Accessed 12 June 2023
363. World Bank (2022b) Circular economy approaches in solid waste management—final report—summary recommendations. Available from: https://documents1.worldbank.org/curated/en/099200012132212911/pdf/P173141084571c0f3084a307acc29708a0e.pdf. Accessed 12 June 2023
364. World Bank (2023) Diagnostic analysis for circular economy interventions in Romania. © World Bank. Available from: https://openknowledge.worldbank.org/server/api/core/bitstreams/7d6f7c52-d302-4c86-baad-d5cdb4cb9703/content. Accessed 5 February 2024
365. World Circular Economy Forum 2023 (n.d.) World Circular Economy Forum 2023. Available from: https://wcef2023.com/. Accessed 22 August 2023
366. Xrobb l-G.a.in Nature Park and Sustainable Development Centre. https://naturetrustmalta.org/what-we-do/natural-parks/xrobb-l-ghagin/

367. ZWS (n.d.) Zero Waste Scotland homepage. Edinburgh, Scottish Government. Available from: https://www.zerowastescotland.org.uk. Accessed 21 Aug 2025
368. Zagreb Airport (2021) Zagreb Airport achieves ACI airport carbon accreditation level 3—optimisation. Available from: https://www.zagreb-airport.hr/business/newsroom/press-releases/zagreb-airport-achieves-aci-airport-carbon-accreditation-level-3-optimisation/705. Accessed 14 Sept 2023

Chapter 6
Standards and Frameworks Supporting Circular Construction

Ruben Paul Borg, Genesis Camila Cervantes Puma, Faye Sciberras, Andrea Francesca Bellia, Jan Pesta, and Luís Bragança

Abstract Circular Economy is a crucial factor in addressing sustainability. It is also on top of the agenda of the European Union and at a global scale. For Circular Economy to be successful in the construction industry, however, standardisation is a vital requirement. This document refers to an overview on the need for standardisation in Circular Economy. It further refers to national level standards and standardisation, together with a review of European and International level standards. To this effect, following a review of the state of the art in standardisation, key components were identified, and a questionnaire was designed and conducted across 19 countries with experts and practitioners in Circular Economy in Construction. The objective of the questionnaire was to assess and understand the level of implementation and standardization of Circular Economy across different countries. The assessment conducted was intended as a review of the perception of standards and their implementation across Europe by the participants of the survey as practitioners in Circular Economy. A review of standards in Circular Economy at a European and International level was carried out to define gaps to be addressed, together with opportunities in standardisation, to support Circularity in the Construction Industry. Case studies in implementation of standardisation are presented with respect to Deconstruction, Recycled Aggregate and Reuse of products.

Keywords Circular economy · Standardisation · Sustainability · Construction · Demolition waste

---

R. P. Borg (✉) · F. Sciberras · A. F. Bellia
Faculty for the Built Environment, University of Malta, Msida, Malta
e-mail: ruben.p.borg@um.edu.mt

G. C. C. Puma · L. Bragança
Department of Civil Engineering, ISISE, ARISE, University of Minho, Guimarães, Portugal

J. Pesta
University Centre for Energy Efficient Buildings, Czech Technical University in Prague, Třinecká, Buštěhrad, Czech Republic

L. Bragança et al. (eds.), *Shaping Circular Transitions in the Built Environment*,
Springer Tracts in Civil Engineering, https://doi.org/10.1007/978-3-032-02834-1_6

6.1 Overview of the Need for Standardization

The transition to a circular economy is crucial for the European Union's sustainable, efficient, and competitive economy. The value of materials and resources must be kept in the economy for as long as possible, in order for this to be achieved. Recent studies show wider benefits of circular economy, including reducing carbon dioxide emissions. This aligns with the key EU priorities, including jobs, investments, climate and energy, the social and industrial innovation, and global drive on sustainable development [1]. In the context of the built environment, standardisation plays a crucial role in promoting and operationalising circular economy practices. By establishing common frameworks, guidelines, and protocols, standardisation ensures consistency, interoperability, scalability, and efficiency in the implementation of circular practices. It provides a structured approach to adopting circular economy principles in various aspects of the built environment, including in the construction of new buildings, building refurbishment and the building end of life. For instance, key aspects of circular economy refer to standardisation at product level (e.g. standardised approach for the use of recycled aggregates in concrete), or at the building level (e.g. standardised approach for design for disassembly and adaptability).

The widespread adoption of circular economy practices in the construction sector has become a global priority to address environmental challenges and promote long-term sustainability. In this context, standardisation emerges as a key element that can catalyse this transition towards more circular and resource-efficient models. Standardisation can be essential in driving the conversion towards to a low-carbon, sustainable, resource-efficient, and competitive economy. By establishing common frameworks, guidelines, and protocols, standardisation enables the consistent and efficient implementation of circular practices in the built environment. This contributes to the achievement of EU priorities, global sustainable development goals, and the development of a more sustainable and environmentally friendly future.

A number of studies and analyses have shed light on the importance of standardisation in different aspects of sustainable construction, providing valuable insights into how it can drive the adoption of circular practices across the industry. A few of these studies are discussed below:

The framework proposed in "A New Framework for Circular Refurbishment of Buildings to Operationalize Circular Economy Policies" provides detailed guidance for the implementation of circular economy policies in building refurbishment. This structured approach ranges from the initial mapping of buildings to the monitoring of their operation to ensure circularity, highlighting the importance of standardising processes to achieve consistent and sustainable results [2].

On the other hand, the study on the "Implementation Stage for Circular Economy in the Danish Building and Construction Sector" highlights the potential of circular economy practices in the Danish construction industry to improve sustainability. It proposes three key approaches for the implementation of circular practices: the use of waste as building materials, the reuse of building materials and design for disassembly. These strategies, supported by standardised standards and processes, can contribute significantly to reducing resource consumption and minimising waste in construction [3].

In addition, the review on the "Application of circular economy principles in buildings" offers a wide range of strategies to maximise the reuse of materials and components, optimise building design and promote a product-as-a-service model. These recommendations, supported by standardised standards and processes, can serve as a solid framework for the effective implementation of circular economy principles in the built environment [4].

Furthermore, the study on "Drivers and barriers towards circular economy in the building sector: Stakeholder interviews and analysis of five European countries policies and practices" provides a comprehensive overview of the challenges and opportunities related to the adoption of the circular economy in the building sector. Through field interviews with stakeholders in five European countries, the study highlights the need for common standards and regulations that can facilitate the shift towards more circular practices [5].

Finally, the publication on "Use of Recycled Aggregates in Concrete: Opportunities for Upscaling in Europe" highlights the crucial role of standardisation in promoting the use of recycled aggregates in concrete. By addressing challenges and opportunities throughout the supply chain, this analysis highlights how standards can facilitate the integration of recycled materials into conventional construction practices, thus promoting greater circularity in the construction industry [6].

The references provided, along with others, add to our understanding of the significance of standardization in circular economy practices within the built environment. By establishing common frameworks, guidelines, and protocols, standardization ensures consistency, interoperability, scalability, quality and efficiency in the application of circular practices, leading to a more sustainable and resource-efficient built environment. It also gives a clearer picture that standardisation can play a crucial role in promoting and implementing circular economy practices within the built environment. As outlined in the scope of ISO TC323 for Circular Economy [7], standardization in the field of Circular Economy contributes towards the development of frameworks, provide supporting tools and addresses requirements for the implementation of activities to maximize the contribution to Sustainable Development. Standardization is an important process in advancing the Circular Economy, by providing a framework to assess the quality and sustainability of materials, products and processes. Standards are a requirement in building trust in the market where good quality is recognised. This will lay the foundation for a smooth transition to the circular model. It should also be noted that the quality of the circular product is assessed based on material composition or properties of input materials (e.g., EN 206 for concrete). However, this does not reflect the suitability of a circular product for its final use, so its performance for intended function should be analysed.

In conclusion, standardisation plays a vital role in promoting and implementing circular economy practices in the built environment. These efforts contribute to the shift towards a more sustainable and resource-efficient built environment.

6.1.1 Adoption of Performance-Based Analysis

As outlined above to ensure quality of a circular product to build trust in the market where good quality is recognised, performance-based analysis (PBA) should be implemented as a key perspective ensuring consumer´s satisfaction. The level of implementation of PBA differs among regions and is affected by local authorities or local standardization bodies. This highlights the importance of having standards and uniformity on an international level which then translates to local level.

The main organizations contributing to development of Performance-Based Analysis approach in Civil Engineering include [8]:

1. ATC: Applied Technology Council.
2. BSI: British Standards Institute.
3. Caltrans: California Department of Transportation.
4. CSA: Canadian Standards Association.
5. CEN/CENELEC: European Standards Organizations.
6. FEMA: Federal Emergency Management Agency.
7. UNEP: United Nations Environment Programme.
8. USEPA: United States Environmental Protection Agency.
9. TAC: Transportation Association of Canada.

According to Easa et al. the level of implementation of standards varies across different countries and regions [8]. Their paper highlights that North American countries, particularly the United States and Canada, are leading in promoting and creating specifications and codes for the use of Performance-Based Analysis (PBA). These countries have made significant progress in implementing PBA in various civil engineering fields [8].

On the other hand, Asian countries like China, Japan, Iran, and India are also showing an emerging focus on PBA, although to a lesser extent compared to North America. Australia and European countries such as France, Greece, and the United Kingdom have comparatively fewer publications in the PBA areas, indicating a potentially lower level of implementation in these regions [8].

The paper also emphasizes the need for formal PBA procedures in transportation and environmental engineering, following the example of structural engineering, suggesting that the level of implementation may vary across different civil engineering disciplines [8].

Overall, the level of implementation for standards related to Performance-Based Analysis in civil engineering, as discussed in the paper, appears to be more advanced in North America compared to other regions, with varying degrees of progress in different countries and fields within the discipline [8].

6.1.2 Good Practice for the Adoption of Standards: Lessons Learnt from Electronic Industry

The electronics industry presents an opportunity for an understanding of the application of circular economy and can be referred to in a review to assess approaches in that sector; it is referred to as an example of good practice for considerations in the field of the built environment and a circular building industry. Regulations and standards play a vital role in promoting sustainability in the power electronics industry by setting guidelines and requirements that support the growth of environmentally friendly products. This can be used as an example of good practice also for built environment and circular building industry. Here are some ways in which regulations and standards contribute to promoting sustainability in the power electronics industry [9]:

- **Environmental Performance Requirements**: Regulations and standards can impose requirements on the environmental performance of power electronics beyond just energy efficiency levels. By considering factors such as material efficiency, lifecycle impacts, and end-of-life performance, regulations can push manufacturers to design products that reduce negative environmental impacts throughout the lifespan of the product.
- **Mitigating Pollution**: Regulations and directives aim to limit or forbid the use of toxic materials in the manufacturing progress of power electronics. By restricting the use of hazardous materials, regulations contribute to reducing pollution and promoting a cleaner production process.
- **Guiding Sustainable Design**: Regulations and standards provide guidance to designers and manufacturers on developing power electronics products that are durable, energy-efficient, repairable, recyclable, and incorporate recycled materials. By outlining specific criteria for sustainable product development, regulations help drive innovation towards more sustainable practices.
- **Circular Economy Perspective**: Standards in the circular economy perspective offer best practices for designing products that comply with material efficiency requirements. By aligning with circular economy principles, power electronics manufacturers can adopt sustainable practices that encourage resource efficiency and waste reduction.
- **Compliance and Implementation**: Understanding and complying with regulations and standards is essential for implementing sustainable product development in the power electronics industry. By adhering to specific criteria and adopting relevant standards, designers can ensure that their products meet sustainability requirements and contribute to a more environmentally friendly industry.

The progress of implementation of standards and policies can be depicted in the level of development of energy efficiency, which became an important pillar of the EU´s climate and energy policy. In the beginning, the initial product policies of the EU were established during the late 1970s and have since developed into a cohesive array of measures within framework directives, aimed at standardizing and enhancing the regulatory framework. Following years of insufficient execution, significant strides have been made in terms of both scope and ambition over the past decade. By 2020, mandatory product measures encompassed 50% of the EU's total final energy consumption, resulting in energy savings of 46 Mtoe [10].

6.1.3 Barriers to Implementation of Standards

As discussed above the implementation of Standards at a National and International level are important as they create a framework to assess the quality and sustainability of materials, products and processes. They also lay the foundations for a smooth transition to the circular model on a National and International level and rise up the level of trust in the purchasing and use of these materials. Below are some barriers which can be faced when implementing standards:

- **Cost of Compliance**: Adhering to standards often involves costs related to training, certification, testing, and documentation. Companies may view these costs as barriers to full compliance with standards, especially if they perceive limited benefits in return.
- **Lack of Resources**: Small and medium-sized enterprises (SMEs) may lack the resources, expertise, and dedicated staff needed to implement and comply with standards effectively. Limited resources can hinder the adoption of standards in smaller companies.
- **Resistance to Change**: Resistance from employees, management, or stakeholders within an organization can impede the implementation of standards. Resistance to change and unwillingness to integrate new practices can be significant barriers to compliance.
- **Inconsistency in Interpretation**: Different stakeholders may interpret standards differently, leading to inconsistencies in implementation across the industry. Lack of clarity and uniformity in interpreting standards can hinder effective compliance.
- **Regulatory Uncertainty**: Rapid changes in regulations and standards can create uncertainty for companies, making it challenging to keep up with evolving requirements. Uncertainty about future standards can delay or complicate implementation efforts.
- **Compatibility Issues**: Standards from different organizations or regions may conflict or overlap, creating compatibility issues for companies operating in multiple markets. Ensuring alignment and consistency across various standards can be a barrier to implementation.
- **Training and Capacity Building**: Companies may lack the necessary training and capacity to understand, implement, and monitor compliance with standards effectively. Investing in training programs and building internal capacity can help overcome this barrier.

- **Supply Chain Challenges**: Ensuring that suppliers and partners also comply with relevant standards can be a challenge. Lack of visibility and control over the entire supply chain can hinder efforts to implement standards consistently.
- **Monitoring and Enforcement**: Monitoring and enforcing compliance with standards can be resource-intensive and require robust systems for tracking and reporting. Without effective monitoring mechanisms in place, companies may struggle to ensure ongoing compliance.

6.2 National Level Standards and Standardisation

6.2.1 National Level Standards

National standards play a crucial role in implementing circular economy principles in specific contexts and regions. These standards provide guidelines and requirements tailored to local conditions, industries and regulations, facilitating the implementation of sustainable practices and promoting resource efficiency.

In the construction sector, for example, national standards transposing European Standards, such as EN 206 on concrete production and conformity, ensure that materials meet certain quality and performance criteria. However, as the circular economy gains traction, there's a growing recognition of the need to go beyond traditional quality assessments and consider the suitability of circular products for their intended use.

In this context, in the case of construction materials, national standards play a crucial role with reference to the local and regional context and scenario, key products used and building typologies, and in setting up characterisation and classification frameworks, which are essential in transforming industrial byproducts into recycled products for construction.

National standards can address this by incorporating criteria that evaluate not only the material composition, but also the performance of circular products in real-world applications, and with reference to an intend use. By setting clear benchmarks for performance and durability, these standards help to build confidence in circular products and technologies, thereby accelerating their uptake in the market.

In addition, national standards can address specific challenges and opportunities within different sectors and industries. For example, in the construction sector, standards may focus on promoting the use of recycled materials, designing for disassembly, or optimising resource efficiency throughout a building's lifecycle.

By aligning with international frameworks and guidelines, national standards also ensure consistency and compatibility across borders, facilitating trade and cooperation on circular economy initiatives. This harmonisation of standards not only improves market access for circular products, but also promotes innovation and knowledge sharing on a global scale.

6.2.2 National Level Standardisation

At a national level, standardization is overseen by organizations like the British Standards Institution (BSI) [11], German Institute of Standardization ("Deutsches Institut für Normung", DIN), the Bureau of Indian Standards (BIS) [12] and others in respective countries. These groups create and release national standards covering various construction aspects, including design codes, safety regulations, material usage, and quality control.

National standardization involves various steps, beginning with identifying standardization requirements in response to industry needs, technological advancements, and regulatory demands [13]. Technical committees or working groups are then formed to develop standards for specific areas of interest. These committees comprise experts from industry, academia, government, and other relevant stakeholders who work together to create guidelines, specifications, and best practices.

Consensus building is a vital step in the development of national standards. Stakeholders review and provide input on draft standards to ensure that different perspectives are considered and the standards reflect the industry's collective expertise and interests [8]. Once the standards are finalized, they are published and made available to the public through various channels, such as online platforms, libraries, and specialized publications.

National governments adopt and integrate these standards into their legal frameworks, making compliance mandatory for construction activities. Enforcement mechanisms are implemented to ensure adherence to these standards.

6.2.3 Research Methodology

A study was conducted with the objective of assessing the understanding and perception of standardisation in Circular Economy in different countries, among practitioners and stakeholders active in the field. The assessment refers to an overview on the need for standardisation in Circular Economy. It further refers to national level standards and standardisation, together with a review of European and International level standards.

To this effect, following a review of the state of the art in standardisation, key components were identified, and a questionnaire was designed and conducted across 19 countries with experts and practitioners in circular Economy in Construction. The objective of the questionnaire was to assess and understand the level of implementation and standardization of Circular Economy across different countries. The assessment conducted was intended as a review of the perception of standards and their implementation across Europe by the participants of the survey as practitioners in Circular Economy.

The methodology selected to address the review of standards and policies at national and international level across EU, based on a questionnaire with practitioners and stakeholders in different countries, allowed for information to be gathered addressing the objectives outlined. The survey was structured in different sections to address key aspects related to standardisation. These sections included an introduction to each country, information related to the National standards body, as well as the mechanism and standards developed in each country. The standards identified for each country were also addressed, identifying which of them are implemented and to what extent, along with any corresponding guidelines and recommendations.

In addition, reference was made to European and International standard, including how they are applied, the national parameters set, an overview of the standards and objectives, whether their adoption is mandatory or voluntary, the level/extent of implementation, how each standard is referenced at the national level (partially or fully), as well as any regulations to promote their implementation. The effectiveness, limitations and target audience of these standards were also assessed, as well as any gaps, revisions or references to implementation.

Reference was also made to each National standard/guideline, including the language in which it is written, the key objectives, how it is adopted, publishing entity, the level/extent of implementation, its efficiency and merits, how it is referenced in national regulations (partially or fully), the reference data sets, boundaries and justification, target audience, gaps, revisions and reference to its implementation.

Further the survey identified two areas of focus as key example, specifically: the dismantling of buildings and the use of recycled aggregates. For each of these areas, the relevant standards, reference data sets, the limits set in the guidelines and their justification, how they are implemented and to what extent, and any gaps were identified.

6.2.4 Survey of Standards

The objective of the survey was to address key aspects pertaining to standardisation, specifically to address critical aspects, including reviewing the need for standardisation, identifying national level standards and standardisation, and identifying International and European standards adopted in each country. The questionnaire was structured in various sections as follows:

1. Introduction to country, national standards entity and link, and mechanism and standards developed per country,
2. Standards referred to in each country—and which of them are implemented, along with guidelines and recommendations,

3. Reference to each international European standard mentioned—and how these are enforced, parameters set, overview, objectives, mandatory or voluntary adoption, the level/extent of implementation, how each standard is referenced nationally and whether partially or fully, along with any regulation to enact such standards, the effectiveness, limits, and to whom standards are addressed, as well as any gaps, revisions and references to implementation,
4. Reference to each national standard/guideline—including the language in which it is written, key objectives, how it is adopted, by whom it is published, the level/extend to which it is implemented, its efficiency and merits, how it is references in national regulations and whether partially or fully, its efficiency and merits, referenced data sets, limits and justification, to whom it is addressed, gaps, revisions, and reference to its implementation,
5. Focus area: building disassembly—relevant standards, reference data sets, limits set in the guidelines and their justification, how it is implemented and to what extent, and any gaps,
6. Focus area: recycled aggregate—relevant standards, reference data sets, limits set in the guidelines and their justification, how it is implemented and to what extent, and any gaps.

The countries participating in the survey for which responses where obtained are the following:

1.	2.
3.	4.
5.	6.
7.	8.
9.	10.
11.	12.
13.	14.
15.	16.
17.	18.
19.	

1. Belgium
2. Croatia
3. Czech Republic
4. Cyprus
5. Georgia
6. Ireland
7. Italy
8. Kosovo

(continued)

(continued)

9. Latvia
10. Malta
11. North Macedonia
12. Poland
13. Portugal
14. Romania
15. Serbia
16. Spain
17. The Netherlands
18. Turkey
19. UK

The survey referred to primarily European Union members states but also other countries and territories in Europe and beyond. The survey was presented in the English language for all countries.

The Questionnaire. The survey was divided into 6 sections of different interest for standardization at the national and international level in each country. Table 6.1 outlined the different section and, includes the areas covered in the survey.

The data gathered through the questionnaire was clustered for each country, edited and organised in a structured template whilst taking into account all the information provided including key references and links to standard platforms and associated documents. The key challenge in this process referred to national standards which were published in languages other than English as is the case for non-English speaking regions, which required translation of key components to identify the objectives of respective documents. This was not the case for International Standards which are generally in English and some National Standards in specific countries. The structure used for clustering and analysis is presented in Table 6.2.

Standard Language Adopted in different Countries. When it comes to national standards/guidelines, the default language for most countries is English. Those countries where standards are presented fully in English are the UK, Ireland, North Macedonia (only few standards are translated to national language), Republic of Serbia and Malta. The countries whose standards are mostly in English are: Croatia (Croatian is also used), Cyprus (the decision is taken by the stakeholders, otherwise Greek is used as the national language), Kosovo (Albanian and Serbian are also used) and Latvia (English and translated also into Latvian. Occasionally German is used). Countries that do not use English include Brazil and Portugal (Portuguese), Czech Republic (Czech), Georgia (Georgian), Italy (Italian), Spain (Spanish), and Turkey (Turkish). Respondents for Belgium, Poland, Romania and the Netherlands did not provide clear information on the standard language adopted.

Table 6.1 Questions used in the survey

Section 1: General information
• Where are you from?
• Reference link to the National Standards Entity
• Through which mechanisms are standard developed in your country?
Section 2: List standards referred to in your country
• Which International Standards in Circular Economy are referred to and/or implemented in your country? (List the Standards)
• Which International Standards in Circular Economy are referred to and/or implemented in your country? (List the reference links)
• Which European Standards in Circular Economy are referred to and/or implemented in your country? (List the Standards)
• Which European Standards in Circular Economy are referred to and/or implemented in your country? (List the reference links)
• Which National Standards in Circular Economy are referred to and/or implemented in your country? (List the Standards)
• Which National Standards in Circular Economy are referred to and/or implemented in your country? (List the References links)
• Which Guidelines in Circular Economy are referred to and/or implemented in your country?
• Which Recommendations in Circular Economy are referred to and/or implemented in your country?
Section 3: With reference to each international European standards mentioned
• How is the standard referenced or enforced at National Level? Is it through a regulation or a policy framework? Does it cover the entire standard or parts/specific clauses only?
• How are the National Determined parameters set in the standard? (Example percentage limits on the use of recycled aggregate in structural concrete set at national level through national regulation)
• Provide a short overview of the different sections of the standard
• List the key objectives of the Standard
• How is the standard adopted in your country? Is it mandatory or voluntary?
• How is the Standards implemented and to which level/extent?
• How is each standard referenced in national regulations? Does it cover the entire standard or parts of it/specific clauses? Which Regulation if any enacts the standard? Provide link to regulation
• How effective is the standard and which are its merits in the National territory?
• Which are the limits set in the guidelines—what is their justification?
• Which Regulation if any enacts the standard?
• For whom is the standard addressed?
• Are there gaps which are not covered by the standard?
• Are there any revisions required?
• Provide Reference to the Implementation of the standard in your country

(continued)

Table 6.1 (continued)

Section 4: With reference to each national standard/guidelines listed
• What is the Standard Language?
• List the key objectives of the Standard
• How is the standard adopted in your country (voluntary or business case through incentives and disincentives, or regulation enforcing a standard or part of it)? Is it mandatory or voluntary?
• Who published the document: industry or ministry or standards entity or other stakeholders?
• How is the Standards implemented and to which level/extent?
• How effective is the standard and which are its merits in the National territory?
• How is each standard referenced in national regulations? Does it cover the entire standard or parts of it/specific clauses? Which Regulation if any enacts the standard? Provide link to regulation
• How effective is the standard and which are its merits ?
• What is the reference data set used for the standard?
• Which are the limits set in the guidelines—what is their justification?
• For whom is the standard addressed?
• Are there gaps which are not covered by the standard?
• Are there any revisions required?
• Provide Reference to the Implementation of the standard in your country
Section 5: Focus area—building disassembly
• List relevant standard covering this area
• What is the reference data set used for the standard?
• Which are the limits set in the guidelines—what is their justification?
• How is it implemented and to which extent?
• Which are the gaps in the standard?
Section 6: Focus area—recycled aggregate
• List relevant standards covering this area
• What is the reference data set used for the standard?
• Which are the limits set in the guidelines—what is their justification?
• How is it implemented and to which extent?
• Which are the gaps in the standard?

6.2.5 *Country Factsheets*

This section gives a summary of the information given by each respondent from each country via the survey above. The questionnaire in the survey is not exhaustive and may contain gaps. The purpose of the questionnaire was to assess the perception of practitioners on standards at national, European and International level, and how Standards are implemented in the country, from their perception.

Standards generally aim to define terminology, methodologies, and certification bases, providing technical measurements and supporting new technologies. They

Table 6.2 Structure used to cluster and analyze national survey responses

1. National standard entities and mechanism/s
a. Summary of national standard entities and mechanism through which standards are developed per country
2. Standards referred to
a. National, International, and European Standards referred to and/or implemented
b. Guidelines and recommendations in Circular Economy referred to and/or implemented
3. Reference to international European standards
a. Key objectives of the Standard
b. Overview of different sections of the standard
c. Standard reference or enforcement
d. How National Determined parameters in the standard are set
e. How the standard is adopted and implemented, and to what level/extent
f. How each standard is referenced in national regulations, and which Regulation (if any) enacts the standard
g. Effectiveness of the standard and its merits in the National territory
h. Which are the limits set in the guidelines - what is their justification?
i. Regulation (if any) which enacts the standard
j. For whom is the standard addressed?
k. Are there gaps which are not covered by the standard?
l. Are there any revisions required?
m. Provide Reference to the Implementation of the standard in your country
4. Reference to national standard guidelines
a. Objectives of the Standard
b. Publisher
c. Adoption
d. Effectiveness and Merits
e. Implementation
f. How is each standard referenced in national regulations? Does it cover the entire standard or parts of it/specific clauses? Which Regulation, if any enacts the standard? Provide links to regulation
g. Standard's effectiveness and merits
h. Standard's reference data set
i. Limits and justification
j. For whom is the standard addressed?
k. Are there gaps which are not covered by the standard?
l. Are there any revisions required?
m. Provide references to the implementation of the standard in your country
5. Focus area building disassembly
6. Focus area recycled aggregate

include sections on scope, normative references, terms and definitions, and specific clauses relevant to the standard's title and scope.

The stages of the development process are the: proposal, committee formation, drafting, public consultation, approval, publication, and periodic review. Committees and technical groups comprising industry, academia, government, and other stakeholders collaborate to create and revise standards based on technological advancements, regulatory changes, and market needs.

Belgium. The Belgian Bureau for Standardisation (NBN) [14] is the central entity responsible for developing standards in Belgium. NBN collaborates with various stakeholders, including industry experts, government bodies, and academic institutions. The development process involves consultation, drafting, and consensus-building to ensure that standards are comprehensive, practical, and aligned with national and international objectives.

Croatia. The key body responsible for developing and maintaining standards is the Croatian Standards Institute (CSI/HZN) [15]. HZN's involvement in international and European standardization organizations ensures that such standards are accessible on a national level, allowing representatives to participate in the development of national standards at international and European levels.

Croatia refers to and implements various national, international, and European standards. The assessment of the sustainability of construction works using the framework provided by HRN EN 15643:2021 [16] is one such example. Other standards relevant to circular economy include EN 303 808 V1.1.1:2023 [17] and EN 17861:2023 [18]. National standards are accessible through HZN's repository [19]. The development of a program for circular management of space and buildings for 2021–2030 has provided guidelines [20] to promote CE principles in urban planning and construction. This aims at extending buildings' life cycles, encouraging the reuse of spaces, and reducing construction waste.

Standards may be either mandatory or voluntary and are designed for a broad audience, including industry professionals, businesses, government agencies, and consumers. These standards provide technical specifications and best practices for various stakeholders to ensure compliance and promote sustainability. While some standards, particularly those regulated at the EU level, are obligatory, others, especially those related to sustainability in construction, are not mandatory. Implementation typically occurs at the project design, material level, and through testing. Such standards that contribute to sustainable development do so by promoting circularity in construction and urban planning. However, information on the effectiveness and specific merits of these standards in the national territory is limited.

Croatia currently has no specific standard for building disassembly. However, the Development Program for Circular Management of Space and Buildings [20] includes principles relevant to this area. Similarly, while no specific standard is available solely for recycled aggregates, standard HR 1128 [21] addresses concrete requirements, incorporating elements relevant to the use of recycled materials in construction.

Cyprus. In Cyprus, standards development is managed by the national standardisation body and technical committees. The country currently lacks published international standards related to the CE, although the first ones are expected from ISO/TC 323 [16] in early 2024 and respective complimentary European standards are currently being developed by CEN/TC 473. National standards are limited, with one in progress for certifying circular plastic products. CE guidelines in Cyprus are mainly driven by private initiatives, such as assessments for specific industries like hotels, conducted by private companies and associations. The Ministries of Energy, Commerce and Industry, and Agriculture, Rural Development and Environment, and Deputy Ministry of Research, Innovation and Digital Policy are coordinating the national action plan on CE and supporting funding schemes CE transition.

The effectiveness of standards is monitored through market surveillance and conformity reports from random checks. National regulations usually refer to entire standards, as seen in the regulation on fuel specifications (Κ.Δ.Π. 252/2015 [22]).

All standards are voluntary unless they are harmonized European standards, which become de facto mandatory and provide CE marking. These harmonized standards are referenced in national regulations and tenders. National regulations often directly reference standards or develop national annexes to supplement European standards. Adoption of standards in Cyprus is voluntary by default, with European standards adopted nationally by law and published in the Official Journal of the Cyprus Government. Implementation varies, with harmonized standards becoming mandatory if they support EU Directives or Regulations.

For building disassembly, the relevant standard ISO 20887:2020 [23] is not yet implemented in Cyprus. There are no existing standards for recycled aggregates, though studies are ongoing to evaluate their inclusion in future standards.

Czech Republic. In the Czech Republic, the national standards are developed by the Czech Office for Standards, Metrology, and Testing (UNMZ) [24] in collaboration with the Czech Standardization Agency. This process begins with a government request for a standard, followed by public procurement for document preparation. The country utilizes various international and European standards, including ISO 14044 [25], EN 17861 [26], ETSI EN 303 808 V1.1.1 [27], EN 17410 [28], EN 15804 [29], EN 15978 [30], EN 13108-8 [31], and EN 15342-7 [32]. However, there are no specific national standards for Circular Economy at present. Guidelines are based on the European CD Waste protocol and national guidelines for asbestos management, while recommendations include the State Health Institute's methodology for ensuring the safety of children and young people in public facilities.

Standards on Life Cycle Assessment are adopted, too. These standards are comprehensive in describing LCA and related tools, but still voluntary, with limited actual application. They are primarily referenced in national regulations, impacting environmental consultants. The effectiveness of these standards is generally assessed through voluntary market implementation, with a noted lack of enforceable regulations. They are aimed at various stakeholders but have limited effectiveness due to their non-mandatory nature.

For building disassembly, no specific standards are mentioned. Regarding recycled aggregates, relevant standards include ČSN EN 1744-6 [33], ČSN EN 1097-6 [34], ČSN EN 1367-4 [35], ČSN EN 1744-5 [36], ČSN EN 1744-1+A1 [37], and ČSN EN 933-11 [38]. These standards cover various testing methods for chemical, mechanical, and physical properties of aggregates. However, the extent of their implementation and any existing gaps are not detailed.

Georgia. In Georgia, standards are primarily developed through laws or strategic documents under the Ministry of Economy and Sustainable Development. The Waste Management Code, adopted in accordance with EU directives, serves as a key international standard for waste management, aiming to prevent waste, increase re-use, and ensure environmentally safe treatment.

The Waste Management Code's objectives include waste prevention, environmentally safe treatment, and efficient resource use, covering various types of waste except for specific exclusions like radioactive waste. The code aims to protect the environment and human health through effective waste management mechanisms. Enforcement of the Waste Management Code is through the law and by-laws, addressing business and state entities. The code's effectiveness and merits are ensured through legal enforcement, although specific limits and justifications are not outlined. The English version of the code is available for reference [39].

There is limited information available regarding national standard guidelines in Georgia, with no details on objectives, publisher, adoption, effectiveness, or implementation. The code primarily targets waste management but lacks specific information on building disassembly or recycled aggregate standards.

Ireland. The National Standards Authority of Ireland (NSAI) [40] is the body that oversees the development and implementation of standards [41], including those related to the circular economy. The NSAI National Committee on Circular Economy, along with its subcommittees and working groups, plays a crucial role in developing standards in this area. These standards include those developed under ISO and CEN committees, addressing various aspects of circular economy principles and frameworks.

Internationally, Ireland is engaged in developing standards under ISO/TC323 and CEN/TC350, focusing on circular economy principles, business models, and circularity measurement frameworks. The Circular Economy and Miscellaneous Provisions Act 2022, signed into law, underpins Ireland's transition towards a more sustainable production and consumption model, aiming to reduce greenhouse gas emissions and promote resource efficiency.

The key objectives of the Circular Economy Act include shifting Ireland away from a linear economy, incentivizing the use of reusable and recyclable alternatives, tackling illegal dumping and littering, and promoting waste prevention and recycling. The Act introduces various measures, including a mandatory segregation regime for commercial waste, support for environmental and circular economy projects, and the establishment of circular economy policies.

Implementation of the Circular Economy Act is to be enforced, with measures such as mandatory segregation and incentivized charging for commercial waste.

The Act also establishes legal requirements for the Circular Economy Strategy and National Food Loss Prevention Roadmap. While details about national standard guidelines are limited, the Act signifies a significant step towards promoting circular economy practices in Ireland.

Italy. The development of technical standards in Italy is overseen by the National Standards Body (UNI) [42], which has been actively involved in creating voluntary technical standards for over a century. These standards codify the state of the art of products, services, processes, and professions, often in collaboration with thousands of experts across various sectors. Additionally, emerging themes at a European level are addressed through CEN Workshop Agreements, with pre-standardization documents processed by restricted author tables.

Internationally, Italian technical standards are developed with consideration for the work of ISO/TC 323 "Circular economy" working groups, resulting in standards like ISO 59004 [43], ISO 59010 [44], and ISO 59020 [45]. Nationally, standards such as UNI/TS 11820:2022 [46] and UNI/TR 11821:2023 [47] provide methods, indicators, and best practices for measuring circularity and analyzing circular economy applications.

Guidelines and recommendations in the circular economy are provided by entities like the Ministry of Ecological Transition [48] and the CAM (Minimum Environmental Criteria) [49]. These resources offer strategic guidance and criteria for implementing circular practices.

The key objectives of standards such as UNI/TS 11820:2022 [46] and UNI/TR 11821:2023 [47] include evaluating the circularity of organizations and analyzing good practices of circular economy applications. These standards outline criteria for measurement, evaluation, and analysis, addressing different aspects of circular economy principles and methodologies.

Adoption of these standards is voluntary, with implementation primarily at the company and organization level. While details regarding enforcement, effectiveness, and regulatory references are limited due to the recent nature of these standards, their adoption by 37 companies and organizations during the experimental phase indicates growing interest and participation in circular economy initiatives in Italy.

The standards and guidelines serve as valuable resources for companies and organizations seeking to adopt circular economy practices, providing frameworks, indicators, and best practices for evaluating and implementing circularity in various contexts.

Kosovo. The standardization process is facilitated by the Kosovo Standardization Agency (KSA) and operates within the framework of European integration, particularly through the Stabilization and Association Process Tracking Mechanism (STM) [50]. International [51, 52], European [53], and national [54] standards are referenced and implemented, with the adoption of construction codes equivalent to European standards being a key objective. However, Kosovo is not a member of international standardization bodies like ISO and is working towards cooperation with these organizations.

The Law on Standardization in Kosovo mirrors European and international systems, providing procedural rules for drafting, approving, and applying standards. While adoption of Kosovar standards is voluntary, they can become mandatory through legislation. The KSA oversees standardization activities, with any entity eligible to propose standards. Despite effective implementation, there are ongoing efforts to develop national annexes and revise existing standards, particularly in civil engineering.

The KSA, under the Ministry of Industry, oversees the approval and harmonization of standards, contributing to economic development and regional competition. The effectiveness of standards is supervised by the Ministry of Trade, with Kosovo standards identified as "SK" and harmonized with European counterparts. The country is striving to align its laws and regulations with EU standards, promoting regional integration and economic growth.

In specific focus areas like building disassembly and recycled aggregate, relevant standards and legal frameworks exist, but implementation challenges persist. These challenges include an underdeveloped market for repurposed goods, inefficient waste management practices, and gaps in record-keeping and operational procedures. Addressing these gaps is crucial for Kosovo's transition towards a more circular economy and sustainable waste management practices.

Latvia. In Latvia, national standardization is facilitated by the National Standardization Institution "Latvijas Standarts" (LVS) [55] which operates through standardization technical committees composed of stakeholders in specific areas. Latvia, as a member state of the European Union and the European Standardization Organization, adopts international and European standards, such as EN 50614 [56] and EN 50625 [57]. However, national standards specific to the circular economy have not yet been developed or approved.

The adoption of European and international standards is initiated by the industry or the relevant industry ministry, with "Latvijas Standards" taking them over after recommendation by standardization technical committees. While the application of standards is generally voluntary, their fulfilment becomes mandatory if referenced in regulatory acts. Latvian regulatory acts may refer to standards or specific clauses, with mandatory application determined by law or Cabinet of Ministers regulations.

Efforts to translate standards referenced in regulatory acts into Latvian are underway, with the objective of ensuring compliance. The application of standards typically applies to the applicator of the relevant normative act, with mandatory application not extending to other persons. Although specific information regarding the effectiveness and merits of standards in Latvia is not provided, the adoption and implementation process aligns with European and international practices.

In focus areas such as building disassembly and recycled aggregate, relevant standards and guidelines are not explicitly mentioned, indicating potential gaps in standardization efforts in these domains. Overall, while Latvia actively adopts international and European standards, there is room for further development and implementation of standards tailored to specific national priorities, such as the circular economy.

Malta. Malta is active in Standardisation through the MCCAA (Malta Communication and Consumer Affairs Authority) [58] and participates in Technical Committees for standards on Circular Economy, including CEN TC350/SC1 Circular Economy in the Construction sector [59]. In Malta, the transition to circular economy is an ongoing development.

The main government institutions in charge of implementing the circular economy in Malta is the Ministry for the Environment, Energy and Enterprise [60]. The Circular Economy Malta (CEMalta) [61] is the unit set up by Government to administer Circular Economy in Malta under the Environment Protection Act, Malta CAP 549: Circular Economy Malta [62]. CEMalta was established through Subsidiary Legislation SL.595.28 [61].

SL. 623.08 Construction Site Management Regulations [63] refer to the standard SM810 for Recycling oriented Deconstruction and Classification of Waste [64] 65, therefore incorporating the standard within the regulatory framework. Compliance with the standards is considered as an essential requirement before an executable Development Permit is issued. Any deconstruction or excavation works which is carried out on a site will need address sections five (5), seven (7) and eight (8) of the national standard SM 810:2022 'Recycling-oriented Deconstruction, Controlled Excavation Works and Classification of Waste' [65].

Furthermore, apertures in a residential building are required to follow the standardisation of dimensions of internal and external apertures as specified in the Technical Document titled 'standardisation of Apertures for Residential Buildings in Malta'.

The objectives of the standards SM810 [65] in Malta was to:

- establish best practice for (de)construction, aimed at reducing the generation of construction and demolition waste and purifying the resulting waste streams;
- establish a classification for construction and demolition waste by type, material, composition and weight to encourage on-site separation as well as improve the quality of waste streams for subsequent reuse or recycling;
- establish standards for appropriate excavation work, with the aim of reusing excavated rock for the purposes of construction.
- establish standards for the dimensions of internal and external apertures of residential dwellings aimed at encouraging the re-use of fittings as well as reducing diversity, bringing about economies of scale.

These standards are merged within the regulatory framework. It is an essential requirement to comply with these standards before an executable Development Permit is issued. In order to improve the quality of the waste, the above standards aim to classify construction and demolition waste according to the material type and composition. The standards were organized into: two standards for C&D Waste (SM810: Recycling-oriented Deconstruction, Controlled Excavation Works and Classification of Waste [65]; SM820 Classification of Recycled Aggregate (In consultation phase) [66]); and a technical guideline for apertures.

The Recycling Oriented Deconstruction of Buildings Standard SM810:2022 (The Recycling-oriented Deconstruction, Controlled Excavation Works and Classification of Waste) was published in 2022 by MCCAA as an important and fundamental

National Standard, following a public Consultation [65]. The standard is incorporated within the regulatory framework through SL. 623.08 Construction Site Management Regulations. Compliance [67] with the standards is considered as an essential requirement prior to the issuance of an executable Development Permit.

North Macedonia. The Standardization Institute of the Republic of North Macedonia (ISRSM) is the official organization responsible for standardization in North Macedonia [68]. ISRSM is key to creating, promoting, and spreading national standards. To address specific standardization areas, ISRSM sets up technical committees. These committees consist of experts from various sectors, including industry, government, academia, and other relevant fields.

The country has a system of implementation where either European Standards are adopted or local standards are created. In both cases, the standards need to pass a public hearing before implementation. North Macedonia follows a number of standards related to sustainable development, waste management and energy efficiency. Their national guidelines are available in English. No information was given regarding standards related to Circular Economy and Building Disassembly. With reference to the focus area of Recycled Aggregate, two standards were mentioned, namely: MKC EN 1744-6:2010/AC:2010 [69] and MKC EN 933-11:2010 [70]. Standard "MKC EN 197-6:2024—Cement—Part 6: Cement with recycled building materials" was adopted to control the production of cement from recycled materials.

Poland. Significant detailed information on Circular Economy and Standardisation was not provided by the Polish respondents. The key information provided refers to the national standardisation entity platform and standardisation which is in Polish [71].

Portugal. Portugal also has a website in Portuguese regarding national standard entities and mechanisms through which standards are developed [72]. It was also explained that they have technical committees which mirror the ISO and CEN committees.

With regards to Circular Economy, they use a number of guidelines and standards. They refer to the *Circular Economy Action Plan: The European Green Deal* [73] from the international level, while *A new Circular Economy Action Plan: For a cleaner and more competitive Europe* [73] is implemented from the European level and at a national level they have the standard *Plano de Ação para a Economia Circular* [74]. A number of key objectives used for the standards were outlined. These are: to decouple economic growth from resource consumption, extract and consume resources more efficiently, improve waste management, share goods and services, reduce GHG emissions and other environmental impacts, from the perspective of analyzing the energy cycle life and close product life cycles.

Portugal has a circular economy policy framework which is implemented on a voluntary basis and they also have a circular economy action plan which is reflected in the National Standards.

Romania. Romania's only information provided was the national standard entity—ASRO [75]. This is the national platform for the development and adoption of European and International standards. At European level, the standardization activity is regulated by Regulation 1025_2012 [76].

Serbia. The Serbian Standards Institute (ISS) [77] is the national body responsible for standardization in Serbia. ISS plays a central role in the development, promotion, and dissemination of national standards. Technical committees are established by the Serbian Standards Institute to focus on specific areas of standardization.

These committees are composed of experts from relevant industries, government agencies, academia, and other stakeholders. They work together to develop and update standards in their respective fields. Serbian standards (and related documents) may be original or may be developed based on international, European, and other regional standards and related documents, as well as national standards and related documents of other countries in accordance with agreements signed with the national standardization bodies of those countries. Standards are developed and defined through a process of knowledge sharing and best practices, and they are built upon a consensus reached among experts representing stakeholders in standardization committees. Serbian standards are created and published while adhering to the principles and rules established by international and European standardization organizations. Draft standards or revisions are typically subject to public consultation, allowing stakeholders, including the public, to provide feedback and input on proposed standards. This inclusive process helps ensure that standards reflect the needs and perspectives of various sectors. Involving all stakeholders in the process of adopting Serbian standards is possible in Step 1—Submitting proposals for a new project to the relevant standards committee and in Step 4—Public consultation on the draft Serbian standard. In the other stages of development, the standard is only accessible to members of the standards committee until its publication. The Serbian respondent gave very detailed answers to the survey.

Serbia has a number of standards they refer to and implement which are ISO/DIS standards for Circular Economy, CE standards which are in the consideration and adoption phase and National standards which seem to be identical to the ISO and EN standards. It should be noted that the standards listed for the National and International section are the same links. Serbia has a website which is used for guidelines and recommendations for circular economy [78].

With reference to the two areas of focus which are Building Disassembly and Implementation and extent of implementation, one will notice that according to the Serbian respondent, *ISO 20887: Sustainability in building and civil engineering works* [79] is mentioned which for implementation, there is a low level of implementation and although the standard is published (but not mentioned in the survey) it is not generally applied.

With reference to another focus area mentioned in the survey—Recycled Aggregate, the Serbian respondent mentioned a number of standards, namely:

- SRPS EN 206 Concrete—Specification, performance, production and conformity [80].

- SRPS EN 12620 Aggregates for concrete [81].
- SRPS EN 13055 Lightweight aggregates [82].
- SRPS EN 933 series of standards "Tests for geometrical properties of aggregates" [83].
- SRPS EN 197-6 Cement—Part 6: Cement with recycled building materials—only draft version, still not in application. Identical to corresponding EU standards.

The respondent also outlines that the application of fine recycled aggregates is still being researched. The implementation requirements are outlined in the guidelines Rulebook on Technical Requirements for the composition of concrete determined by the standards SRPS EN 206 [78] and SRPS U.M1.206 [84] and the application of aggregates in accordance with SRPS EN 12620 [81] or SRPS EN 13055 [82].

Spain. Spain has a national entity called the UNE [85] and implements standards from an International, European and National level. Spain has guidelines and recommendations for Circular Economy and is referred to in the document Estrategia Española de Economía Circular, España [86]. The Spanish Circular Economy Action plan for 2021–2021 includes 116 measures proposed by eleven ministries aimed at achieving the objectives for the year 2030, while also maintaining coherence with initiatives and policies undertaken at the community level.

The main objectives in Spain for International standards are the following: Protection and improvement of the environment, preventive action, decarbonisation of economy, "Polluter pays" principle, health protection, rationalization and efficiency, cooperation and coordination between public entities, public participation, sustainable development, solidarity between people and territories, integration of environmental conscience in decision making, increasingly competitive economy, generation of quality employment.

On a national level, the objectives for standards differ to the international ones. Here are the objectives:

- Production: the production of a product from conception to manufacture keeping in mind it's lifespan, repair and repurpose to create less waste.
- Consumption: promote sustainable consumption and promote recycling where possible.
- Waste management: when facing the reality that raw materials worldwide are becoming expensive and scarce, it is imperative for action to be taken to reuse, recover and recycle
- Secondary raw materials: the implementation of secondary raw materials gives the possibility of better usage of natural resources
- Water reuse and purification: Water is an important resource in the Iberian Peninsula. It has an important contribution to the Spanish economy and work is done to address obtaining second raw materials.

The Spanish also have a Spanish Strategy for Circular Economy called, España Circular 2030 [86]. España Circular 2030 created a production and consumption model platform for the reduction of waste and better usage of resources within the economy. In this way, waste is recycled and reused as long as possible before becoming unavoidable waste.

Although España Circular 2030 has a long-term vision, it is to be implemented in the short and medium term by means of successive action plans. The regulation which enacts the standard is Directive 2018/851/UE. Such standards are addressed to construction, farming, fishing and forestry, industry in general, consumption goods, tourism, and textile and garment sector.

The Netherlands. The Dutch respondent had very little information to share in the survey. NEN [87] is the standard entity through which standards are developed. More information (in Dutch) on how standards are developed is available on the site [88]. Guidelines for circular economy are in development. For four years there are progress studies by the Planbureau voor de Leefomgeving (PBL) which will be developed into Guidelines and Standards. To date all guidelines and standards are on a voluntary basis. There are also policy frameworks, strictly in the building industry for new buildings. This is not yet applicable for existing buildings and the circular economy as most is energy-driven.

Turkey. The national standard entities and mechanism to develop standards are carried out by the Ministry for Environment, Urbanism and Climate Change via ISO 59000 (under publication). A number of standards at National, European and International level are implemented in Turkey. The adoption of international European standards is mandatory, with implementation being required at national and company levels. All implemented actions/projects are available in guidelines called *Technical Assistance for Assessment of Turkey's Potential on Transition to Circular Economy*.

United Kingdom. The Department for the Environment, Food and Rural Affairs provides information regarding the national standard entities and mechanism [89]. The implementation of standards at European and international level refer to: BS EN 18177 Circular Economy in the construction sector—framework, principles and definitions; BS ISO 59040:2025 Circular Economy—Product circularity Data Sheet; BS ISO 59014:2024 Environmental management and circular economy. Sustainability and traceability of the recovery of secondary materials. Principles, requirements and guidance; BS ISO 59004:2024—Circular Economy. Vocabulary, principles and guidance for implementation; BS ISO 59020:2024 Circular Economy. Measuring and assessing circularity performance; BS ISO 59010:2024 Circular economy. Guidance on the transition of business models and value networks. Standards and guidelines at national level primarily include the BS8001 Circular Economy standards [90] as well as recommendations through the IStructE guideline for reuse [91].

In the UK, the key objectives of international European standards are for circular economy for new buildings, where reference/enforcement, along with adoption and how national determined parameters in the standard are set, are voluntary. The same can be said for how each standard is references in national regulations at this stage, where no regulations enact the standard. As for implementation of the standard however, it follows BS 8001 environmental benefits through improved resource use.

The UK also as yet does not have any measurements for the effectiveness of the standard so far. However, the goal for net zero has been changed.

The standard requires revisions and is currently addressed to building construction, with gaps not covered by the standard including existing building stocks, and various applications. When it comes to implementing the standard, the UK references BS 8001 environmental benefits through improved resource use.

In the national standard guidelines, which are in English, the objectives of the standard are determined from BS 8001 [90] environmental benefits through improved resource use. At this level, standards are published by the standards entity, with adoption on a voluntary basis. The current effectiveness of the standard is stated as being good. As for implementation, this is carried out for large construction projects and is addressed to construction in general.

6.2.6 *Country Factsheets: Discussion*

The review of CE standards and standardisation across Europe and beyond, and the perception of their implementation, provides some key insights in the current state of standardisation in Circular Economy. The survey outcomes suggest that there is an understanding of the need for standardisation across most countries, to support policy and regulations. The level of implementation of European standards and International Standards, however varies across States, as does the perception of standardisation as identified through the questionnaire. While generally there is a voluntary framework of standards, these are not necessarily covering all aspects of the industry. However, there is a degree of implementation in various countries, through policy, regulations frameworks and legislation which refer to the standards. The survey conducted suggests that the level of implementation varies, with different recent and current developments, and with a perception of key ongoing actions towards standardisation to further support policy and implementation.

6.3 International and European Level Standards and Standardisation

Standardization refers to the creation, publishing, and implementation of standards in different sectors, including the construction industry. This process occurs at both national and international levels. Two prominent organizations that play a crucial role in standardization efforts are the International Organization for Standardization (ISO) [92] and the European Committee for Standardization (CEN) [93].

6.3.1 International Level Standardisation

In today's globalized world, standardization has become crucial in numerous industries, including the construction sector. At the international level, standardization is facilitated by organizations such as the International Organization for Standardization (ISO) [92] and the European Committee for Standardization (CEN) [93]. These organizations combine national standardization bodies from various countries to develop and publish international standards [13]. The primary objective is to promote harmonization, interoperability, and compatibility across borders.

The process of international standardization involves the formation of technical committees that address specific areas of standardization. Experts from participating countries collaborate to develop global standards. The draft standards are reviewed to ensure accuracy, clarity, and relevance. They are then circulated among participating countries for voting and approval. The standards are approved if a significant majority of the participating countries support them. Once approved, the international standards are published and made available to the public. National standardization bodies adopt these standards and incorporate them into their national frameworks.

The implementation of standards involves various aspects, including the acceptance of international standards as national ones and their referencing through laws and regulations. In this way, compliance with the standards becomes mandatory. Authorized bodies often verify compliance through inspections, audits, and certifications. Educational initiatives and training programs are conducted to raise awareness and build the capacity of professionals in the construction industry. This ensures they have the necessary information and skills to implement and conform with the standards.

The construction industry, including contractors, engineers, architects, and suppliers, plays a vital role in implementing standards. They incorporate the standards into their processes, designs, and specifications to ensure compliance and quality. Standards are periodically reviewed and updated to reflect advancements in technology, changes in regulations, and emerging industry practices. Monitoring and feedback mechanisms are established to gather information on the effectiveness and relevance of the standards, leading to continuous improvement.

6.3.2 Key Players in European Standardisation

The process of European standardization involves numerous parties and aims to reach a consensus. Standards development is mainly driven by market demands, with the industry playing a crucial role in initiating it. Three European Standards Organizations, namely the European Committee for Standardization (CEN) [93], the European Committee for Electrotechnical Standardization (CENELEC) [94], and the European Telecommunications Standards Institute (ETSI) [95], are responsible for the development of European standards [96].

ESOs—The European Standardization Organizations. The European Standardization Organizations, namely CEN, CENELEC, and ETSI, are officially recognized as providers of European standards by Regulation (EU) No 1025/2012. These organizations have cooperated with the European Commission since 1984, when the first agreement was signed. The agreement was later revised in 2003 and provides general guidelines for collaboration between the parties involved [97].

CEN. The European Committee for Standardization, also known as CEN, is an organization that brings together the national standardization bodies of 33 European countries. Its primary purpose is to create a space for developing European standards and other technical documents related to a wide range of products, materials, services, and processes, including air, space, chemicals, construction, and more [93].

CENELEC. CENELEC, the European Committee for Electrotechnical Standardization, is in charge for standardization in electro-technical engineering. The voluntary standards developed by CENELEC help to enable trade between different countries, access new markets, reduce compliance costs, and encourages the growth of the EU single market. In addition, CENELEC collaborates closely with the International Electrotechnical Commission (IEC) to create market opportunities at the international level [94].

ETSI. The European Telecommunications Standards Institute (ETSI) creates international applicable standards for information and communications technology (ICT). This covers fixed, mobile, radio, converged, broadcast, and internet technologies. ETSI's primary aim is to develop and preserve the technical standards required by its members. With ETSI, the industry can directly participate in standards development. However, to access CEN and CENELEC, the sector can only involve themselves through their respective national standardization bodies [95].

Other stakeholders involved in European standardization [97]: EU and EFTA countries' National Standardization Bodies (NSBs) are responsible for developing a European consensus. Small and medium-sized enterprises (SMEs) are represented by the non-profit organization Small Business Standards, which reinforces their representation. Consumer, trade union, and environmental interests are represented by ANEC, ETUC, and ECOS, respectively. They are essential for accountability. Public authorities develop standards-receptive regulation, issue standardization directives and public procurement, and provide funding to the European Standardization Organizations (ESOs) and the NSBs.

6.3.3 Review of International Standardization Bodies

Procedures for the reporting, quantification, assessment and guarantee of sustainable buildings have been established in many countries. There was, however, the need to develop systems and standards for the assessment of better insertion of life cycle analysis methodologies with regards to environmental, social and economic aspects

[59]. Standardization is crucial in facilitating the transition to a circular economy, where the goal is to move from a linear throwaway society to a circular model. This transition involves using raw materials as long and frequently as possible and managing natural resources within a closed circle without consuming new resources [98].

In this section a brief overview will be given of the main International and European standardisation bodies which are working towards creating common standards in the Circular Economy for buildings and construction. These standardisation bodies are namely CEN/TC 350, ASTM International, ISO Technical Committee 323 and CEN-CLC/JTC 10.

European Committee for Standardization (CEN). CEN, the European Committee for Standardization, holds a distinguished position as one of the three European Standardization Organizations officially recognized by the European Union as well as the European Free Trade Association. Its primary function is to develop and define voluntary standards at the European level. Boasting a membership of national standardization bodies from 33 European countries, CEN plays a crucial role in promoting the creation of European standards across various sectors, such as construction [99].

CEN has established standards in the circular economy realm designed to encourage sustainable practices, enhance resource efficiency, and embrace circular construction principles. These criteria offer direction and specifications for utilizing recycled materials, assessing circularity, and carrying out circular economy strategies within the construction industry [93].

CEN/TC350 provides a framework of coordinated European standards for the analysis of environmental, social and economic performance of structures. These are based on a life cycle method as a contribution to applying sustainable construction. The aim of these standards is to grow and implement a mechanism of sustainability assessment tools for buildings that address the assessment of the environmental, social and economic performance as well as the technical quality and functionality/serviceability of the building. In this European standard, the building including its foundations and the external environment footprint of the building site, is the area of assessments and the basis of impacts over the life cycle of the property. It is important to note, that in order to assess the environmental, economic and social performance of the building, the footprint of the site does not take into consideration the impacts of non-building appliances [59].

When assessing the results and carrying out comparisons between alternative design options at design stage or when comparing finished buildings, a "fair basis" must always be considered. For this reason, the concept of "functional equivalent" was introduced. The functional performance of buildings should form the basis of the comparison. These requirements effect the results of the assessment. For this reason, they need to be considered during any assessment. It is clear that the functional equivalent must always be communicated together with the results of the assessment. Should the National, Regional or Local entity prefer, it is allowed to communicate the results with more collected groups of information and/or indications after the valuation. The results, however, must be reported separately from the standardised assessment results making it clear that this is additional information [59].

This European system of assessments strives to have truly transparent performance-based standards and for this reason, CEN/TC 350 only considers performance aspects and impacts that are reported with the quantitative indicators. In the framework level standards, these indicators establish the requirements for the building and product level being use. At the building level, the assessment communicates that the descriptive model of the building is defined by the requirements for technical and functional performance of the building [59].

In CEN/TC 350 the sustainability assessment of buildings as a player in the implementation process of sustainability of construction works is a multi-level process. The framework for the assessment of social and economic performance has working groups developing these processes. The descriptive and assessment of sustainability aspects consider the economic, environmental and social performance. This performance creates a relationship between the technical performance and functionality aspects and thus one can say that the two concepts, performance-based buildings and sustainability in the building construction are emerging as a complementary tool. The addition of economic and social performance has encouraged to transition from an assessment of environmental impacts to a wider assessment of sustainability characteristics. This assessment is significant in the discussion of the quality of the building and the built environment [59].

CEN has formed specialized working groups, including CEN/TC 350/SC 1, dedicated to promoting circular economy principles within the construction industry. These groups strive to create standards that encourage the adoption of circular construction practices, focusing on certifications for products containing recycled materials and guidelines for future life cycles [100]. CEN ensures coordination and alignment with existing construction committees supporting the circular economy, such as ISO/TC 323 and CEN-CLC/JTC 10 [100].

ASTM International. ASTM International is a well-known organization that has been actively developing standards for the circular economy. The organization has been mainly focused on sustainability and circularity and has created specialized technical committees and workgroups to define frameworks, principles, and sector-specific applications. These efforts have led to the development of various methods to measure sustainability and circularity, making ASTM International a pivotal contributor to the development of circular economy standards worldwide.

ASTM International has formed technical committees and workgroups to promote the circular economy concept. The workgroups comprise experts from diverse industries collaborating to create standards that facilitate circular practices. For instance, Subcommittee E60.13 on Sustainable Manufacturing concentrates on devising design principles for the circular economy [101].

ASTM International has developed a roadmap for technical standards to improve production efficiencies and reduce waste materials. The roadmap identifies openings for new and existing standards in relation to circular economy in manufacturing [101].

ASTM International actively partners with key players in the manufacturing industry to pinpoint the necessary standards for prolonging the lifespan and improving the recovery of materials and goods through repair, reuse, remanufacturing, and recycling. Moreover, the organization spearheads roundtables and workshops to cultivate circular economy principles and establish standards that bolster circularity. By spearheading the creation of circular economy standards, ASTM International emphasizes its dedication to promoting sustainable initiatives, optimizing resource utilization, and mitigating waste across diverse sectors.

International Organization for Standardization (ISO). Another organisation and technical committee to consider is ISO Technical Committee 323—Circular Economy. ISO/TC 323 is a committee of experts from several countries, including industry, academia, and government representatives. ISO TC 323 was set up to standardize the field of CE by proposing a number of international texts, guidelines and tools for all public and private organizations with the aim of contributing to the field of circular economy and sustainable development. TC 323 aims to propose some transversal texts but not targeted towards a specific activity. If specific activities need some personalised text, then these will be written baked on the TC 323 experience. The scope is to develop frameworks, guidelines, supporting tools and requirements for the application of activities of all organisations involved in this sector, to amplify the contribution to Sustainable Development [7].

The purpose of ISO TC 323 is to contribute to an answer of the collective emergencies by proposing an alternative environmental emergency. These are mainly, resource depletion, water generation, climate change and biodiversity losses. There are also social emergencies to consider which are closely linked to the environmental ones, such as resource access and social inequalities. By promoting a new mindset, one can initiate more collaborative partnerships and create more value networks, to develop concrete, operational texts for all user types, including small and medium enterprises from all over the world [7].

This committee is tasked with developing frameworks, guidance, supporting tools, and requirements to promote implementing circular economy practices that contribute to sustainable development [102]. ISO plans to introduce new international standards for the circular economy in 2024, providing guidance and requirements for transitioning to a circular economy model [92]. The scope of the standards is to cover principles, terminology, frameworks, and implementation guidance for the circular economy, ensuring consistent practices [103]. For that purpose, the committee collaborates with other ISO technical committees responsible for developing standards related to the circular economy, such as sustainable procurement, quality management, and environmental management [103].

The primary goal of ISO/TC 323 is to address the pressing need to shift to a circular economy. This is in response to resource depletion, climate change, and growing inequalities. ISO aims to develop standards that provide organizations with a global vision and a framework for adopting circular economy practices [92].

Some of the opportunities considered as part of the circular economy mindset are:

- integrated systematic eco design to manage one's products/services/solutions

- consider waste as a resource as far as possible to reinforce material valorisation
- promote collaboration and partnership
- integrate long term vision by balancing the economic/financial results/indicators with the environmental and social indicators to better change the mindset [7].

TC323 decided to propose a package to better understand the implementation of Circular Economy. Four main International Standards are under development, namely:

- ISO 59004—to give definitions/principles/framework for implementation and areas of action.
- ISO 59010—to shift from linear to circular business models to help organizations implement a new business model.
- ISO 59020—to measure if a business is really circular. This standard identifies which set of indicators to use and which data framework is required.
- ISO 59040—to focus on a product along the value chain and provide circularity at this level [7].

A standard on secondary material is also under development to propose guidelines to manage the material and include recommendations for traceability mainly for waste.

CEN-CENELEC Joint Technical Committee 10 on Energy-related products—Material Efficiency Aspects for Eco-design. The last technical committee highlighted for discussion is CEN-CENELEC [59] Joint Technical Committee 10 on Energy-related products—Material Efficiency Aspects for Eco-design (CEN-CLC/JTC 10). This was created after the publication of the Circular Economy Action Plan in 2015. The European Commission asked the three European Standardization Organizations—CEN, CENELEC and ETSI to create standards on material efficiency to establish future eco-design requirements on durability, reparability and recyclability of products as well as other things. This introduced a new focus on material efficiency during the design phase [102].

The CEN-CLC/JTC 10 created a group of eight standards with principles to consider when dealing the material efficiency of energy-related products. Some of the factors considered were extending the product lifetime, the ability to reuse components or recycle material from products at end-of-life and use of reused components and/or recycled materials in products. The standards developed by CEN-CLC/JTC 10 were used by Technical Committees to process product-specific or product group standards addressing material efficiency features [102].

International Standardisation: General Conclusions

Circular Economy is a complex challenge that needs to consider not only the environmental question but also the social one. The question one needs to ask has to deal with more than the value chain with the value network and also consider the holistic/systemic/life cycle thinking perspective relationship. One needs to rethink the long-term vision to balance economic/financial indicators with environmental/

social ones. Thus, one needs to change the mode of production and consumption at individual/organisational/country level [7].

For products or materials to be designed intelligently, one has to also consider the efficient use of the resources in the production processes as inefficiency can reduce business opportunities and substantial waste generation. One of the problems experienced by operators showing interest in using secondary raw materials is the vagueness of the quality of the materials in hand. Due to the lack of EU-wide standards, it will be hard to assess the impurity levels or suitability of high-grade recycled products. By creating quality standards, the overall trust in secondary raw materials and in recycled materials will increase. This ensures better support and take-up of this market [102].

The facilitation of cross-border circulation of secondary raw materials ensures that trading of these materials can be facilitated across the EU. With the use of electronic data exchange, the cross-border process can be simplified. By enhancing the accessibility of data on secondary raw materials, the EU will further grow the Raw Materials Information Systems and support the research on raw material flows across the European Union [102].

The biggest sources of waste across Europe are construction and demolition surplus. Most of the materials generated in this sector are recyclable and can be reused. There is however a barrier in this process since the rates to reuse and recycling vary widely across the EU. The construction industry plays an essential role in the environmental performance and infrastructure, throughout the lifespan of the building. Considering buildings are permanent structures that last for decades, it is imperative to promote design improvements that will mitigate and reduce the negative environmental footprint of buildings and enhance the durability and recyclability of their materials [102].

Innovation and creativity will play an important role in this systemic change. Europe and the world need to rethink the way of production and consumption. To transition material waste into products that have a high value-added quality, new technologies, processes, frameworks, services and business models are needed to determine the future of our economy and society. Research and innovation are major contributors in encouraging this shift as they provide knowledge which leads to trust in the investment. Research and innovation will also shift the EU industry towards competitiveness and modernisation. Reliable indicators will be important to understand the progress of this transition towards a more circular economy and the impact of these actions at EU and National level [102].

6.3.4 Identification of Challenges at International, European and Regional Levels

In considering the discussion on standards and standardisation above, different challenges and gaps are identified.

In the working groups of ISO/TC 323, a critical evaluation of the standards was evaluated and the following was drawn up [7]:

- the ambition level is very low,
- no holistic interpretation in definition of principles. points such as service life extension, design, upstream innovation were not considered,
- no guidance for circularity,
- the concepts were in part only of little help when transferred to company and product level.

The possible negative effects were also identified:

- Discrepancy with goals of Green Deal and CEAP due to carrying interpretation of circular economy concepts;
- Increasing fragmentation of the EU internal market in Circular Economy standardization;
- Linear concepts were declared as Circular Economy without the necessary transformation process;
- Problem of availability of critical raw material in the EU was insufficiently addressed;
- In particular, the principles of established environmental management systems such as ISO 14001 et al. were insufficiently used for evaluation of options for action.

The negative effects regarded in ISO/TC 323 Circular Economy can be absorbed by mirroring at European level. The option of European annexes would make it possible to adapt the forthcoming international standards to the needs of the European internal markets. The call for the European mirroring of ISO/TC 323 is for the following reasons [7]:

- Fundamental issues of the Circular Economy are defined which implications for all industries and elementary components of the transformation;
- In the legislative projects, relevant to the industry such as the Eco-design regulations, the Green Claims Directive, the Critical Raw Materials Acts and the Digital Product Passport, ISO/TC323 standards, especially in the area of definitions risk not being compatible with the main objectives of the legislative projects unless they undergo European adaptation [7].

In CEN/TC350, the standards address just the analytical part of the building assessment system but does not address the valuation part. It should also be noted that the standards do not offer valuation methods, nor do they set levels, classes or benchmarks to understand the performance of the building when they are used. Thus CEN/TC 350 standards assume that the assessment and weighting factors in the performance of buildings have to be addressed by other assessment systems, or as part of national assessment systems. It should be noted however, that initiatives like LEnSE and UNEP = SBCI discuss these assessments and approach to weighting and benchmarking, both on an international level as well as on a European level [102].

The CEN/TC350 *Final Gap analysis Consultation report and findings*, drawn up to summarize the gap analysis work carried out by CEN/TC 350 work groups and the feedback received from National Standards Bodies and other key stakeholders. Nine focus areas were developed in this document. Each area outlines the feedback given by the National Standard Bodies and other stakeholders and the feedback was drawn up in a table for each focus area. Below is a list of the nine areas, a description of what the area encompasses and a concluding summary of the gaps and action required as outlined in the document [104].

6.3.5 Framework and Definitions

Framework and definitions are required to guide and document the standard development process in the field of circular economy in the construction sector [104]. In this way, a common language for all stakeholders involved is established which addresses and encourages stakeholders to define what they need in the standardization process [104]. The framework will establish the boundaries of standardization work and the required interfaces to regulatory systems.

This focus area is addressed by the activities of WG1. This means that this WG will be addressing the feedback as part of their work.

6.3.6 Circular Indicators, Measurements and Assessment

Circularity criteria can be measured using indicators which give quantitative and qualitative insights on how well a product can perform in the context of a circular economy. These indicators were developed by "Ellen MacArthur Foundation" and the European Commission is actively supporting the circular economy in EU States. Towards this goal, an online monitoring tool launched by Eurostat gives users an overview of different circular indicators across EU countries.

From the feedback received it was clear that this is an area which is not aligned and consensus was not given from the stakeholders concerned. There was not any great agreement on the measurement and assessment needed to support circularity, where the gaps are and what methods and data would therefore deliver the requirements. Given the lack of agreement on what should be measured and how, it is clear that more work needs to be done in this area and a dedicated working group should be created to address these requirements. Although one would not ignore the current assessment methods, there is no agreement that the existing approaches and methods are yielding the results needed to assess circularity in relation to costs and benefits [104].

6.3.7 Circular Data, Data Storage and Maintenance and Product/Building Passports

Circular data documentation is an important means to achieve more circular construction in the construction sector. It promotes the reuse of materials, products and elements at a project level while also reducing waste, consumption of raw materials and reduction in the negative environmental impact. The objective for additional documentation on a product and on a project is to "support sustainable production, to enable the transition to a circular economy, to provide new business opportunities to economic actors, to support consumers in making sustainable choices and to allow authorities to verify compliance with legal obligations." (The Circular Economy Action Plan's Sustainable Product Initiative).

By developing a standardization product passport, it is felt that the knowledge of the product content used in the building can be tracked and facilitate the maintenance of the building, as well as the repair and renovation. The passport can also increase the worth of the materials and products when it comes to reuse and recycling. The product passport would also support the goal of the Green Deal and in particular the Circular Economy Action Plan's Sustainable Product Initiative, the EU's Digital Europe Programme and the EU Data strategy. This can be achieved by improving product sustainability and digitalization, boosting material and energy efficiency, enabling new business models and circular value extraction based on data sharing.

In the circular construction world, having the ability to compare and exchange digital representations of objects in construction is essential. Thus, it is imperative that developers of the passports make agreements in this respect, as otherwise, the desired outcome will not be achieved. To facilitate this, harmonized development of digital passports in the EU is essential by defining the format, scope, digitalization, data generation and access for product passports.

It should be noted that the Circularity requirements for EU Taxonomy request digital tools to support enhancing and extending the service life and future adaptations and reuse of materials or buildings. No existing standards to support implementation exist to date.

From the feedback received it seems there is a general agreement that product and building data should be available in digital form to facilitate circularity in construction. Currently there are a lot of activities being carried out at EU level which needs to be considered in this context especially in relation to CPR AQUIS revision, EU taxonomy, Digital Building logbook. There is also a need to further define the data needs in order to then standardise such templates [104].

6.3.8 *Planning and Design and Regulations for Circularity (Product/Building/CEW)*

The main objective of Circular Economy is to keep materials in a closed loop and extend their lives through the reuse, repair, remanufacture and recycle approaches. Buildings are like material banks in which the ease and ability to deconstruct and reconstruct a building plays a key role in the Circular Economy. Construction waste from one building should be considered as building material for another building.

There needs to be a focus on design building components that include at least the 5 "design for R options". This means: design for reduce, design for reuse, design for repair, design for remanufacture and design for recycling. This means that the using, managing, operation and renovation of a building needs to be rethought.

To avoid gaps and misunderstandings between researchers and industry, Standardization is essential. By having standardized parts for buildings together with standardized systems, will ease the disassembly and repair activities. If buildings are built modular, it will facilitate the repair/replacement/change/upgrade of components without the need to demolish the whole building. To achieve Circular Economy, the technical based obstacles as well as the hurdles in business models and social embracement need to be addressed.

This section outlines a number of gaps which most stakeholders agreed on overall. There are, however, many comments related to the scope and nature of standardization which need to be addressed. It was also highlighted that there are needs for better assessment and pre-demolition auditing to facilitate better use of structure and resources. Having design rules, guidance and indicators related to future disassembly, adaptability and reusability were also highlighted as priorities. From the comments presented in the document, the gap in this area could be split into 2 with one area focused on reuse and the other designing circularity [104].

6.3.9 *Construction Site and Resources and Regulations in Context of Circular Economy*

Construction and demolition waste is known to make-up for more than a third of all waste which is generated in the EU. Thus Reuse of construction products is an important part of the circular economy process. The waste is made up of a wide variety of materials such as concrete, bricks, wood, glass, minerals, wool, metals and plastics. To reap the full potential of this waste one needs to consider the following steps:

- construction site management of waste,
- resume of residual and cut-off materials,
- take-back for reuse on other sites,
- take-back to reintroduce into the production process,

- planning phase,
- circularity assessment as designed.

Despite the potential of REUSE of construction waste, the level of recycling and material recovery varies greatly across Europe, ranging from less than 10% to over 90% reuse. Since different EU countries apply different definitions of waste, it makes cross-country comparisons difficult. This could be addressed through standardization.

In the document there was a medium to high recognition that further clarification and guidance to support the optimal reuse of structures and products is essential. It also highlighted the importance of supporting recycling and end of waste criteria. References were made to the need to develop technical details related to both reuse and recycling. It was however made clear that issues to address the resume were more urgent than the issues of recycling. Recycling is dealing with materials while reuse is more relevant as an asset, product or component level. Ways to address the concerns of the insurance sector like carrying out non-destructive tests were also highlighted [104].

6.3.10 Operation and Maintenance/Repair/Refurbishment and Regulations in the Context of Circular Economy

Three methods to extend the lifespan of a building or material is maintenance, repair and refurbishment. Maintenance and repair of constructions are effective for extending the technical lifespan and reducing the need of reconstruction. Refurbishment of used construction products on the other hand led to extended lifespan of a product. For effective maintenance, repair and the consecutive use of construction materials in more than one construction-cycle, one needs to design these construction products and elements for this purpose.

Circular CPR AQUIS needs harmonized technical standards to function in such design aspects as intended lifespan, maintainability, reparability and deconstruction ability. Testing is also essential to ensure functionality, health and safety of the products in the next life cycle.

A number of stakeholders referred to draft standard *EN 17680 Sustainability of construction work—Evaluation of the potential for sustainable refurbishment of buildings.* Thus, this standard should be considered before any further gaps related to refurbishment are addressed [104].

6.3.11 Built Environment End-of-Life and Maximizing Resource Retention (Building Product, Material)

This area considered the end-of-life aspects of products and structures in the built environment. The aspects to be considered include process and data needed to support maximizing the value of assets and materials. In the London Circular Economy Statement guidance policy document, a 'waste hierarchy' is illustrated and this can be adapted to convey the relative priority. Another approach is to consider buildings as Material banks and should be used to the highest value possible thus reducing environmental impacts of construction through displacing primary products and materials. This method helps reduce the impact of managing the demolition waste, avoid landfill/disposal and can yield economic benefits.

Dealing with end-of-life materials arising from demolition has been the emphasis for many years. There has been a change with the introduction of pre-demolition/redevelopment auditing and retention of existing assets before demolition commences. Considering the better management of demolition waste has been a property in a number of Member States national policy and EU policy, there is substantial guidance, standards, protocol, tools and policies to consider.

The feedback from the stakeholders had many comments and proposals for additional gaps relating to the End-of-life phase. There is a need for further clarification and refinement of standards especially for pre-demolition audits/appraisals and decision to maximise retention. There were also many gaps raised from previous focus areas, namely in aspects of definitions, decisions relating to refurbish vs demolition, reuse and recycling. The focus should thus be on standardisation related to pre-demolition audit, selective demotion and end-of-waste criteria [104].

6.3.12 Circular Business Models

In order to achieve a circular economy, one needs to develop circular business models to drive it. In linear economy models, the mindset is to 'take, make, use, dispose/recycle'. In a circular economy, however, one needs to shift away from this mindset. The alternative approach for a better alignment with the circular economy is related to the way construction products and materials are supplied.

The first shift in mindset is to see buildings as material banks which need to be 'mined' instead of the reliance on new products (including recycled materials). These banks could be from deconstruction of existing assets or acquiring from digital marketplaces often called material exchanges. As already mentioned in previous areas above, the reuse of existing materials has its multiple benefits as well as issues to consider. Suppliers could also consider a shift from supplying products to services or at least offer to retrieve the product back for reuse, remanufacturing or other recovery. If a supplier is permanently tied to its product, it will be in the supplier's interest to design products which are easily modified and maintain their value.

The stakeholders did not support the gaps in this focus area mainly due to national policy and markets dictating the activities discussed in this area. Reference was also made to other focus areas like circular procurement and standardization gaps for promoting reuse and recycling of products and materials [104].

6.3.13 Circular Procurement

This area deals with different procurement processes within the European Commission such as the Green Public Procurement (GPP), The Socially Responsible Public Procurement (SRPP) and the ISO 20400: 2017 Sustainable Procurement. The procurement process is a key component meant for Organisations to interact with their supply chain. This can drive change in the product ecosystem. The procurement process and collaboration therein are essential to achieve change from linear to circular economy.

The following documents are considered important in relation to circular procurement in Europe. There are namely:

1. The Level(s) framework—this document focuses on life cycle environmental performance, health and comfort, cost, value and risk.
2. The EU GPP—this document aims to create a good balance between environmental performance, cost consideration, market availability and ease of verification.
3. Planform CB '23 Guide "Measuring Circularity in the Construction Sector 2.0"—this document proposes three goals for circular construction which are:
   - environmental protection,
   - to protect stocks of material,
   - value retention.

The guide also addresses social fairness. This area scored low for all the preliminary gaps. It was uncertain if this was something that should be done outside of the national boundaries. Since certain areas are still evolving at EU level with Level(s) and EU Taxonomy, this area is not considered a priority to date. It was felt that Standardization would be needed prior to specifying the context of circular procurement [104].

6.4 Case Studies in Standardisation

6.4.1 Recycled Aggregate

An extensive overview of standards describing the use of recycled aggregate in concrete is presented in the study by Pacheco et al. [105]. According to this study, the requirements can be divided into two groups based on the aggregate use. The first group is addressed by standard, EN 12620 which defines the quality of the recycled aggregate [106]. The second group is described in EN 206 [107] and deals with the incorporation ratio of RA in concrete. Both of these standards have their specifications at national levels. Therefore, RA's maximum allowed incorporation ratios for structural use differ among European countries. For example, in Denmark, for use in low to moderate exposure requirements, fine aggregates can be used up to 30% of the incorporation ratio ($\leq$ C30/37) and coarse aggregate up to 100% ($\leq$ C40/50) [108].

In France, the standard NF EN 206-1/CN describes different conditions based on recycled aggregate size and allows use of up to 30% of RA for low-demanding exposure classes. In Italy, the use of RA is described based on Ministry Decree NTC:2018, and its incorporation rate depends on the origin of the RA and the strength class (e.g. 60% for non-demanding structural concrete (C20/25) and moderate quality RA; 30% for C30/37 and high-quality RA). Requirements are implemented using national specification LNEC E471 in Portugal, where high-quality coarse aggregate from waste can be used up to 25% of total aggregates (fine + coarse) ($\leq$ C40/50) and even in most environmental exposure classes.

Recycled aggregate was standardised in Malta using the SM820—Classification of Recycled Aggregate, which describes technical parameters for the classification of waste aggregate [64].

The Czech version of the European Standard (CSN EN 206, 2014) outlines the exposure classes for various environmental actions along with the recommended technical limits for concrete composition and strength classes [109]. Within the X0 environmental classes, there are sub-classes that specify the humidity levels in different environments. Each class must adhere to the requirements for a minimum amount of cement, a maximum water-cement ratio, and a minimum compressive strength class. Recycled aggregate in concrete must comply with the geometrical, physical, and chemical properties outlined in (EN 12620, 2002; CSN EN 12620+A1, 2008) [106, 110]. Some crucial properties are detailed in (CSN EN 206-1/Z3, 2008) and the updated Czech European Standard (CSN EN 206, 2014) [111].

6.4.2 Deconstruction and Reuse of Elements

A key element for proper waste management of construction materials is sorting Construction Demolition Waste (CDW) at a deconstruction site. Therefore, multiple documents tend to standardise this approach. Historically, the example of good practice is Austrian standard ÖNORM B 3151 Dismantling of buildings as a standard method for demolition, which describes the standardised approach for each deconstruction where more than 100 t of CDW will be produced and the scope of information depends on the volume of the building (greater or smaller than 3500 m^3) [112]. Nevertheless, pre-demolition audits are mandatory in projects where waste production is estimated to be higher than 750 t. It describes the pre-demolition audit and prescribes an assessment of dangerous compounds and materials interfering with the future recycling process in the demolished construction and how to eliminate their occurrence in the waste flows. Furthermore, prescribed forms for the declaration of materials are presented in the document appendix. It also suggests what material flows should be declared in the pre-demolition audit and recommends waste management scenarios. The standard, however, is focused on main materials such as asphalt, concrete, excavated soil, wood, and metals.

At European standardisation level, the key document is the EU Construction and Demolition Waste Protocol and Guidelines 2018 [113], which is currently under revision. In the revised version, renovation is recommended as a preferred option of how to deal with a building after its end of life. To support this, the guideline stressed the importance of digital building logbooks (DBL), which provide building data in a more interoperable manner. This approach opens space for the implementation of standards related to building renovation passports, building material passports, and digital product passports, which should support the user in better maintaining products and services and thus prolong the service life of building products and buildings. Furthermore, this guideline standardises the steps of a pre-demolition audit, such as the definition of the scope of the audit, study before inspection, field survey, inventory of materials, resource and waste management recommendations, reporting, and further traceability. In addition, it describes the requirements for auditors. This guideline also standardises the procedure for selective deconstruction, which should be performed in the following steps: interruption of supply and drainage lines, removal of furniture, removal of hazardous substances and materials, removal/dismantling of non-load bearing structures (ceilings, floor coverings, doors, windows etc.), demolition of load-bearing structures, demolition/removal of foundations and ground covers.

In relation to waste management, JRC's Best Environmental Management Practices provides guidelines and recommendations in relation to CDW management and suggests the use of a life cycle perspective in consideration of waste management scenarios [114].

For instance, in the Czech Republic, there is a guideline for pre-demolition audits, which is not mandatory. The EU CDW protocol focuses mainly on the separation of dangerous waste and, similarly to Austrian standards, suggests protocols that can help auditors identify the main material flows. Currently, a new standard is being prepared, which also focuses on recommended reuse and recycling approaches and combines information among the circular value chain from pre-demolition audit to selective deconstruction and waste management.

In Malta, the procedure of selective deconstruction is standardised by SM810 for recycling-oriented Deconstruction and Classification of Waste [64], 65. Together with SM820, standards were developed to cover deconstruction and use of aggregates with focus on proper classification to support the implementation of circular principles and minimize consumption of primary resources [64]. SM810 covers information related to planning, classification of waste and conducting demolition operations. SM810 mentions also practice based on BS 6187:2011, which prescribes demolition practice in Great Britain [115].

In other European countries, the threshold for performing a pre-demolition audit differs. For example, in Sweden, pre-demolition audits are always required [116]. In some countries, such as Germany, the pre-demolition audit is mandatory with a focus on the identification of dangerous materials, also considering the possible risks for workers at building sites [117]. At least an inventory of materials present in the buildings is mandatory in Flanders for non-residential buildings with volumes higher than 1000 m^3 [116]. Inventory and intended waste management scenarios are mandatory in the Netherlands for each building demolition, where more than 10 m^3 is produced [117]. Among other construction and demolition waste flows, asbestos waste is a dangerous type of waste that must be audited in most of the countries in the European Union [118].

6.4.3 Reuse of Products

Besides the implementation of standards for deconstruction, there is a growing emphasis on the concept of re-utilisation or reuse of products. This circular strategy is, according to the hierarchy of waste management, the most prioritised one as it results in the keeping of the value of the product. According to the ISO standard focusing on design for disassembly, reuse is defined as the use of products or components more than once for the same or other purposes without reprocessing [119]. Therefore, efforts are being made to establish and broaden markets for re-utilisation and to devise structures that support it. This includes defining the status of reused products, setting responsibility guidelines, and adopting quality standards and certificates for construction products. In addition, standards should be implemented to reduce legislative hurdles.

Furthermore, to make reuse more accessible, new physical and digital platforms and networks are being developed to link the providers and consumers of materials from deconstruction. These platforms can play a role in all stages of a project, from planning to execution [120, 121]. For instance, reclaimed materials from the regions of France, Belgium, and the Netherlands are well described at Project Opalis, which provides technical documentation on the most common construction products [121].

In addition, Finland has introduced Materialitori, a waste information platform that also serves as a marketplace for reusable products, waste by-products, and recyclable materials [122]. Similarly, the Netherlands is promoting re-utilisation in the construction sector through the use of material passports and the Concrete Agreement. This agreement brings stakeholders together to collaborate on enhancing the sustainability of concrete, including the re-utilisation of concrete [123].

Reuse is generally used as a term for any activity that leads to the re-utilisation of products and prolongs their service life. However, based on the final use, two types of reuses can be identified: (1) reuse for the same purpose and (2) reuse for different purposes (re-purposing). The second option leads to finding a new place in new construction, where the product will have a different function. The advantage of this approach can be that the product does not need to fulfil technical requirements for the original function, and its new function is, for example, rather aesthetical. On the other hand, this approach can result in just time-limited use, where the product is exposed to conditions that it is not intended to use. Thus, it can lead to a shorter service life and may be considered in some situations as only the scenario expanding the time before landfilling. To maximise the reuse of products and find the proper function, several standards have been implemented. The important perspective on the assessment of the quality of reclaimed materials is described in the guideline, which focuses on best practices in France, Belgium, and the Netherlands [124].

Acknowledgements The authors of this chapter would like to thank the European Union and COST (European Cooperation in Science and Technology) for supporting the COST Action CircularB CA21103. www.circularb.eu.

This chapter was developed with the support of, and within the framework of the CircularB research project. The following Circular Economy experts and practitioners are also acknowledged for their participation in the questionnaire. Their responses to our survey provided key insights into the implementation of circular economy standards across Europe and beyond: Daniel-Viorel Ungureanu (Romania), Jan Pešta (Czech Republic), Vanessa Tavares (Portugal), Ivana Carevic and Ivan Lukačević (Croatia), Mercedes del Rio (Spain), Meri Cvetkovska (North Macedonia), Marilena de Simone (Italy), Olivera Bukvić (Serbia), Visar Krelani (Kosovo), Bengü Güngör (Turkey), Ekaterine Kardava (Georgia), Simon Duindam (The Netherlands), Sandra Gusta (Latvia), Haitham AbuGhaida (Belgium), Ruben Paul Borg (Malta), Paul Archbold (Ireland), Jan Kazak (Poland), and Peter Griffiths (United Kingdom).

References

1. ISO/TC 323 Circular Economy_English by Bundesverband der Deutschen Industrie e.V.—Issuu
2. Fernandes J, Ferrão P (2023) A new framework for circular refurbishment of buildings to operationalize circular economy policies. Environments 10(3):51. https://doi.org/10.3390/environments10030051
3. Ottosen LM et al (2021) Implementation stage for circular economy in the Danish building and construction sector. Detritus 16:26–30. https://doi.org/10.31025/2611-4135/2021.15110
4. Akhimien NG, Latif E, Hou SS (2021) Application of circular economy principles in buildings: a systematic review. J Build Eng 38:102041. https://doi.org/10.1016/j.jobe.2020.102041
5. Giorgi S, Lavagna M, Wang K, Osmani M, Liu G, Campioli A (2022) Drivers and barriers towards circular economy in the building sector: Stakeholder interviews and analysis of five European countries policies and practices. J Clean Prod 336:130395. https://doi.org/10.1016/j.jclepro.2022.130395

6. European Commission. Joint Research Centre., Use of recycled aggregates in concrete: opportunities for upscaling in Europe. LU: Publications Office, 2023. Accessed 16 Mar 2024. [Online]. Available: https://data.europa.eu/doi/https://doi.org/10.2760/144802
7. ISO Technical Committee 323 Circular Economy. November 2021. Chair:Catherine Chevauché (SUEZ); Committee Manager: Clarisse Issanes (AFNOR); TPM: Monja Korter
8. Easa SM, Yan WY (2019) Performance-based analysis in civil engineering: overview of applications. Infrastructures 4(2):28. ISSN 2412-3811. https://doi.org/10.3390/infrastructures4020028
9. Fang L, Turkbay Romano T, Rio M, Mélot J, Crébier JC (2024) Enhancing sustainability in power electronics through regulations and standards: a literature review. Sustainability 16(3):1042. ISSN 2071-1050. https://doi.org/10.3390/su16031042
10. Gonzalez-torres M, Bertoldi P, Castellazzi L, Perez-lombard L (2023) Review of EU product energy efficiency policies: what have we achieved in 40 years? J Clean Prod 421:138442. ISSN 0959-6526, 1879-1786. https://doi.org/10.1016/j.jclepro.2023.138442
11. BSI Group. Accelerating progress towards a sustainable world. [Online]. Available: https://www.bsigroup.com/en-GB/. Accessed 13 Mar 2024
12. Bureau of Indian Standards. [Online]. Available: https://www.bis.gov.in/. Accessed 13 Mar 2024
13. Krigsvoll G, Fumo M, Morbiducci R (2010) National and international standardization (International Organization for Standardization and European Committee for Standardization) relevant for sustainability in construction. Sustainability 2(12):3777–3791. https://doi.org/10.3390/su2123777
14. Belgian Bureau for Standardisation (NBN). [Online]. Available: https://www.nbn.be/
15. Hrvatski zevod za norme (HZN)—Croatian Standards Institute. [Online]. Available: https://www.hzn.hr/default.aspx?id=435
16. HZN. HRH EN 15643:2021. [Online]. Available: https://repozitorij.hzn.hr/search/norms
17. HZN. EN 303 808 V1.1.1:2023. [Online]. Available: https://repozitorij.hzn.hr/norm/HRN+EN+303+808+V1.1.1%3A2023
18. HZN. EN 17861:2023. [Online]. Available: https://repozitorij.hzn.hr/norm/HRN+EN+17861%3A2023
19. HZN Norms Search Platform. [Online]. Available: https://repozitorij.hzn.hr/search/norms
20. Ministry of Spatial Planning of Croatia (2021) Development program for the circular management of space and buildings for the period 2021 to 2030. [Online]. Available: https://mpgi.gov.hr/UserDocsImages/14146
21. HZN. HR 1128: Concrete—National requirements to HRN EN 206:2020. Croatian Standards Institute, Zagreb
22. Ministry of Energy, Commerce and Industry, Cyprus. ΟΗ ΠΔΡΗ ΠΡΟΓΗΑΓΡΑΦΩΝ ΠΔΣΡΔΛΑΗΟΔΗΓΩΝ ΚΑΗ ΚΑΤʹΗΜΩΝ ΝΟΜΟΗ ΣΟΤ 2003 ΔΩ 2015. [Online]. Available: https://energy.gov.cy/assets/entipo-iliko/2015_1_252.pdf
23. ISO. (2020). ISO 20887:2020—Sustainability in buildings and civil engineering works—design for disassembly and adaptability—principles, requirements and guidance. International Organization for Standardization
24. Czech Office for Standards, Metrology, and Testing (UNMZ). [Online]. Available: https://www.unmz.cz/en/
25. ISO. ISO 14044—Environmental management—life cycle assessment—requirements and guidelines. [Online]. Available: https://csnonline.agentura-cas.cz/Detailnormy.aspx?k=77067
26. ČAS (2023) ČSN EN 17861—Resilient, textile, laminate and modular mechanical locked floor coverings—circular economy—terms and definitions. [Online]. Available: https://csnonline.agentura-cas.cz/Detailnormy.aspx?k=518001
27. ČAS (2023) ČSN ETSI EN 303 808 V1.1.1—Environmental engineering—applicability of EN 45552 to EN 45559 methods for assessment of material efficiency aspects of ICT network infrastructure goods in the context of circular economy. [Online]. Available: https://csnonline.agentura-cas.cz/Detailnormy.aspx?k=517355

28. ČAS (2022) ČSN EN 17410—Plastics—Controlled loop recycling of PVC-U profiles from windows and doors. [Online]. Available: https://csnonline.agentura-cas.cz/Detailnormy.aspx?k=513744
29. ČAS. ČSN EN 15804+A2—Sustainability of construction works—environmental product declarations—core rules for the product category of construction products. [Online]. Available: https://csnonline.agentura-cas.cz/Detailnormy.aspx?k=513904
30. ČAS (2012) ČSN EN 15978—Sustainability of construction works—assessment of environmental performance of buildings—Calculation method. [Online]. Available: https://csnonline.agentura-cas.cz/Detailnormy.aspx?k=91565
31. ČAS. ČSN EN 13108-8 ed. 2—Bituminous mixtures—Material specifications—Part 8: Reclaimed asphalt. [Online]. Available: https://csnonline.agentura-cas.cz/Detailnormy.aspx?k=502819
32. ČAS (2008) ČSN EN 15342—Plastics—recycled plastics—Characterisation of polystyrene (PS) recyklates. [Online]. Available: https://csnonline.agentura-cas.cz/Detailnormy.aspx?k=81552
33. ČAS. ČSN EN 1744-6—Zkoušení chemických vlastností kameniva—Část 6: Stanovení vlivu výluhu z recyklovaného kameniva na počátek tuhnutí cementu
34. ČAS. ČSN EN 1097-6—Zkoušení mechanických a fyzikálních vlastností kameniva—Část 6: Stanovení objemové hmotnosti zrn a nasákavosti
35. ČAS. ČSN EN 1367-4—Zkoušení odolnosti kameniva vůči teplotě a zvětrávání—Část 4: Stanovení smršťování
36. ČAS. ČSN EN 1744-5—Zkoušení chemických vlastností kameniva—Část 5: Stanovení chloridových solí rozpustných v kyselině
37. ČAS. ČSN EN 1744-1 + A1—Zkoušení chemických vlastností kameniva—Část 1: Chemický rozbor
38. ČAS. ČSN EN 933-11—Zkoušení geometrických vlastností kameniva—Část 11: Klasifikace složek hrubého recyklovaného kameniva
39. Parliament of Georgia (2014) Waste management code. Legislative Herald of Georgia. [Online]. Available: https://matsne.gov.ge/en/document/view/2676416?publication=12
40. National Standards Authority of Ireland (NSAI). [Online]. Available: https://www.nsai.ie
41. NSAI. How are indigenous standards made? National Standards Authority of Ireland. [Online]. Available: https://www.nsai.ie/standards/about-standards/indigenous-standards/
42. UNI—Ente Italiano di Normazione. National Standards Body (UNI). [Online]. Available: https://www.uni.com/
43. ISO. ISO 59004—Circular economy—framework and principles for implementation. [Online]. Available: https://www.iso.org/standard/80648.html
44. ISO. ISO 59010—Circular economy—guidelines on business models and value chains. [Online]. Available: https://www.iso.org/standard/80649.html
45. ISO. ISO 59020—Circular economy—measuring circularity framework. [Online]. Available: https://www.iso.org/standard/80650.html
46. UNI. UNI/TS 11820:2022—Measuring circularity—methods and indicators for measuring circular processes in organizations. [Online]. Available: https://store.uni.com/uni-ts-11820-2022
47. UNI. UNI/TR 11821:2023—Collection and analysis of circular economy best practices. [Online]. Available: https://store.uni.com/uni-tr-11821-2023
48. Ministry of Ecological Transition (2022) National strategy for the circular economy. [Online]. Available: https://www.mase.gov.it/sites/default/files/archivio/allegati/PNRR/SEC_21.06.22.pdf
49. Ministry for Ecological Transition (2023) Minimum environmental criteria (Criteri Ambientali Minimi—CAM). [Online]. Available: https://gpp.mite.gov.it/CAM-vigen-ti
50. U.S. Department of Commerce, International Trade Administration (2024) Kosovo—country commercial guide—standards for trade. [Online]. Available: https://www.trade.gov/country-commercial-guides/kosovo-standards-trade

51. U.S. Department of Commerce, International Trade Administration (2024) Kosovo—country commercial guide—waste management and recycling. [Online]. Available: https://www.trade.gov/country-commercial-guides/kosovo-waste-management-and-recycling
52. European External Action Service (EEAS) (2022) EU-German funded project helps develop a joint waste management system in Pristina region. [Online]. Available: https://www.eeas.europa.eu/delegations/kosovo/eu-german-funded-project-helps-develop-joint-waste-management-system-pristina-region_en?s=321
53. Government of the Republic of Kosovo—Ministry of Environment. (2023). Circular economy roadmap of Kosovo. [Online]. Available: https://mmphi.rks-gov.net/MMPHIFolder/DocumentsFiles/2023_8171e270-643b-4de3-9387-91794084eb8b.pdf
54. Kosovo Standardisation Agency. National Standardisation Portal. [Online]. Available: https://mint.rks-gov.net/page.aspx?id=2,59
55. Latvijas Standarts (LVS). [Online]. Available: https://www.lvs.lv/
56. CENELEC. EN 50614—Requirements for the preparation for re-use of waste electrical and electronic equipment (WEEE)
57. CENELEC. EN 50625—Collection, logistics & treatment requirements for WEEE
58. Malta Competition and Consumer Affairs Authority (MCCAA). (2024). [Online]. Available: https://mccaa.org.mt. Accessed 19 May 2024
59. Ilomäki A, Lützkendorf T, Trinius W. Sustainability assessment of buildings in CEN/TC350 "Sustainability of construction works". In: Proceedings of the world conference SB08
60. Government of Malta. Sostenibilità—National sustainability portal. [Online]. Available: https://sustainability.gov.mt/. Accessed 18 May 2024
61. CE Malta. Home. [Online]. Available: https://www.cemalta.gov.mt/. Accessed 18 May 2024
62. Government of Malta. LEĠIŻLAZZJONI MALTA—Environmental Protection Act (Cap. 549). [Online]. Available: https://legislation.mt/eli/cap/549/eng/pdf. Accessed 18 May 2024
63. Building and Construction Authority (BCA). Official website. [Online]. Available: https://bca.org.mt/. Accessed 19 May 2024
64. Borg RP (2024) Standards for recycling oriented deconstruction and classification of waste. In: Bragança L, Cvetkovska M, Askar R, Ungureanu V (eds) Creating a roadmap towards circularity in the built environment. Springer Nature Switzerland, Cham, pp 433–444. https://doi.org/10.1007/978-3-031-45980-1_35
65. MCCAA (2022) SM810—Recycling-oriented deconstruction, controlled excavation works and classification of waste—requirements for planning and execution. Malta Competition and Consumer Affairs Authority
66. MCCAA (2022) SM820—Classification of recycled aggregate (Consultation Document). Malta Competition and Consumer Affairs Authority
67. Government of Malta (2022) Construction management site regulations—subsidiary legislation S.L. 623.08. Legal Notice 340 of 2022. Published: December 16, 2022
68. The Standardization Institute of the Republic of North Macedonia (ISRSM). [Online]. Available: https://isrsm.gov.mk/en/
69. Buildwise. EN 1744-6: Tests for chemical properties of aggregates—Part 6: Determination of the influence of recycled aggregate extract on the initial setting time of cement. [Online]. Available: https://www.buildwise.be/en/standards-and-regulations/standards-regulations/en-1744-6-en/. Accessed 16 May 2024
70. GlobalSpec. CEN—PREN 933-11—Tests for geometrical properties of aggregates—Part 11: Classification test for the constituents of coarse recycled aggregate. [Online]. Available: https://standards.globalspec.com/std/14375601/pren-933-11. Accessed 16 May 2024
71. Polish Committee for Standardization (PKN). Polish standards. [Online]. Available: https://www.pkn.pl/
72. IPQ—Instituto Português da Qualidade (2024) [Online]. Available: https://www.ipq.pt/. Accessed 16 May 2024
73. European Commission. Circular economy action plan. [Online]. Available: https://environment.ec.europa.eu/strategy/circular-economy-action-plan_en. Accessed 16 May 2024

74. Governo de Portugal. Economia circular—Novo Plano de Ação para a Economia Circular. [Online]. Available: https://economiacircular.gov.pt/pt/recursos/noticias/133-new-eu-circular-economy-action-plan. Accessed 16 May 2024
75. ASRO. About—Organismul Național de Standardizare din România. [Online]. Available: https://www.asro.ro/en/about-asro/. Accessed 16 May 2024
76. European Commission. Regulation (EU) No 1025/2012 on European standardisation. [Online]. Available: https://eur-lex.europa.eu/eli/reg/2012/1025/oj. Accessed 16 May 2024
77. Institute for Standardization of Serbia (ISS). [Online]. Available: https://iss.rs/en/. Accessed 16 May 2024
78. Kamberović S. Sector for strategic planning and projects. Ministry of Environmental Protection, Serbia. [Internal report]
79. ISO (2020) ISO 20887:2020—Sustainability in buildings and civil engineering works—design for disassembly and adaptability—Principles, requirements and guidance. [Online]. Available: https://www.iso.org/standard/69370.html. Accessed 16 May 2024
80. GlobalSpec. CEN—EN 206—Concrete—Specification, performance, production and conformity. [Online]. Available: https://standards.globalspec.com/std/14462892/en-206. Accessed 16 May 2024
81. Institute for Standardization of Serbia (ISS). SRPS EN 12620:2010. [Online]. Available: https://iss.rs/sr_Cyrl/project/show/iss:proj:26477. Accessed 16 May 2024
82. Institute for Standardization of Serbia (ISS). SRPS EN 13055:2017. [Online]. Available: https://iss.rs/sr_Cyrl/project/show/iss:proj:54616. Accessed 16 May 2024
83. Institute for Standardization of Serbia (ISS). SRPS EN 933-1:2013. [Online]. Available: https://iss.rs/en/project/show/iss:proj:40709. Accessed 16 May 2024
84. Institute for Standardization of Serbia (ISS). SRPS U.M1.206:2023. [Online]. Available: https://iss.rs/en/project/show/iss:proj:65149. Accessed 16 May 2024
85. UNE—Spanish Association for Standardization. [Online]. Available: https://www.en.une.org/. Accessed 16 May 2024
86. Ministerio para la Transición Ecológica y el Reto Demográfico. Estrategia Española de Economía Circular y Planes de Acción. [Online]. Available: https://www.miteco.gob.es/es/calidad-y-evaluacion-ambiental/temas/economia-circular/estrategia.html. Accessed 16 May 2024
87. NEN—Netherlands Standardization Institute. Home. [Online]. Available: https://www.nen.nl/en/. Accessed 16 May 2024
88. NEN—Netherlands Standardization Institute. Normontwikkeling. [Online]. Available: https://www.nen.nl/en/normontwikkeling. Accessed 16 May 2024
89. GOV.UK. Resources and waste strategy: at a glance. [Online]. Available: https://www.gov.uk/government/publications/resources-and-waste-strategy-for-england/resources-and-waste-strategy-at-a-glance. Accessed 16 May 2024
90. BSI Group. BS 8001—Circular economy. [Online]. Available: https://www.bsigroup.com/en-SG/Standards/Benefits-of-using-standards/Becoming-more-sustainable-with-standards/bs-8001---circular-economy-_new/. Accessed 16 May 2024
91. The Institution of Structural Engineers. Circular economy and reuse: guidance for designers. [Online]. Available: https://www.istructe.org/resources/guidance/circular-economy/. Accessed 16 May 2024
92. ISO—Organización Internacional de Normalización. [Online]. Available: https://www.iso.org/es/home. Accessed 13 Mar 2024
93. CEN-CENELEC. About CEN. [Online]. Available: https://www.cencenelec.eu/about-cen/. Accessed 13 Mar 2024
94. CEN-CENELEC. CEN-CENELEC. [Online]. Available: https://www.cencenelec.eu/. Accessed 13 Mar 2024
95. ETSI. Welcome to the world of standards! [Online]. Available: https://www.etsi.org/. Accessed 13 Mar 2024
96. European Commission. Key players in European standardisation. [Online]. Available: https://single-market-economy.ec.europa.eu/single-market/european-standards/key-players-european-standardisation_en. Accessed 13 Mar 2024

97. European Commission. Work programs of standardization bodies. [Online]. Available: https://single-market-economy.ec.europa.eu/single-market/european-standards/notification-system/work-programmes-standardisation-bodies_en. Accessed 13 Mar 2024
98. National Standards Authority of Ireland (NSAI). The role of standards in the circular economy. [Online]. Available: https://www.nsai.ie/standards/sectors/circular-economy-standards/the-new-cen-commission-on-circular-construction.pdf. Accessed 13 Mar 2024
99. CEN-CENELEC. Sustainability, safety, and accessibility. [Online]. Available: https://www.cencenelec.eu/areas-of-work/cen-sectors/construction/sustainability-safety-and-accessibi lity/. Accessed 13 Mar 2024
100. ASTM International. Standards for the Circular Economy. ASTM standardization news. [Online]. Available: https://sn.astm.org/features/standards-circular-economy.html. Accessed 13 Mar 2024
101. ISO. ISO/TC 323—Circular economy. [Online]. Available: https://www.iso.org/committee/7203984.html. Accessed 13 Mar 2024
102. European Commission (2015) Communication from the commission to the European parliament, the council, the European economic and social committee and the committee of the regions: closing the loop—an EU action plan for the circular economy. Brussels, 2 December 2015. COM(2015) 614 final
103. CEN/TC 350/SC 1. Circular economy in the construction sector—final gap analysis consultation report and findings, Version 3, 21st July 2023
104. Pacheco JN, de Brito J, Lamperti Tornaghi M (2023) Use of recycled aggregates in concrete—opportunities for upscaling in Europe. Publications Office of the European Union, Luxembourg
105. Pacheco J, de Brito J, Lamperti Tornaghi M (2023) Use of recycled aggregates in concrete: opportunities for upscaling in Europe. JRC Publications Repository. https://doi.org/10.2760/144802
106. CEN. EN 12620:2013—Aggregates for concrete. iTeh Standards. [Online]. Available: https://standards.iteh.ai/catalog/standards/cen/aef412e6-36ce-49d3-afaa-5200d721ff84/en-12620-2013. Accessed 23 May 2024
107. CEN. EN 206:2013+A2:2021—Concrete—specification, performance, production and conformity. iTeh Standards. [Online]. Available: https://standards.iteh.ai/catalog/standards/cen/4f3d2008-978a-47ec-bc12-5660aa40e04d/en-206-2013a2-2021. Accessed 23 May 2024
108. Danish Standards. DS 2426—EN 206-1:2011. [Online]. Available: https://webshop.ds.dk/en/standard/M253247/ds-2426-en-206-1-2011. Accessed 23 May 2024
109. ČAS (2018) ČSN EN 206 + A1—Beton—Specifikace, vlastnosti, výroba a shoda. Czech office for standards, metrology, and testing
110. ČAS (2008) ČSN EN 12620+A1—Kamenivo do betonu. Czech office for standards, metrology, and testing
111. ČAS (2018) ČSN EN 206+A1—Concrete: specification, performance, production and conformity (in Czech). Prague
112. Austrian Standards Institute (2014) ÖNORM B 3151—Dismantling of buildings as a standard method for demolition. [Online]. Available: https://www.ris.bka.gv.at/Dokumente/Bundesnor men/NOR40187245/II_290_2016_OeNORM_B_3151.pdf. Accessed2 May 2024
113. European Commission (2018) EU construction and demolition waste protocol and guidelines. [Online]. Available: https://ec.europa.eu/growth/content/eu-construction-and-demoli tion-waste-protocol-0_en. Accessed 10 Jan 2020
114. European Commission. Best environmental management practice. joint research centre. [Online]. Available: https://joint-research-centre.ec.europa.eu/scientific-activities-z/best-env ironmental-management-practice_en. Accessed 23 May 2024
115. BSI. BS 6187:2011—Code of practice for full and partial demolition. [Online]. Available: https://www.en-standard.eu/bs-6187-2011-code-of-practice-for-full-and-partial-demoli tion/. Accessed 23 May 2024
116. Wahlström M, Zu Castell-Rüdenhausen M, Hradil P, Hauge-Smith K, Oberender A, Ahlm M et al (2019) Improving quality of construction & demolition waste—requirements for pre-demolition audit. Nordisk Ministerråd

117. CLOSER—Close to resources recovery. Guia Português de Auditorias Pré-Demolição. [Online]. Available: http://closer.lnec.pt/guide/R4_Guia%20portugues%20auditoria%20pre%20demolicao.pdf. Accessed 23 May 2024
118. Generaldirektoratet for Miljø (European Commission), Akelytė R, Chiabrando F, Camboni M, Ledda C, Vencovsky D et al (2024) Study on asbestos waste management practices and treatment technologies. Den Europæiske Unions Publikationskontor
119. ISO. ISO 20887:2020—Sustainability in buildings and civil engineering works—design for disassembly and adaptability—principles, requirements and guidance. [Online]. Available: https://www.iso.org/standard/69370.html. Accessed 23 May 2024
120. Cyrkl. Digitální odpadové tržiště. [Online]. Available: https://www.cyrkl.com/cs. Accessed 23 May 2024
121. Opalis. Homepage. [Online]. Available: https://opalis.eu/en. Accessed 23 May 2024
122. Ympäristöministeriö (2022) Kierrätyksestä kiertotalouteen: Valtakunnallinen jätesuunnitelma vuoteen 2027. [Online]. Available: https://julkaisut.valtioneuvosto.fi/handle/10024/163978. Accessed 23 May 2024
123. Ministerie van Infrastructuur en Waterstaat (2021) Afvalpreventieprogramma Nederland—Rapport. Rijksoverheid. [Online]. Available: https://www.rijksoverheid.nl/documenten/rapporten/2021/02/18/afvalpreventieprogramma-nederland. Accessed 23 May 2024
124. Opalis. Insurance and reuse—learning from case studies and perspectives. [Online]. Available:https://opalis.eu/sites/default/files/2024-04/Prospective%20report%20on%20reuse%20and%20insurance%20-%20EN.pdf. Accessed 23 May 2024

Part III
Harmonising Language and Concepts

Chapter 7
Glossary

Ana Andabaka, Martina Basarac Sertić, Petra Palić, Inês Burmester, and Marco Frazão Pedroso

Abstract This chapter introduces a comprehensive glossary developed as part of the COST Action CircularB, aimed at standardising terminology related to the circular economy and the built environment. Acknowledging the conceptual complexity and terminological inconsistencies across various disciplines and practices, this glossary serves as an essential reference to enhance communication, research, and implementation. It encompasses over 400 terms and definitions sourced from European regulations, ISO standards, and expert practices, providing clarity and consistency for stakeholders, including policymakers, practitioners, researchers, and industry professionals. The glossary was designed with the overarching goal of promoting a shared language that fosters collaboration and accelerates the transition toward a sustainable and circular built environment.

Keywords Circular economy · Built environment · Terminology harmonization · Sustainability glossary · COST Action CircularB

A. Andabaka (✉)
Faculty of Economics and Business, University of Zagreb, Zagreb, Croatia
e-mail: ana.andabaka@efzg.unizg.hr

M. Basarac Sertić
Economic Research Division, Croatian Academy of Sciences and Arts, Zagreb, Croatia

P. Palić
Department of Sociology, Catholic University of Croatia, Zagreb, Croatia

I. Burmester
BUILT CoLAB Porto, Porto, Portugal

M. F. Pedroso
CERIS, Técnico Lisbon, University of Lisbon, Lisbon, Portugal

L. Bragança et al. (eds.), *Shaping Circular Transitions in the Built Environment*, Springer Tracts in Civil Engineering, https://doi.org/10.1007/978-3-032-02834-1_7

7.1 Introduction

To aid in the development of the COST Action "Implementation of Circular Economy in the Built Environment (CircularB)", a glossary of terms was suggested at the kick-off meeting in Brussels in 2022 and adopted by WG4. During the Action, we realized that dealing with so many terms can sometimes be confusing, and therefore this publication allows its readers to discover the concepts behind many terms related to a sustainable built environment. This is a live document for the collection and agreeing of terminology which is used in the built environment and in the circular economy. All COST Action participants are encouraged to use the terms below to avoid confusion, and to aid with the dissemination of knowledge in this subject. However, we strongly believe that sharing the document beyond the scope of this Action will raise awareness and understanding of a wider audience. This glossary grows as our Action progresses, already covering more than 400 terms. Its mission is to encompass as many relevant concepts as possible and to allow easier navigation through the circular built environment. Moreover, it strives to reflect the changes in the regulatory framework and the progress achieved by the leading companies and institutions that promote the implementation of circular economy and sustainability initiatives.

7.1.1 Purpose

The glossary aims to provide a clear understanding of terms and concepts related to the implementation of the circular economy in the built environment. It serves as a foundational resource for professionals, researchers, and practitioners in sustainable construction and design.

7.1.2 Instruction Notes for Users

The glossary is organized alphabetically for easy lookup. The emphasis is on a user-friendly approach providing users with a wider context of the described terms. Numbering is used when multiple definitions are provided, with the latest definition taking the first place. This can also reflect the changes in the documents such as Guidelines or Protocols, allowing for the comparison of previous and current definitions of the terms used.

Feel free to share it with your colleagues to establish a common understanding of key concepts in the circular built environment. This publication was designed for electronic publication and we appreciate your commitment to the conservation of resources by using this publication in digital format.

7.2 Glossary of Terms

7.2.1 Abiotic

Physical rather than biological, i.e. not derived from living organisms [1].

7.2.2 Accessibility

1. The ability to allow for easy access to building components for disassembly, refurbishment, replacement, or upgrade [2].
2. Ability of a space to be entered with ease [3].

7.2.3 Accreditation

An attestation by a national accreditation body that a conformity assessment body meets the requirements set by harmonised standards and, where applicable, any additional requirements including those set out in relevant sectoral schemes, to carry out a specific conformity assessment activity [4].

7.2.4 Adaptability

Ability to be changed or modified to make suitable for a particular purpose [3].

7.2.5 Adaptable Building

Building that is designed in such a way that, over time, it can be readily transformed to accommodate uses for which it was not originally conceived [1].

7.2.6 Adaptation to Climate Change

Collective effort that is needed to devise new strategies, policies, technologies, materials and methods to assist the human race to cope with the effects of climate change [1].

7.2.7 Add Value

Process to increase the value of the object of consideration [5].

7.2.8 Alternative Material

Material which substitutes a primary material [1].

7.2.9 Anaerobic Digestion

A controlled biological process where microorganisms break down organic material without oxygen, producing biogas, carbon dioxide and, normally, nutrient-rich digestate [5].

7.2.10 Annual Cost

Costs that represent the sum of running costs and periodic costs or replacement costs paid in a certain year [6].

7.2.11 Artificial Intelligence (AI)

A disruptive technology consisting of a machine that through Artificial Neural Networks mimics human cognitive functions, like problem-solving, pattern recognition, and learning. It has recently acquired ground-breaking capabilities thanks to important progresses made in computational power, and whose impact can stretch through the whole lifecycle of a building [7].

7.2.12 Assembly

Set of related components attached to each other [8].

7.2.13 Assembly Sequences

The sequences in assembly create dependencies between building elements, by locking elements together. The way we assemble a building sets the mirror image of the building during its disassembly and transformation phase. Two major assembly/disassembly sequences can be distinguished: a parallel sequence and a sequential sequence [9].

7.2.14 Assessment Boundary

Boundary where the delivered and exported energy are measured or calculated [10].

7.2.15 Auditor

1. The expert or the team of experts (auditors´ team) performing the pre-demolition or pre-renovation audit. The auditor or the team of auditors can be appointed by the owner of the construction works or consultant (e.g. an architect or structure engineer) acting on behalf of the owner [4].
2. The expert or the team of experts (auditing team) performing the waste audit. It can be represented by the building owner or consultant (e.g. an architect or structure engineer) acting on behalf of the owner [11].

7.2.16 Augmented Reality (AR)

A digital view of a real and physical environment/object whose elements are augmented (or supplemented) by computer-generated sensory input [7].

7.2.17 Authorised Representative

Any natural or legal person established in the European Union that has received a written mandate from the manufacturer to act on the manufacturer's behalf in relation to specified tasks with regard to the manufacturer's obligations under the Regulation (EU) 2024/1781 (Establishing a framework for the setting of ecodesign requirements for sustainable products) [12].

7.2.18 Authority

The national or regional administration responsible for granting the demolition or renovation permits and supervision of the demolition or renovation process [4].

7.2.19 Backfilling

Any recovery operation where suitable non-hazardous waste is used for purposes of reclamation in excavated areas or for engineering purposes in landscaping. Waste used for backfilling must substitute non-waste materials, be suitable for the aforementioned purposes, and be limited to the amount strictly necessary to achieve those purposes [13].

7.2.20 Base Element

A building product (system/component/element) is a carrier of a specific building's functions. Each assembled product represents a cluster of elements that are carriers of sub-functions. To provide independence for elements within two clusters, each cluster should define its base element, which integrates all surrounding elements of that cluster. This element functions as intermediary between elements as well as between clusters. Such an intermediary share element functions on two levels: (i) it connects elements within the cluster and (ii) performs as an intermediary with other clusters. Base elements/intermediaries can be found at all levels of technical composition of a building [9].

7.2.21 Baukultur (Literally "Building Culture")

German term for a concept describing all human activity that changes spaces and the living environment in a positive way. It includes existing buildings, including monuments and other elements of built heritage, as well as the design and construction of contemporary buildings, infrastructure, public spaces and landscapes embedded in and relating to the natural environment. Baukultur also refers to planning procedures for building projects, infrastructures, cities, villages and open landscapes, and to traditional and innovative building techniques and methods. There is a broad implicit societal understanding of quality as well as tools, procedures and practices, hence the selective use of the word 'culture' in the term [20].

7.2.22 Benchmark

Reference point against which comparisons can be made [14].

7.2.23 Benchmarking

Process of collecting, analysing and relating performance data of comparable buildings or other types of construction works [14].

7.2.24 Best Practice

Level representing best available real performance [14].

7.2.25 Brownfield Land (Brownfield Site)

Land available for development which has previously been built on or used for purposes that have altered the condition of the soil [1].

7.2.26 Bio-effluents

A variety of pollutants produced during metabolic processes that take place within the human body. These pollutants generally include CO_2, ammonia, hydrocarbons, alcohols, ketones and aldehydes and their increase in concentration is associated with a decrease in perceived indoor air quality. CO_2 would normally be used as a "marker" for bio-effluents [15].

7.2.27 Biodiversity

The variability among living organisms arising from all sources including terrestrial, marine and other aquatic ecosystems and the ecological complexes of which they are part and includes diversity within species, between species and of ecosystems [16].

7.2.28 *Biogenic Carbon*

Carbon derived from biomass [17].

7.2.29 *Biological Cycle*

Cycle(s) through which biological nutrients are utilized by living organisms and subsequently restored into or within the biosphere in a way that rebuilds ecosystem resilience and natural capital and enables the regrowth of renewable resources [5].

7.2.30 *Biomass*

Material of biological origin excluding material embedded in geological formations and material transformed to fossilized material, excluding peat [17].

7.2.31 *Biosphere*

Part of the environmental system that is capable of supporting life [5].

7.2.32 *Building*

Construction works that has the provision of shelter for its occupants or contents as one of its main purposes, usually partially or totally enclosed and designed to stand permanently in one place [8].

7.2.33 *Building Automation and Control System*

A system comprising all products, software and engineering services that can support energy efficient, economical and safe operation of technical building systems through automatic controls and by facilitating the manual management of those technical building systems [18].

7.2.34 Building Component

A construction product manufactured as a distinct unit to serve a specific function or functions [2].

7.2.35 Building Fabric

All construction works that are fixed to the building in a permanent manner, so that the dismantling or replacement of the product constitute construction operations [2].

7.2.36 Building Information Modelling (BIM)

A digital form of construction and asset operations. It brings together technology, process improvements and digital information to radically improve client and project outcomes and asset operations. BIM is a strategic enabler for improving decision making for both buildings and public infrastructure assets across the whole life-cycle. It applies to new build projects; and crucially, BIM supports the renovation, refurbishment and maintenance of the built environment which have the largest share of the sector [7].

7.2.37 Building Owners (or Clients)

The persons or organisations which own the building and/or are managing the financial resources for conducting construction and demolition operations [19].

7.2.38 Built Environment

The existing space that surrounds people, which they actively shape and which in turn impacts on people's life and behaviour. It includes the living environment [20].

7.2.39 Built Environment Sector

All of the activities that encompass the making of environments for human occupants and activities, including the construction of homes, commercial buildings, streets, highways, infrastructure and zoning [21].

7.2.40 Built Environment Process

All of the activities, processes, and decisions that encompass the making of environments for human occupants and activities, across the life cycle of building projects from extraction to design construction and end of life, including the construction of homes, commercial buildings, streets, highways, infrastructure and zoning [21].

7.2.41 Carbon Dioxide Equivalent (CO_2e)

1. An equal green-house gas emissions quantity that represents the number of metric tons of (CO_2) emissions with the equivalent global warming potential as another greenhouse gas. It is commonly used, since it is the primary component in greenhouse gas emissions that result from the use of fossil-based energy and emissions from biological materials, waste and chemical reactions [21].
2. Unit for comparing the radiative forcing of a greenhouse gas to that of carbon dioxide [17].

7.2.42 Carbon Capture

In the context of buildings and construction, carbon capture refers to active processes of removing carbon from the atmosphere through processes such as plant photosynthesis and carbonation in cementitious materials [21].

7.2.43 Carbon Carrying Capacity

Amount of carbon able to be stored in a forest ecosystem under prevailing environmental conditions and natural disturbance regimes, but excluding anthropogenic disturbance [1].

7.2.44 Carbon Cycle

Biogeochemical cycle by which carbon is exchanged among the biosphere, pedosphere, geosphere, hydrosphere, and atmosphere of the earth [1].

7.2.45 Carbon Footprint (Whole Life Carbon Measurement)

Sum of greenhouse gas emissions and removals in a product system, expressed as CO_2 equivalents and based on a life cycle assessment using the single impact category of climate change [17].

7.2.46 Carbon Loophole

The greenhouse gas loophole that is created by global trade, whereby production of materials are relocated and emissions allocated to contexts with lower production costs and historic emission patterns [21].

7.2.47 Carbon Offsets

Instruments that allow companies to demonstrate net zero greenhouse gas emissions by paying for an activity outside of its organization that verifiably reduces greenhouse gas emissions. They are measured in units of one metric ton of CO_2–equivalent emissions, and they have an "additionality" requirement, which means that they must come from a process that is verified to be "additional" to what would happen under typical "business-as-usual" scenarios [21].

7.2.48 Carbon Offsetting

Compensation for unavoidable emissions by helping fund projects that deliver an equivalent CO_2 saving elsewhere [21].

7.2.49 Carbon Storage

1. In the context of buildings and construction, carbon storage refers to how carbon is kept within the building material itself over time [21].
2. Carbon removed from the atmosphere and stored as carbon in a product [17].

7.2.50 Cascading

Repeated use of a usually starting at a level of high value with decreasing quantity and quality at each subsequent stage or cycle, depending on the processes used [5].

7.2.51 Cascading (Re)use

Reusing a construction material or product for an application with lower technical requirements than the original one [19].

7.2.52 CE Marking

A marking by which the manufacturer indicates that the relevant product is in conformity with the applicable requirements set out in European Union harmonisation legislation providing for its affixing [12].

7.2.53 CE Marking for Construction Products

Indicates that manufacturers take responsibility for the conformity of their products with the declared performance as well as the compliance with all applicable requirements of Regulation (EU) 2024/3110 of the European Parliament and of the Council laying down harmonised rules for the marketing of construction products and repealing Regulation (EU) No 305/2011 and in other relevant Union harmonisation legislation providing for its affixing. CE marking should be affixed to all construction products for which the manufacturer has drawn up a declaration of performance and conformity in accordance with the Construction Products Regulation [4].

7.2.54 Cement Sustainability Initiative (CSI)

Global effort by 24 major cement producers with operations in more than 100 countries, which is part of the World Business Council for Sustainable Development (a unique, CEO-led, global association of some 200 companies dealing exclusively with business and sustainable development) [1].

7.2.55 Certification

A procedure by which a third party gives written assurance that a product, process or service is in conformity with certain standards. Certification can be seen as a form of communication along the supply chain. The certificate demonstrates to the buyer that the supplier complies with certain standards, which might be more convincing than if the supplier itself provided the assurance [4].

7.2.56 Certified Emission Reduction

Output of clean development mechanism (CDM) projects, as defined by the Kyoto Protocol [1].

7.2.57 Certified Environmental Profiling (CEP)

Method to identify and assess the environmental effects associated with a building material over its life cycle (extraction, processing, use and maintenance and eventual disposal) [1].

7.2.58 Circular Buildings Coalition

The circular buildings coalition is an initiative by Metabolic, Circle Economy, World Business Council for Sustainable Development, World Green Building Council, Ellen McArthur Foundation and Arup to accelerate the transition to a circular economy in the built environment [1].

7.2.59 Circular Building

A circular building optimises the use of resources whilst minimising waste throughout its whole lifecycle. The building's design, operation and deconstruction maximise value over time using: durable products and services made of secondary, non-toxic, sustainably sourced, or renewable, reusable or recyclable materials; space efficiency over time through shared occupancy, flexibility and adaptability; longevity, resilience, durability, easy maintenance and reparability; disassembly, reuse or recycling of embedded material, components and systems; lifecycle assessment (LCA), lifecycle costing (LCC) and readily available digital information (such as building material passports) [22].

7.2.60 Circular Construction

Developing, using and reusing buildings, construction components or products and materials, areas, and infrastructure without unnecessarily depleting natural resources, polluting the living environment and affecting ecosystems. Carrying out construction such that it is economically justifiable and contributes to the welfare of people and animals, here and there, now and later [23].

7.2.61 Circular Economy

1. An economic system whereby the value of products, materials and other resources in the economy is maintained for as long as possible, enhancing their efficient use in production and consumption, thereby reducing the environmental impact of their use, minimising waste and the release of hazardous substances at all stages of their life cycle, including through the application of the waste hierarchy [24].
2. Economy that is restorative and regenerative by design, and which aims to keep products, components and materials at their highest utility and value at all times, distinguishing between technical and biological cycles [25].
3. An economy that uses a systems-based approach in order to maintain the lifespan and/or circulation of materials, products, and services for as long as possible. "Circularity" is a common term that refers to production processes and economic models that are restorative and regenerative, enabling resources to maintain their highest value, firstly by "avoiding" their extraction in the first place through "improving" design processes to eliminate waste and material overconsumption, and secondly by "shifting" to renewable and recycled materials [21].
4. Economic system that uses a systemic approach to maintain a circular flow of resources, by recovering, retaining or adding to their value, while contributing to sustainable development [5].

7.2.62 Circular Design

Creating products and services that no longer have a lifecycle with a beginning, a middle and an end. The purpose is to design products that can be used, reused and recycled, resulting in less resource consumption, less waste and more value added to the ecosystem. Circular design focuses on creating products and services for the circular economy [20].

7.2.63 Circular Flow of Resources

Systematic cycling of the provision and use of resources within multiple technical or biological cycles [5].

7.2.64 Circular Material Use Rate (Circularity Rate)

The ratio of the circular use of materials to the overall material use. This indicator measures the share of material recycled and fed back into the economy—thus saving extraction of primary raw materials—in overall material use [26].

7.2.65 Circular Public Procurement

A process by which public authorities purchase works, goods or services that seek to contribute to closed energy and material loops within supply chains, whilst minimising, and in the best case avoiding, negative environmental impacts and waste creation across their whole life-cycle [27].

7.2.66 Circularity Aspect

Element of an organization's activities or solutions that interacts with the circular economy (e.g., durability, recyclability, reusability, repairability, recoverability). Circularity aspects should be considered in relation to the principles, as well as the organization's objectives, goals and actions, for the implementation of a circular economy [5].

7.2.67 Circularity Impact

Change to economic, social and environmental systems, whether adverse or beneficial, including possible consequences, wholly or partially resulting from an organization's circularity aspects [5].

7.2.68 Circularity Performance

Degree to which a set of circularity aspects align with the objectives and principles for a circular economy [5].

7.2.69 Circularity Measurement

Process to help determine the circularity performance through collection, calculation or compilation of data or information [5].

7.2.70 Circularity Assessment

Evaluation and interpretation of results and impacts from a circularity measurement [5].

7.2.71 Circularity Indicator

Metric used to measure one or more circularity aspects [5].

7.2.72 City Metabolism

Concept which considers a city as a constantly changing, evolving system with flows of various essential elements such as water, food, energy, telecoms, waste etc. [1].

7.2.73 *Climate Change Adaptation*

The process of adjustment to actual and expected climate change and its impacts [24].

7.2.74 *Climate Change Mitigation*

The process of holding the increase in the global average temperature to well below 2 °C and pursuing efforts to limit it to 1.5 °C above pre-industrial levels, as laid down in the Paris Agreement [24].

7.2.75 *Closed Loop Recycling*

Recycling a waste product into the same product [1].

7.2.76 *Closed Loop System*

System by which products or resources are used and then recovered and turned into new products or recovered resources, without losing their inherent properties [5].

7.2.77 *CO_2 Reduction Potential*

Potential of a replacement technology, material or process that is used in construction to reduce the CO_2 emissions as compared to the emissions arising from the technology, material or process it is replacing [1].

7.2.78 *Cogeneration*

Use of a heat engine or a power station to simultaneously generate both electricity and useful heat [1].

7.2.79 *Collection of Waste*

The gathering of waste, including the preliminary sorting and preliminary storage of waste for the purposes of transport to a waste treatment facility [4].

7.2.80 *Combined Indicator*

Indicator that includes information on more than one aspect of operations, management conditions or impacts [28].

7.2.81 *Component*

1. A product intended to be incorporated into another product [12].
2. Product manufactured as a distinct unit to serve a specific function or functions [3].

7.2.82 *Composting*

Aerobic biological process usually carried out under controlled conditions, which converts organic material into a normally nutrient-rich, humus-like, material [5].

7.2.83 *Consumer Product*

Any product, excluding components and intermediate products, primarily intended for consumer [12].

7.2.84 *Constructed Asset*

Anything of value that is constructed or results from construction operations [29].

7.2.85 *Conformity Assessment*

The process demonstrating whether the ecodesign requirements set out in the relevant delegated acts adopted pursuant to Article 4 of Regulation (EU) 2024/1781 have been fulfilled [12].

7.2.86 *Conformity Assessment Body*

A body that performs conformity assessment activities including calibration, testing, certification and inspection [12].

7.2.87 *Construction and Demolition Waste (CDW)*

1. Waste generated by construction and demolition activities [30].
2. Any waste generated in the activities of companies belonging to the construction sector and included in category 17 of the European List of Wastes. The category 17 provides for codes for several individual materials that can be collected separately from a construction or demolition site. It includes waste streams (hazardous and non-hazardous; inert, organic and inorganic) resulting from construction, renovation and demolition activities. C&D waste originates at sites where construction, renovation or demolition takes place. Construction waste contains several materials, often related to cut-offs or packaging waste. Demolition waste comprises all materials found in constructions. Renovation waste can contain both construction-related materials and demolition-related materials [4].

7.2.88 *Construction Material*

Material used in the construction industry to create buildings and structures, e.g. steel, timber, aggregates, plaster, concrete and plastic products as well as manufactured products [1].

7.2.89 *Construction Product*

1. Any formed or formless physical item, including 3D-printed products, or a kit that is placed on the market, including by means of supply to the construction site, for incorporation in a permanent manner into construction works or parts

thereof with the exception of items that need first to be integrated into a kit or another construction product prior to being incorporated in a permanent manner into construction works [31].
2. Item manufactured or processed for incorporation in a building [32].

7.2.90 Construction Works

1. Everything that is constructed or results from construction operations [32].
2. Buildings and civil engineering works whether over or in the ground or water, including but not limited to roads, bridges, tunnels, pylons and other facilities for transport of electricity, communication cables, pipelines, aqueducts, dams, airports, ports, waterways, and installations which are the bases for the rails of railways [31].

7.2.91 Convertibility

Ability to accommodate a substantial change(s) in user needs by making modifications [33].

7.2.92 Co-processing

The term used when introducing alternative fuels and raw materials into a standard production process, rather than using conventional fuels and raw materials [4].

7.2.93 Corporate Social Responsibility

Concept whereby companies integrate social and environmental concerns in their business operations and in their interaction with their stakeholders [1].

7.2.94 Cost Benefit Analysis

Financial tool which allows a decision-maker to consider a number of complicated environmental, social and economic factors to permit a cost optimum choice to be made [1].

7.2.95 *Customer*

A natural or legal person that purchases, hires or receives a product for their own use whether or not acting for purposes which are outside their trade, business, craft or profession [12].

7.2.96 *Cradle to Cradle*

Approach to the design of all building components that mimics nature so that, at the end of their useful life, they become a resource for future materials, creating no waste at any stage [1].

7.2.97 *Critical Raw Materials*

The following 34 raw materials shall be considered critical: antimony, arsenic, bauxite, baryte, beryllium, bismuth, boron, cobalt, coking coal, copper, feldspar, fluorspar, gallium, germanium, hafnium, helium, heavy rare earth elements, light rare earth elements, lithium, magnesium, manganese, natural graphite, nickel—battery grade, niobium, phosphate rock, phosphorus, platinum group metals, scandium, silicon metal, strontium, tantalum, titanium metal, tungsten, and vanadium [34].

7.2.98 *Data Carrier*

A linear barcode symbol, a two-dimensional symbol or other automatic identification data capture medium that can be read by a device [12].

7.2.99 *Davos Baukultur Quality System*

A set of instruments that enables the definition and assessment of 'Baukultur' qualities in places incorporating and weighing up social, emotional and cultural values equally to technical and functional aspects, with the help of the following eight criteria: governance, functionality, environment, economy, diversity, context, sense of place and beauty [20].

7.2.100 Daylight Design

Specific approach to the design of a building that ensures that a comfortable, acceptable indoor environment is achieved using natural light (and natural ventilation), thus reducing the demand for artificial lighting and its related energy and climate change impact [1].

7.2.101 Decarbonisation

Phasing-out of dependence on carbon-containing fossil fuels, and carbon embedded in other societal and industrial processes (travel, manufacturing) [1].

7.2.102 Deconstruction (See Selective Deconstruction)

1. Removal by disassembly of a building in the reverse order in which it was constructed [35].
2. Taking a building apart normally using the reverse of the methods by which it was built [19].

7.2.103 Decontamination

The removal of hazardous substances and materials [4].

7.2.104 Degradation Indicator

Measurement which estimates the impact of human activities on land and soil [1].

7.2.105 Delivered Energy

Energy, expressed per energy carrier, supplied to the technical building systems through the assessment boundary, to satisfy the uses taken into account or to produce the exported energy (Note: delivered energy can be calculated for defined energy uses or it can be measured) [10].

7.2.106 Demolish (See Demolition)

Remove and legally dispose of off-site [35].

7.2.107 Demolition

1. Removal by destructive methods [33].
2. The process or action of removing or clearing a building or part of a building, usually using heavy machinery, often by the most expedient means. Demolition often includes a soft strip, which is the segregation of materials intended to be recycled or reused. Demolition is undertaken by demolition contractors [19].

7.2.108 Design for Adaptability

Creating buildings that can be easily modified and repurposed over their lifespan. This approach aims to extend the building's useful life, reduce waste, and improve its long-term value. It involves considering factors like material selection, structural design, and service integration to facilitate future changes [36].

7.2.109 Design for Circularity (DfC)

Design and development based on the circular economy principles [5].

7.2.110 Design for Deconstruction

Approach to the design of a product or structure that facilitates deconstruction at the end of its useful life, in such a way that components and parts can be reused, recycled or recovered for further economic use or, in some other way, diverted from the waste stream [75].

7.2.111 Design for Disassembly

Approach to the design of a product or constructed asset that facilitates disassembly at the end of its useful life, in such a way that enables components and parts to be reused, recycled, recovered for energy or, in some other way, diverted from the waste stream [33].

7.2.112 *Design Life*

Service life intended by the designer [17].

7.2.113 *Design Ventilation Airflow Rate*

According to EN 16798-1, it is the ventilation rate that the ventilation system is able to provide in design conditions (including boost, weather and loads) [15].

7.2.114 *Destructive*

Characteristic of a process to obtain recoverable resources without any intent of preserving the functionality or form of the original object [5].

7.2.115 *Digital Building Logbook*

A dynamic tool that allows a variety of data, information and documents to be recorded, accessed, enriched and organised under specific categories. It represents a record of major events and changes over a building's life cycle, such as change of ownership, tenure or use, maintenance, refurbishment and other interventions [37].

7.2.116 *Digital Construction Platforms*

Virtual or physical platforms gathering private and public stakeholders, aiming to support the integration of digital technologies in the construction sector. These can also take the shape of "associations". National digital platform related to construction play a role in helping implementing construction policies by e.g. coordinating the digitalisation of public and private stakeholder initiatives, providing space for dialogue [7].

7.2.117 Digital Product Passport (DPP)

A tool to electronically register, process and share product-related information amongst supply chain businesses, authorities and consumers [34].

7.2.118 Digital Product Passport Service Provider

A natural or legal person that is an independent third-party authorised by the economic operator which places the product on the market or puts it into service and that processes the digital product passport data for that product for the purpose of making such data available to economic operators and other relevant actors with a right to access those data under the Regulation (EU) 2024/1781 or other European Union law [12].

7.2.119 Digital Twin

The real-time digital representation of the physical building or infrastructure. Usually, data is gathered by on-site sensors that continuously monitor changes in the building and in the environment and update the BIM model with the most recent data and measurement [7].

7.2.120 Direct Land Use Change (dLUC)

Change in human use or management of land within the product system being assessed [17].

7.2.121 Dismantling

The process or action of carefully taking apart, during a demolition, a building to save reusable elements [19].

7.2.122 *Disassembly*

1. The taking apart of a constituent element of a building or assembly at the end of its useful life in such a way that allows components and parts to be re-used, recycled or recovered [38].
2. Non-destructive taking apart of a construction works or constructed asset into constituent materials or components [39].

7.2.123 *Disposal Cost*

Cost of removing or getting rid of refuse or unwanted materials left over from a manufacturing process [1].

7.2.124 *Distance Selling*

The offer for sale, hire or hire purchase of products, online or through other means of distance sales, whereby the potential customer cannot physically access the product [12].

7.2.125 *Distributed Generation*

Any electricity generating technology installed by a customer or independent electricity producer that is connected at the distribution system level of the electric grid [1].

7.2.126 *Distributor*

Any natural or legal person in the supply chain, other than the manufacturer or the importer, that makes a product available on the market [12].

7.2.127 *Do not Significant Harm*

Not supporting or carrying out economic activities that do significant harm to any environmental objective, where relevant, within the meaning of Article 17 of Regulation (EU) 2020/852 [24].

7.2.128 *Downcycling*

A process in which the material's quality, potential for future reuse, and economic value decreases during the conversion process [40].

7.2.129 *Durability*

1. The ability of a product to maintain over time its function and performance under specified conditions of use, maintenance and repair [12].
2. The ability of a building or any of its components to perform its required functions in its service environment over a period of time without unforeseen cost for maintenance or repair [41].

7.2.130 *Ecodesign*

1. The integration of environmental sustainability considerations into the characteristics of a product and the processes taking place throughout the product's value chain [12].
2. Design and development based on a life cycle perspective aimed at supporting sustainable development. Also referred to by ISO as Environmentally conscious design (ECD), Design for environment (DfE), Green design, Environmentally sustainable design [5].

7.2.131 *Ecodesign Requirement*

A performance requirement or an information requirement aimed at making a product, including processes taking place throughout the product's value chain, more environmentally sustainable [12].

7.2.132 *Economic Operator*

The manufacturer, the authorised representative, the importer, the distributor, the dealer and the fulfilment service provider [12].

7.2.133 Economic System

System by which a society organizes and allocates resources [5].

7.2.134 Ecosystem

A dynamic complex of plant, animal, and micro-organism communities and their non-living environment interacting as a functional unit [24].

7.2.135 Ecosystem Services

The direct and indirect contributions of ecosystems to the economic, social, cultural and other benefits that people derive from those ecosystems [24].

7.2.136 Eco-management and Audit Scheme (EMAS)

A voluntary environmental management tool for companies and other organisations to evaluate, report and improve their environmental performance. It is the most credible and robust environmental management tool on the market, adding several elements on top of the requirements of the international standard for Environmental Management Systems EN ISO 14001:2004 [42].

7.2.137 Elemental Cost Plan

The total construction cost for the project divided into building elements, estimated based on the actual quantities of work and materials that will be required to construct the project [6].

7.2.138 Elementary Flow

Material or energy entering the system being studied that has been drawn from the environment without previous human transformation, or material or energy leaving the system being studied that is released into the environment without subsequent human transformation [43].

7.2.139 *Embodied Carbon*

A term commonly used in the built environment industries to denote the amount of CO_2 that is emitted as a result of all the energy that goes into a material's production (extraction, manufacture) construction, maintenance, refurbishment and end-of-use (demolition, incineration, landfill, etc.) across the life cycle of the built environment process [21].

7.2.140 *End-of-Life*

The life cycle stage that begins when a product is discarded and ends when the waste material of the product is returned to nature or enters another product's life cycle [12].

7.2.141 *End of Life (Product)*

Point in time when a product is taken out of use and its resources are either recovered for processing or it is disposed of [5].

7.2.142 *End of Life (Resource)*

Point in time when a resource is taken out of use and is disposed of [5].

7.2.143 *Energy Carrier*

Substance or phenomenon that can be used to produce mechanical work or heat or to operate chemical or physical processes [10].

7.2.144 *Energy Efficiency Improvement*

The more efficient use of energy at all the stages of the energy chain from production to final consumption [24].

7.2.145 Energy Need

For domestic hot water: heat to be delivered to the needed amount of domestic hot water to raise its temperature from the cold network temperature to the prefixed delivery temperature at the delivery point without the losses of the domestic hot water system.

For heating or cooling: heat to be delivered to or extracted from a thermally conditioned space to maintain the intended space temperature conditions during a given period of time.

For humidification or dehumidification: latent heat in the water vapour to be delivered to or extracted from a thermally conditioned space by a technical building system to maintain a specified minimum or maximum humidity within the space [10].

7.2.146 Energy Recovery

Generation of useful energy through direct and controlled transformation of recovered resources [5].

7.2.147 Energy Source

Source from which useful energy can be extracted or recovered either directly or by means of a conversion or transformation process [10].

7.2.148 Energy-Related Product

Any product that has an impact on energy consumption during use [12].

7.2.149 Energy Use

- For lighting: electrical energy input to a lighting system [10].
- For other services: energy input to appliances providing services not included in the EPB services [10].
- For space heating or cooling or domestic hot water: energy input to the heating, cooling or domestic hot water system to satisfy the energy need for heating, cooling (including dehumidification) or domestic hot water respectively [10].

- For ventilation: electrical energy input to a ventilation system for air transport and heat recovery [10].

7.2.150 Enterprise

An enterprise is considered to be any entity engaged in an economic activity, irrespective of its legal form. This includes, in particular, self-employed persons and family businesses engaged in craft or other activities, and partnerships or associations regularly engaged in an economic activity [44].

7.2.151 Environmental Impact

1. For a product: Any change to the environment, whether adverse or beneficial, wholly or partially resulting from a product during its life cycle [12].
2. Any change to the environment, whether adverse or beneficial, wholly or partially resulting from an organization's environmental aspects [45].

7.2.152 Environmental Footprint

A quantification of the environmental impacts resulting from a product throughout its life cycle, whether in relation to a single environmental impact category or an aggregated set of impact categories based on the Product Environmental Footprint method established by Recommendation (EU) 2021/2279 or other scientific methods developed by international organisations, widely tested in collaboration with different industry sectors and adopted or implemented by the European Commission in other European Union law [12].

7.2.153 Environmental Performance

Quantification of the performance of a product, service or organisation in terms of its environmental impact [1].

7.2.154 Environmental Product Declaration (EPD)

A document that is meant to transparently communicate the environmental performance/impact of a product or material across its life cycle [21].

7.2.155 Environmental Quality of Life

Personal satisfaction (or dissatisfaction) with the environmental conditions in which people live [1].

7.2.156 Environmental Sustainability

Approach which ensures that biological systems remain diverse and productive through responsible and active management and exploitation by humans over time [1].

7.2.157 Environmental Tax

Tax intended to promote environmentally sustainable activities via economic incentives [1].

7.2.158 Environmental System

Systems of the natural environment that interact, encompassing biotic and abiotic components [5].

7.2.159 Estimated Service Life

The service life that a building or an assembled system would be expected to have in a set of specific in-use conditions, determined from reference service life data after taking into account any differences from the reference in-use conditions [2].

7.2.160 Expandability

Ability of a design or the characteristic of a system to accommodate a substantial change that supports or facilitates the addition of new space, features, capabilities and capacities [33].

7.2.161 Exported Energy

Energy, expressed per energy carrier, supplied by the technical building systems through the assessment boundary [10].

Note 1: It can be specified by generation types (e.g. combined heat and power, photovoltaic) in order to apply different weighting factors [10].

Note 2: Exported energy can be calculated or it can be measured [10].

7.2.162 Extended Contractor Responsibility

Contractor or material supplier has responsibility for maintenance for a specified time range after project delivery and/or take back of the materials at end of use, which incentivises the choice of durable, recoverable materials [46].

7.2.163 Extended Producer Responsibility (EPR)

1. An environmental policy approach in which a producer's responsibility for a product is extended to the post-consumer stage of a product's life cycle [47].
2. An environmental policy approach whereby producers take over the financial and/or organisational responsibility for collecting or taking back used goods, as well as sorting and treatment for their recycling [48].

7.2.164 Extended Producer Responsibility Scheme (EPR Scheme)

Set of measures taken by Member states to ensure that producers of products bear financial responsibility or financial and organisational responsibility for the management of the waste stage of a product's life cycle [30].

7.2.165 Externality (External Effect)

Consequence of an activity that affects stakeholders (interested parties) other than the organization undertaking the activity, for which the organization is neither compensated nor penalized through markets or regulatory mechanisms [49].

7.2.166 Forced Labour

All work or service that is extracted from any person under the threat of a penalty and for which the person has not offered themself voluntarily [21].

7.2.167 Façades-as-a-Service (FaaS)

A model for building envelope procurement in which the building envelope (i.e. façade) builder, and/or the supply consortium behind him, assumes responsibility for the construction, cleaning, maintenance, updating, replacement, decommissioning, and circular material reprocessing of the building envelope and its constituent materials throughout its service life [50].

7.2.168 Final Recycling Process

The recycling process which begins when no further mechanical sorting operation is needed and waste materials enter a production process and are effectively reprocessed into products, materials or substances [4].

7.2.169 Finite Materials

Materials that are non-renewable on timescales relevant to the economy, i.e. not geological timescales [51].

7.2.170 Fixation Materials

Non-structural materials (all materials apart from aggregates) [4].

7.2.171 Fly Ash

Residue generated from the combustion of coal at high temperature in power stations, which can be used as an alternative constituent in cement production [1].

7.2.172 Fossil Carbon

Carbon which is contained in fossilised material [17].

7.2.173 Functional Decomposition

Functional decomposition defines the level of integration of functions within elements and the dependence of functions. It embraces the functional separation and the functional autonomy. Functional autonomy defines the dependency between independent functions of an element (e.g. structure, finishing, technical elements, piping, etc.) by planned or unplanned interpenetration of components having different functions. This means that relocation or resizing of components that have one function influences the integrity of other components that have other functions [75].

7.2.174 Functional Equivalent

Quantified functional requirements and/or technical requirements for a building or other types of construction works for use as a reference basis for comparison [14].

7.2.175 Functional Unit

Quantified performance of a product system for use as a reference unit [17].

7.2.176 Functionality

Suitability or usefulness for a specific purpose or activity [29].

7.2.177 Geometry of the Connections

The geometry of the product edge and the standardization of the connections, which will affect the ease of disassembly as well as the assembly sequence [75].

7.2.178 Geothermal Energy System

Extraction of power from heat stored in the earth [1].

7.2.179 Global Cost

Sum of the present value of the initial investment costs, annual running costs and replacement costs (referred to the starting year) as well as disposal costs if applicable [6].

7.2.180 Global Reporting Initiative (GRI)

Organisation that seeks to ensure that disclosure on economic, environmental, and social performance is as commonplace and comparable as financial reporting and important to organisation success [1].

7.2.181 Global Warming

Result of the greenhouse effect by which the average global temperatures are increasing at such a rate that they provoke significant climate change resulting in risks to future generations [1].

7.2.182 Global Warming Potential (GWP)

Characterization factor describing the radiative forcing impact of one mass-based unit of a given greenhouse gas relative to that of carbon dioxide over a given period of time [17].

7.2.183 Governance

Principles, policies and framework by which an organization is directed and controlled [52].

7.2.184 Green Belt

Zone of land that encircles a city and on which development is forbidden [1].

7.2.185 Green Facade

A vertical structure where climbing or hanging plants are directly or indirectly supported to grow up or down the structure. The growing media for these plants is ground-based for climbing plants and plant-trough-based for hanging plants [53].

7.2.186 Green Wall

A vertical structure that is normally fitted with modular, pre-planted panels that contain wall-bound growing media [53].

7.2.187 Greenfield Land

Land on which no previous development has taken place [1].

7.2.188 Greenhouse Effect

Heating of the surface of a planet or moon due to the presence of an atmosphere containing gases that absorb and emit infrared radiation [1].

7.2.189 Greenhouse Gas (GHG)

Gaseous constituent of the atmosphere, both natural and anthropogenic, that absorbs and emits radiations at specific wavelengths within the spectrum of infrared radiation emitted by the Earth's surface, the atmosphere, and clouds [17].

7.2.190 Greenhouse Gas Emission

1. Mass of a greenhouse gas released to the atmosphere [17].
2. The GHG Protocol classifies greenhouse gas emissions into three different scopes: Scope 1 emissions are direct emissions that occur from a source that can be controlled by an organization. Scope 2 emissions refer to indirect emissions that result from the generation of purchased energy. Scope 3 emissions are indirect emissions that occur in the value chain of the reporting entity (company, municipality, community, etc.) including both upstream and downstream emissions, but that are not in the control of the reporting entity [21].

7.2.191 Greenhouse Gas Emission Factor

Mass of a greenhouse gas emitted relative to an input or output of a unit process or a combination of unit processes [17].

7.2.192 Greenhouse Gas (GHG) Protocol

A framework established by the World Resources Institute and the World Business Council for Sustainable Development to assist stakeholders across the built environment process such as governments, businesses, industry associations, non-governmental organizations and other entities, to identify, measure, manage and report the greenhouse gas emissions that result from their activities, according to standard guidelines [21].

7.2.193 Greenhouse Gas Sink

Process that removes a greenhouse gas from the atmosphere [17].

7.2.194 Green Public Procurement (GPP)

A process whereby public authorities seek to procure goods, services and works with a reduced environmental impact throughout their life cycle when compared to goods, services and works with the same primary function that would otherwise be procured [17].

7.2.195 Green Public Procurement Criteria (GPP Criteria)

Environmental criteria that address environmental impacts and are designed to be used in procurement procedures [54].

7.2.196 Green Roof

1. A vegetated landscape installed on a roof surface in a loose-laid or modular format. Green roofs may be further classified as extensive, semi-intensive or intensive depending on the depth of the substrate installed and vegetation planted [53].
2. Roof of a building that is partially or completely covered with seeded soil and vegetation laid over a waterproofing membrane [1].

7.2.197 Greenwash

Practice of companies disingenuously spinning their products and policies as environmentally friendly, such as by presenting cost cuts as reductions in use of resources [1].

7.2.198 Greywater

Domestic wastewater, excluding wastewater from WCs and urinals. This term is normally applied to wastewater from sinks, wash basins, showers, baths, washing machines and dishwashers. A sub-division into "light greywater" may be used for domestic wastewater from showers, baths and wash-basins, due to their inherently lower organic loads [53].

7.2.199 Groundwater

Water which is below the surface of the ground in the saturation zone and in direct contact with the ground or subsoil [1].

7.2.200 Hazardous Waste

1. A waste that due to its (intrinsic) chemical -or other- properties poses a risk to the environment and/or human health. In the European List of Waste, hazardous wastes are marked with an asterisk (*)[11].
2. Waste which displays one or more of the hazardous properties listed in Annex III of the Directive 2008/98/EC of the European Parliament and of the Council of 19 November 2008 on waste and repealing certain Directives [13].

7.2.201 Hazardous Construction and Demolition Waste

The 2016 EU Construction & Demolition waste management protocol defines hazardous construction and demoliltion waste as debris that has hazardous properties and that may prove to be harmful to human health or the environment. This comprises contaminated soil and dredging spoil, materials and substances that may include adhesives, sealants and mastic (flammable, toxic or irritant), tar (toxic, carcinogenic), asbestos-based materials in the form of respirable fibre (toxic, carcinogenic), wood treated with fungicides, pesticides, etc. (toxic, ecotoxic, flammable), coatings of halogenated flame retardants (ecotoxic, toxic, carcinogenic), equipment with PCBs (ecotoxic, carcinogenic), mercury lighting (toxic, ecotoxic), systems with CFCs, insulation containing CFCs, containers for hazardous substances (solvents, paints, adhesives, etc.) and the packaging of likely contaminated waste [55].

7.2.202 Heat Pump

The part of a heating system that generates a temperature difference allowing heat supply to the space or process to be heated, using an electric vapour compression cycle [34].

7.2.203 Heating System

A combination of the components required to provide a form of indoor air treatment, by which the temperature is increased [18].

7.2.204 High Level Construction Forum (HLCF)

The HLCF is an initiative that has evolved from the Construction 2020 Strategy with the purpose to co-create the green, digital and resilient transition pathway for the EU construction industry ecosystem [56].

7.2.205 Hierarchical Relations Between Elements

The type of relations between individual elements within a building has a significant impact on the disassembly potential of a structure. The number of links between the different elements, as well as how the different clusters or functions connect to each other, will have an impact on the ease of disassembly. When the relations (read connections) between different clusters of functions or sub-assemblies is limited to the relation with the base element of this sub-assembly, the disassembly of the different sub-assemblies will be facilitated as well as the adaptation within one sub-assembly. When relations are made between the different elements (other than the base element) of different sub-assemblies, the disassembly and adaptations will be more elaborate [75].

7.2.206 Importer

Any natural or legal person established in the European Union that places a product from a third country on the Union market [12].

7.2.207 In-Use Condition

Any circumstance that can impact the performance of a building or assembled system under normal use [2].

7.2.208 Independence

Quality that allows parts, components, modules and systems to be removed or upgraded without affecting the performance of connected or adjacent systems [33].

7.2.209 Independent Operator

A natural or legal person that is independent of the manufacturer and is directly or indirectly involved in the refurbishment, repair, maintenance or repurposing of a product, and includes waste management operators, refurbishers, repairers, manufacturers or distributors of repair equipment, tools or spare parts, as well as publishers of technical information, operators offering inspection and testing services and operators offering training for installers, manufacturers and repairers of equipment [12].

7.2.210 Indicator

1. Quantitative, qualitative or descriptive measure [39].
2. Quantitative, qualitative or binary variable that can be measured or described, representing the status of operations, management, conditions or impacts [49].

7.2.211 Indirect Land Use Change (iLUC)

Change in the use or management of land which is a consequence of direct land use change but which occurs outside the product system being assessed [17].

7.2.212 Inert Waste

Waste that does not undergo any significant physical, chemical or biological transformations (e.g., concrete, bricks, masonry, tiles). Inert waste will not dissolve, burn or otherwise react physically or chemically, biodegrade or adversely affect other matter with which it comes into contact in a way likely to give rise to environmental pollution or harm human health [4].

7.2.213 Information Requirement

An obligation for a product to be accompanied by information [12].

7.2.214 *Initial Investment Cost*

Cost incurred up to the point when the building (or the building element) is delivered to the customer, ready to use. These costs include design, purchase of building elements, connection to suppliers, installation and commissioning process [6].

7.2.215 *Integrated Waste Management Plans and Strategies*

A geographically-based plan that promotes and supports the C&D waste management [4].

7.2.216 *Intelligent Building*

Building that provides a productive and cost-effective environment based on three basic elements: (1) people, (2) products and (3) processes and the interrelationships between them [1].

7.2.217 *Intermediate Product*

1. A product that requires further manufacturing or transformation such as mixing, coating or assembling to make it suitable for end-users [12].
2. Product that has undergone a partial processing and is used as raw material in a successive productive step [1].

7.2.218 *Internet of Things (IoT)*

The concept of connecting to the internet household appliances, devices, sensors, vehicles, etc. Thus, allowing for communication, remote control, exchange of data, etc. IoT is, currently, closely related to sensors as in most cases it requires some form of data provided by the sensors. However, as mentioned earlier, IoT could also be coupled with drones and 3D scanners which would lead to a better monitoring and inspection of construction projects [7].

7.2.219 *Irrigation*

The practice of supplying water to an area of land through pipes, sprinklers or channels so that vegetation will grow [53].

7.2.220 *Key Performance Indicator (KPI)*

Indicator of performance deemed by an organization to be significant and giving prominence and attention to certain aspects of operations, management, conditions or impacts [49].

7.2.221 *Kyoto Protocol*

International agreement linked to the United Nations Framework Convention on Climate Change, the major feature of which is that it sets binding targets for 37 industrialised countries and the European Community for reducing greenhouse gas (GHG) emissions [1].

7.2.222 *Labelling*

A certification label is a label or symbol indicating that compliance with standards has been verified. Use of the label is usually controlled by the standard-setting body. Where certification bodies certify against their own specific standards, the label can be owned by the certification body [4].

7.2.223 *Landfill*

Means a waste disposal site for the deposit of the waste onto or into land (for instance underground), including: internal waste disposal sites (for instance own waste disposal carried out by the producer of waste at the place of production), and a permanent site (older than one year) which is used for temporary storage of waste.

But excluding: facilities where waste is unloaded in order to permit its preparation for further transport for recovery, treatment or disposal elsewhere, and storage of waste prior to recovery or treatment for a period less than 3 years as a general rule, storage of waste prior to disposal for a period less than 1 year [4].

7.2.224 LCI (Lowest Concentration of Interest)

According to EN 16516, LCI is the Lowest Concentration of Interest, which is a substance-specific value and is quoted in terms of mass concentration in the air of the reference room, for health-related evaluation of emission levels from construction products [15].

7.2.225 Lead Market

Innovation-friendly market for creating new innovative products and services in areas that are promising but are currently constrained due to regulatory or other obstacles [1].

7.2.226 Lean Construction

Way to design production systems to minimise waste of materials, time, and effort in order to generate the maximum possible amount of value [1].

7.2.227 Level(s)

1. A voluntary reporting framework to improve the sustainability of buildings. Using existing standards, Level(s) provides a common EU approach to the assessment of environmental performance in the built environment [57].
2. An assessment and reporting tool, developed by the European Commission, for sustainability performance of buildings, firmly based on circularity [58].

7.2.228 Life Cycle

1. The consecutive and interlinked stages of a product's life, consisting of raw material acquisition or generation from natural resources, pre-processing, manufacturing, storage, distribution, installation, use, maintenance, repair, upgrading, refurbishment and reuse, and end-of-life [12].
2. All consecutive and interlinked stages in the life of the object under consideration. For consideration of environmental impacts and environmental aspects, the life cycle comprises all stages, from raw material acquisition or generation of natural resources to end-of-life. For consideration of economic impacts and economic

aspects, in terms of costs, the life cycle comprises all stages from construction to decommissioning [39].

7.2.229 Life Cycle Assessment (LCA)

1. Compilation and evaluation of the inputs, outputs and the potential environmental impacts of a product system throughout its life cycle [43].
2. A procedure for tabulating and reporting the environmental impact of a material, product or service over its lifetime. The life-cycle assessment process includes the following procedures: (1) goal and scope definition, (2) inventory analysis, (3) impact assessment and (4) interpretation [21].

7.2.230 Life Cycle Co-ordination in Assembly/Disassembly

Co-ordination of elements, with those that have a long life cycle and greatest dependencies in assembly, should be assembled first and disassembled last taking into account use, technical and size aspects [75].

7.2.231 Life Cycle Cost (LCC)

1. Cost of an asset or its parts throughout its life cycle, whilst fulfilling its performance requirements [6].
2. Quantification of all the costs incurred during the lifetime of a good or service including planning, design, acquisition, operation, maintenance and disposal [1].

7.2.232 Life Cycle Costing (LCC)

1. Methodology for systematic economic evaluation of life cycle costs over a period of analysis, as defined in the agreed scope [29].
2. Method to appraise not just the capital costs of a project, but also the manufacturing costs of the construction materials, operation and maintenance costs [1].

7.2.233 *Life Cycle Impact Assessment (LCIA)*

Phase of a life cycle assessment aimed at understanding and evaluating the magnitude and significance of the potential environmental impacts for a product system throughout the life cycle of the product [1].

7.2.234 *Life Cycle Impact Category Indicator*

Quantifiable representation of an environmental issue of concern (impact category) to which life cycle inventory results may be assigned [1].

7.2.235 *Life Cycle Inventory*

Mass and energy information about the functional unit in question which is used to quantify the inputs and outputs of a unit process, regardless of whether the collected data is measured, calculated or estimated [1].

7.2.236 *Life Cycle Inventory Analysis*

1. Phase of a life cycle assessment involving the compilation and quantification of inputs and outputs for product throughout its life cycle [38].
2. Phase of a life cycle assessment involving the compilation and quantification of inputs and outputs for a product throughout its life cycle [1].

7.2.237 *Life Cycle Inventory Analysis Result*

Outcome of a life cycle inventory analysis that includes the flows crossing the system boundary and provides the starting point for life cycle impact assessment [1].

7.2.238 *Life Cycle Perspective (Life Cycle Thinking)*

Consideration of the circularity aspects relevant to a solution during its life cycle which includes consideration of the relevant environmental, social and economic impacts [59].

7.2.239 Lifespan (Lifetime)

The period of time from when a product is released for use after manufacture to the moment it becomes obsolete beyond recovery at product level [1].

7.2.240 Light to Solar Gain Ratio

Method to evaluate the relationship between visible light transmittance and heat gain [1].

7.2.241 Limit Value

Upper or lower acceptable performance level on a performance scale [14].

7.2.242 Linear Economy

An economy in which finite resources are extracted to make products that are used, generally not to their full potential, and then thrown away ('take-make-waste') [51].

7.2.243 Living Environment

A balanced composition of the artificial and natural environments, which includes both outdoor and indoor spaces. The planning and creation of a sustainable and integrated living environment (urban or, more broadly, spatial design) include planning and designing activities (including spatial planning, architecture, landscape architecture), construction activities and the spatial coordination of other space-related decisions (such as traffic and mobility possibilities, utility services, agriculture etc.), with the goal of improving the living environment for as many people as possible in the long term through spatial decisions [20].

7.2.244 Locally Sourced Material

Material obtained from a defined radius around a project site, helping to support the local economy and reducing transportation costs and energy [1].

7.2.245 Losses

Unmanaged outflows of a resource from the system in focus that are not recovered [5].

7.2.246 Low-Carbon Buildings Programme

National member state scheme where grants are currently provided towards the cost of installing the following technologies: solar photovoltaic's; wind turbines; small scale hydro; solar hot water; ground source heat pumps; air source heat pumps; biomass [1].

7.2.247 Low-Carbon Energy Source

Source of power which produces fewer greenhouse gases than traditional means of power generation [1].

7.2.248 Low or Zero Carbon Technology (LZC)

Technology which emits low levels of CO2 emissions, or no net CO_2 emissions [1].

7.2.249 Maintenance

1. Combination of all technical and associated administrative actions during the service life to retain a building or an assembled system (part of works) in a state in which it can perform its required functions [17].
2. One or more actions carried out to keep a product in a condition where it is able to fulfil its intended purpose [12].

7.2.250 Maintenance Cost

Total of necessarily incurred labour, material or other costs incurred to retain a building or its parts in a state in which it can perform its required functions [6].

7.2.251 *Making Available on the Market*

Any supply of a product for distribution, consumption or use on the European Union market in the course of a commercial activity, whether in return for payment or free of charge [12].

7.2.252 *Manufacturer*

Any natural or legal person that manufactures a construction product or that has such a product designed or manufactured, and markets that product under their name or trademark [60].

7.2.253 *Material Footprint*

The total amount of raw materials extracted to meet final consumption demands [12].

7.2.254 *Material Passport*

Digital document listing all the materials that are included in a product or a construction during its life cycle, in order to facilitate strategizing circularity decisions in supply chain management. Passports generally consists of a set of data describing defined characteristics of materials in products, which enables the identification of value for recovery, recycling, and re-use [61].

7.2.255 *Material Separation*

An operation to separate materials, including mechanical, chemical or thermal processes (e.g. shredding, smelting, sorting etc.) other than dismantling or disassembly [2].

7.2.256 *Materiality*

Information related to circular economy that is essential for decision-making and can be applied to identify issues that reflect an organization's environmental and social impacts, as well as information that supports interested party and strategic decision-making [59].

7.2.257 Materiality Assessment

Method to identify and prioritize the issues most important to an organization and its interested parties, and relevant to its circular economy strategy [59].

7.2.258 Mechanical Ventilation

According to EN 16798-1, a ventilation system where air is supplied or extracted from the building or both by a fan using air terminal devices, ducts and roof/wall devices [15].

7.2.259 Microgeneration

Zero or low-carbon generation of heat and power by individuals, small businesses and communities to meet their own needs [1].

7.2.260 Mixed Construction and Demolition Waste

Mixture of different fractions of construction and demolition waste [4].

7.2.261 Modular

Composed of modules for easy construction or arrangement and adaptation or disassembly [33].

7.2.262 Module

Set of standardised parts or independent units [33].

7.2.263 *Natural Capital*

Extension of the economic notion of capital (manufactured means of production) to environmental goods and services [1].

7.2.264 *Natural Resource*

Resource occurring in nature [5].

7.2.265 *Natural Ventilation*

According to EN 16798-1, where ventilation is provided by thermal, wind, or diffusion effects through doors, windows, or other intentional devices in the building designed for ventilation. Natural ventilation may be manually or automatically controlled [15].

7.2.266 *Nature-Based Solutions (NBS)*

Solutions that are inspired and supported by nature, which are cost-effective, simultaneously provide environmental, social and economic benefits and help build resilience. Such solutions bring more, and more diverse, nature and natural features and processes into cities, landscapes and seascapes, through locally adapted, resource-efficient and systemic interventions [62].

7.2.267 *Nearly Zero-Energy Building*

1. According to the Energy performance of Buildings Directive, a building that has a very high energy performance, as determined in accordance with Annex I. The nearly zero or very low amount of energy required should be covered to a very significant extent by energy from renewable sources, including energy from renewable sources produced on-site or nearby [10].
2. Building that, by virtue of the materials from which it is constructed and of the fact that it produces surplus energy from renewable sources, ensures that, over its design life, it compensates for all carbon emissions associated with the construction and use of the building [1].

7.2.268 Net Present Cost

Sum of the discounted future costs [6].

7.2.269 Net Zero Energy Building

Building where, as a result of its very high level of energy efficiency, the overall annual primary energy consumption is equal to the energy (produced from renewable energy sources) supplied to the building [1].

7.2.270 Nominal Cost

Expected price that will be paid when a cost is due to be paid, including estimated changes in price due to, for example, forecast changes in efficiency, inflation or deflation and technology [6].

7.2.271 Non-destructive

Characteristic of a process to obtain recoverable resources with the intent of preserving the entire functionality or form of the original object to the extent possible [5].

7.2.272 Non-hazardous Waste

According to the Waste Framework Directive, it means a waste that does not display one or more of the hazardous properties listed in Annex III (to the same Directive) [55].

7.2.273 Non-recoverable Resource

Resource that cannot be recovered and used again after it has been processed or used [5].

7.2.274 Non-renewable Primary Energy Factor

Non-renewable primary energy for a given energy carrier, including the delivered energy and the considered energy overheads of delivery to the points of use, divided by the delivered energy [10].

7.2.275 Non-renewable Energy

Energy resource that is not replaced or is replaced only very slowly by natural processes [1].

7.2.276 Non-renewable Resource

Natural resource that exists in a fixed amount and which cannot be replenished on a human time scale [1].

7.2.277 Obsolescence

Loss of ability of an item to perform satisfactorily due to changes in performance requirements [29].

7.2.278 Offsetting

Mechanism for compensating for all or for a part of the carbon footprint through the prevention of the release of, reduction in, or removal of an amount of greenhouse gas emissions in a process outside the boundary of the product system [17].

7.2.279 Onsite (in the Context of Delivered Energy)

Means energy produced on the premises and the parcel of land on which the building is located and the building itself [10].

Note: onsite is defining a strong link between the energy source (localization and interaction) and the building [10].

7.2.280 *Operation Cost*

Cost incurred in running or managing a facility or the built environment, including administration support services [6].

7.2.281 *Operational Carbon*

Emissions generated through the function and maintenance of the building [21].

7.2.282 *Operational Energy Use*

Energy use of the building-integrated technical systems during use and operation of the building [17].

7.2.283 *Operational Water Use*

Water use of the building-integrated technical systems and of the user, as needed for the technically and functionally defined operation of the building [17].

7.2.284 *Periodic Cost*

Substitute investment that is necessary for age reasons; that corresponds to cost replacement of all components (or system) according to their lifespan that occur on the anticipate year of wear out. Periodic costs include all replacement costs for any components or system that occur the same year [6].

7.2.285 *Performance*

Ability to fulfil required functions under intended use conditions [8].

7.2.286 *Performance Level*

Value indicating the relative performance required (or provided) for a particular attribute on a relative scale, from the level of the least (performance) to the level of the most (performance). For some attributes, such as adaptability, the level may be expressed with help of criteria e.g., an 'A level' achieved when 80% of criteria is fulfilled, or a B level when only 60% of criteria is fulfilled [14].

7.2.287 *Performance Requirement (Performance Criterion)*

1. Minimum acceptable level of a critical property [29].
2. A quantitative or non-quantitative requirement for or in relation to a product to achieve a certain performance level in relation to a product parameter referred to in Annex I [12].

7.2.288 *Place-Based Approach*

An approach that targets the specific circumstances of a place and engages the community and a broad range of local organisations from different sectors as active participants in their development and implementation [20].

7.2.289 *Placing on the Market*

The first making available of a product on the Union market [12].

7.2.290 *Pollutant*

A substance, vibration, heat, noise, light or other contaminant present in air, water or land which may be harmful to human health or the environment, which may result in damage to material property, or which may impair or interfere with amenities and other legitimate uses of the environment [24].

7.2.291 Pollution

The direct or indirect introduction of pollutants into air, water or land as a result of human activity [24].

(a) the direct or indirect introduction of pollutants into air, water or land as a result of human activity [24];
(b) in the context of the marine environment, pollution as defined in point 8 of Article 3 of Directive 2008/56/EC [24];
(c) in the context of the water environment, pollution as defined in point 33 of Article 2 of Directive 2000/60/EC [24].

7.2.292 Post-consumer Material (PCM)

Material generated by households or by commercial, industrial or institutional facilities in their role as end-users of the product which can no longer be used for its intended purpose [59].

7.2.293 Potable Water

Water, in its original state or after treatment, that is wholesome and clean and therefore suitable for human consumption for drinking, cooking, food preparation and other domestic purposes. The term can be considered as synonymous with "drinking water" [53].

7.2.294 Premature Obsolescence

A product design feature or subsequent action or omission resulting in the product becoming non-functional or performing less well without such changes of functionality or performance being the result of normal wear and tear [12].

7.2.295 Preparing for Re-use

According to the Waste Framework Directive, it means checking, cleaning or repairing recovery operations, by which products or components of products that have become waste are prepared so that they can be re-used without any other pre-processing [2].

7.2.296 Pre-demolition Audit

A preparatory activity with the purpose of (1) collecting information about the qualities and quantities of the C&D waste materials that will be released during the demolition or renovation works and (2) giving general and site-specific recommendations regarding the demolition process [4].

7.2.297 Pre-development Audit

Pre-development audits include demolition and refurbishment assessments of what can be reused from deconstruction and stripped out, respectively. These should also inform the potential to reuse products and materials in subsequent construction and/or fit out (of refurbishment) [63].

7.2.298 Prefabrication

Prefabrication refers to the process of manufacturing building components in a controlled environment before transporting them to the construction site for assembly. This method enhances efficiency, reduces waste, and minimizes on-site construction time. Prefabricated components can include wall panels, roofs, and modular buildings. The use of prefabrication allows for greater precision in manufacturing and often results in cost savings for construction projects [64].

7.2.299 Prevention

Measures taken before a substance, material or product has become waste that reduces:

(a) the quantity of waste, including through the reuse of products or the extension of the life span of products;
(b) the adverse impacts of the generated waste on the environment and human health; or
(c) the content of hazardous substances in materials and products [12].

7.2.300 Primary Energy

Energy from renewable and non-renewable sources which has not undergone any conversion or transformation process [17].

7.2.301 (Total) Primary Energy Factor

For a given energy carrier, non-renewable and renewable primary energy divided by delivered energy, where the primary energy is that required to supply one unit of delivered energy, taking account of the energy required for extraction, processing, storage, transport, generation, transformation, transmission, distribution, and any other operations necessary for delivery to the building in which the delivered energy will be used [17].

7.2.302 Performance Requirement

Minimum acceptable level of a critical property [29].

7.2.303 Preparing for Re-use

Checking, cleaning or repairing waste materials for recovery operations. The waste, products or components of products, that have been collected by a recognised re-use operator or deposit-refund scheme, are prepared so that they can be re-used without any further pre-processing [4].

7.2.304 Product

1. Physical-based object designed for or utilized with a purpose [5].
2. Any physical goods that are placed on the market or put into service [12].

7.2.305 Product Group

A set of products that serve similar purposes and are similar in terms of use, or have similar functional properties, and are similar in terms of consumer perception [12].

7.2.306 Product Presenting a Risk

A product that, by not complying with an ecodesign requirement set in or pursuant to this Regulation other than those listed in Article 71(1), could adversely affect the environment or other public interests protected by that requirement [12].

7.2.307 *Product Presenting a Serious Risk*

A product presenting a risk for which, based on an assessment, the degree of the relevant non-compliance or the associated harm is considered to require rapid intervention by the market surveillance authorities, including cases where the effects of the non-compliance are not immediate [12].

7.2.308 *Professional Repairer*

A natural or legal person that provides professional repair or maintenance services for a product, irrespective of whether that person acts within the manufacturer's distribution system or independently [12].

7.2.309 *Property Owner*

The owner of the building or infrastructure, the developer or the party stated by the national legislation as the original waste holder [4].

7.2.310 *Provider of an Online Market*

Means a provider of an intermediary service using an online interface which allows customers to conclude distance contracts with economic operators for the sale of products covered by delegated acts adopted pursuant to Article 4 [12].

7.2.311 *Putting into Service*

The first use, for its intended purpose, in the Union, of a product [12].

7.2.312 *Rainwater Harvesting*

The collection and storage of rainwater to partially or fully meet current and future water demand. Rainwater is normally collected from roofs but may also be collected from other impermeable or pervious ground surfaces depending on the risk of contamination and the intended end use [53].

7.2.313 *Raw Material*

A substance in processed or unprocessed state used as an input for the manufacturing of intermediate or final products, excluding substances predominantly used as food, feed or combustion fuel [34] (see critical and strategic raw materials).

7.2.314 *Raw Materials Supply Chain*

All activities and processes of the raw materials value chain up to the point where a raw material is used as an input for the manufacturing of intermediate or final products [34].

7.2.315 *Raw Materials Value Chain*

All activities and processes involved in the exploration, extraction, processing and recycling of raw materials [34].

7.2.316 *Real Cost*

Cost expressed as a value at the base date, including estimated changes in price due to forecast changes in efficiency and technology [6].

7.2.317 *Reclamation*

To carefully salvage, deconstruct, dismantle, demolish or otherwise take apart a building for the purpose of maximising the amount of material saved for reuse. This activity encompasses the purchase, salvage, processing, storage and sale of reusable materials and products [19].

7.2.318 *Reclamation Audit*

The action of undertaking an assessment of potentially reusable building materials and products, after the decision has been taken to demolish a building. The audit results in the reclamation inventory, which is a list of reclaimable products or materials. The level of detail may vary but will normally include factors such as quantity, dimensions, constituent material, and condition. A reclamation audit may also assess comparative carbon content, ease of removal, technical characteristics, and general reclamation advice [19].

7.2.319 Reclamation Dealers (Salvage Dealers)

1. Companies that are specialised in the reclamation of construction products. They source construction products released from demolition works, process and store them, before putting them back on the market. Some dealers are only trading very specific materials, while others are much more diversified. They usually provide a range of services that facilitate the reuse of the products [19].

2. Economical actor who acquires and sells reclaimed construction elements. Most salvage or reclamation dealers are equipped to undertake specific operations to prepare reclaimed construction elements for reuse. This involves, for example, sorting, cleaning, dimensioning, documenting, advertising and shipping. Some salvage or reclamation dealers combine the sale of reclaimed products with an activity as deconstructors or demolition contractors [19].

7.2.320 Reclaimed Building Material or Product

Material and product derived from reclamation [19].

7.2.321 Reclamation Inventory

A list of reclaimable products or materials which normally includes factors such as quantity, dimensions, constituent material, and condition [19].

7.2.322 Reclamation Sector

A narrow segment of the reuse sector. It includes mainly salvage dealers and dismantling contractors, runners, restorers [19].

7.2.323 Recoverable Resource

Resource that can be recovered and used again after it has already been processed or used [5].

7.2.324 Recovered Resource (Secondary Resource)

Resource that is obtained from one that has already been processed or used [5].

7.2.325 Recovery

1. Any operation the principal result of which is waste serving a useful purpose by replacing other materials which would otherwise have been used to fulfil a particular function, or waste being prepared to fulfil that function, in the plant or in the wider economy [65].
2. Any activity carried out for the purposes of reclaiming, recycling or reusing the waste [11].

According to the Waste Framework Directive, it means any operation the principal result of which is waste serving a useful purpose by replacing other materials which would otherwise have been used to fulfil a particular function, or waste being prepared to fulfil that function, in the plant or in the wider economy [2].

7.2.326 Recyclable

Characteristic of a product or associated component, that can be diverted from the waste stream through available processes and programmes, and can be collected, processed and returned to use in the form of raw materials or products [66].

7.2.327 Recyclability

1. The ability of a waste product to be recycled, based on actual practices [2].
2. The ability of component parts, materials or both to be separated and reprocessed from products and systems and subsequently used as material input for the same or different use or function [33].

7.2.328 Recycling

1. Any recovery operation by which waste materials are reprocessed into products, materials or substances, whether for the original or other purposes. It includes the reprocessing of organic material but does not include energy recovery and the reprocessing into materials that are to be used as fuels or for backfilling operations [65].
2. Activities to obtain recovered resources for use in a process or a product, excluding energy recovery [5].

7.2.329 Reference Service Life

The service life of construction product which is known to be expected under a particular reference set of in-use conditions and which may form the basis for estimating the service life under other in-use conditions [2].

7.2.330 Reference Study Period

Period over which the time-dependent characteristics of the object of assessment are analysed [17].

7.2.331 Reference Value

Performance level on a performance scale that represents state of the art or best practice [14].

7.2.332 Refurbish (Recondition)

Restore an item, during its expected service life, to a useful condition for the same purpose with at least similar quality and performance characteristics [5].

7.2.333 Refurbishability

The ability to restore the aesthetic and functional characteristics of a product, building or other constructed asset to a condition suitable for continued use [33].

7.2.334 Refurbishment

1. The modification and improvements to an existing building or civil engineering works in order to bring it up to an acceptable condition [8].
2. Actions carried out to prepare, clean, test, service and, where necessary, repair a product or a discarded product in order to restore its performance or functionality within the intended use and range of performance originally conceived at the design stage at the time of the placing of the product on the market [12].

7.2.335 Regenerate

Improve or restore a degraded ecosystem [5].

7.2.336 Regenerative Production

Regenerative production provides food and materials in ways that support positive outcomes for nature, which include but are not limited to: healthy and stable soils, improved local biodiversity, improved air and water quality [52].

7.2.337 Releases

Managed emissions to air and discharges to water or land from the system in focus [5].

7.2.338 Reliability

The probability that a product functions as required under given conditions for a given duration without an occurrence which results in a primary or secondary function of the product no longer being performed [12].

7.2.339 Remanufacturability

The ability of a product to be disassembled and refabricated at the end of its useful life in a manner that provides restoration to a condition suitable for resale [33].

7.2.340 Remanufacture

Return an item to a like-new condition from both a quality and performance perspective using an industrial process [5].

7.2.341 Remanufacturing

1. The process of returning a used product to, at least, its original performance with a warranty that is equivalent to or better than that of the newly manufactured product [67].
2. Actions through which a new product is produced from objects that are waste, products or components and through which at least one change is made that substantially affects the safety, performance, purpose or type of the product [12].

7.2.342 Renewable Energy

Means energy from renewable non-fossil sources, namely wind, solar, aerothermal, geothermal, hydrothermal and ocean energy, hydropower, biomass, landfill gas, sewage treatment plant gas and biogases [10].

7.2.343 Renewable Primary Energy Factor

Renewable primary energy for a given distant or nearby energy carrier, including the delivered energy and the considered energy overheads of delivery to the points of use, divided by the delivered energy [10].

7.2.344 Renovation

Work that involves the structural alteration of buildings, the substantial replacement of main services or finishes and/or the substantial changed use of floor space whilst at the same time including associated redecoration and repair works on the one hand and related new building on the other. Renovation covers all the work done to existing buildings as the four R's: renovation, rehabilitation, restoration and remodelling. Renovation is addressed from a broad perspective, including residential, historical and commercial buildings owned and managed by private/public companies or authorities [4].

7.2.345 Repair

1. Returning a product, component, assembly or system to an acceptable condition by renewal or replacement of worn, damaged, or degraded parts [8].
2. Returning an item to an acceptable condition through the renewal, replacement or mending of worn, damaged or degraded parts [17].
3. One or more actions carried out to return a defective product or waste to a condition where it fulfils its intended purpose [12].

7.2.346 Repurpose

Adapt a product or its component parts for use in a different function than it was originally intended for, without making major modifications to its physical, chemical or mechanical structure [5].

7.2.347 Repurposing

Using an obsolete item considered by its owner as a waste, with another use totally different from the initial one [68].

7.2.348 Required Service Life

Service life required by the client or through regulations [17].

7.2.349 *Resilience*

1. Ability to resist, adapt to, or quickly recover from potentially disruptive events or conditions, whether natural or anthropogenic, in order to maintain or restore the intended service [32].
2. Ability to face economic, social and environmental shocks or persistent structural changes in a fair, sustainable and inclusive way [69].

7.2.350 *Resource Productivity*

Resource productivity is a measure of the total amount of materials directly used by an economy [measured as domestic material consumption (DMC)] in relation to GDP. It provides insights into whether decoupling between the use of natural resources and economic growth is taking place [70].

7.2.351 *Reusability*

1. Ability of a material, product, component or system to be used in its original form more than once and maintain its value and functional qualities during recovery to accommodate reapplication for the same or any purpose [71].
2. The ability to restore the aesthetic and functional characteristics of a product, building or other constructed asset to a condition suitable for continued use. The refurbishing of products can reduce the consumption of natural resources. Depending on the intended design life of the construction works, refurbishability can also help reduce operating and maintenance [33].

7.2.352 *Reuse*

1. Any operation by which products or components that are not waste are used again for the same purpose for which they were conceived [65].
2. Another use, at the end of its current life, of a construction product or material. Ideally, the installation of the reused material or product should be reversible, allowing for future uses of the same product. Reuse includes handmade and low-tech recrafting, repurposing or remanufacturing, such as the sawing of reclaimed timber beams into planks for use as flooring, or in making new furniture, for example [19].

7.2.353 Reuse Expert

A relatively new professional profile that offers assistance for achieving reuse ambitions (which are often coupled with a larger view on circular economy). Depending on the missions at stake, they can be contracted directly by the client or team up with the architects (in the context of an architectural contest, for instance). In some cases, these reuse assistants are former architects who have built up such specific know-how on reuse that it becomes a new service which they are able to provide [19].

7.2.354 Reuse Sector

Refers to all parties actively involved in reuse including clients, the reclamation sector, construction professionals, specifiers and end users. It may also include interested parties such as environmental and building conservation organisations, regulators, government and local authorities [19].

7.2.355 Reverse Logistics

1. Supply chains dedicated to the reverse flow of products and materials for the purpose of maintenance, repair, reuse, refurbishment, remanufacture, recycling, or regenerating natural systems [52].
2. Process of managing, collecting and moving products from their current location after the end of use for the purpose of recovering or retaining value through proper handling such as remanufacturing, repair, recycling or other treatment [5].

7.2.356 Reversible Connection

Connection that can be disconnected and/or disassembled for easy alterations and additions to structures [33].

7.2.357 Robotics

Robotics in the construction sector consists of the use of devices with robotic arms which operate repetitive processes such as laying floor tiles or bricks, lifting heavy objects and placing them in exact coordinates [7].

7.2.358 Salvage

1. Removal of disassembled building materials for the purpose of reuse or recycling [35].
2. Saving elements of a building, construction or landscape for the purpose of reuse. Synonym is reclamation, although salvage usually refers to more modern materials and products [19].

7.2.359 Same-Site Reuse

Use of an element dismantled from a building in a new construction or renovation project happening on the same site. It may require storage, either on site or elsewhere. The expression in situ reuse is used in France and Belgium [19].

7.2.360 Scavenging

The activity of identifying usable materials that takes place after demolition; in this context, particularly re-usable and recyclable materials [4].

7.2.361 Scenario

Collection of assumptions and information concerning an expected sequence of possible future events [17].

7.2.362 Secondary Materials (Non-virgin Materials)

Materials that have been previously used. This includes: materials in products that have been reused, refurbished or repaired; components that have been remanufactured; and materials that have been recycled [52].

7.2.363 Selective Deconstruction (See Deconstruction)

Disassembly and removal of selected portions of building or structure [35].

7.2.364 Selective Demolition

Sequencing the demolition activities to allow the separation and sorting of building materials [4].

7.2.365 Self-regulation Measure

A voluntary agreement or a code of conduct, concluded by economic operators on their own initiative, which they are responsible for enforcing [12].

7.2.366 Sensor

Any device that offers the possibility to collect data and monitor the performance of an aspect of a construction project (site, building, machine, etc.) during the whole building lifecycle, including operation and maintenance [7].

7.2.367 Separated Collection

Collection where a waste stream is kept separately by type and nature so as to facilitate a specific treatment [4].

7.2.368 Service

Activity designed or executed with a purpose [5].

7.2.369 Service Life

Period of time after installation during which a building or an assembled system (part of works) meets or exceeds the technical requirements and functional requirements [17].

7.2.370 Simplicity

Quality of an assembly or system that is designed to be straightforward, easy to understand and meet performance requirements with the least amount of customization [33].

7.2.371 Site-to-Site Reuse

Expression forged as part of the FCRBE project to refer to situations in which a reclaimed material or product travels directly from dismantling to end-user, avoiding the stockholding (dealer) network [19].

7.2.372 Small Enterprise

Within the SME category, a small enterprise is defined as an enterprise which employs fewer than 50 persons and whose annual turnover and/or annual balance sheet total does not exceed EUR 10 million [44].

7.2.373 Social System

System by which human beings are expected to undertake different types of tasks in order to achieve common goals within a society [5].

7.2.374 Solution

Product or service, or a combination thereof, that fulfils a need of an interested party [5].

7.2.375 Soil

The top layer of the Earth's crust situated between the bedrock and the surface, which is composed of mineral particles, organic matter, water, air and living organisms [24].

7.2.376 Sphere of Influence

Range or extent of political, contractual, economic or other relationships through which an organization has the ability to affect the decisions or activities of other individuals or organizations [5].

7.2.377 Stakeholder (Interested Party)

Person or organization that can affect, be affected by, or perceive itself to be affected by a decision or activity [5].

7.2.378 Stockpiling Location

A platform for storing waste that can be moved [4].

7.2.379 Strategic Raw Materials

The following raw materials shall be considered strategic: bismuth, boron—metallurgy grade, cobalt, copper, gallium, germanium, lithium—battery grade, magnesium metal, manganese—battery grade, natural graphite—battery grade, nickel—battery grade, platinum group metals, rare earth elements for magnets (Nd, Pr, Tb, Dy, Gd, Sm, and Ce), silicon metal, titanium metal, and tungsten [34].

7.2.380 Stripping

The activity of removing valuable materials from a site, installation or building that takes place before demolition [4].

7.2.381 Supply Chain

All upstream activities and processes of the product's value chain, up to the point where the product reaches the customer [12].

7.2.382 Supply Disruption

The unexpected significant decrease in the availability of a raw material or significant increase in the price of a raw material [34].

7.2.383 Substance of Concern

Means a substance that [12]:

(a) meets the criteria laid down in Article 57 of Regulation (EC) No 1907/2006 and is identified in accordance with Article 59(1) of that Regulation;
(b) is classified in Part 3 of Annex VI to Regulation (EC) No 1272/2008 in one of the following hazard classes or hazard categories:

  (i) carcinogenicity categories 1 and 2;
  (ii) germ cell mutagenicity categories 1 and 2;
  (iii) reproductive toxicity categories 1 and 2;
  (iv) endocrine disruption for human health categories 1 and 2;
  (v) endocrine disruption for the environment categories 1 and 2;
  (vi) persistent, mobile and toxic or very persistent, very mobile properties;
  (vii) persistent, bioaccumulative and toxic or very persistent, very bioaccumulative properties;
  (viii) respiratory sensitisation category 1;
  (ix) skin sensitisation category 1;
  (x) hazardous to the aquatic environment—categories chronic 1 to 4;
  (xi) hazardous to the ozone layer;
  (xii) specific target organ toxicity—repeated exposure categories 1 and 2;
  (xiii) specific target organ toxicity—single exposure categories 1 and 2;

(c) is regulated under Regulation (EU) 2019/1021; or
(d) negatively affects the reuse and recycling of materials in the product in which it is present.

7.2.384 Sustainability Indicator

Indicator related to economic, environmental or social impacts [72].

7.2.385 Sustainable Construction

Application of sustainable development principles to the design and construction process, i.e.: use of fewer virgin materials, less energy in construction, less energy in use, less pollution and less waste; 'whole life' approach to design, construction and life use; and providing safe places and work with acceptable social conditions integrated into sustainable communities [1].

7.2.386 Sustainable Development

Development that meets the environmental, social and economic needs of the present without compromising the ability of future generations to meet their own needs [5].

7.2.387 Sustainable Public Procurement

A process by which public authorities seek to achieve the appropriate balance between the three pillars of sustainable development—economic, social and environmental—when procuring goods, services or works at all stages of the project [54].

7.2.388 System

Set of interrelated or interacting elements [73].

7.2.389 Systemization

The grouping of sub-systems and the constituent components in a building to encourage open/independent strands of construction/layers; includes structure and material levels and clustering aiming to limit the number of disassembly sequences [75].

7.2.390 Technosphere

Sphere or realm of human technological activity which results in a technologically modified environment [5].

7.2.391 Upcycling

A process in which the material's quality, potential for future reuse, and economic value is increased during the conversion process [40].

7.2.392 Unique Facility Operator

A unique string of characters for the identification of locations or buildings involved in a product's value chain or used by actors involved in a product's value chain [12].

7.2.393 Unique Operator Identifier

A unique string of characters for the identification of an actor involved in a product's value chain [12].

7.2.394 Unique Product Identifier

A unique string of characters for the identification of a product that also enables a web link to the digital product passport [12].

7.2.395 Unsold Consumer Product

Means any consumer product that has not been sold including surplus stock, excess inventory and deadstock and products returned by a consumer on the basis of their right of withdrawal in accordance with Article 9 of Directive 2011/83/EU or, where applicable, during any longer withdrawal period provided by the trader [12].

7.2.396 Upgrading

Actions carried out to enhance the functionality, performance, capacity, safety or aesthetics of a product [12].

7.2.397 3D Printing (Additive Manufacturing)

The process of creating an object by solidifying a material (e.g. plastic, metal, wood or concrete) under the control of a computer using a Computer-Aided Design (CAD) or BIM file to guide the 3D printer nozzle. In the construction sector, 3D printing can be used to create building components or to print entire buildings. It has been used for buildings, bridges, printed moulds or architectural models [7].

7.2.398 3D Scanning

The process of creating a 3D model of a real-world object or construction by scanning it from all possible angles. This process can be used in the construction sector for creating 3D models of existing buildings and infrastructures for which there is no digital information [7].

7.2.399 Technical Building System

Technical equipment for space heating, space cooling, ventilation, domestic hot water, built-in lighting, building automation and control, on-site electricity generation, or a combination thereof, including those systems using energy from renewable sources, of a building or building unit [18].

7.2.400 Technical Cycle

1. Cycle(s) within the social system through which resources are used, recovered, restored and utilized within existing or new solutions [5].
2. The processes that products and materials flow through in order to maintain their highest possible value at all times. Materials suitable for these processes are those that are not consumed during use, such as metals, plastics, and wood [52].

7.2.401 Total Volatile Organic Compounds (TVOCs)

According to EN 16516, TVOC is the sum of the concentrations of the identified and unidentified volatile organic compounds as defined in 3.1.3.11 (of EN 16516), calculated by summing the reference room concentrations of every individual compound (target and non-target, identified and unidentified) eluting between n-hexane and n-hexadecane inclusively using the specified column, and calculated using the TIC response factor for toluene after subtracting the blank values and after excluding compounds calculated to be below 5 $\mu g/m^3$ in the air of the reference room using the TIC response factor for toluene, additionally all compounds listed in Annex G (of EN 16516) are included even if they elute after n-hexadecane or before n-hexane under the specific test conditions [15].

7.2.402 Technical Specification

A document that prescribes technical requirements to be fulfilled by a product, process or service [12].

7.2.403 Total Primary Energy Factor

Sum of renewable and non-renewable primary energy factors for a given energy carrier [10].

7.2.404 Type of Connections

The type of connection refers to the way in which elements are connected to each other using, for example, dry or wet connections, as well as the type of these connections and the (planned) design [75].

7.2.405 Usable Space

Area of rooms on all floors of a building, either assigned to or available for assignment to an occupant or specific use or necessary for the general operation [33].

7.2.406 Value

Gain(s) or benefit(s) from satisfying needs and expectations, in relation to the use and conservation of resources [5].

7.2.407 Value Chain

1. Set of organizations that provide a solution that results in value for them [5].
2. All activities and processes that are part of the life cycle of a product, as well as its possible remanufacturing [12].

7.2.408 Value Creation Model (Business Model)

Organization's chosen system of interconnected and interdependent decisions and activities that determines how it creates, delivers and captures value [5].

7.2.409 Value Network

Network of interlinked value chains and interested parties [5].

7.2.410 Ventilation

According to EN 16798-1, ventilation is the process of providing outdoor air by natural or mechanical means to a space or building [15].

7.2.411 Ventilation Rate

According to EN 16798-1, the ventilation rate is the magnitude of outdoor air flow to a room or building through the ventilation system or device [15].

7.2.412 Ventilation System

According to EN 16798-1, a ventilation system is a combination of appliances or building components designed to supply indoor spaces with outdoor air and/or to extract polluted indoor air. The system may be mechanical (e.g. using a combination of air handling units, ducts and terminals), natural (e.g. achieving air flow via temperature differences and wind via façade grills) or a hybrid combination of both mechanical and natural aspects [15].

7.2.413 Versatility (Flexibility)

Ability to accommodate different functions with minor system changes [32].

7.2.414 Virgin Resource (Primary Resource)

Natural resource or energy that is used as a resource for the first time as input in a process or for creating a solution [5].

7.2.415 Virtual Reality (VR)

A detailed virtual model of the project and places the user directly inside the virtual environment, so that it is possible to experience the building before or during its realization [5].

7.2.416 Waste

1. Resource that is no longer considered to be an asset as it, at the time, provides insufficient value to the holder [59].
2. According to the EU Waste Framework Directive, it means any substance or object that the holder discards or is required to be discarded. Notable exemptions to the scope of the Waste Framework Directive set in Article 2 are: (i) land (in situ) including unexcavated contaminated soil and buildings permanently connected with land; and (ii) uncontaminated soil and other naturally occurring material excavated in the course of construction activities where it is guaranteed that the material will be used for the purposes of construction in its natural state on the site from which it was excavated [55].
3. Any substance or object which the holder discards or intends or is required to discard [65].

7.2.417 Waste Audit

According to the 2018 EU Guidelines for waste audits before demolition and renovation works of buildings, it means a qualitative and quantitative assessment of waste that will be produced from the construction, demolition/deconstruction or refurbishment activities including residual waste that is not part of the building. An important part of the waste audit is also the identification and removal of materials/components [55].

7.2.418 Waste Collection

The gathering of waste, including the preliminary sorting and preliminary storage of waste for the purposes of transport to a waste treatment facility [34].

7.2.419 Waste Hierarchy

A priority order in waste prevention and management legislation and policy [65]:

(a) prevention,
(b) preparing for re-use,
(c) recycling,
(d) other recovery, e.g. energy recovery, and
(e) disposal.

7.2.420 Waste Holder

According to the EU Waste Framework Directive, it means the waste producer or the natural or legal person who is in possession of the waste [55].

7.2.421 Waste Management

The collection, transport, recovery and disposal of waste, including the supervision of such operations and the after-care of disposal sites, and including actions taken as a dealer or broker [65].

7.2.422 Waste Management Plan

According to the 2016 EU Construction and Demolition waste management protocol, it means a plan that contains information about how the different steps of the demolition/renovation/construction will be performed, by whom they will be performed, which materials will be collected selectively at source, where and how they will be transported, what will be the recycled, re-use of final treatment and how to follow up [55].

7.2.423 Waste Producer

According to the EU Waste Framework Directive, it means anyone whose activities produce waste (original waste producer) or anyone who carries out pre-processing, mixing or other operations resulting in a change in the nature or composition of this waste [55].

7.2.424 Waste Transfer Station

Any site, location, tract of land, installation, or building that is used or intended to be used primarily for the purpose of transferring solid wastes [4].

7.2.425 Waste Treatment

Recovery or disposal operations, including preparation prior to recovery or disposal [4].

7.2.426 Whole Life-Cycle Assessment

A method that quantifies the carbon impact of a material or process across the entire building life cycle, from embodied to operational carbon and end of use. Whole life-cycle assessment requires rigorous standardised methodology so that the scope and benchmarks of assessments can be transparently communicated and evaluated [21].

Acknowledgements The editors would like to highlight that this glossary is a collection of terms and concepts defined mainly by the EU institutions and bodies, and the International Organization for Standardization (ISO). We would also like to acknowledge the contributions of leading organizations and pioneering companies in the field of circular economy implementation and environmental protection such as Ellen MacArthur Foundation and the United Nations Environment Programme.

Finally, we would like to thank the European Union and COST (European Cooperation in Science and Technology) for supporting the COST Action CircularB CA21103 (www.circularb.eu), and Philip Griffiths, WG4 leader, for his contribution and support.

References

1. Architects' Council of Europe, European Concrete Platform, European Economic and Social Committee (2011) Let's speak sustainable construction: multilingual glossary. European Economic and Social Committee. Accessed: 23 June 2025. [Online]. Available: https://data.europa.eu/doi/10.2864/61857
2. Donatello S, Dodd N, Cordella M (2021) Level(s) indicator 2.1: Bill of Quantities, materials and lifespans user manual: introductory briefing, instructions and guidance (Publication version 1.1). Accessed: 23 June 2025. [Online]. Available: https://susproc.jrc.ec.europa.eu/product-bureau/sites/default/files/2021-01/UM3_Indicator_2.1_v1.1_34pp.pdf
3. ISO (2020) ISO 6707–1:2020—Buildings and civil engineering works—Vocabulary—Part 1: general terms. International Organization for Standardization, Geneva. https://www.iso.org/obp/ui/#iso:std:iso:6707:-1:ed-6:v1:en. Last accessed 23 June 2025
4. European Commission, Directorate-General for Internal Market, Industry, Entrepreneurship and SMEs, Oberender A, Fruergaard Astrup T, Frydkjær Witte S, Camboni M et al (2024) EU construction & demolition waste management protocol including guidelines for pre-demolition and pre-renovation audits of construction works: updated edition 2024. Publications Office of the European Union. Accessed: 23 June 2025. [Online]. Available: https://data.europa.eu/doi/10.2873/77980
5. ISO (2024) ISO 59004:2024—Circular economy—Vocabulary, principles and guidance for implementation. International Organization for Standardization, Geneva. https://www.iso.org/obp/ui/en/#iso:std:iso:59004:ed-1:v1:en. Last Accessed 23 June 2025
6. Dodd N, Donatello S, Cordella M (2021) Level(s) indicator 6.1: Life cycle costs user manual: introductory briefing, instructions and guidance (Publication version 1.1). Accessed: 23 June 2025. [Online]. Available: https://susproc.jrc.ec.europa.eu/product-bureau/sites/default/files/2021-01/UM3_Indicator_6.1_v1.1_21pp.pdf
7. European Construction Sector Observatory (2021) Analytical report: digitalisation in the construction sector. Accessed: 22 June 2025. [Online]. Available: https://ec.europa.eu/docsroom/documents/45547/attachments/1/translations/en/renditions/native
8. ISO (2017) ISO 6707-1:2017—Buildings and civil engineering works—Vocabulary—Part 1: General terms. International Organization for Standardization, Geneva. https://www.iso.org/obp/ui/fr/#iso:std:iso:6707:-1:ed-5:v1:en:en. Last accessed 23 June 2025
9. Durmisevic E (2018) Reversible Building Design (WP3)—BAMB Reversible building design guidelines. Available from: https://www.bamb2020.eu/wp-content/uploads/2018/12/Reversible-Building-Design-guidelines-and-protocol.pdf. Accessed 23 June 2023
10. Dodd N, Donatello S, Cordella M (2021) Level(s) indicator 1.1: Use stage energy performance user manual: introductory briefing, instructions and guidance (Publication version 1.2). Accessed: 22 June 2025. [Online]. Available: https://susproc.jrc.ec.europa.eu/product-bureau/sites/default/files/2021-11/UM3_Indicator_1.1_v1.2_34pp_clean_15.07.2021.pdf
11. European Commission (2018) Guidelines for the waste audits before demolition and renovation works of buildings. Accessed: 22 June 2025. [Online]. Available: https://ec.europa.eu/docsroom/documents/31521/
12. European Parliament and Council of the European Union (2024) Regulation (EU) 2024/1781 of 13 June 2024 establishing a framework for the setting of ecodesign requirements for sustainable products, amending Directive (EU) 2020/1828 and Regulation (EU) 2023/1542 and repealing Directive 2009/125/EC. Official Journal of the European Union, L 1781, 28 June 2024, pp 1–89. https://eur-lex.europa.eu/eli/reg/2024/1781/oj/eng. Last accessed 23 June 2025
13. European Parliament and Council of the European Union (2024). Consolidated text: Directive 2008/98/EC of 19 November 2008 on waste and repealing certain Directives. Official Journal of the European Union.http://data.europa.eu/eli/dir/2008/98/2024-02-18/eng. Last accessed 23 June 2025
14. ISO (2020) ISO 21678:2020—Sustainability in buildings and civil engineering works—Indicators and benchmarks—Principles, requirements and guidelines. International Organization for

Standardization, Geneva. https://www.iso.org/obp/ui/en/#iso:std:iso:21678:ed-1:v1:en. Last accessed 23 June 2025
15. Dodd N, Donatello S, Cordella M, (2021) Level(s) indicator 4.1: Indoor air quality user manual: introductory briefing, instructions and guidance (Publication version 1.1). Accessed: 22 June 2025. [Online]. Available: https://susproc.jrc.ec.europa.eu/product-bureau/sites/default/files/2021-02/UM3_Indicator_4.1_v1.1_37pp.pdf
16. European Parliament and Council of the European Union (2020) Regulation (EU) 2020/852 of 18 June 2020 on the Establishment of a Framework to Facilitate Sustainable Investment, and Amending Regulation (EU) 2019/2088. Official Journal of the European Union, L 198, 22 June 2020, pp 13–43. https://eurlex.europa.eu/eli/reg/2020/852/oj/eng
17. Dodd N, Donatello S, Cordella M (2020) Level(s) indicator 1.2: life cycle global warming potential (GWP) user manual: overview, instructions and guidance (Publication version 1.0). Accessed: 22 June 2025. [Online]. Available: https://susproc.jrc.ec.europa.eu/product-bureau/sites/default/files/2020-10/20201013%20New%20Level(s)%20documentation_Indicator%201.2_Publication%20v1.0.pdf, https://susproc.jrc.ec.europa.eu/productbureau/sites/default/files/2021-01/UM3_Indicator_1.2_v1.1_37pp.pdf
18. European Parliament and Council of the European Union (2018) Directive (EU) 2018/844 of 30 May 2018 on the energy performance of buildings. Official Journal of the European Union. https://eur-lex.europa.eu/eli/dir/2018/844/oj/eng. Last accessed 23 June 2025
19. Deweerdt M, Mertens M, Vrijders J, Poncelet F, Bernair C, Scherrier N et al (2020) A guide for identifying the reuse potential of construction products (Working draft version). Accessed: 23 June 2025. [Online]. Available: https://vb.nweurope.eu/media/10132/en-fcrbe_wpt2_d12_a_guide_for_identifying_the_reuse_potential_of_construction_products.pdf
20. Council of the European Union (2021) Council conclusions on culture, high-quality architecture and built environment as key elements of the New European Bauhaus initiative. Brussels, 30 Nov 2021, ST 14534/21 INIT. Accessed: 22 June 2025. [Online]. Available: https://data.consilium.europa.eu/doc/document/ST-14534-2021-INIT/en/pdf
21. United Nations Environment Programme, Yale Center for Ecosystems + Architecture (2023) Building materials and the climate: constructing a new future. Accessed: 23 June 2025. [Online]. Available: https://wedocs.unep.org/20.500.11822/43293
22. Bates A, Kelly P, Schoonhoven J, Riis-Tolman T, and Snabe K (2021) The business case for circular buildings: exploring the economic, environmental and social value. World Business Council for Sustainable Development. Accessed: 23 June 2025. [Online]. Available: https://www.wbcsd.org/wp-content/uploads/2023/10/The-business-case-for-circular-buildings-Exploring-the-economic-environmental-and-social-value-.pdf
23. Platform CB'23 (2020) Circular construction lexicon. Accessed: 23 June 2025. [Online]. Available: https://platformcb23.nl/wp-content/uploads/PlatformCB23_Lexicon-circular-construction.pdf
24. European Parliament and Council of the European Union (2020) Regulation (EU) 2020/852 on the establishment of a framework to facilitate sustainable investment. Official Journal of the European Union. https://eur-lex.europa.eu/eli/reg/2020/852/oj/eng. Last accessed 23 June 2025
25. ISO (2017) ISO 20400:2017—Sustainable procurement—Guidance. International Organization for Standardization, Geneva. https://www.iso.org/obp/ui#iso:std:iso:20400:ed-1:v1:en. Last accessed 23 June 2025
26. Eurostat (2025) Circular material use rate (cei_srm030). https://ec.europa.eu/eurostat/cache/metadata/en/cei_srm030_esmsip2.htm. Last accessed 23 June 2025
27. European Commission (2025) Green business—Biblioteca. https://circabc.europa.eu/ui/group/44278090-3fae-4515-bcc2-44fd57c1d0d1/library/b1af4b10-e76c-4e7c-816a-6d9644a5691b?p=1&n=-1&sort=name_ASC. Last accessed 23 June 2025
28. ISO (2021) ISO 14031:2021—Environmental management—environmental performance evaluation—guidelines. International Organization for Standardization, Geneva. https://www.iso.org/obp/ui/#iso:std:iso:14031:ed-3:v1:en. Last accessed 23 June 2025

29. ISO (2011) ISO 15686–1:2011—Buildings and constructed assets—service life planning—Part 1: general principles and framework. International Organization for Standardization, Geneva. https://www.iso.org/obp/ui/#iso:std:iso:15686:-1:ed-2:v1:en. Last accessed 23 June 2025
30. European Parliament and Council of the European Union (2018) Directive (EU) 2018/851 of 30 May 2018 amending Directive 2008/98/EC on waste. Official Journal of the European Union. http://data.europa.eu/eli/dir/2018/851/oj/eng. Last accessed 23 June 2025
31. European Union (2024) Regulation (EU) 2024/3110 of the European Parliament and of the Council. Official Journal of the European Union. https://eur-lex.europa.eu/eli/reg/2024/3110/oj/eng. Last accessed 23 June 2025
32. ISO (2022) ISO 21931–1:2022—Sustainability in buildings and civil engineering works—Framework for methods of assessment of the environmental, social and economic performance of construction works as a basis for sustainability assessment—Part 1: Buildings. International Organization for Standardization, Geneva. https://www.iso.org/obp/ui/en/#iso:std:iso:21931:-1:ed-2:v1:en. Last accessed 23 June 2025
33. ISO (2020) ISO 20887:2020—Sustainability in buildings and civil engineering works—Design for disassembly and adaptability—Principles, requirements and guidance. International Organization for Standardization, Geneva. https://www.iso.org/obp/ui/en/#iso:std:iso:20887:ed-1:v1:en. Last accessed 23 June 2025
34. European Commission (2023) Proposal for a Regulation of the European Parliament and of the Council establishing a framework for ensuring a secure and sustainable supply of critical raw materials and amending Regulations (EU) 168/2013, (EU) 2018/858, 2018/1724 and (EU) 2019/1020. https://eur-lex.europa.eu/legal-content/EN/TXT/?uri=CELEX%3A52023PC0160. Last accessed 23 June 2025
35. Guy B, Ciarimboli N (2008) DfD: Design for Disassembly in the Built Environment: A Guide to Closed-Loop Design and Building. Hamer Center for Community Design, The Pennsylvania State University. Accessed 20 June 2025. [Online]. Available: https://www.lifecyclebuilding.org/docs/DfDseattle.pdf
36. RIBA (2020) RIBA 2020 Plan of work: a roadmap for thoughtful architecture. https://www.architecture.com/knowledge-and-resources/resources-landing-page/riba-plan-of-work. Last accessed 23 June 2025
37. European Commission: Executive Agency for Small and Medium-sized Enterprises, Volt J, Toth Z, Glicker J, De Groote M, Borragán G, De Regel S, et al. (2020) Definition of the digital building logbook: Report 1 of the study on the development of a European Union framework for buildings' digital logbook. Publications Office of the European Union, Luxembourg. https://data.europa.eu/doi/10.2826/480977. Last accessed 23 June 2025
38. Dodd N, Donatello S, Cordella M (2021) Level(s) indicator 2.4: Design for deconstruction user manual: introductory briefing, instructions and guidance (Publication version 1.1). Accessed: 22 June 2025. [Online]. Available: https://susproc.jrc.ec.europa.eu/product-bureau/sites/default/files/2021-01/UM3_Indicator_2.4_v1.1_18pp.pdf
39. ISO (2019) ISO 15392:2019—Sustainability in buildings and civil engineering works—General principles. International Organization for Standardization, Geneva. https://www.iso.org/obp/ui/#iso:std:iso:15392:ed-2:v1:en. Last accessed 23 June 2025
40. Chini AR, Bruening SF (2003) Report 10—deconstruction and materials reuse in the United States. CIB Publication, University of Florida, Gainesville. https://www.irbnet.de/daten/iconda/CIB1303.pdf. Last accessed 22 June 2025
41. ISO (2017) ISO 17738–1:2017—Thermal insulation products—Exterior insulation and finish systems—Part 1: Materials and systems. International Organization for Standardization, Geneva. https://www.iso.org/obp/ui/#iso:std:iso:17738:-1:ed-1:v1:en. Last accessed 23 June 2025
42. European Commission (2025) Join EMAS. https://green-forum.ec.europa.eu/emas/join-emas_en. Last accessed 23 June 2025
43. ISO (2006) ISO 14040:2006—Environmental management—life cycle assessment—principles and framework. International Organization for Standardization, Geneva. https://www.iso.org/obp/ui#iso:std:iso:14040:ed-2:v1:en. Last accessed 26 March 2025

44. Commission of the European Communities (2003) Commission Recommendation of 6 May 2003 concerning the definition of micro, small and medium-sized enterprises (2003/361/ EC). Official Journal of the European Union, L124:36–41. http://data.europa.eu/eli/reco/2003/ 361/oj. Last accessed: 22 June 2025
45. ISO (2004) ISO 14001:2004—Environmental management systems—requirements with guidance for use. International Organization for Standardization, Geneva. https://www.iso.org/obp/ ui/#iso:std:iso:14001:ed-2:v1:en. Last accessed 23 June 2025
46. ICLEI Europe, Eurocities (2020) Public procurement of circular construction materials: key takeaways from the Big Buyers Initiative working group (lessons learned report). Accessed: 22 June 2025. [Online]. Available: https://circulars.iclei.org/wp-content/uploads/2020/09/BBI-CCM-lessons-learned.pdf
47. ISO (2022) ISO 24161:2022—Waste collection and transportation management—vocabulary. International Organization for Standardization, Geneva. https://www.iso.org/obp/ui/en/#iso: std:iso:24161:ed-1:v1:en. Last accessed 23 June 2025
48. Bourguignon D (2016) Closing the loop: new circular economy package. European Parliamentary Research Service. Accessed: 22 June 2025. [Online]. Available: https://www.europarl.eur opa.eu/RegData/etudes/BRIE/2016/573899/EPRS_BRI%282016%29573899_EN.pdf
49. ISO (2020) ISO 14050:2020—Environmental management—vocabulary. International Organization for Standardization, Geneva. https://www.iso.org/obp/ui/en/#iso:std:iso:14050:ed-4: v1:en. Last accessed 23 June 2025
50. Azcárate-Aguerre JF (2023) Facades-as-a-Service: a cross-disciplinary model for the (re)development of circular building envelopes. Doctoral thesis, Delft University of Technology. A+BE | Architecture and the Built Environment, No. 11. https://doi.org/10.7480/abe. 2023.11. Last accessed 22 June 2025
51. Ellen MacArthur Foundation (2021) Circular economy glossary. Ellen MacArthur Foundation, Cowes. https://www.ellenmacarthurfoundation.org/topics/circular-economy-introduct ion/glossary. Last accessed 19 August 2025
52. ISO (2017) ISO 21505:2017—Project, programme and portfolio management—guidance on governance. International Organization for Standardization, Geneva. https://www.iso.org/obp/ ui/en/#iso:std:iso:21505:ed-1:v2:en. Last accessed 23 June 2025
53. Donatello S, Dodd N, Cordella M (2021) Level(s) indicator 3.1: Use stage water consumption user manual: introductory briefing, instructions and guidance (publication version 1.1). Accessed: 22 June 2025. [Online]. Available: https://susproc.jrc.ec.europa.eu/product-bureau/ sites/default/files/2021-01/UM3_Indicator_3.1_v1.1_31pp.pdf
54. European Commission (2025) Green public procurement. https://green-forum.ec.europa.eu/ green-public-procurement_en. Last accessed 23 June 2025
55. Donatello S, Dodd N, Cordella M (2021) Level(s) indicator 2.2: Construction and Demolition waste and materials user manual: introductory briefing, instructions and guidance (Publication version 1.1) Accessed: 22 June 2025. [Online]. Available: https://susproc.jrc.ec.europa.eu/pro duct-bureau/sites/default/files/2021-01/UM3_Indicator_2.2_v1.1_40pp.pdf
56. European Commission (2025) High Level Construction Forum—Meeting report: reporting from the 1st meeting of the Digital Cluster Group. Accessed: 22 June 2025. [Online]. Available: https://single-market-economy.ec.europa.eu/system/files/2021-11/HLCF_1st_Resilienc eClusterGroup_MeetingReport.20.pdf
57. European Commission (2025) Level(s). https://green-forum.ec.europa.eu/levels_en. Last accessed 23 June 2025
58. European Commission (2022) Proposal for a Regulation of the European Parliament and of the Council laying down harmonised conditions for the marketing of construction products, amending Regulation (EU) 2019/1020 and repealing Regulation (EU) 305/2011. COM(2022) 144 final. Accessed: 22 June 2025. [Online]. Available: https://ec.europa.eu/docsroom/docume nts/49315/attachments/2/translations/en/renditions/native
59. ISO (2024) ISO 59010:2024—Circular economy—guidance on the transition of business models and value networks. International Organization for Standardization, Geneva. https:// www.iso.org/obp/ui/en/#iso:std:iso:59010:ed-1:v1:en. Last accessed 23 June 2025

60. European Parliament and Council of the European Union (2011) Regulation (EU) No 305/2011 of 9 March 2011 laying down harmonised conditions for the marketing of construction products and repealing Council Directive 89/106/EEC. Official Journal of the European Union. http://data.europa.eu/eli/reg/2011/305/oj/eng. Last accessed 23 June 2025
61. Interreg Central Europe (2023) Materials Passports. https://www.interreg-central.eu/news/materials-paasports/. Last accessed 26 June 2025
62. European Commission (2025) Nature-based solutions. https://research-and-innovation.ec.europa.eu/research-area/environment/nature-based-solutions_en. Last accessed 23 June 2025
63. Hobbs G, Adams K (2017) Reuse of building products and materials—barriers and opportunities. International HISER conference on advances in recycling and management of construction and demolition waste. Delft University of Technology, Delft, pp 109–113. Accessed: 22 June 2025. [Online]. Available: https://www.bamb2020.eu/wp-content/uploads/2017/07/Reuse-of-building-products-and-materials-barriers-and-opportunities.pdf
64. Westman Lumber (2025) Prefabrication—glossary. https://www.westmanlumber.com/glossary/prefabrication. Last accessed 24 June 2025
65. European Parliament and Council of the European Union (2008) Directive 2008/98/EC of 19 November 2008 on waste and repealing certain Directives. Official Journal of the European Union. http://data.europa.eu/eli/dir/2008/98/oj/eng. Last accessed 24 June 2025
66. ISO (2016) ISO 14021:2016—Environmental labels and declarations—self-declared environmental claims (Type II environmental labelling). International Organization for Standardization, Geneva. https://www.iso.org/obp/ui/#iso:std:iso:14021:ed-2:v1:en:sec:7.7.1. Last accessed 24 June 2025
67. European Remanufacturing Network (2025) What is remanufacturing? https://www.remanufacturing.eu/about-remanufacturing.php. Last accessed 24 June 2025
68. Cruz PJS (ed) (2013) Structures and architecture: concepts, applications and challenges. Proceedings of the second international conference on structures and architecture, 24–26 July 2013, Guimarães, Portugal. CRC Press, Boca Raton
69. European Parliament and Council of the European Union (2021) Regulation (EU) 2021/241 of 12 February 2021 establishing the Recovery and Resilience Facility. Official Journal of the European Union. http://data.europa.eu/eli/reg/2021/241/oj/eng. Last accessed 24 June 2025
70. Eurostat (2025) Glossary: resource productivity. https://ec.europa.eu/eurostat/statistics-explained/index.php?title=Glossary:Resource_productivity. Last accessed 24 June 2025
71. European Commission. Joint Research Centre. Level(s), a common EU framework of core sustainability indicators for office and residential buildings. Part 3, How to make performance assessments using level(s) (Beta v1.0). LU: Publications Office, 2017. Accessed: 24 June 2025 [Online]. Available: https://data.europa.eu/doi/10.2760/95143
72. ISO (2011) ISO 21929–1:2011—Sustainability in building construction—sustainability indicators—Part 1: framework for the development of indicators and a core set of indicators for buildings. International Organization for Standardization, Geneva. https://www.iso.org/standard/46599.html. Last accessed 24 June 2025
73. ISO (2015) ISO 9000:2015—Quality management systems—fundamentals and vocabulary. International Organization for Standardization, Geneva. https://www.iso.org/obp/ui/#iso:std:iso:9000:ed-4:v1:en. Last accessed 24 June 2025
74. ISO (2014) ISO 6707–1:2014—Buildings and civil engineering works—vocabulary—Part 1: general terms. International Organization for Standardization, Geneva. https://www.iso.org/obp/ui/#iso:std:iso:6707:-1:ed-4:v1:en. Last accessed 23 June 2025
75. European Commission (2020) Circular Economy—Principles for Buildings Design. Available from: https://ec.europa.eu/docsroom/documents/39984. Accessed 23 June 2023

MIX
Papier aus verantwortungsvollen Quellen
Paper from responsible sources
FSC® C105338

If you have any concerns about our products,
you can contact us on
ProductSafety@springernature.com

In case Publisher is established outside the EU,
the EU authorized representative is:
Springer Nature Customer Service Center GmbH
Europaplatz 3, 69115 Heidelberg, Germany

Printed by Libri Plureos GmbH
in Hamburg, Germany